AF692644

Advances in Differential Dynamical Systems with Applications to Economics and Biology

Advances in Differential Dynamical Systems with Applications to Economics and Biology

Editors

Mihaela Neamţu
Eva Kaslik
Anca Rădulescu

MDPI • Basel • Beijing • Wuhan • Barcelona • Belgrade • Manchester • Tokyo • Cluj • Tianjin

Editors
Mihaela Neamțu
West University of Timişoara
Romania

Eva Kaslik
West University of Timişoara
Romania

Anca Rădulescu
State University of New York
at New Paltz
USA

Editorial Office
MDPI
St. Alban-Anlage 66
4052 Basel, Switzerland

This is a reprint of articles from the Special Issue published online in the open access journal *Mathematics* (ISSN 2227-7390) (available at: https://www.mdpi.com/si/mathematics/Differential_Dynamical_Systems_Applications_Economics_Biology).

For citation purposes, cite each article independently as indicated on the article page online and as indicated below:

LastName, A.A.; LastName, B.B.; LastName, C.C. Article Title. *Journal Name* **Year**, *Volume Number*, Page Range.

ISBN 978-3-0365-5865-3 (Hbk)
ISBN 978-3-0365-5866-0 (PDF)

© 2022 by the authors. Articles in this book are Open Access and distributed under the Creative Commons Attribution (CC BY) license, which allows users to download, copy and build upon published articles, as long as the author and publisher are properly credited, which ensures maximum dissemination and a wider impact of our publications.
The book as a whole is distributed by MDPI under the terms and conditions of the Creative Commons license CC BY-NC-ND.

Contents

About the Editors . . . vii

Eva Kaslik, Mihaela Neamţu and Anca Rădulescu
Preface to the Special Issue on "Advances in Differential Dynamical Systems with Applications to Economics and Biology"
Reprinted from: *Mathematics* **2022**, *10*, 3561, doi:10.3390/math10193561 . . . 1

Alexander Musaev, Andrey Makshanov and Dmitry Grigoriev
Evolutionary Optimization of Control Strategies for Non-Stationary Immersion Environments
Reprinted from: *Mathematics* **2022**, *10*, 1797, doi:10.3390/math10111797 . . . 5

Alexander Musaev, Andrey Makshanov and Dmitry Grigoriev
Numerical Studies of Channel Management Strategies for Nonstationary Immersion Environments: EURUSD Case Study
Reprinted from: *Mathematics* **2022**, *10*, 1408, doi:10.3390/math10091408 . . . 23

Hashem Najafi, Sina Etemad, Nichaphat Patanarapeelert, Joshua Kiddy K. Asamoah, Shahram Rezapour and Thanin Sitthiwirattham
A Study on Dynamics of CD4$^+$ T-Cells under the Effect of HIV-1 Infection Based on a Mathematical Fractal-Fractional Model via the Adams-Bashforth Scheme and Newton Polynomials
Reprinted from: *Mathematics* **2022**, *10*, 1366, doi:10.3390/math10091366 . . . 43

Ahmed M. Elshenhab, Xingtao Wang, Omar Bazighifan and Jan Awrejcewicz
Finite-Time Stability Analysis of Linear Differential Systems with Pure Delay
Reprinted from: *Mathematics* **2022**, *10*, 1359, doi:10.3390/math10091359 . . . 75

Alexander Musaev, Andrey Makshanov and Dmitry Grigoriev
Statistical Analysis of Current Financial Instrument Quotes in the Conditions of Market Chaos
Reprinted from: *Mathematics* **2022**, *10*, 587, doi:10.3390/math10040587 . . . 85

Mădălina Sofia Paşca, Olivia Bundău, Adina Juratoni and Bogdan Căruntu
The Least Squares Homotopy Perturbation Method for Systems of Differential Equations with Application to a Blood Flow Model
Reprinted from: *Mathematics* **2022**, *10*, 546, doi:10.3390/math10040546 . . . 101

Liviu Popescu, Daniel Militaru and Gabriel Tică
Lie Geometric Methods in the Study of Driftless Control Affine Systems withHolonomic Distribution and Economic Applications
Reprinted from: *Mathematics* **2022**, *10*, 545, doi:10.3390/math10040545 . . . 115

Irina Badralexi, Andrei-Dan Halanay and Ragheb Mghames
Stability Analysis of Equilibria for a Model of Maintenance Therapy in Acute Lymphoblastic Leukemia
Reprinted from: *Mathematics* **2022**, *10*, 313, doi:10.3390/math10030313 . . . 135

Alexander Musaev and Dmitry Grigoriev
Numerical Studies of Statistical Management Decisionsin Conditions of Stochastic Chaos
Reprinted from: *Mathematics* **2022**, *10*, 226, doi:10.3390/math10020226 . . . 155

Santiago Alonso-Quesada, Manuel De la Sen and Raúl Nistal
An SIRS Epidemic Model Supervised by a Control System for Vaccination and Treatment Actions Which Involve First-Order Dynamics and Vaccination of Newborns
Reprinted from: *Mathematics* **2022**, *10*, 36, doi:10.3390/math10010036 . . . 169

Eva Kaslik, Mihaela Neamţu and Loredana Flavia Vesa
Global Stability Analysis of a Five-Dimensional Unemployment Model with Distributed Delay
Reprinted from: *Mathematics* **2021**, *9*, 3037, doi:10.3390/math9233037 . . . . **201**

Xin Jiang
Global Dynamics for an Age-Structured Cholera Infection Model with General Infection Rates
Reprinted from: *Mathematics* **2021**, *9*, 2993, doi:10.3390/math9232993 . . . . **217**

Giacomo Aletti, Alessandro Benfenati and Giovanni Naldi
Graph, Spectra, Control and Epidemics: An Example with a SEIR Model
Reprinted from: *Mathematics* **2021**, *9*, 2987, doi:10.3390/math9222987 . . . . **237**

Mohamed M. Mousa and Fahad Alsharari
A Comparative Numerical Study and Stability Analysis for a Fractional-Order SIR Model of Childhood Diseases
Reprinted from: *Mathematics* **2021**, *9*, 2847, doi:10.3390/math9222847 . . . . **251**

Luis M. Abia, Óscar Angulo, Juan Carlos López-Marcos and Miguel Ángel López-Marcos
Computational Study on the Dynamics of a Consumer-Resource Model: The Influence of the Growth Law in the Resource
Reprinted from: *Mathematics* **2021**, *9*, 2746, doi:10.3390/math9212746 . . . . **263**

Cristian Lăzureanu
Integrable Deformations and Dynamical Properties of Systems with Constant Population
Reprinted from: *Mathematics* **2021**, *9*, 1378, doi:10.3390/math9121378 . . . . **279**

About the Editors

Mihaela Neamțu

Ph.D., is currently Full Professor in the Department of Economics and Business Administration at the West University of Timişoara, Romania. Here, she received both her B.S. degree and Ph.D. degree in mathematics, in 1994 and 2001, respectively. She received her habilitation in mathematics from this same university, in 2019. Her current field of research is nonlinear dynamics, economic modeling, differential equations, stability analysis, biomathematics, numerical simulation and mathematical modeling. She has been the director and investigator of several national and international projects. Professor Neamtu is the author or co-author of six books, and of various scientific papers, of which 110 are articles in peer-reviewed scholarly journals. She has been the director of two national projects, and member of more than 10 national and international projects. She is a member of the Romanian Mathematical Society (R.M.S.). Professor Neamtu has been a visiting scientist at the Mathematical Institute of the University of Bonn, Germany; Nottingham Trent University, UK; University of Perpignan Via Domitia and Universitatea La Rochelle, France.

Eva Kaslik

Ph.D., is currently Full Professor in the Department of Mathematics and Computer Science at the West University of Timişoara, Romania. Here, she received both her B.S. degree in pure Mathematics and M.S. degree in applied Mathematics, in 2002 and 2004, respectively. She received her Ph.D. Degree in 2006 from Université Paris Nord, France, and the West University of Timişoara and received habilitation in mathematics from the West University of Timişoara, in 2015. She has been the director of two national projects, Management Committee Member of the EU-funded program COST Action CA15225 and member of more than 10 national and international projects. Professor Kaslik has been invited to give talks and presentations at more than 30 national and international conferences. Her current field of research is dynamical systems, delay differential equations, fractional-order differential equations, dynamic networks and mathematical modeling in engineering, neuroscience, medicine, biology, economy, and aerospace science.

Anca Rădulescu

Ph.D., is currently Associate Professor in the Department of Mathematics at the State University of New York at New Paltz, USA. She received her B.S. from the University of Bucharest in 1998 and her Ph.D. degree in mathematics from Stony Brook University, in 2005. She was Visiting Assistant Professor at City University of New York, Brooklyn College and Instructor at the University of Colorado Boulder, USA. Since 2014, Dr. Rădulescu has worked with the State University of New York at New Paltz. She has been an invited speaker and served as co-organizer and session chair at many international conferences in her research field. Her main research interest is in problems from complex dynamics that have applications to various fields in the natural sciences. Part of her work is purely theoretical, focused around identifying and understanding new phenomena in discrete random complex dynamics, and in complex dynamic networks. The other part centers around using dynamical systems methods and results to derive and analyze models in a variety of fields, among which mathematical neuroscience, medicine epidemiology, climate, and the environment.

mathematics

MDPI

Editorial

Preface to the Special Issue on "Advances in Differential Dynamical Systems with Applications to Economics and Biology"

Eva Kaslik [1,2,*,†], Mihaela Neamţu [2,3,*,†] and Anca Rădulescu [4,*,†]

1 Department of Mathematics and Computer Science, West University of Timişoara, 300223 Timişoara, Romania
2 Institute for Advanced Environmental Research, West University of Timişoara, 300223 Timişoara, Romania
3 Department of Economics and Business Administration, West University of Timişoara, 300223 Timişoara, Romania
4 Department of Mathematics, State University of New York at New Paltz, New Paltz, NY 12561, USA
* Correspondence: eva.kaslik@e-uvt.ro (E.K.); mihaela.neamtu@e-uvt.ro (M.N.); radulesa@newpaltz.edu (A.R.)
† These authors contributed equally to this work.

Citation: Kaslik, E.; Neamţu, M.; Rădulescu, A. Preface to the Special Issue on "Advances in Differential Dynamical Systems with Applications to Economics and Biology". *Mathematics* **2022**, *10*, 3561. https://doi.org/10.3390/math10193561

Received: 23 September 2022
Accepted: 26 September 2022
Published: 29 September 2022

Publisher's Note: MDPI stays neutral with regard to jurisdictional claims in published maps and institutional affiliations.

Copyright: © 2022 by the authors. Licensee MDPI, Basel, Switzerland. This article is an open access article distributed under the terms and conditions of the Creative Commons Attribution (CC BY) license (https://creativecommons.org/licenses/by/4.0/).

In recent research on natural processes, mathematical modeling has become a very useful tool. It is often the case that in fields such as economics and biology, a temporal lag between cause and effect must often be taken into consideration. In modeling, a natural and practical implementation of this phenomenon is through the use of distributed delays. This is because they illustrate the situation where temporal lags arise in certain ranges of values for certain related probability distributions, taking into account the variables' entire history of behavior. Another mathematical tool that allows for the memory and inherited properties of systems to be encompassed in a model is the replacement of integer-order derivatives with fractional derivatives. To address realistic conditions, stochastic perturbation framed by a stochastic differential delay system can be used to explain the ambiguity about the context in which the system operates.

This Special Issue comprises 16 scientific contributions and focuses on the dynamical analysis of mathematical models arising from economy and biology and innovative developments in mathematical techniques for their applications.

Musaev et al. [1] explored the evolutionary self-organization of control techniques using the example of speculative trading in a non-stationary immersion market environment. Because of the extreme volatility and non-stationarity of the observation series, it is particularly challenging to employ adaptive computational algorithms. The authors suggest a strategy based on evolutionary modeling that provides a control model with structural and parametric self-organization.

In a second paper [2], the same authors take into account the short-term forecasting of a process that is an output signal of a nonlinear system seen against an additive noise background. The authors show that it is fundamentally possible to make profit, even in fields with complicated dynamics and sudden changes in the process under consideration, suggesting updated channel strategies and outlining key methods for boosting their efficiency.

In the paper by Najafi et al. [3], the dynamics of $CD4^+$ T-cells under the influence of HIV-1 infection are studied for the first time in the context of a generalized fractal-fractional structure by using a new mathematical model. Analytical and numerical results reveal stability properties and a lack of a discernible order in the early stages of the illness' dynamics.

In a more theoretical study, Elshenhab et al. [4] studied the finite-time stability of nonhomogeneous systems of second-order linear delay differential equations. The results are applicable to all singular, non-singular and arbitrary matrices.

Using multidimensional statistical analysis, Musaev et al. [5] explored the problem of evaluating the current value of financial instruments, examining various methods for building computational schemes for regression. The chaotic nature of the observation series, which is caused by the instability of the starting data's probabilistic structure, is a key aspect of this issue; hence, the primary approach for examining the effectiveness of forecasting and analytical algorithms is numerical experimentation.

Paşca et al. [6] propose the least-squares homotopy perturbation as a straightforward and accurate method for obtaining approximate analytical solutions for systems of ordinary differential equations. The technique is used to resolve a problem involving the laminar flow of a viscous fluid in a semi-porous channel, which may be used to simulate blood flow through a blood vessel while taking the effects of a magnetic field into account.

In the paper by Popescu et al. [7], the Pontryagin Maximum Principle and Lie geometric methods are employed to study two optimal control problems at the level of the Lie algebroid. It is demonstrated that the cotangent bundle is not the best framework for finding the best solutions to some driftless control affine systems with holonomic distributions. In this context, a financial application is also presented.

Badralexi et al. [8] analyzed the processes of erythropoiesis and leukopoiesis in the context of maintenance therapy for acute lymphoblastic leukemia by considering two mathematical models expressed by delay differential equations. The stability of every equilibrium point is examined either analytically or numerically. The mathematical results are interpreted from a biological point of view.

Another study by Musaev and Grivoriev [9] examined the applicability of conventional statistical management decision-making methods under stochastic chaos. Compositional algorithm variations are proposed, aimed at adjusting statistical methods to the non-deterministic circumstances brought on by the peculiarities of chaotic processes.

Alonso-Quesada et al. [10] investigate an SIRS epidemic model involving the immunization of susceptible individuals and the treatment of contagious individuals, which are both governed by a designed control system for which its inputs are the subpopulations of the epidemic model. Additionally, newborn vaccinations are also taken into account and control strategies are proposed to eradicate the infectious disease.

In the paper of Kaslik et al. [11], a five-dimensional mathematical model for analyzing the labor market was proposed, with a particular emphasis on the number of open positions, migration, fixed-term contractors, full-time employment and unemployment. The rate of change of open positions, which depends on historical regular employment levels, takes the distributed time delay into account.

By incorporating age structures and overall infection rates into a cholera model, Jiang [12] examined the model's global dynamics, the existence and point dissipativeness of the orbit and asymptotical smoothness. Next, they focus on the existence and local stability of equilibria and also discuss uniform persistence, followed by numerical simulations.

Aletti et al. [13] present an epidemiological SEIR population-based model with many groups, which may represent a geographically limited population or a social subpopulation with similar tendencies while also taking into account the heterogeneity in the weighting of contacts between two groups. In order to minimize the sum of the economic and social costs, they suggest a straightforward control algorithm in which connection weights are optimized.

The aim of the paper of Mousa et al. [14] was to examine the dynamics of a fractional-order susceptible-infectious-recovered (SIR) model that simulates epidemiological diseases. An efficient numerical method based on the Grunwald–Letnikov fractional derivative was proposed and a qualitatively stability analysis was also carried out.

Abia et al. [15] investigate the dynamics of a particular consumer-resource model for Daphnia magna from a numerical perspective. Malthusian, chemostatic, and Gompertz growth laws are taken into account in this study for the evolution of the resource population, and the ensuing global dynamics of the model are contrasted as various model parameters vary. The numerical simulations demonstrate that the biological effect of resource scarcity is a real reduction in the size of the consumer population's members.

The paper of Lăzureanu [16] considers systems of three autonomous first-order differential equations such that the sum of the three variables is constant in time. Hamilton–Poisson formulations and integrable deformations are presented, and the case of Kolmogorov systems is also analyzed. As an application, the author examines the three-dimensional Lotka–Volterra system with constant populations from the perspective of Poisson geometry.

As Guest Editors, we are grateful to all authors who contributed to the success of this Special Issue, and to all reviewers for their constructive comments that helped improve initial submissions.

This Special Issue's objective was to draw original contributions in the area of "Advances in Differential Dynamical Systems with Applications to Economics and Biology". We expect that the international scientific community will find this collection of research papers influential and that they will spur additional investigations on diverse applications with respect to dynamical systems in all scientific areas.

Funding: This research received no external funding.

Conflicts of Interest: The authors declare no conflict of interest.

References

1. Musaev, A.; Makshanov, A.; Grigoriev, D. Evolutionary Optimization of Control Strategies for Non-Stationary Immersion Environments. *Mathematics* **2022**, *10*, 1797. [CrossRef]
2. Musaev, A.; Makshanov, A.; Grigoriev, D. Numerical studies of channel management strategies for nonstationary immersion environments: EURUSD case study. *Mathematics* **2022**, *10*, 1408. [CrossRef]
3. Najafi, H.; Etemad, S.; Patanarapeelert, N.; Asamoah, J.K.K.; Rezapour, S.; Sitthiwirattham, T. A study on dynamics of CD4+ T-cells under the effect of HIV-1 infection based on a mathematical fractal-fractional model via the Adams-Bashforth scheme and Newton polynomials. *Mathematics* **2022**, *10*, 1366. [CrossRef]
4. Elshenhab, A.M.; Wang, X.; Bazighifan, O.; Awrejcewicz, J. Finite-time stability analysis of linear differential systems with pure delay. *Mathematics* **2022**, *10*, 1359. [CrossRef]
5. Musaev, A.; Makshanov, A.; Grigoriev, D. Statistical Analysis of Current Financial Instrument Quotes in the Conditions of Market Chaos. *Mathematics* **2022**, *10*, 587. [CrossRef]
6. Paşca, M.S.; Bundău, O.; Juratoni, A.; Căruntu, B. The Least Squares Homotopy Perturbation Method for Systems of Differential Equations with Application to a Blood Flow Model. *Mathematics* **2022**, *10*, 546. [CrossRef]
7. Popescu, L.; Militaru, D.; Tică, G. Lie Geometric Methods in the Study of Driftless Control Affine Systems with Holonomic Distribution and Economic Applications. *Mathematics* **2022**, *10*, 545. [CrossRef]
8. Badralexi, I.; Halanay, A.D.; Mghames, R. Stability Analysis of Equilibria for a Model of Maintenance Therapy in Acute Lymphoblastic Leukemia. *Mathematics* **2022**, *10*, 313. [CrossRef]
9. Musaev, A.; Grigoriev, D. Numerical studies of statistical management decisions in conditions of stochastic chaos. *Mathematics* **2022**, *10*, 226. [CrossRef]
10. Alonso-Quesada, S.; De la Sen, M.; Nistal, R. An SIRS Epidemic Model Supervised by a Control System for Vaccination and Treatment Actions Which Involve First-Order Dynamics and Vaccination of Newborns. *Mathematics* **2021**, *10*, 36. [CrossRef]
11. Kaslik, E.; Neamţu, M.; Vesa, L.F. Global Stability Analysis of a Five-Dimensional Unemployment Model with Distributed Delay. *Mathematics* **2021**, *9*, 3037. [CrossRef]
12. Jiang, X. Global Dynamics for an Age-Structured Cholera Infection Model with General Infection Rates. *Mathematics* **2021**, *9*, 2993. [CrossRef]
13. Aletti, G.; Benfenati, A.; Naldi, G. Graph, spectra, control and epidemics: An example with a SEIR model. *Mathematics* **2021**, *9*, 2987. [CrossRef]
14. Mousa, M.M.; Alsharari, F. A Comparative Numerical Study and Stability Analysis for a Fractional-Order SIR Model of Childhood Diseases. *Mathematics* **2021**, *9*, 2847. [CrossRef]
15. Abia, L.M.; Angulo, Ó.; López-Marcos, J.C.; López-Marcos, M.Á. Computational Study on the Dynamics of a Consumer-Resource Model: The Influence of the Growth Law in the Resource. *Mathematics* **2021**, *9*, 2746. [CrossRef]
16. Lăzureanu, C. Integrable Deformations and Dynamical Properties of Systems with Constant Population. *Mathematics* **2021**, *9*, 1378. [CrossRef]

mathematics

Article

Evolutionary Optimization of Control Strategies for Non-Stationary Immersion Environments

Alexander Musaev [1,2], Andrey Makshanov [3] and Dmitry Grigoriev [4,*]

1 St. Petersburg State Technological Institute (Technical University), 190013 St. Petersburg, Russia; amusaev@technolog.edu.ru
2 St. Petersburg Institute for Informatics and Automation of the Russian Academy of Sciences, 199178 St. Petersburg, Russia
3 Department of Computing Systems and Computer Science, Admiral Makarov State University of Maritime and Inland Shipping, 198035 St. Petersburg, Russia; makshanov@oogis.ru
4 Center of Econometrics and Business Analytics (CEBA), St. Petersburg State University, 199034 St. Petersburg, Russia
* Correspondence: d.a.grigoriev@spbu.ru

Abstract: We consider the problem of evolutionary self-organization of control strategies using the example of speculative trading in a non-stationary immersion market environment. The main issue that obstructs obtaining real profit is the extremely high instability of the system component of observation series which implement stochastic chaos. In these conditions, traditional techniques for increasing the stability of control strategies are ineffective. In particular, the use of adaptive computational schemes is difficult due to the high volatility and non-stationarity of observation series. That leads to significant statistical errors of both kinds in the generated control decisions. An alternative approach based on the use of dynamic robustification technologies significantly reduces the effectiveness of the decisions. In the current work, we propose a method based on evolutionary modeling, which supplies structural and parametric self-organization of the control model.

Keywords: chaotic processes; control strategies; non-stationary environment; channel strategies; observation series; numerical studies; dynamic stability

MSC: 37M20; 37M10; 90C90

Citation: Musaev, A.; Makshanov, A.; Grigoriev, D. Evolutionary Optimization of Control Strategies for Non-Stationary Immersion Environments. *Mathematics* **2022**, *10*, 1797. https://doi.org/10.3390/math10111797

Academic Editors: Mihaela Neamţu, Eva Kaslik and Anca Rădulescu

Received: 31 March 2022
Accepted: 17 May 2022
Published: 24 May 2022

Publisher's Note: MDPI stays neutral with regard to jurisdictional claims in published maps and institutional affiliations.

Copyright: © 2022 by the authors. Licensee MDPI, Basel, Switzerland. This article is an open access article distributed under the terms and conditions of the Creative Commons Attribution (CC BY) license (https://creativecommons.org/licenses/by/4.0/).

1. Introduction

The problem of effective control in non-stationary immersion environments appears in a large number of applications. They are meteorology, the control of turbulent hydrodynamic flows, the stabilization of the state of non-stationary technological processes, asset management in capital markets, etc. The intricacy here is due to the unstable nature of the observed processes described by models of nonlinear chaotic dynamics [1–10]. Non-stationarity means that different observations have means, variances, and covariances that change over time. The non-stationary behavior can be a trend, a cycle, a random walk, or a combination of these. Non-stationary data is generally unpredictable and cannot be modeled or forecasted. Results obtained using non-stationary time series can be misleading because they may indicate a relationship between two variables when in fact there is none. For a better introduction to the subject, see the following Investopedia article [10].

High volatility and low predictability of chaotic processes significantly complicate the effective use of well-known control stabilization techniques based on traditional methods of adaptation and robustification. In particular, adaptive technologies turn out to be too inertial to close the feedback loop in time. Increasing the sensitivity of feedback leads to a high level of statistical errors of type II ("false alarms"). Robust control methods [11–15] focused on solutions with reduced sensitivity to statistical and dynamic variations of the observed process in the conditions of a basic model significantly lose effectiveness compared to optimal a posteriori versions.

A control strategy based on sequential dynamic optimization of the control model in chaotic dynamics, as will be shown in this paper, also faces a number of well-known problems. Here we have, for example, bruteforcing control parameter values with a given step that determines the accuracy of the solution, exponential increasing amounts of calculations with an increase in the number of model parameters. Taking into account the high variability and unsteadiness of the observed process, the number of parameters of the optimized control model should not exceed five or six even for modern techniques. Hence, there is a need to study new techniques of sequential dynamic optimization of the control model based on suboptimal computational schemes.

In this paper, we use an observation series of currency pairs in the Forex market as a testing ground for studying the effectiveness and stability of control algorithms in chaotic immersion environment. Dynamic chaos in an observation series violates the fundamental premise of the repeatability of experiments under identical conditions. In particular, two geometrically similar observation segments in conditions of chaos may have completely different aftereffects [16,17]. As a result, traditional statistical data analysis technologies and control algorithms based on them are ineffective. In our studies, the choice of channel control strategies is determined primarily by their accessibility and interpretability.

This article considers the possibility of applying sequential optimization based on the method of evolutionary modeling in conditions of stochastic market chaos [18–21] etc. The method is based on the Darwinian concept of evolutionary self-organization and the theory of random search.

2. Methods

2.1. Observation Model and Problem Statement

A significant difference between the presented work and traditional studies in the field of asset management is their focus on the Wald's additive observation model [22–24]

$$y_k = x_k + v_k, k = 1, \ldots, n \tag{1}$$

where x_k, $k = 1, \ldots, n$ is the system component used in the process of making management decisions (i.e., open, close or retain current position), and v_k, $k = 1, \ldots, n$ is the noise.

Currently, the prevailing point of view is that in market situations, the system component x_k in (1) is modeled as an output signal of a nonlinear system observed in the conditions of non-stationary and non-Gaussian interference v_k (dynamic chaos model) [5–13]. Lyapunov functions [25] and identification methods based on higher-order spectra [26] are used in order to substantiate such problem statements. There is a large area of research on the direct reconstruction of stochastic differential equations [27,28] for the model (1). Other points of view are based on nonlinear transformations of the y_k process, and, for example, on investigating fractal properties of the process trajectory [29–31].

Sequential filtering of the initial observation series y_k, $k = 1, \ldots, n$ is usually used to isolate the system component x_k, $k = 1, \ldots, n$ from (1) in real time. For this purpose, we utilize an exponential filter [32]:

$$x_k = \alpha y_k + (1 - \alpha) y_{(k-1)} = x_{(k-1)} + \alpha(y_k - x_{(k-1)}), k = 2, \ldots, n \tag{2}$$

with a discounting coefficient $\alpha \in [0.01, 0.3]$. This range is our empirical finding for the Forex market. Simultaneously, it is possible to select α adaptively so it better corresponds to features of considered data. This filtering technique is not the best one, because an increase in the smoothing effect with a decrease of α leads to a significant bias of the generated estimates. However, it produces a satisfactory result for the considered examples, while also providing simplicity of interpretation of the extracted system component.

Note that the conventional statistical observation model is based on the assumption that its system component is an unknown deterministic process, and its noise component is a stationary random process with independent increments [33]. Such a model was used in a wide class of trend analysis-based management strategies [34–37], though all of them are not resistant to possible dynamic variations in the process of changing quotes. One of the reasons for the unfeasible effectiveness of these trend indicators is the intrinsic inadequacy of the statistical approach for a chaotic observation series.

The main difference of the proposed model (1) is that its system component x_k, $k = 1, \dots, n$ is modeled as an oscillatory non-periodic process with a large number of local trends. This description indicates the possibility of interpreting this process as an implementation of the dynamic chaos model [5–13]. Its second distinctive feature is the noise component v_k, $k = 1, \dots, n$ being interpreted as a non-stationary random process described by an approximate Gaussian model with fluctuating parameters. In particular, correlations and spectral characteristics of this process change significantly over time [23,24].

The indicated features of observation series described by (1) violate the conditions of applicability of traditional statistical methods. In this case, numerical studies are the only approach to analyze the effectiveness of the developed management strategies. The statement of the asset management problem essentially coincides with the traditional formalization of the task to maximize the gain in the process of trading or investing capital.

Let $y_k = x_k + v_k$, $k = 1, \dots, n$ be a sequence of observations corresponding to a given time interval of asset management $T = n\Delta t$, where Δt is the selected interval between time counts. During the specified time, M operations are carried out in the trading process, each being determined by their start and finish $(k_{open}, k_{close})_j$, $j = 1, \dots, M$.

The trading problem can be thus formulated as follows: select a management strategy S and construct a sequence of actions u_j, $j = 1, \dots, M$ to obtain maximum profit G(S):

$$G(S) = \sum_{j=1}^{M} u_j(k_{close}) - u_j(k_{open}) = \max \quad (3)$$

In the simplest case, each management strategy is defined by the rules that determine the time of opening and closing a position $(k_{open}, k_{close})_j$, $j = 1, \dots, M$, and, in some cases, the lot size. The sum of the operation results at the k-th step $G_k(S)$ becoming smaller than the trader's available deposit G_0 means the management process resulted in complete loss.

The approach to determining the start time of an operation (the so-called "position opening") is the defining characteristic of a management strategy. The operation can be finished (a "position" can be "closed") when a specified profit level (TP, "take profit") or loss level (SL, "stop loss") is reached, or according to some different rules that could be more flexible.

This paper examines the technique of sequential evolutionary self-organization of the management strategy in the conditions of market chaos. Due to this, the simplest control schemes are used as basic control strategies, which makes it easy to visualize and interpret the obtained results. In particular, we used the so-called channel strategies [4,34–37]. Let us consider the simplest approaches to constructing such a strategy.

2.2. Channel Asset Management Strategies

We have a series of the trading asset's quote observations being modeled by (1). Let us define a "channel" as a range of observation values constrained by $y_k = x_k \pm B$, $k = 1, \dots, n$ [33]. Variations inside the channel $|y_k - x_k| = |\delta y_k| \leq B$, $k = 1, \dots, n$, are fluctuations that do not contain an obvious trend, in which case, the process can be referred to as a sideways trend or a flat. Channel width B can be selected depending on various considerations. It usually lies in the range from s_y to $3s_y$, where s_y is the estimate of the standard deviation (SD) δy_k, $k = 1, \dots, n$. In general, the choice of channel width depends on the nature of the data and the specificity of the selected management strategy. In some cases, the channel width may be some variable value $B_k = B_k(y_k)$, $k = 1, \dots, n$.

The observation series value y_k, $k = 1, \dots, n$ that breaks out of the channel is interpreted as the emergence of a trend in some management strategies. In the case of managing assets according to the trend direction, such events give rise to a recommendation to open a position in accordance with the sign of the channel boundary. Due to strong variability, a trend is often considered to be present when the system component x_k, $k = 1, \ldots, n$ formed by the exponential filter (2) with a given level of smoothing quits the channel. The values of the model parameters α, B, TP, SL are optional. Their selection depends on the knowledge and intuition of the trader, and they fully determine the management effectiveness. But it is often the case that intuition and other abilities of a human person appear to be ineffective in trading. Therefore, there is a need for strictly formalized and mathematically sound solutions.

We named the strategy of moving corresponding to the trend CSF (channel strategy forward). The idea behind it is simple: open a position up or down when the process breaks through the upper or lower bound of the channel respectively. The management algorithm contains two rules: Open Up position at $y_k > x_k + B$ or Open Dn at $y_k < x_k - B$. Otherwise, a position will be opened at each step outside the channel. In this regard, the more often used rules are based on determining the time of crossing the channel boundary $(y_{k-1} \leq x_{k-1} + B)$ & $(y_k > x_k + B)$ or $(y_{k-1} \geq x_{k-1} - B)$ & $(y_k < x_k - B)$, $k = 1, \ldots, n$.

In the simplest case, a position is closed either when the $y_{close} > y_{open} + TP$ or $= y_{close} < y_{open} - SL$ levels are reached (at Open Up) or when $y_{close} < y_{open} - TP$ or $y_{close} > y_{open} + SL$ (at Open Dn).

A more flexible control scheme allows for the obtaining of a model in which the upper and lower bounds of the channel are evaluated separately. An example of the implementation of the CSF management strategy for the Euro/Japanese Yen (EURJPY) currency pair on an observation interval of 1440 min counts (one day) is shown in Figure 1. In this case, the control model was defined by the parameters $\alpha = 0.04$, B1 = 15, B2 = 15, TP = 15, SL = 15. Blue diamonds indicate opening a position up, red diamonds denote opening a position down, and circles are positions being closed.

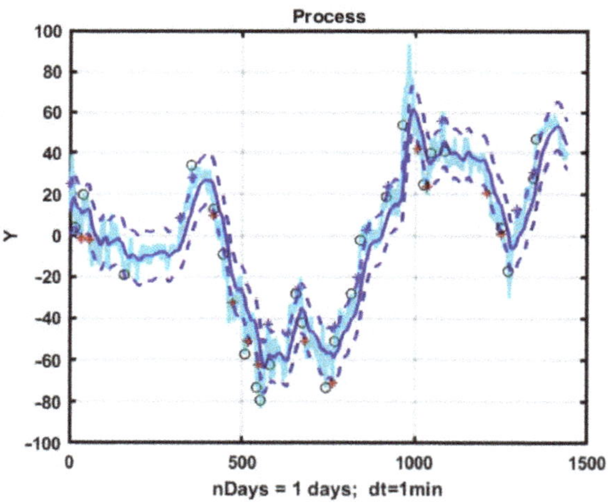

Figure 1. Example implementation of CSF on an observation interval of 1440 min counts.

Figure 2 presents management (3) effectiveness fluctuations in utilizing the channel strategy for the selected example. For the parameters that were used for the control model, the result turned out to be close to zero, and most of time during which trading was being carried out, it was negative. Note that in most cases of direct use of the CSF strategy, the result was negative.

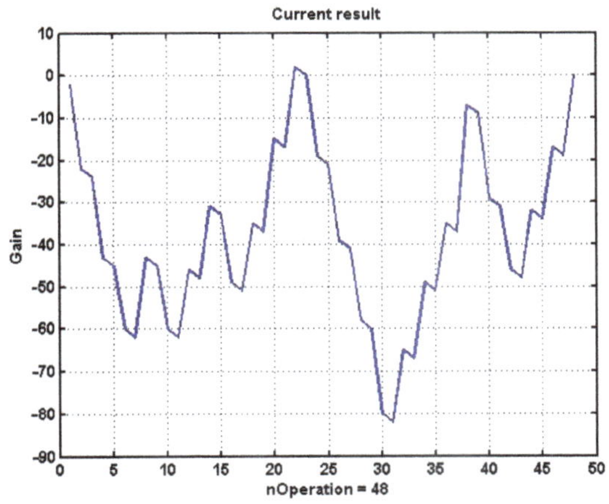

Figure 2. Performance of the CSF strategy for example in Figure 1.

Is it possible to achieve profit if there is a reliable forecast of the development of the process? To answer this question, we use bruteforce posterior optimization of the parameters of the management model M = {α, B1, B2, TP, SL}.

For each of the parameters, 15 iterations were carried out, starting from the values of M_0 = {0.01, 10, 10, 10, 10}. The values of the iteration step, respectively, were Step = {0.01, 1, 1, 1, 1}.

An example of the implementation of the CSF strategy with optimized parameter values α* = 0.05, B1* = 10, B2* = 12, TP* = 18, SL* = 18 at the same observation interval is presented in Figure 3. Note that this is the same EURJPY price history as in Figure 1, but the opening and closing moments are different. Figure 4 presents the change in performance during the use of CSF with optimized parameters. The management result for the day was G = 248 p.

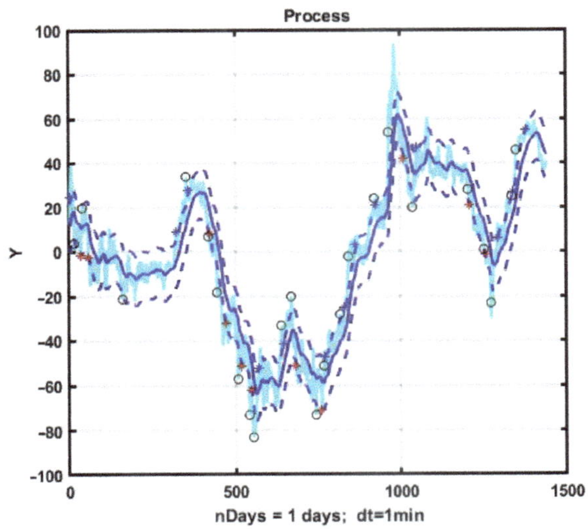

Figure 3. Example implementation of CSF with optimized parameters.

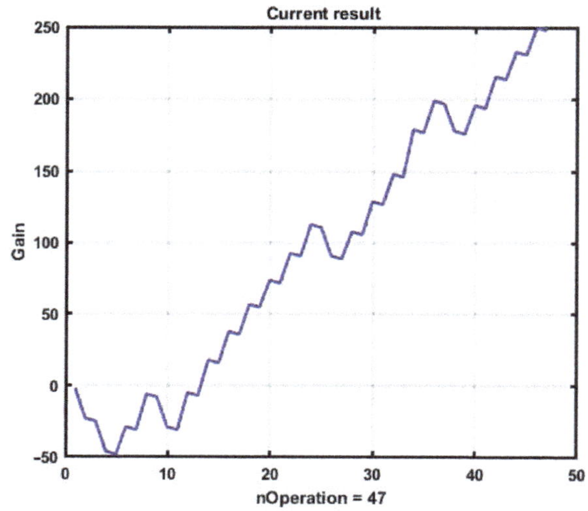

Figure 4. Performance of the CSF strategy with optimized parameters.

It is not difficult to see that this result is unstable. The results of applying the CSF strategy with the optimal parameters found for the next 10 days of management is shown in Figure 5. The obtained values completely confirm the instability of the found solution. None of the following nine days of observation produced a profitable result.

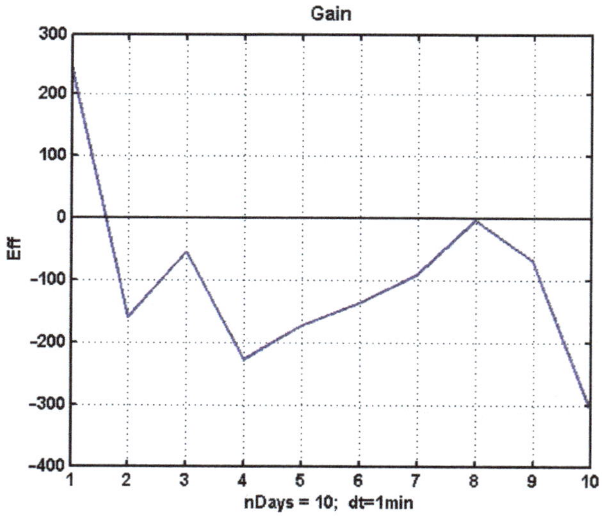

Figure 5. Performance of applying the CSF strategy with parameters optimal for the first day to the next nine days of management.

Comparing the results presented in Figure 6 with the dynamics of quotations of a financial instrument at the same observation interval, we can draw another conclusion. The structure of dynamics is more important than the degree of variability of the process. The minimum loss corresponds to the eighth day of observation, the most similar in structure to the first day for which the parameters of the control model were optimized. The worst result corresponds to the last, 10th day, with its strongly pronounced trend.

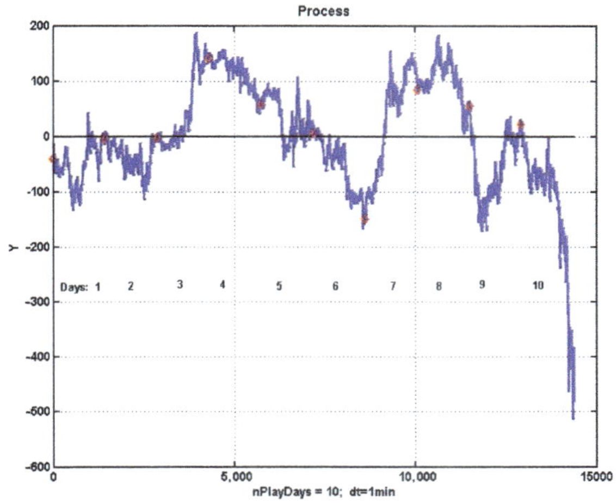

Figure 6. Dynamics of quotations during the indicated 10 days, corresponding to the performance of the CSF strategy in Figure 5.

Thus, in order to preserve, even if not the best, but at least positive result, it is necessary to significantly reduce the size of the time shift at which the optimal parameters are reassessed. This requirement significantly reduces the feasibility of using optimization based on bruteforcing optional parameters. The problem is that the number of calculations shows an exponential increase along with an increase in the number of optimized parameters. It takes about 15 min on a 2.5 GHz 6-core INTEL Core i5 to optimize the model with five parameters by bruteforce on an observation interval of one day. A model with six parameters will take about 1.5 h.

Hence, there is a need to switch to suboptimal computational schemes for sequential optimization, which significantly reduces the calculation time to a level that makes possible real-time sequential data processing.

2.3. Features of Evolutionary Optimization for Chaotic Immersion Environments

In this paper, an algorithm based on the method of evolutionary modeling is proposed as a suboptimal computational scheme for optimizing the management strategy [38]. Modern computing techniques and applications based on this method can be found in [39–46]. Unlike evolutionary modeling, evolutionary optimization of management strategies is not interested in the degree of similarity of the mathematical model to the real data obtained via monitoring the managed object and the parameters of the immersion environment. Its task is to choose a management model that produces the best solution according to (3). Genetic algorithms decrease the volume of computations by about 40% according to estimates given in a number of referenced works.

At the same time, the control strategy itself, as a set of decision rules, can also be modified. The implementation of nonparametric mutations of the management strategy consists in choosing the structure of the model and management rules from the a priori knowledge bank. As a set of decision rules is selected, the list of parameters to be changed, their critical values and ranges of changes are modified. The advantage of this approach is its feasibility. However, at the same time, the arbitrariness of machine choice is limited, and there is no possibility of obtaining radically new strategies that are not provided by the programmer. Any regularization and any set of restrictions can block access to unexpected original solutions.

Moreover, the complete removal of restrictions in the process of random modifications of the structure of strategies leads to a huge number of meaningless decision rules. Waiting for any reasonable solution to appear will take time comparable with real biological evolution. At the same time, the question of artificial generation of management strategies remains open.

Evolutionary technology, like the entire probabilistic-statistical paradigm, is implicitly oriented towards a comfortable hypothesis about the repeatability of experiments in unchanging or slowly changing conditions. The transition to non-stationary, and even chaotic processes, inevitably destroys the optimality of statistical solutions, including those constructed via evolutionary modeling. However, chaos, in general, contains regularizing effects that reduce the degree of total uncertainty. If evolutionary technology can identify, at least not explicitly, and use such hidden patterns, then the task of constructing a winning strategy may be feasible. In addition, using an evolutionary computational scheme will help us answer the question of the fundamental admissibility of particular classes of management strategies.

The paper uses the basic evolutionary modeling algorithm described in [31]. The return to the original version of this concept is due to the fact that it does not introduce additional restrictions on the mechanism of variability and leaves a wide range of opportunities for its formation. For example, genetic algorithms are focused on the model of bisexual reproduction, which is very important in the implementation of variability in biology. However, for the models under consideration, there is no need to limit the process of variability to the mechanisms of gene exchange. The same can be said about the method of differential evolution. The formation of a new genome, as a mutant vector formed from other parental genomes, also introduces unnecessary restrictions for this case.

In econometric models, one can make any modifications to the structure of genes if they do not contradict common sense and the laws of the market. Therefore, it is reasonable to use the computational scheme of evolutionary optimization, which corresponds to the traditional concept of evolutionary modeling. At the same time, the mechanism of variability is based on the well-known mechanism for extracting a random variable from the range of permissible variations in genome parameters.

2.4. Algorithm of Evolutionary Optimization of the Management Model

Consider a set of ancestor strategies $S_A = \{S_{A1}, \ldots, S_{ANa}\}$ with N_a elements, each of which is defined by its structure R (the decision-making rule) and a set of corresponding numerical parameters a, i.e., $S = \{R, a\}$. The effectiveness of a strategy $Eff(s)$ is assessed via applying it to the time series of observations Y(t), which together form an experimental retrospective dataset. We introduce two nonlinear operators.

1. The operator of variability and multiplication of strategies:

$$Var(S_A) : S_A = \{S_{A_1}, \ldots, S_{AN_a}\} \Rightarrow \{S_{D_1}, \ldots, S_{DN_D}\} : N_D = k_b N_A, \ k_b > 1$$

Here $\{S_{D_1}, \ldots, S_{DN_D}\}$ is a set of descendant strategies, each of which is created by modifying one of the ancestor strategies, N_D is the number of descendant strategies in one generation, and $k_b > 1$ is the multiplication coefficient of strategies. The union of ancestor strategies and descendant strategies is a generation of size $N_g = N_a + N_d = N_a\ (1 + k_b)$:

$$S_G = \left\{S_{G_1}, \ldots, S_{GN_g}\right\} = \{S_{A_1}, \ldots, S_{AN_a}\} \cup \{S_{D_1}, \ldots, S_{DN_D}\}$$

2. A selection operator that selects the "surviving" strategies from the generation $S_G = \left\{S_{G_1}, \ldots, S_{GN_g}\right\}$ that become the ancestors of the next generation:

$$Sel\left(S_1, \ldots, S_{N_g}\right) = \left\{S_1, \ldots, S_{N_g}\right\} \Rightarrow \left\{S_{\langle 1\rangle}, \ldots, S_{\langle N_a\rangle} : Eff\left(S_{\langle 1\rangle}\right) \geq \ldots \geq Eff\left(S_{\langle N_a\rangle}\right) \geq Eff(S_j), \forall j > N_a\right\}$$

Let $S_0 = \{K_0,\ p_0, a_0\}$ be a particular management strategy with the specified parameters, adopted as the basic "parent strategy". Then evolutionary optimization is reduced to a cyclic execution of an operator sequence:

$$\begin{array}{c} S_0 \Rightarrow Var(S_0) = S_A = \{S_1, \ldots,\ S_{N_A}\} \ \Rightarrow \ Var(S_A) = S_D = \{S_1, \ldots,\ S_{N_D}\} \\ \Uparrow \\ Sel\left(S_1, \ldots,\ S_{N_g}\right) = \left\{S_1, \ldots,\ S_{N_g}\right\} \Leftarrow \ S_G = (S_A) \cup (S_D) \end{array} \quad (4)$$

where arrows denote the sequential order of operators. Since selection is carried out by superiority, the optimality of the final solution is not guaranteed. However, it will be the best of the whole set created during the implementation of evolutionary modeling.

The process of evolutionary optimization is obviously converging to more effective strategies by virtue of its very construction. This is due to the fact that the new generation always includes ancestor strategies in its composition. Consequently, the most effective strategies in principle cannot be discarded by the accepted selection and selection procedure. However, a high convergence rate cannot be expected due to the randomness of the modification process. The convergence rate will be close to the convergence rate of a random search, and depends on the size N_g of the generation being formed. It can be assumed that the convergence rate will be higher if the multiplication coefficient k_b is made variable so that the number of descendant strategies N_d depends on the effectiveness of parent strategies, i.e., $N_d = k(Eff(S_a)), k > 1$. In other words, a more effective ancestor can produce more offspring. However, this statement requires additional verification.

Other regularization methods aimed at increasing the convergence rate of evolutionary optimization are also possible.

2.5. Computational Aspects of the Evolutionary Optimization Algorithm

The functional structure of the algorithm of evolutionary optimization of management strategies is shown in Figure 7. The sequence of evolution is represented by a diagram of a process developing from the bottom up.

Let us consider the presented algorithm. At the preliminary stage, some basic strategy (prototype) is formed. Its structure is chosen either randomly or based on existing a priori experience in asset management. With the help of the variability mechanism, the basic prototype strategy is modified, giving rise to parent strategies.

Then, in accordance with (4), the program loops for the number of successive generations, which are usually called epochs.

At the first step of this loop, the first generation of descendant strategies is formed using the variability generator, which, combined with their parents, form the first generation of strategies S_G. Furthermore, each of the first-generation strategies undergoes a testing by being applied to a set of retrospective observations {Y(t), Y(t − Y)}, where T is the size of the validation dataset. The created strategies are ranked by their effectiveness $Eff(S_i)$, $i = 0, \ldots, N_g$ and a specified number N_a of "surviving" strategies that are allowed for further "reproduction" (modification) are selected. The selected strategies are the parents of a new set of modified descendant strategies and together with them form the second generation.

Furthermore, the cycle is repeated until either a specified number of generations is reached, or, according to some other criterion, for example, when the effectiveness of management does not improve during a given number of epochs.

The variability mechanism used in the program randomly selects changes made to the strategy from the following set of options:

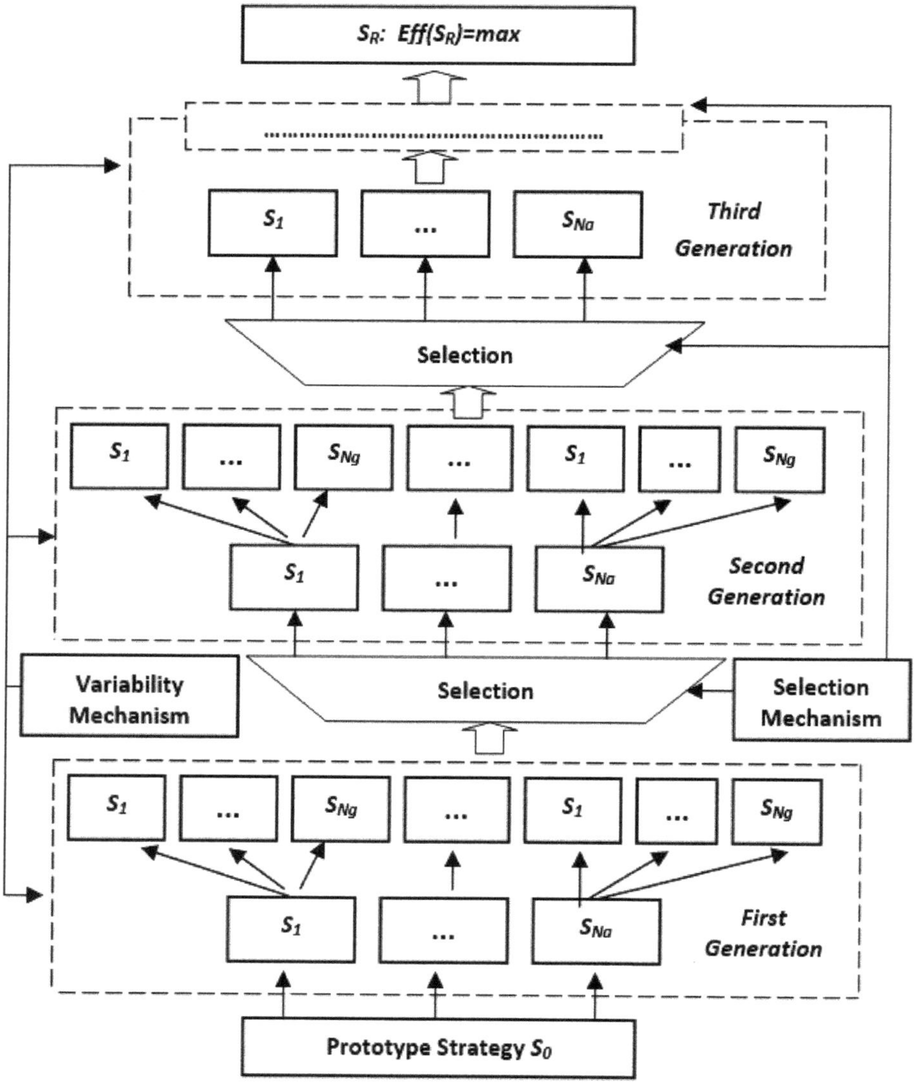

Figure 7. General functional structure of the evolutionary optimization algorithm.

1. Small single changes. In the strategy undergoing modification, relatively small changes are made to only one parameter (gene) selected by a random draw. The size of the change range depends on the parameter. This one-time change does not usually exceed 10% of the original value. The choice of the parameter is carried out by a random draw, similar to how it happens in the Monte Carlo method.
2. Small group changes. They are carried out similarly to the previous case, but are made to several gene parameters at once instead. Their amount and their numbers are selected via a random draw.
3. Strong single mutation or parametric mutation. The gene number is selected via a random draw. Usually, the number of mutations in a generation is small, and the probability of their occurrence does not exceed 2–3%. The size of the change field

also depends on the parameter, usually a single mutation can reach 30–50% of the original value.

4. Strong group changes are similar to the previous case, but are made immediately in several parameters, as in case 2.
5. Structural (nonparametric) mutations. The parent strategy with some relatively small probability (usually less than 0.01) may undergo nonparametric mutation. In this case, the number of genes in the original genome may change, or, in a more radical case, the management strategy itself may be completely modified. The most rational way in this case consists of randomly choosing a management strategy from an a priori created knowledge base.

It should be expected that such mutations will be quite constructive, since the strategies included in the data bank, to one degree or another, have already been pre-selected by at least the common sense of their developer. For example, there is a transition from a channel strategy to management based on trend analysis, etc. However, this approach limits the evolution to the level of the programmer's constructive imagination.

A more radical approach consists in stochastic synthesis of new strategies using artificial intelligence technologies. In this case, the evolution program gets out of the strict control of the developer not only at the parametric, but also at the structural level. This approach, in theory, enables the generation of completely new, unexpected solutions. However, in most cases, such mutations will generate ineffective strategies that will be immediately eliminated by the selection mechanism, without generating variants of management strategies in the next generation. In this regard, it may make sense to deliberately preserve such mutations for a given number of epochs and allow them to generate variants of descendants in a mutant zoo.

The selection mechanism, as already noted, ranks a set of strategies according to their effectiveness, and selects among them the best strategies that become the parents of the next generation. It is important to note the principle of incomplete or open solutions, meaning that at each cycle of selection, not the only best option is selected, but a group of strategies is as well. This approach allows us to come to the best solution along a chain consisting of intermediate options other than optimal, which is especially important when searching for effective management strategies in chaotic immersion environments.

3. Results

As an example of the implementation of the evolutionary optimization method, consider the asset management task based on the CSF management strategy. In order to be able to compare with the best solution found by bruteforce, we will consider the same one-day observation interval as in the examples shown in Figures 1–4. As optimization parameters (the genome) we will use M = {α, B1, B2, TP, SL}. From the above optimization example (Figures 3 and 4) it follows that the best parameters of the control model are $\alpha^* = 0.05$, $B1^* = 10$, $B2^* = 12$, $TP^* = 18$, $SL^* = 18$.

Figure 8 shows an example of CSF implementation with evolutionary optimization of the management model for 10 epochs. The best result was 177 points with the parameters of the model being $\alpha^* = 0.05$, $B1^* = 12$, $B2^* = 12$, $TP^* = 15.5$, $SL^* = 14.8$. The convergence to the found result is shown in Figure 9.

The dependence of the effectiveness of the CSF strategy during the epoch is shown in Figure 10.

The plots show that the obtained result, as expected, is inferior to the gain obtained by bruteforce. However, this search only took 0.26 s with the same CPU and with the observation interval the same as in the example with bruteforce optimization, and half of this time was spent on the implementation and output of graphics.

The question arises of how much the quality of management can be improved by increasing the number of epochs. Figures 11 and 12 show changes in the effectiveness of the best versions of the management model obtained during, 50 and 200 iterations, respectively.

The gain in the above examples reached, 197 p. and 203 p., respectively, did not achieve the effectiveness of bruteforce (248 p). However, the time spent was 0.203 s in the first case, and 0.541 s in the second. A further search for the best option was quite slow, and this can be seen from the section of the horizontal line in Figure 12, starting from about 70 epochs. Increasing the number of epochs to 500 increased the profit to only 209 p., and spending to 1.17 s.

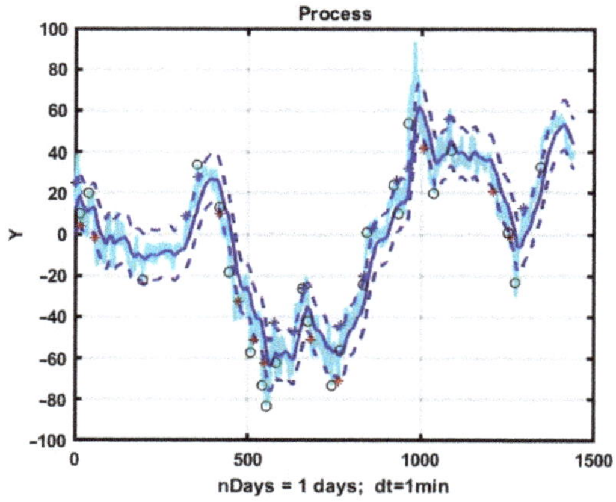

Figure 8. An example implementation of the CSF strategy with parameters obtained via evolutionary optimization.

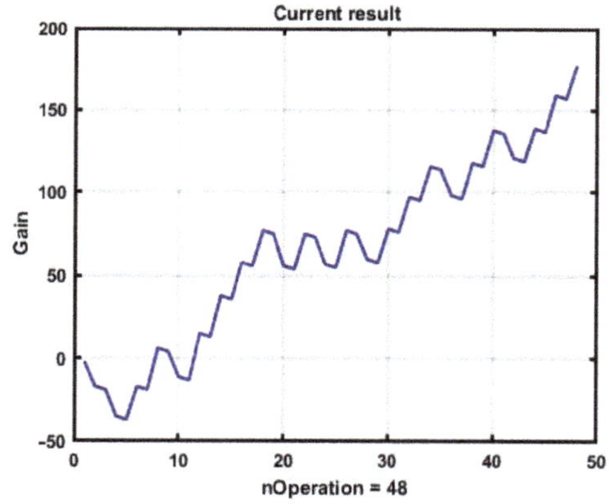

Figure 9. Effectiveness of the CSF strategy with evolutionary optimization.

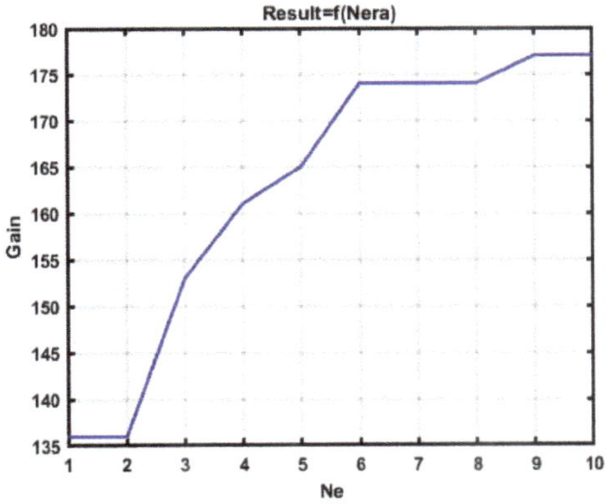

Figure 10. Effectiveness of the CSF strategy with evolutionary optimization for 10 epochs.

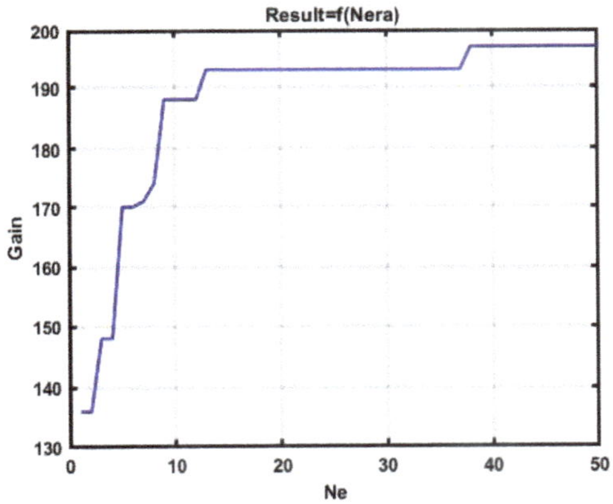

Figure 11. Effectiveness of evolutionary optimized CSF strategy, 50 epochs.

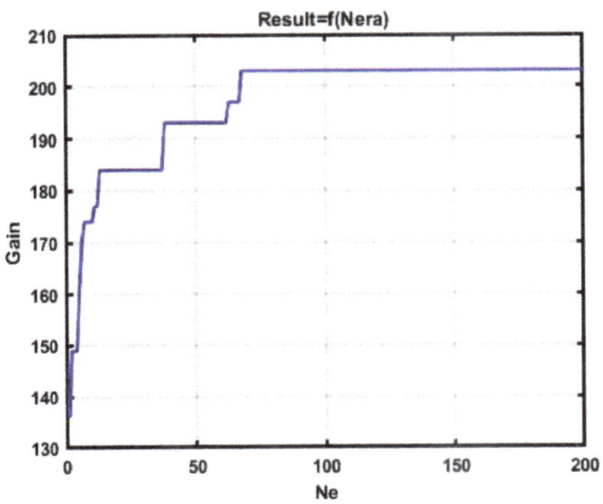

Figure 12. Effectiveness of evolutionary optimized CSF strategy, 200 epochs.

Note that in the considered examples, a simplified version of the variability operator was used. For example, structural mutation was not used at all. However, even in the proposed case, it becomes possible to conclude that this algorithm is suitable for sequential optimization with new data arriving every 0.5–1 s.

4. Discussion

The chaotic dynamics of quotations, characteristic for electronic capital markets, obstruct the effective use of traditional methods of increasing the stability of management strategies. As shown in the examples in Figures 5 and 6, the parameters that are optimal for a given day of observing the dynamics of quotations on the very next day lead to an unpredictable loss. In this regard, the management model should be dynamically corrected as new results of monitoring the market situation become available.

At the same time, the most powerful optimization tool based on bruteforce incurs significant time costs even for modern high-performance processors. This is due to the fact that the time complexity here is described by a power function with an increase in the number of iteration steps, and by an exponential function with an increase in the number of model parameters. As a result, even with a relatively simple model with five to six parameters, the required computational resources exceed the capabilities of modern processors. At the same time, increasing the time interval between successive calculations of the best parameters is unacceptable, given that the dynamics of market assets is absolutely non-inertial and can be discontinuous even within one minute of observation. Due to this, we proposed an optimization of the management model based on evolutionary modeling.

We analyzed the effectiveness of this approach using the example of real data obtained by monitoring the currency exchange market. The results of numerical studies have shown that maximum gain decreased by about 15–20% compared to the result obtained by bruteforcing parameter values. At the same time, the number of operations decreased to such a level that model parameters can be optimized even with a one-second interval of monitoring results.

It is important to note that the potential of evolutionary optimization is by no means exhausted by the version of its implementation considered in the paper. First of all, the above example did not use structural mutations. In essence, we are talking about utilizing qualitatively different strategies, which can give a result that significantly exceeds the mechanistic optimization based on bruteforcing parameter values. Moreover, new opportunities arise, such as artificial intelligence technologies: the computer independently generates management strategies that are not contained in the a priori knowledge base.

The second point related to the advantage of using evolutionary optimization is due to the fact that the drawing of parameter values is carried out on a scale of counts close to continuous. The discreteness is essentially determined only by the mantissa of the bit grid. This means that the conducted adjustment is finer than that obtained with a fixed iteration step. The advantage of fine-tuning is due to the extremely high sensitivity of the management result to variations in the values of the model parameters, which is generally characteristic of a chaotic series of observations.

It should be noted that we used a very simple channel strategy as a demonstration example in this work. This was done in order to ensure the clarity and interpretability of the results. However, in practice, such strategies are not used as is, because in most cases they only lead to loss.

In [33] it was shown that for almost any variations of the observed dynamic process, profitable solutions exist for this class of strategies. However, their implementation requires fine-tuning, minor deviations from which, as follows from the theory of dynamic chaos, lead to loss. Nevertheless, with a small time shift of observation series relative to the optimization tuning interval, the positive gain on average is preserved. This means that it is fundamentally possible to build a dynamic self-adjusting asset management system. One of the variants of such a solution is considered in this paper.

Other formulations of asset management tasks that also use genetic algorithms for optimization can be seen, for example, in [47,48]. At the same time, various algorithms based on random search in some cases lead to completely unexpected and, often, anti-intuitive results [49,50].

A development of these studies being focused on constructing a dynamic self-adjusting asset management system with an increased level of resistance to variations in the dynamic and statistical characteristics of observation series implies considering the following points:

- the study of potential characteristics of self-adjusting asset management systems for various sets of dynamic properties of observation intervals of chaotic processes;
- the development of a knowledge base of management strategies and its application for implementing structural mutations of the management model in the mechanism of variability of the evolutionary optimization algorithm;
- the development of randomized synthesis of management strategies using multi-expert data analysis [51];
- the use of composite algorithms combining the capabilities of robustification and adaptation in management decision-making.
- the effectiveness of the application of evolutionary optimization in markets and periods that differ in the degree of market efficiency within the Efficient Market Hypothesis (EMH) [52]. It is supposed that the greatest profit can be made in a highly inefficient market. At the same time various exchange markets all have multifractal structural properties with different levels in the sample and sub-samples that cause inefficiency with different levels in these foreign exchange markets [30]. Another work reveals that the efficiency of the cryptocurrency markets varies over time, which is consistent with adaptive market hypothesis (AMH) [53]. The question about the level of current market inefficiency which is acceptable for self-adjusting asset management systems needs to be investigated.
- the use of external add-ons that carry information exogenous to technical analysis on expected trends of the considered financial instrument and market mood in general.
- The outlined issues constitute the subject of our further research.

Author Contributions: Conceptualization, Methodology, A.M. (Andrey Makshanov); Validation, Writing, A.M. (Alexander Musaev); Review and Editing, Investigation, Programming, Visualization, Administration, Scientific Discussions, Supervision, Funding Acquisition, D.G. All authors have read and agreed to the published version of the manuscript.

Funding: The research of Alexander Musaev described in this paper is partially supported by the state research FFZF-2022-0004. The research of Dmitry Grigoriev for this paper is supported by Saint Petersburg State University, project ID: 93024916.

Institutional Review Board Statement: Not applicable.

Informed Consent Statement: Not applicable.

Data Availability Statement: The source of data is Finam.ru (https://www.finam.ru/, accessed on 21 April 2022).

Acknowledgments: The authors are grateful to participants at the Center for Econometrics and Business Analytics (ceba-lab.org, CEBA) seminar series for helpful comments and suggestions.

Conflicts of Interest: The authors declare that they have no conflict of interest.

References

1. Layek, G.C. *An Introduction to Dynamical Systems and Chaos*; Springer: New Delhi, India, 2015; Volume 449.
2. Kim, E.J. Intermittency and self-organisation in turbulence and statistical mechanics. *Entropy* **2019**, *21*, 574. [CrossRef]
3. Inglada-Perez, L. A comprehensive framework for uncovering non-linearity and chaos in financial markets: Empirical evidence for four major stock market indices. *Entropy* **2020**, *22*, 1435. [CrossRef]
4. Gregory-Williams, J.; Williams, B.M. *Trading Chaos: Maximize Profits with Proven Technical Techniques*, 2nd ed.; John Wiley & Sons: New York, NY, USA, 2004; p. 251.
5. Feldman, D. *Chaos and Dynamical Systems*; Princeton University Press: Princeton, NJ, USA, 2019.
6. Bensaïda, A.; Litimi, H. High level chaos in the exchange and index markets. *Chaos Solitons Fractals* **2013**, *54*, 90–95. [CrossRef]
7. Klioutchnikov, I.; Sigova, M.; Beizerov, N. Chaos theory in finance. *Procedia Comput. Sci.* **2017**, *119*, 368–375. [CrossRef]

8. Gu, E.G. On the existence of chaos in a discontinuous area-preserving map arising in financial markets. *Int. J. Bifurc. Chaos* **2018**, *28*, 1850177. [CrossRef]
9. Sarkar, A.; Chakrabarti, G.; Sen, C. Volatility, long memory, and chaos: A discussion on some "Stylized facts" in financial markets with a focus on high frequency data. In *Development and Sustainability*; Springer: New Delhi, India, 2013; pp. 71–101.
10. Iordanova, T. An Introduction to Non-Stationary Processes. Available online: https://www.investopedia.com/articles/trading/07/stationary.asp (accessed on 21 April 2022).
11. Riani, M.; Hubert, M. Editorial, special issue on "Advances in robust statistics". *Metron* **2021**, *79*, 121–125. [CrossRef]
12. Llanos, C.; Galletti, R.C.; Maronna, R. On-line process monitoring using a robust statistics based methodology. *Lat. Am. Appl. Res.-Int. J.* **2019**, *49*, 111–116. [CrossRef]
13. Maronna, R.A.; Martin, R.D.; Yohai, V.J.; Salibián-Barrera, M. *Robust Statistics: Theory and Methods (with R)*; John Wiley & Sons: New York, NY, USA, 2019.
14. Rousseeuw, P.J.; Hubert, M. Anomaly detection by robust statistics. *Wiley Interdiscip. Rev. Data Min. Knowl. Discov.* **2018**, *8*, e1236. [CrossRef]
15. Petrelli, M. Introduction to robust statistics. In *Introduction to Python in Earth Science Data Analysis*; Springer: Cham, Switzerland, 2021; pp. 181–194.
16. Musaev, A.; Grigoriev, D. Machine learning-based cyber-physical systems for forecasting short-term state of unstable systems. In *Cyber-Physical Systems: Intelligent Models and Algorithms. Studies in Systems, Decision and Control*; Springer: Cham, Switzerland, 2022; Volume 417.
17. Musaev, A.; Makshanov, A.; Grigoriev, D. Forecasting multivariate chaotic processes with precedent analysis. *Computation* **2021**, *9*, 110. [CrossRef]
18. Jin, Y.; Wang, H.; Chugh, T.; Guo, D.; Miettinen, K. Data-driven evolutionary optimization: An overview and case studies. *IEEE Trans. Evol. Comput.* **2018**, *23*, 442–458. [CrossRef]
19. Zhang, Y.; Gong, D.W.; Sun, J.Y.; Qu, B.Y. A decomposition-based archiving approach for multi-objective evolutionary optimization. *Inf. Sci.* **2018**, *430*, 397–413. [CrossRef]
20. Abu Khurma, R.; Aljarah, I.; Sharieh, A.; Abd Elaziz, M.; Damaševičius, R.; Krilavičius, T. A review of the modification strategies of the nature inspired algorithms for feature selection problem. *Mathematics* **2022**, *10*, 464. [CrossRef]
21. Kalyuzhnaya, A.V.; Nikitin, N.O.; Hvatov, A.; Maslyaev, M.; Yachmenkov, M.; Boukhanovsky, A. Towards generative design of computationally efficient mathematical models with evolutionary learning. *Entropy* **2020**, *23*, 28. [CrossRef]
22. Boer, K.; Kaymak, U.; Spiering, J. From discrete-time models to continuous-time, asynchronous modeling of financial markets. *Comput. Intell.* **2007**, *23*, 142–161. [CrossRef]
23. Hastie, T.J.; Tibshirani, R.J. *Generalized Additive Models*; Routledge: Abingdon, UK, 2017.
24. Musaev, A.; Grigoriev, D. Analyzing, modeling, and utilizing observation series correlation in capital markets. *Computation* **2021**, *9*, 88. [CrossRef]
25. Bezruchko, B.P.; Smirnov, D.A. *Extracting Knowledge from Time Series: An Introduction to Nonlinear Empirical Modeling*; Springer: New York, NY, USA, 2010.
26. Petropulu, A.P. Higher-order spectral analysis. In *The Digital Signal Processing Handbook, Second Edition: Wireless, Networking, Radar, Sensor Array Processing, and Nonlinear Signal Processing*; CRC Press: Boca Raton, FL, USA, 2009; pp. 18–21.
27. Cremers, J.; Hübler, A. Construction of differential equations from experimental data. *Z. Nat. A* **1987**, *42*, 797–802. [CrossRef]
28. Crutchfield, J.P.; McNamara, B.S. Equations of motion from a data series. *Complex Syst.* **1987**, *1*, 417–452.
29. Cheng, C. Multi-scale Gaussian process experts for dynamic evolution prediction of complex systems. *Expert Syst. Appl.* **2018**, *99*, 25–31. [CrossRef]
30. Han, C.; Wang, Y.; Ning, Y. Comparative analysis of the multifractality and efficiency of exchange markets: Evidence from exchange rates dynamics of major world currencies. *Phys. A Stat. Mech. Appl.* **2019**, *535*, 122365. [CrossRef]
31. Kirichenko, L.; Radivilova, T.; Bulakh, V. Binary classification of fractal time series by machine learning methods. In *International Scientific Conference Intellectual Systems of Decision Making and Problem of Computational Intelligence*; Springer: Cham, Switzerland, 2019; pp. 701–711.
32. Gardner, E.S.J. Exponential smoothing: The state of the art—Part II. *Int. J. Forecast.* **2006**, *22*, 637–666.
33. Musaev, A.; Makshanov, A.; Grigoriev, D. Numerical studies of channel management strategies for nonstationary immersion environments: EURUSD case study. *Mathematics* **2022**, *10*, 1408. [CrossRef]
34. Nti, I.K.; Adekoya, A.F.; Weyori, B.A. A systematic review of fundamental and technical analysis of stock market predictions. *Artif. Intell. Rev.* **2020**, *53*, 3007–3057. [CrossRef]
35. Niederhoffer, V.; Kenner, L. *Practical Speculation*; John Wiley & Sons: New York, NY, USA, 2005.
36. Colby, R.W.; Meyers, T.A. *The Encyclopedia of Technical Market Indicators*; IRWIN Professional Publishing: Burr Ridge, IL, USA, 2012.
37. Chordia, T.; Goyal, A.; Saretto, A. *P-Hacking: Evidence from Two Million Trading Strategies*; Swiss Finance Institute: Zürich, Switzerland, 2018.
38. Lu, Y. Artificial intelligence: A survey on evolution, models, applications and future trends. *J. Manag. Anal.* **2019**, *6*, 1–29. [CrossRef]
39. Jia, L.U.; Shazemeen, N.M.; Martinkute-Kauliene, R. Portfolio decision using time series prediction and multi-objective optimization. *Rom. J. Econ. Forecast.* **2020**, *23*, 118.

40. Eltaeib, T.; Mahmood, A. Differential evolution: A survey and analysis. *Appl. Sci.* **2018**, *8*, 1945. [CrossRef]
41. Mamanis, G. A comparative study on multi-objective evolutionary algorithms for tri-objective mean-risk-cardinality portfolio optimization problems. In *Computational Management*; Springer: Cham, Switzerland, 2021; pp. 277–303.
42. Masich, I.S.; Kulachenko, M.A.; Stanimirović, P.S.; Popov, A.M.; Tovbis, E.M.; Stupina, A.A.; Kazakovtsev, L.A. Formation of fuzzy patterns in logical analysis of data using a multi-criteria genetic algorithm. *Symmetry* **2022**, *14*, 600. [CrossRef]
43. Slowik, A.; Kwasnicka, H. Evolutionary algorithms and their applications to engineering problems. *Neural Comput. Appl.* **2020**, *32*, 12363–12379. [CrossRef]
44. Bansal, J.C.; Singh, P.K.; Pal, N.R. *Evolutionary and Swarm Intelligence Algorithms*; Springer: Cham, Switzerland, 2019; Volume 779.
45. Tian, Y.; Cheng, R.; Zhang, X.; Li, M.; Jin, Y. Diversity assessment of multi-objective evolutionary algorithms: Performance metric and benchmark problems. *IEEE Comput. Intell. Mag.* **2019**, *14*, 61–74. [CrossRef]
46. Cai, X.; Zhao, H.; Shang, S.; Zhou, Y.; Deng, W.; Chen, H.; Deng, W. An improved quantum-inspired cooperative co-evolution algorithm with muli-strategy and its application. *Expert Syst. Appl.* **2021**, *171*, 114629. [CrossRef]
47. Asjad, M.; Khan, S. Analysis of maintenance cost for an asset using the genetic algorithm. *Int. J. Syst. Assur. Eng. Manag.* **2017**, *8*, 445–457. [CrossRef]
48. Sarijaloo, A.; Moradbakloo, A. Asset management using genetic algorithm: Evidence from Tehran stock exchange. *Manag. Sci. Lett.* **2014**, *4*, 221–226. [CrossRef]
49. Fogel, G.B. Evolutionary computation for the inference of natural evolutionary histories. *IEEE Connect.* **2005**, 3, 11–14.
50. Yang, X.S. Nature-inspired optimization algorithms: Challenges and open problems. *J. Comput. Sci.* **2020**, *46*, 101104. [CrossRef]
51. Musaev, A.; Grigoriev, D. Multi-expert systems: Fundamental concepts and application examples. *J. Theor. Appl. Inf. Technol.* **2022**, *100*, 336–348.
52. Singh, J.E.; Babshetti, V.; Shivaprasad, H.N. Efficient market hypothesis to behavioral finance: A review of rationality to irrationality. *Mater. Today Proc.* 2021, *in press*. [CrossRef]
53. Chu, J.; Zhang, Y.; Chan, S. The adaptive market hypothesis in the high frequency cryptocurrency market. *Int. Rev. Financ. Anal.* **2019**, *64*, 221–231. [CrossRef]

mathematics

MDPI

Article

Numerical Studies of Channel Management Strategies for Nonstationary Immersion Environments: EURUSD Case Study

Alexander Musaev [1,2], Andrey Makshanov [3] and Dmitry Grigoriev [4,*]

1 St. Petersburg State Technological Institute (Technical University), 190013 St. Petersburg, Russia; amusaev@technolog.edu.ru
2 St. Petersburg Institute for Informatics and Automation of the Russian Academy of Sciences, 199178 St. Petersburg, Russia
3 Department of Computing Systems and Computer Science, Admiral Makarov State University of Maritime and Inland Shipping, 198035 St. Petersburg, Russia; makshanov@oogis.ru
4 Center of Econometrics and Business Analytics (CEBA), St. Petersburg State University, 199034 St. Petersburg, Russia
* Correspondence: d.a.grigoriev@spbu.ru

Abstract: This article considers a short-term forecasting of a process that is an output signal of a nonlinear system observed on the background of additive noise. Forecasting is made possible thanks to the technique of nonparametric estimation of local trends. The main problem in this case is the instability of the time of the existence of these local trends. The average duration of relatively stable intervals can be estimated from earlier observation history. Such approaches are called channel strategies. The task of constructing such strategies for EURUSD asset management in the conditions of market chaos is considered, as well as the potential capabilities of these management strategies via computational experiments. We demonstrated the fundamental possibility of achieving profit even for areas with complex dynamics with abrupt changes in the considered process. We propose improved channel strategies and also denote the main directions of increasing their effectiveness.

Keywords: nonstationary immersion environment; chaotic processes; channel strategies; series of observations; channel control strategies; numerical studies; dynamic stability; currency asset management

MSC: 37M20; 37M10; 90C90

Citation: Musaev, A.; Makshanov, A.; Grigoriev, D. Numerical Studies of Channel Management Strategies for Nonstationary Immersion Environments: EURUSD Case Study. *Mathematics* **2022**, *10*, 1408. https://doi.org/10.3390/math10091408

Academic Editors: Mihaela Neamţu, Eva Kaslik and Anca Rădulescu

Received: 12 March 2022
Accepted: 18 April 2022
Published: 22 April 2022

Publisher's Note: MDPI stays neutral with regard to jurisdictional claims in published maps and institutional affiliations.

Copyright: © 2022 by the authors. Licensee MDPI, Basel, Switzerland. This article is an open access article distributed under the terms and conditions of the Creative Commons Attribution (CC BY) license (https://creativecommons.org/licenses/by/4.0/).

1. Introduction

Today, financial markets are the object of increased interest from scientists, since the growth of the national economy and the well-being of individuals depend on their stability and effective functioning. Followers of the efficient-market hypothesis (EMH) [1] prefer the "buy and hold" trading strategy, counting on the long-term growth of the selected asset. Within this theory, it is believed that it is impossible to beat stock indexes for a long time. However, a number of fundamental investors, e.g., Warren Buffet, who invest in accordance with the company's current financial performance [2], demonstrate their advantage over the dynamics of the stock index. In contrast to passive investment, active investment involves the purchase and sale of a security according to some condition that develops in the fundamental indicators of the company (fundamental analysis), or according to the current dynamics of quotations (technical analysis).

Fundamental analysis (FA) uses available news and other information to evaluate the direction of price movement in accordance with general economic, political and other considerations [3]. Results of expert FA are regularly published on the websites of brokers and traders in the form of analytical reports and forecasts.

Technical analysis (TA) is associated with a formalized study of the dynamics of quotations and uses the entire arsenal of modern computer mathematics [4]. In practice, traders usually use a combination of both types of analysis.

Thus, at the moment there is no algorithmic approach to making a trading decision that would have a clear advantage over others. Under these conditions, stock analysts prefer to justify their recommendations using channel strategies that are identified on simple calculating rules, insider information, market psychology, and visual interpretation [5].

Various strategies of this type are described in well-known monographs on the management of market assets in investing and trading [5–10]. The main idea of channel strategies is to detect local trends in the development of the observed process. Such trends are assessed based on the localization of the current value of the asset in a certain range of its changes (i.e., a "channel"). Most often, channels are understood as areas of price changes of a priori selected width, parallel to the smoothed quotation average.

The simplicity and visual expressiveness of such strategies makes them extremely attractive, but their effectiveness, in general, is quite low. The main reason for this fact is the chaotic nature of the change in the values of quotations, described, for example, in [11,12]. The presence of dynamic chaos in the series of observations violates the basis of the entire probabilistic-statistical paradigm, i.e., the repeatability of experiments under identical conditions. In essence, the price series are formed by a flow of bifurcation points, for which arbitrarily small perturbations lead to an unpredictable development of the dynamics of the process.

It is difficult to assess the last statement analytically due to the fact that the right-hand sides of the corresponding systems of differential equations are random functions with constantly changing structures. Observations of financial markets demonstrate this: a newly found trend is very likely to reverse in the next moment. Therefore, back-test effective strategies could be effective later on by pure chance. The inability of trends to go on can be explained by the composition of the trading participants being an equilibrium balance. Trend-following traders (momentum traders) impact is compensated by market makers and mean reversion traders [13]. At the same time, according to the inaction inertia theory [14], individual investors who missed the opportunity to buy in at a good price withhold from a trade at a less favorable price, thus halting trend development. However, all of this becomes irrelevant when the fundamental valuation of the asset changes. If the majority of traders shares this opinion, a continuous trend may arise [15,16].

The only available approaches to effectiveness analysis of channel strategies and to investigating the chaotic nature of the observation series of market assets are numerical and statistical big data methods. In the work [5], a large-scale empirical study of the effectiveness of the channel strategy developed by the authors was carried out from 31 July 1990 to 31 July 2010 for more than 3000 stocks of the NYSE and NASDAQ markets. It was shown that using the support and resistance levels (SAR), it is in general impossible to build a steadily profitable trading management system. However, making a decision on the price rollback from the support level works better than at the resistance level. The authors substantiate this fact by the globally growing trend in the stock market during this period.

This paper discusses performance of channel strategies, as well as their enhancement within nonstationary immersion environments, such as the currency market. We show that there are time intervals at which the use of this strategy will be profitable. The issue of the influence of optimization of the parameters of the trading system on its profitability is also considered.

We consider the EURUSD currency pair because it is the most liquid asset at the international financial market. This makes it possible to assess the effectiveness of channel strategies without the influence of such manipulative factors as insider trading or mistakes made by large investors.

2. Methods

As a basic model of market asset quotations, we will use the following additive model (Wald) [10]:

$$y_k = x_k + v_k, \; k = 1, \ldots, n \tag{1}$$

where x_k, k = 1, ... ,n is a system component formed by sequentially smoothing the time series of initial observations y_k, k = 1, ... ,n and used in the process of making management decisions, and v_k, k = 1, ... ,n is the noise.

The traditional observation model for statistical analysis is based on the assumption that the system component is an unknown deterministic process, and the noise component is a stationary random process with independent increments. In the more general approach, based on the Bayesian concept [17–19], the system component is also considered as a random process, and various computational schemes for sequential estimation of the conditional mean are used to identify it. As an example of the Bayesian approach, we can cite the Kalman filtration model [20], in which the system component is modeled by a linearized model of the form $x_{k+1} = \Phi_{k+1/k}x_k + w_k$, k = 1, ... ,n, where w_k is the noise process that imitates the admissible errors in the system.

These traditional models allow us to construct classes of management strategies based on trend analysis [7–9], each of which is effective only in a narrow range of possible variations of the observed process. The underlying reason for the low efficiency of the proposed trend indicators stems from the fundamental discrepancy between the traditional statistical approach and the nature of real observations. The following facts represented in Model (1) are the fundamental differences of the considered market asset observation series:

- Their system component x_k, k = 1, ... ,n is an oscillatory nonperiodic process with a large number of local trends. This description indicates the possibility of interpreting this process as an implementation of the dynamic chaos model [21–29]. However, the proof of this statement requires a strict formalization of the discrimination of deterministic chaotic and non-stationary random processes, which will require additional research.
- The noise v_k, k = 1, ... ,n is a nonstationary random process roughly described by the Gaussian model with fluctuating parameters [11,12]. At the same time, noise deviations contain local trends, and their correlations and spectral characteristics change significantly over time [30].

These characteristics violate the conditions of applicability of traditional statistical methods for effective decision making. Moreover, violation of the condition of repeatability prohibits any analytical assessments of forecasts and corresponding proactive management strategies. In essence, the main method of analyzing the quality of asset management in this case is numerical studies that assess the effectiveness of management algorithms over long observation intervals.

We have a number of financial instrument quote observations described by Model (1). Consider the simplest version of a single-channel management strategy. We define a "channel" as a range of observations limited by a range $Y_k = x_k \pm B$, k = 1, ... ,n. Variations of observations inside the channel $|y_k - x_k| = |\delta y_k| \leq B$ are interpreted as fluctuations that do not contain a pronounced trend, and the process itself is sometimes called a sideways trend or a flat. The choice of the channel width can be driven by various considerations. It usually lies in the range (1–3)s_y, where s_y is the estimate of the standard deviation (SD) δy_k, k = 1, ... ,n. In general, the choice of channel width is an option that depends on the features of the TP management. In some cases, it can be a variable value $B_k = B_k(y_k)$, k = 1, ... ,n.

The current value of observations y_k, k = 1, ... ,n breaking out of the channel is interpreted as the emergence of a trend in some management strategies. In the case of "playing by the trend", this invokes a recommendation to open a position in the direction corresponding to the sign of the channel boundary. The position can be closed when a given level of gain (TP, "take profit") or loss (SL, "stop loss") is reached, or in accordance with other, more flexible rules defined by the management strategy.

The task of a trader or a trading robot is to choose a management strategy S and form a sequence of actions u_j, j = 1, ... , M corresponding to it that provide maximum profit:

$$R(S) = \sum_{j=1}^{M} |y_j(k_{open}) - y_j(k_{close})| = max \quad (2)$$

At the same time, each management strategy determines the moments of opening and closing a position $(k_{open}, k_{close})_j$, j = 1, ... ,M, and, in some cases, the lot size. If the resulting amount at some k-th step turns out to be less than the trader's available deposit R_0, then this means a complete loss. In order to isolate the system component, any technique of sequential filtration can be applied. In the simplest case, an exponential filter is used for this purpose, defined as [31]:

$$x_k = \alpha y_k + (1 - \alpha) y_{(k-1)} = x_{(k-1)} + \alpha(y_k - x_{(k-1)}),\ k = 2, \ldots, n \quad (3)$$

with a smoothing coefficient α, whose value most often lies in the range [0.01,0.3].

The given simplified channel strategy makes it possible to remove many minor details. This makes the problem clear for the terminal task of producing the best proactive management for the selected class of management strategies.

2.1. Channel Strategy Based on Moving by the Trend

Let y_k, k = 1, ... ,n be the observed monitoring process of an asset quotation with a given duration n. The system component x_k, k = 1, ... ,n consists of an exponential filter (3) with a given smoothing coefficient α. The channel width is denoted as B, measured in pipses. Similarly, the levels of profit TP and the level of acceptable loss SL are set. The values of α, B, TP and SL are optional parameters. The choice of these parameters depends on the knowledge and intuition of the trader, as they completely determine the effectiveness of management. However, in trading, intuition and other human abilities often turn out to be ineffective. Thus, a need for strictly formalized and mathematically sound solutions arises.

The management strategy based on moving by the trend, which we have named CSF (channel strategy forward), consists in opening a position up or down when the process exits, respectively, the upper or lower boundary of the channel. The management algorithm consists of two rules: Open Up position at $y_k > x_k + B$ or Open Dn at $y_k < x_k - B$, k = 1, ... ,n.

Note that a stricter formalization of this rule has the form $y_{k-1} \leq (x_{k-1} + B)$ & $(y_k > x_k + B)$ or $y_{k-1} \geq (x_{k-1} + B)$ & $(y_k < x_k + B)$, k = 1, ... ,n. Otherwise, a position will be opened at each step outside the channel. The position is closed either when the $y_k = y_{close} > y_{open} + TP$ or $y_k = y_{close} < y_{open} - SL$ levels are reached (with Open Up) or $y_{close} < y_{open} - TP$ or $y_{close} > y_{open} + SL$ (with Open Dn).

The considered version of the channel strategy is basic and can be easily augmented in various ways, presented, for example, in [5–9]. In particular, it is natural to close a position at the moment of trend reversal, when the sign of the difference $x_k - x_{k-nW}$, k = 1, ... ,n, nW being the size of the sliding observation window, changes to the opposite. However, this article is focused on the analysis of the fundamental features of channel management strategies, and, as a result, omits various complications.

2.2. Channel Strategy Based on Moving against the Trend

We will now consider an alternative interpretation of the observed process exiting the channel boundaries as another well-known channel strategy. In this case, the exit of the quotation from the channel is considered as a random deviation of the asset value, which inevitably triggers market compensation mechanisms that return the process to the limits of normalized fluctuations, i.e., to the limits of the channel. We will call this approach a channel strategy of playing against the trend (CSB, channel strategy back).

As a demonstration example, let us consider a day-long observation segment of EURUSD quotes shown in Figure 1. The strategy consists in opening a position up or down when the process exits, respectively, beyond the lower or upper boundary of the channel.

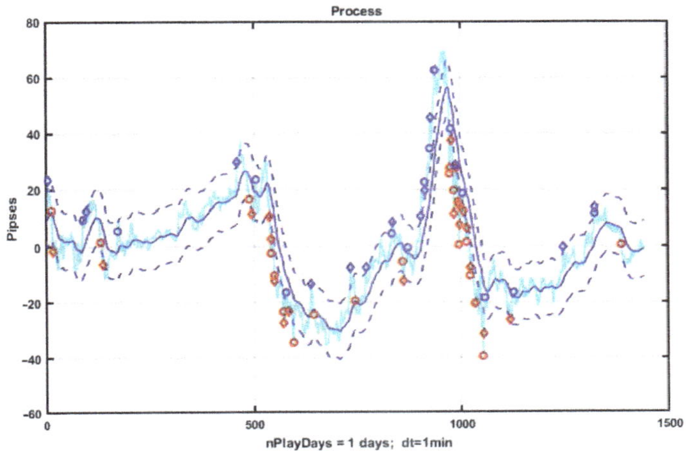

Figure 1. Example of CSF implementation.

Note that opening a position at the moment when the observed process leaves the channel can lead to a large negative drawdown, since the time and level of reversal of the observed process are unknown. Moreover, if the previous hypothesis about the emergence of a trend is true, then such a position opening can lead to a big loss. Therefore, it is more rational to open a position at a reverse intersection of the channel, which increases the stability of management, but, at the same time, makes a more modest gain.

The basic algorithm for controlling the movement against the trend is described by the following rules: Open Up is carried out under the condition $(y_k > x_k - B)$ & $(y_{k-1} < x_{k-1} - B)$, $k = 2, \ldots, n$, and Open Dn $-$ $(y_k < x_k + B)$ & $(y_{k-1} \geq x_{k-1} + B)$, $k = 2, \ldots, n$. The position is closed when either of the levels $y_{close} > y_{open} + TP$; $y_{close} < y_{open} - SL$ are reached with Open Up or $y_{close} < y_{open} - TP$; $y_{close} > y_{open} + SL$ with Open Dn.

An example of an alternative option for closing a position is the moment when the observed process crosses the conditional mean level x_k, $k = 1, \ldots, n$.

It should be pointed out that the presented algorithm is extremely "careful", as in, the opening is carried out only when the asset price returns to the channel, and the expectation of profit relies on the fact that the compensation mechanisms of the market do not act instantly and are prone to a certain inertia. At the same time, the gain turns out to be quite small, but this provides protection against losses in the event of a systemic trend that takes the quotation far beyond the channel.

3. Results

Figure 1 illustrates the characteristics of a basic CSF strategy using a concrete implementation. We consider the quotes of the EURUSD currency pair as the observed process, and the management duration is one day (1440 min counts). The system component x_k, $k = 1, \ldots, n$ is formed by exponential filter (3) with $\alpha = 0.05$. The width of the channel $B = \pm 10$ p is measured in points or pipses (p). The level of profit and the level of acceptable loss are $TP = SL = 10$ p.

On the chart, blue diamonds indicate the detection of an upward trend, red ones indicate a downward trend, and circles of corresponding colors indicate the levels of decision making about the quality of management. Figure 2 shows a plot of changes in the effectiveness of management (3) in the process of applying the channel strategy in the selected example. The result is obviously negative: the loss was -93 p per day.

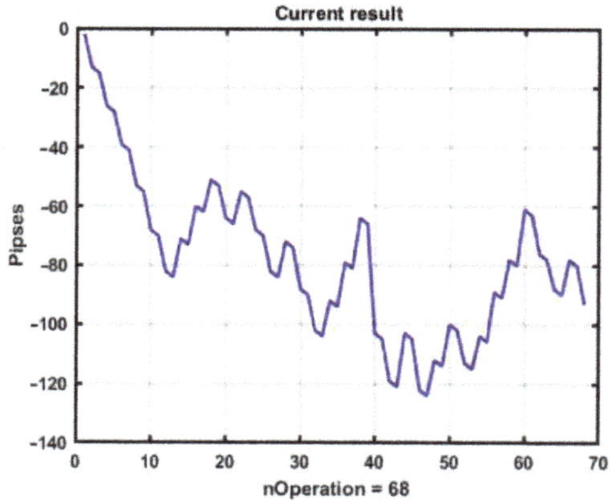

Figure 2. Performance in the case of using CSF.

The reasons for the loss are obvious. The presented forecasting technique can produce results only for areas with a strong, pronounced trend. For areas with weak local trends, this strategy only produces loss. Nevertheless, it allows for improvement via parametric fitting and introducing additional conditions into the management algorithm.

For example, increasing only the channel width up to B = ±18 p can make the management strategy at the same observation area profitable (see Figures 3 and 4).

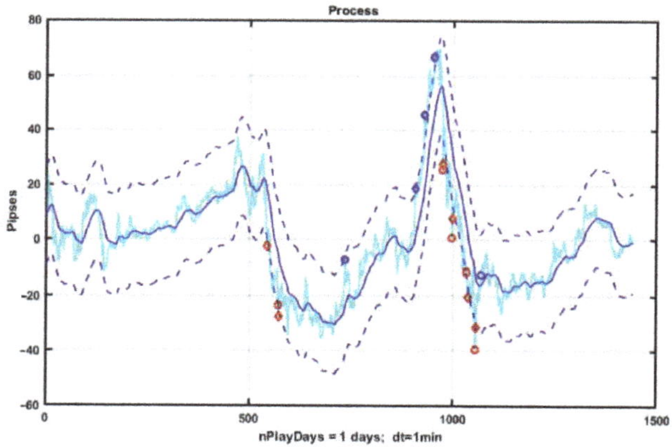

Figure 3. Example 2 of CSF implementation.

To assess the potential gain, we have initially used a brute force search for only two optional parameters in the ranges B = 11:1:25 and TP = 11:1:25. For the given values of the options, the best profit was R = 76 p with B = 19, TP = 22. Three-dimensional plots of the management results for the selected optional parameter ranges are shown in Figure 5. The same plots with network approximation are shown in Figure 6.

It is not difficult to see that the approximating surface is discontinuous with jumps in values even at minor variations in arguments. Thus, it can be concluded that the effectiveness of a channel management strategy based on movement in the direction of the trend is unstable. At the same time, parametric fitting based on retrospective data on a selected observation interval does not guarantee a positive result at the next subsequent time interval with the same parameters. An example of this is shown in Figure 7. When using the optimal parameters for a selected observation interval for the next interval of the same duration, the result turned out to be negative and amounted to R = −29 p.

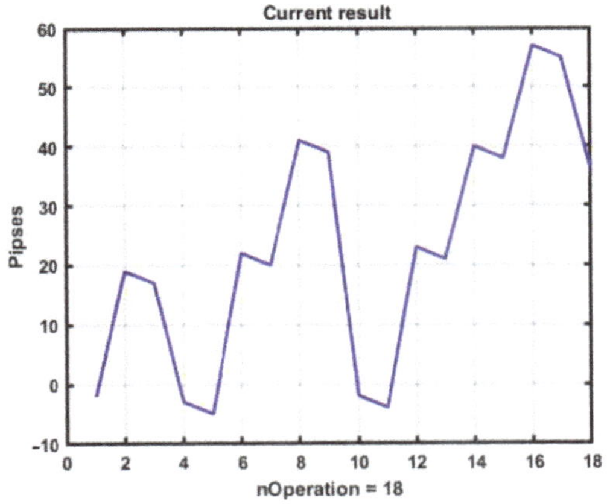

Figure 4. Performance in the case of using CSF in example 2.

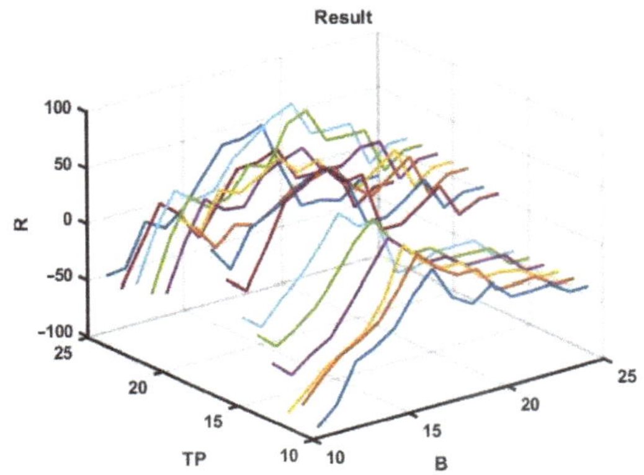

Figure 5. Three-dimensional plots of the management results for the selected ranges of optional parameters.

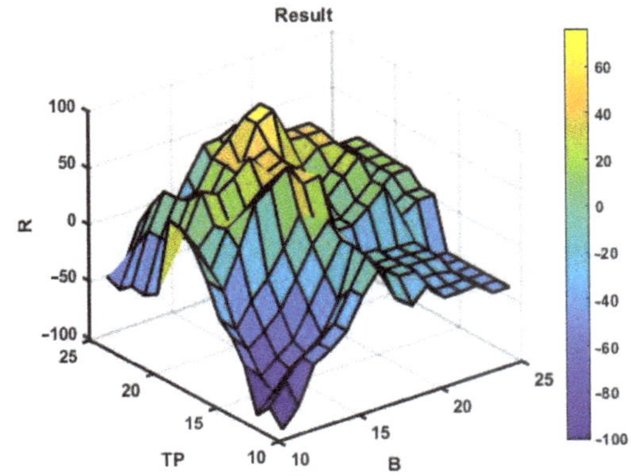

Figure 6. Network approximation of Figure 5.

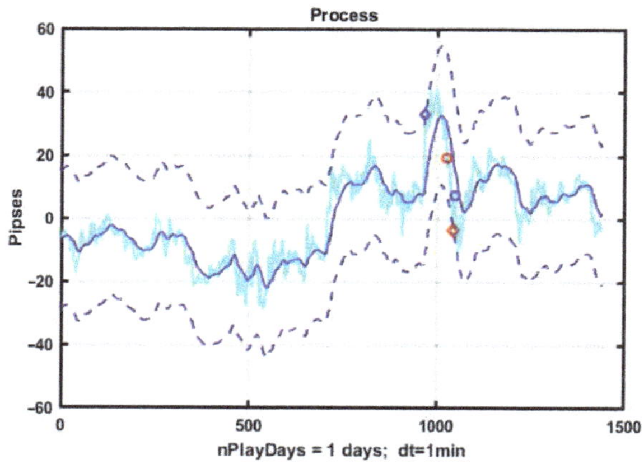

Figure 7. Example of CSF implementation on a subsequent one-day observation interval.

Repeated a priori optimization on the same area produced a result of only R = 5 p with the best option parameters being TP = 12, B = 19.

It should be noted that channel management can undergo various modifications, each of which makes sense for certain types of the observed dynamic process [6–8], but none of them has a margin of stability that makes it possible to obtain a positive result on the whole variety of processes generated by market chaos.

3.1. Evaluation of CSF's Potential Performance

Let us consider typical versions of improving channel management strategies. Figure 8 shows the results of management based on CSF on a three-day observation interval with pronounced areas of strong trends. On the basis of an a priori three-parameter optimization obtained by brute force search, we have identified best values of B = 19, TP = 22 and SL = 22, corresponding to exponential smoothing (1) with the transfer coefficient $\alpha = 0.05$.

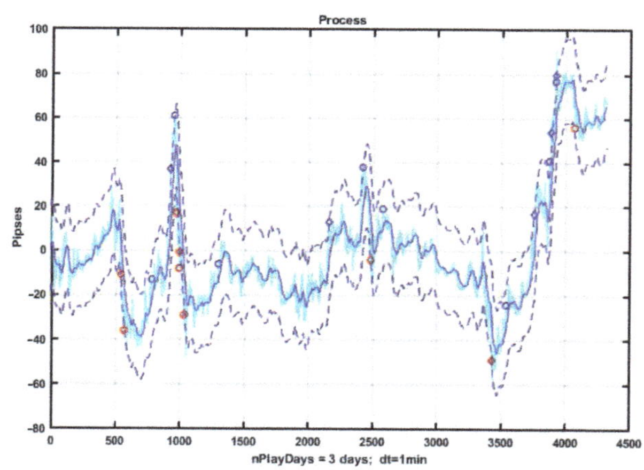

Figure 8. Example CSF implementation with three optimized parameters.

The example is profitable at R = 32 p; however, the result is unstable and even small variations in the structure of observation series definitely lead to a loss. Without a priori optimization, a trend-based management strategy almost always leads to a negative result. The rationale for this is given in studies of the inertia of chaotic processes in [32].

Is it possible to improve the result by extending the vector of optimized parameters and using different values for the upper and lower boundaries of the channel? To solve this problem, we consider five-parameter optimization, including the filter smoothing coefficient α, the lower and upper channel boundaries B_{Dn} and B_{Up}, and the stop parameters TP and SL. Let us set the ranges of the selected parameters to α = 0.01:0.01:0.15, B_{Dn}, B_{Up} = 10:1:15, TP, SL = 10:1:15. For the selected three-day observation segment considered in the previous example, the best result of CSF was R = 166 p with the parameters P* = (α, B_{Dn}, B_{Up}, TP, SL)* = (0.1, 13, 16, 23, 16). The implementation of the management strategy with the specified parameters at the selected observation interval is shown in Figure 9, and the changes in the overall result of management are shown in Figure 10.

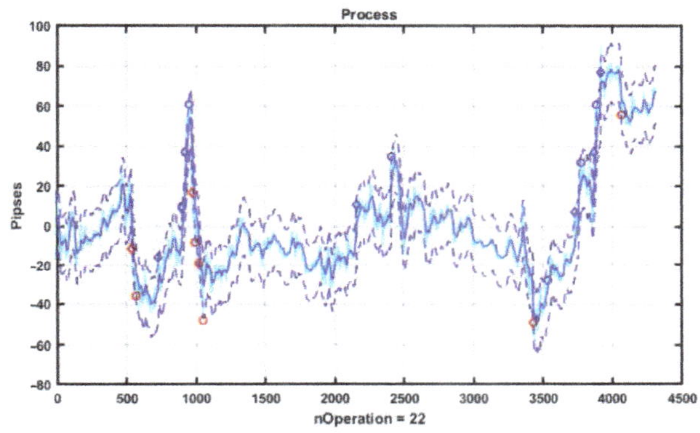

Figure 9. Example CSF implementation with five optimized parameters.

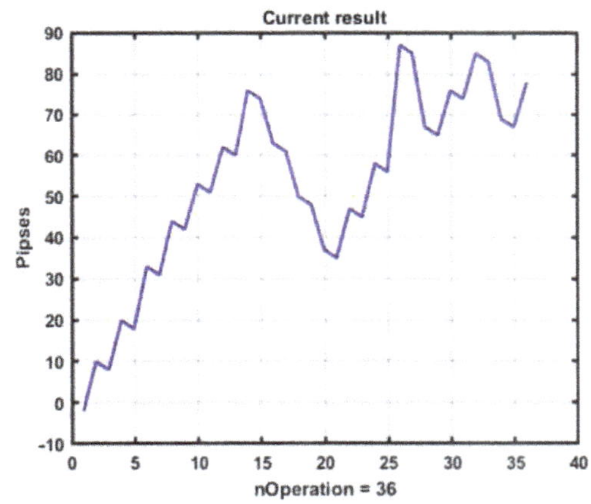

Figure 10. Performance in the case of using CSF in the previous example.

Experiments show that R(P) is unstable: even minor deviations of the vector of optional parameters P from the optimum P* lead to a sharp decrease in the effectiveness.

It should be noted that brute force parameter optimization leads to an exponential growth of arithmetic complexity with an increase in the number of parameters or requirements for the accuracy of estimating optimal parameter values (i.e., decrease of the iteration step). This well-known fact can lead to serious problems in the construction of adaptive management strategies that sequentially estimate the optimal values of P.

Another important point is the asymmetry of management performance: among all the selected combinations, only 26% give a positive management result.

Attempts to increase the effectiveness of this strategy by improving system component isolation, the possibility of which was shown in the previous section, can only have an indirect effect, because during optimization it was compensated for by separately choosing the boundaries B_{Dn}, B_{Up} and the filtration coefficient α. It is possible that this strategy will be more effective with separate dynamic adjustment of the upper and lower boundaries B_{Dn}, B_{Up}, regarded as functions of the rate of change of the system component estimated on the sliding observation window $B_{Dn}, B_{Up} = F(x_k - x_{k-\tau})$, $k = \tau + 1, \ldots, n$. This issue requires additional research.

Figure 11 shows an example of CSB implementation. Similarly to the previous example, the system component x_k, $k = 1, \ldots, n$ is evaluated as the result of the smoothing performed by exponential filter (3) with $\alpha = 0.05$. The initial values of the option parameters are: channel width $B = \pm 10$ p, take profit level TP = 10 p, acceptable loss level SL = 10 p.

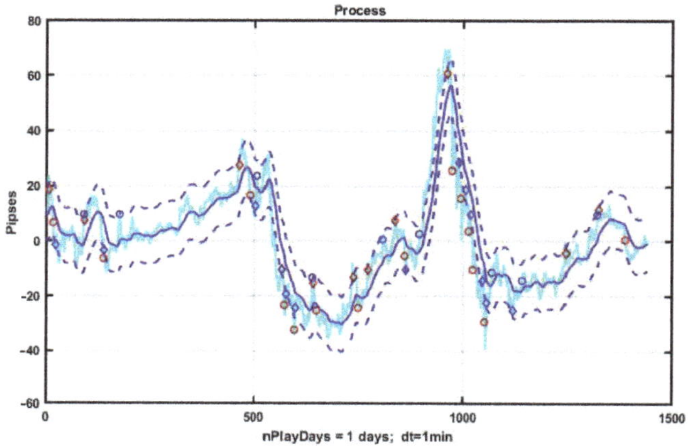

Figure 11. Example implementation of CSB.

From the above figure, as well as from the plots of changes in the result due to the management strategy (Figure 12), it can be seen that the adopted asset management produces stable gain on weak trends (see the first third of the plot). However, when strongly pronounced trends occur, this strategy leads to incorrect decisions.

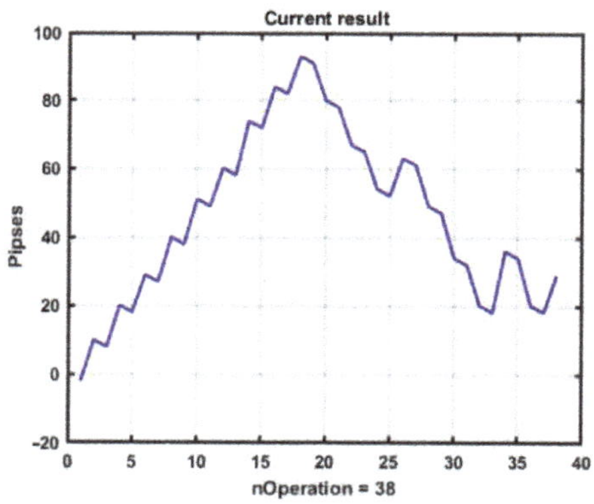

Figure 12. Performance of CSB in the example implementation.

To evaluate the effectiveness potential of the proposed management strategy a posteriori, we apply a brute force search to optional parameters, as discussed above. We iterated over the values of three control parameters in the ranges $\alpha = 0.01{:}0.01{:}0.1$, $B = 10{:}1{:}25$, TP = 10:2:25 on a single-day observation segment shown in Figure 11. As a result of comparing 880 combinations, we found that the optimal values of the option parameters were

α^* = 0.03, B* = 10 and TP* = 10, i.e., the parameters selected in the example were either close to or coincided with the best. However, the probability of a positive outcome was low: out of 880 parameter combinations, only 137 provided a positive result. The optimal profit value for the selected combination option was R* = 78 p.

An example implementation of a CSB management strategy with an optimal combination of optional parameters is shown in Figure 13, and the corresponding plot of changes in performance is shown in Figure 14.

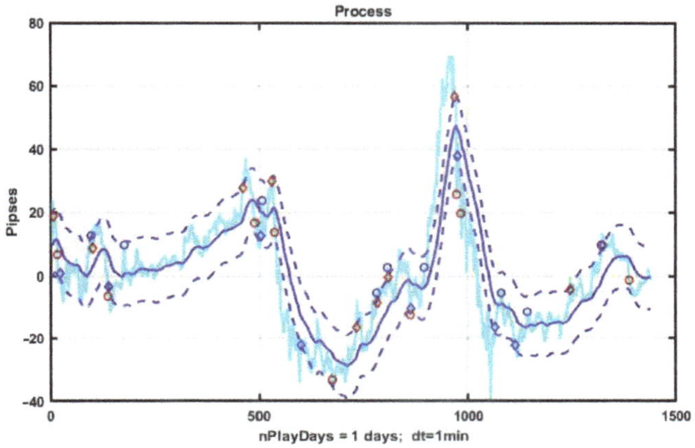

Figure 13. Example implementation of CSB with three optimized parameters.

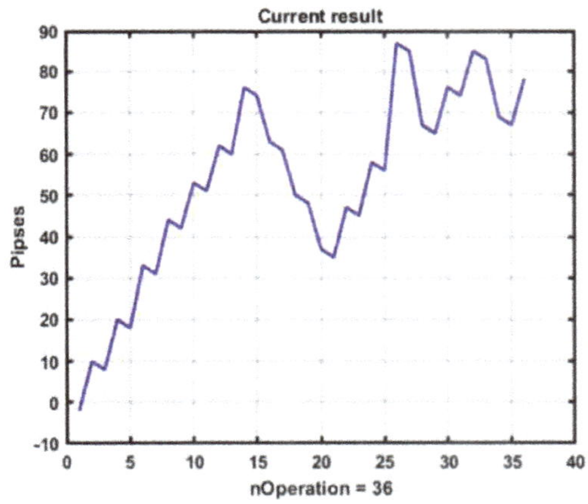

Figure 14. Performance of CSB with three optimized parameters.

It should be noted that a number of losing decisions are explained not by the overall decision quality, but by a significant bias that occurs during the filtration process (3) during the isolation of the system component x_k, k = 1, ... ,n. In Figures 1, 8, 11 and 13, it can be easily seen that when a strong trend occurs, the system component formed by the smoothing filter lags behind the observation process, which leads to a shift of the entire channel and incorrect use of the management strategy concept itself. Hence, natural suggestions arise for improving the management by modifying the sequential filtering procedure and dynamically changing the size of the lower and upper channel boundaries.

3.2. Evaluation of CSB's Potential Performance

Let us compare the estimates of the potential effectiveness of CSF and CSB management strategies on the same three-day observation interval. We will retain the same set of ranges for optional parameters, α = 0.01:0.01:0.15, B_{Dn}, B_{Up} = 5:1:15, TP, SL = 7:1:15. Using a brute force search, the best result of the CSF strategy was R = 166p with P* = (α, B_{Dn}, B_{Up}, TP, SL)* = (0.1, 13, 16, 23, 16). The implementation with the specified parameters at the selected observation interval is shown in Figure 11.

The CSB's best result for the selected observation area was R = 250 p, corresponding to P* = (α, B_{Dn}, B_{Up}, TP, SL)* = (0.03, 8, 5, 17, 17). Its implementation is shown in Figure 15.

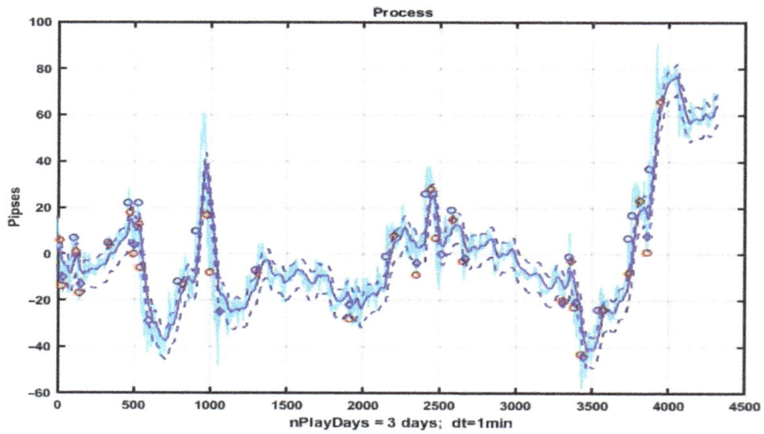

Figure 15. Example implementation of CSB with five optimized parameters.

Comparing the two channel strategies with optimal vectors of optional parameters allows us to draw the following conclusions:

- Both strategies have profitable decisions, but the result in both cases is not stable and small changes in control parameters can lead to a radical decrease in gain;
- Management against the trend needs a higher level of smoothing, which activates the opening of a position on a sideways trend and reduces the frequency of management in areas with a strong trend. On the contrary, playing by the trend requires reducing the degree of smoothness, which makes it possible to more effectively detect a strong trend and activate management in these areas of observation;
- The width of the channel when playing against the trend is significantly, two to three times, smaller than in the opposite case, which increases the frequency of opening in horizontal quotation sections. Obviously, in both cases, the choice of channel width is related to the degree of volatility, as a measure of which, for example, the standard deviation (SD) of observations relative to the system component can be used.

Due to the paper limit, we will illustrate our further research with numerical results based on the use of the CSB management strategy.

3.3. Parametric Stability of Optimal Solutions

In order to check the stability of the solutions produced by the considered a posteriori determination of optimal parameters, we will consider another three-day observation section, following the one presented in Figure 15, containing discontinuous sections of quotation changes.

The result of optimization using CSB was R = 267 p with the parameters P* = (α, B_{Dn}, B_{Up}, TP, SL)* = (0.02, 7, 14, 8, 19).

Now let us evaluate the effectiveness of the same strategy at the same observation segment, but with the parameters optimal in the previous three-day interval. Comparisons of management performance with the CSB strategy in the example shown in Figure 15 with optimal parameters (left) and using the same parameters in the subsequent three-day observation interval (see Figure 16) are shown in Figure 17.

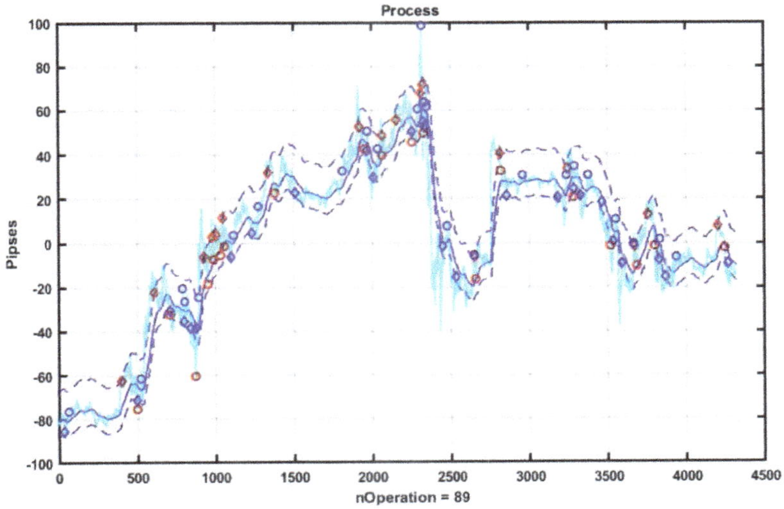

Figure 16. Example implementation of CSB with five optimized parameters on subsequent three-day interval.

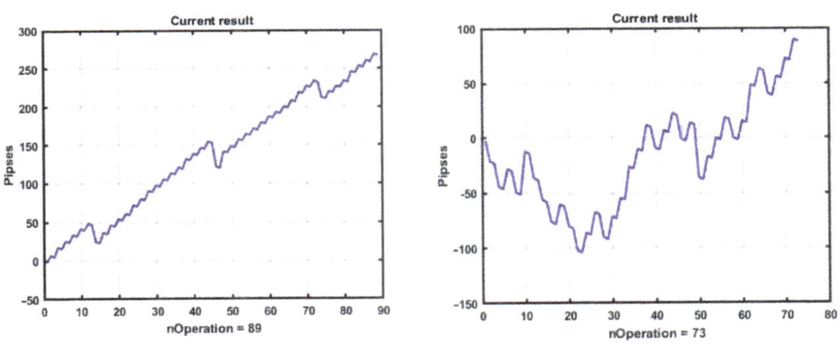

(**a**) (**b**)

Figure 17. Performance of optimized control after applying the CSB strategy in the observation area with optimization (**a**); and in the subsequent interval of the same duration (**b**).

It is easy to see that the gain decreased by about three times (R = 88 p), and in the process of management, the result decreased to a loss of 100 p.

Let us consider the stability of the CSB strategy performance in response to particular variations of individual parameters. To this end, we will evaluate the effectiveness on the same three-day time interval as in the previous experiment, but with deviations of the optional parameters B_{Dn}, B_{Up}, TP and SL from optimal in the range $P^* \pm 0.1\ P^*$, and α in range $\alpha^* \pm 0.5\ \alpha^*$. The range of changes was divided into eight steps, i.e., four steps in each direction from the optimal value of the parameter. Table 1 presents the changes in performance for this strategy.

Table 1. Effectiveness of the management strategy with varying parameters.

k	α	R	B_{Dn}	R	B_{Up}	R	TP	R	SL	R
1	0.0100	50	6.300	244	12.60	170	7.2	250	17.10	229
2	0.0125	80	6.475	260	12.95	203	7.4	250	17.57	229
3	0.0150	202	6.650	270	13.30	241	7.6	250	18.05	246
4	0.0175	240	6.825	263	13.65	247	7.8	250	18.52	246
5	0.0200	267	7.000	267	14.00	267	8.0	267	19.00	267
6	0.0225	238	7.175	268	14.35	258	8.2	272	19.47	267
7	0.0250	203	7.350	261	14.70	248	8.4	272	19.95	267
8	0.0275	148	7.525	262	15.05	245	8.6	272	20.42	263
9	0.0300	96	7.700	257	15.40	238	8.8	272	20.90	263

3.4. Examining the Dynamic Stability of Optimal Solutions

From the above data, it can be seen that the management has a certain margin of stability, i.e., small changes in the values of the option parameters do not lead to an abrupt change in performance. This conclusion leaves hope for the possibility of implementing an adaptive approach, when the optimal setting at the previous stage is used as the set of optional parameters at the subsequent observation site.

However, a more accurate conclusion requires an analysis of the stability of the control with simultaneous variations of groups of parameters. In addition, it is necessary to examine this issue in dynamics, i.e., taking into account the time shift of the observation area.

For clarity, let us consider the process of asset management with the CSB strategy during a single day of observation, shown in Figure 18. With brute forcing, we obtain a vector of optimal parameters $P^* = (\alpha, B_{Dn}, B_{Up}, TP, SL)^* = (0.03, 7, 6, 11, 12)$, the use of which produces gain R = 164 p.

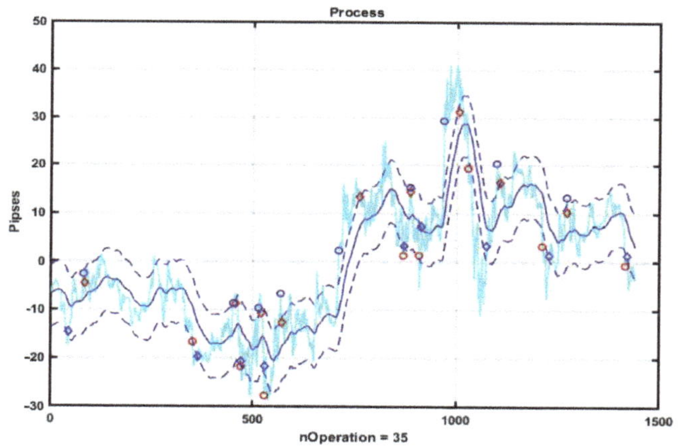

Figure 18. Example implementation of CSB during a single observation day with optimal optional parameters.

Let us consider the question of how stable the obtained vector will be during the next two days of management.

Figure 19 shows graphs of three days of observation (separated by days by red circles), from which it can be easily seen that the relatively smooth dynamics of the first day, on which parametric optimization was carried out, are not at all similar to the dynamic data structures of the two following days. The second day is characterized by a decline followed by a strong positive trend, the third day is characterized by an uneven positive trend (growth).

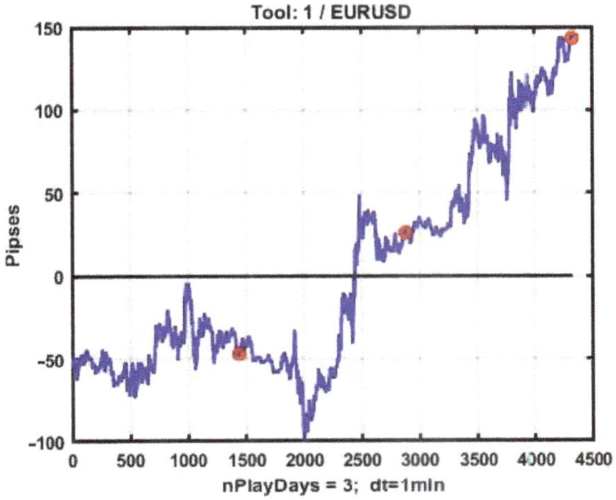

Figure 19. EURUSD quotation chart during three observation days.

Thus, the settings produced on the first day of observation will obviously produce only loss for the next two days. In this example, the loss was R = −90 p on the second day and R = −57 p on the third day. Thus, the assumption about the parametric instability of the channel strategy, which is obvious to all practicing traders, is confirmed.

Note that the first day, which was used for the training, i.e., determining the best values of optional parameters, was quite "comfortable" for a CSB-type management strategy. The corresponding observation segment was a sideways trend with small variations. Naturally, when strong trends occur, management with parameters optimal for a sideways trend turns out to be losing.

Consider a channel management strategy with enhanced resistance to variations in the dynamics of observations: a model with parameters optimal over a large observation interval. In the example, we examine an observation interval of 15 days. This area contains a large variety of dynamics: sideways trends with different heights, areas of slow and abrupt increases and decreases, jumps, etc.

We optimized the parameters via brute force on the following ranges: $\alpha = 0.01{:}0.01{:}0.15$, B_{Dn}, B_{Up} = 5:1:15, TP, SL = 7:1:15. The test result of using CSB was R = 475 p with $P^* = (\alpha^*, B_{Dn}^*, B_{Up}^*, TP^*, SL^*) = (0.02, 6, 8, 17, 21)$. The performance of the management strategy with specified parameters at the selected observation interval is shown in Figure 20.

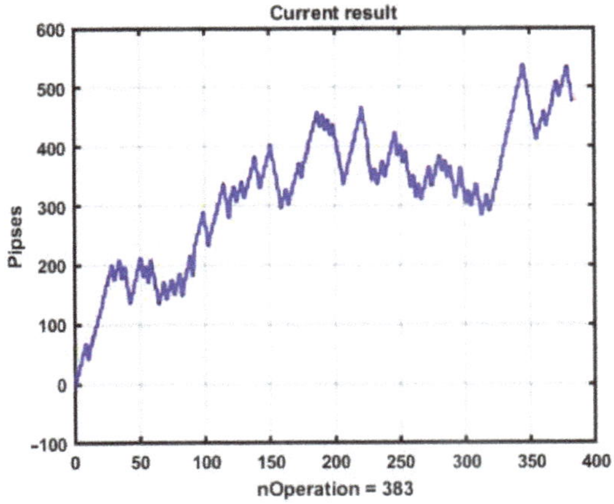

Figure 20. Performance change in the process of using the CSB strategy.

Using these parameters separately on each observation day, we obtained R = 456 p, which is quite close to the previous result of optimizing for the entire 15-day observation area. Let us consider how effective the parameters P* are in the subsequent 10-day observation interval that does not intersect with the training data interval. The total sum of one-day results in this case is R = 51.

We chose the same large observation interval of 15 days as for the CSB strategy, preserving the same ranges for the following parameters: $\alpha = 0.01{:}0.01{:}0.15$, B_{Dn}, $B_{Up} = 5{:}1{:}15$, TP, SL = 7:1:15. Their best values, $P^* = (\alpha^*, B_{Dn}{}^*, B_{Up}{}^*, TP^*, SL^*) = (0.1, 11, 14, 16, 13)$, were obtained by brute force search.

Figure 21 presents the performance of the strategy. The best result was R = 252 p, which is significantly worse than the CSB's result with optimal parameters at the same interval.

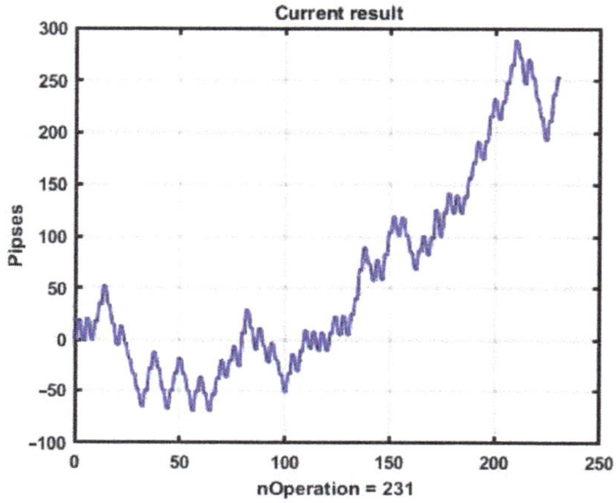

Figure 21. Performance of the CSF strategy with the best parameters.

Following the proposed method of stability analysis of the obtained result, let us consider how effective the found optimal parameters are in the subsequent 10-day observation interval that does not intersect with the training interval (Figure 21). The total sum of one-day results in this case is R = −534, which indicates low efficiency and pronounced dynamic instability of this management strategy.

4. Discussion

Given studies clearly indicate the theoretical applicability of channel strategies to the task of speculative management of financial instruments in conditions of market chaos. However, their practical implementation faces the problem of weak predictability of changes in the value of market assets, which, combined with low statistical stability of management in the conditions of market chaos, leads in most cases to negative results.

The obvious consequence of this conclusion is the need to extensively modify channel strategies to increase their resistance to dynamic variations in the structure of observation series.

The traditional approach to improving the stability of system functioning in conditions of uncertainty is based on various adaptive computing schemes. For systems with a high level of dynamic uncertainty characteristic of market pricing processes, it is natural to use sequential adaptation on a sliding window of application adjacent to the current time moment. At the same time, it is possible to use both nonparametric adaptation, which modifies the structures of the observation model, and traditional parametric adaptation.

It should be noted that the parametric adaptation based on posterior brute force optimization of the values of the optional parameters, in conditions of strong variability characteristic of quotation dynamics, may not be feasible. This is due to the rapid growth of the time complexity with an increase in the number of parameters. Due to this, it is advisable to utilize suboptimal optimization with significantly lesser computational demands.

It should be pointed out that any kind of adaptation of decision-making algorithms in conditions of chaotic dynamics is a very difficult problem. The market information space is an unstable immersion environment. As shown previously [32], similarity of aftereffects does not follow from the geometric or correlational similarity of two multidimensional situations. As a consequence, an effective solution on the previous sliding observation interval does not guarantee a quality solution on the current observation window. Nevertheless, the studies of the stability of solutions presented in the article retain hope for the possibility of creating fast self-organizing systems, for example, based on evolutionary optimization technologies [33], allowing to increase the stability of channel management.

The main alternative to adaptation is reducing the sensitivity of decision-making algorithms to variations in the dynamic and statistical characteristics of observation series. The robustification of forecasting algorithms and management decisions is based on the search for the best options for the least favorable variants of initial data implementation [34]. However, as already noted, there is no similarity of aftereffects for such observation segments in the conditions of stochastic chaos. In this case, a uniform decrease in the quality of management does not follow from a consistent decrease in the quality of estimation and forecasting algorithms. In other words, the quality of the best solution for the worst case of the source data under these conditions will not be stable for other parts of the source data.

Nevertheless, this requires numerical validation, and even if the assumption is confirmed, it is of interest to use robustification in combination with other methods to increase the stability of proactive management strategies.

Another possible solution is to switch to multi-expert decision-making systems (MES), involving both strategies in the management [35]. Such an approach can produce many possible solutions based on consensus and compromise.

As an example, consider an expert arbitrator who has information about the general mood of the market. In particular, such a software expert can be implemented as an automatic text analyzer that evaluates market sentiment by identifying the required knowledge from analytical reports circulating on the web [36]. Based on the decision of the expert, one of the two channel management strategies is selected, the most consistent with the expected dynamics of the quotations.

Apparently, the best results of management in chaos should be expected from composite algorithms and multi-expert systems that widely use external add-ons such as fundamental analysis or the results of technical analysis of alternative markets.

Generalizing the abovementioned conclusions, the following directions of increasing the stability of management seem to be worthy of consideration:

- Dynamic adaptation of decision-making algorithms with the choice of the best parameters on sliding observation windows;
- Robustification, i.e., reduction of the sensitivity of solutions to changes in the dynamic characteristics of the observed processes;
- Improvement of management based on more flexible computational schemes for selecting boundaries of a channel or several channels and decision-making algorithms;
- Structural adaptation, for example, based on switching between management strategies as a reaction to changes in the dynamic properties of the system component of the observation series;
- Self-organization based on artificial intelligence, which independently creates strategies that are not contained in the source code of the management program;
- The use of composite algorithms combining the capabilities of robustification and adaptation in management decision making;
- The use of external add-ons that carry information exogenous to technical analysis on expected trends of the considered financial instrument and market mood in general.

The outlined issues constitute the subject of our further research.

5. Conclusions

Our research strategy is to consistently study the effectiveness and sustainability of management strategies in the conditions of stochastic chaos and develop proposals for their improvement. An approximate plan of the forthcoming research is given in the Discussion.

The materials presented in this article are an introduction to the general problem of management stability in the conditions of market chaos. The choice of primitive channel management strategies at this stage was due to the need to illustrate the problems that arise in the process of managing financial assets [37]. All computational experiments conducted in the article were made using real data obtained by monitoring the currency exchange market.

As a result of the conducted research, the following conclusions were obtained:

1. CSF and CSB strategies are strict alternatives. Nevertheless, in the absence of prior information about the expected dynamics of the observed process, both strategies, as a rule, lead to losses on the same trading intervals;
2. Posterior parametric optimization of both CSF and CSB management strategies has shown that they can be profitable in almost any observation area. This result is quite paradoxical, because these strategies are strict alternatives;
3. With a shift of the observation interval equal to one full day, the strategy optimized on the previous interval becomes losing again. This leads to a fairly obvious conclusion about the priority of the problem of management stability in an unstable immersion environment;
4. Small time shifts of observation series lead to a loss of optimality of the management algorithm parameters; however, the overall positive result of management is preserved in most cases. This means there is feasibility to using computational schemes with sequential parametric adaptation of the management model;
5. Parametric adaptation based on brute forcing the values of optional parameters produces an optimal solution. However, the computational load increases exponentially with the increase in the dimension of the parameter vector. If the time shifts between the observation intervals on which the model is optimized are small, the complexity can go beyond the limits even for high-performance computers. This implies the recommendation to switch to suboptimal computational schemes for adapting the management model;
6. As a variant of suboptimal adaptation of the management model, we propose to use algorithms based on evolutionary modeling [36]. These studies have been completed and are being prepared for publication.

Author Contributions: Conceptualization, Methodology, A.M. (Andrey Makshanov); Validation, Writing, A.M. (Alexander Musaev); Review and Editing, Investigation, Programming, Visualization, Administration, Scientific Discussions, Supervision, Funding Acquisition, D.G. All authors have read and agreed to the published version of the manuscript.

Funding: This research received no external funding.

Institutional Review Board Statement: Not applicable.

Informed Consent Statement: Not applicable.

Data Availability Statement: The source of data is Finam.ru (https://www.finam.ru/, accessed on 21 April 2022).

Acknowledgments: The authors are grateful to participants at the Center for Econometrics and Business Analytics (ceba-lab.org, CEBA) seminar series for helpful comments and suggestions.

Conflicts of Interest: The authors declare no conflict of interest.

References

1. Ying, Q.; Yousaf, T.; Ain, Q.U.; Akhtar, Y.; Rasheed, M.S. Stock investment and excess returns: A critical review in the light of the efficient market hypothesis. *J. Risk Financ. Manag.* **2019**, *12*, 97. [CrossRef]
2. Rajablu, M. Value investing: Review of Warren Buffett's investment philosophy and practice. *Res. J. Financ. Account.* **2011**, *2*, 1–12.
3. Bartram, S.M.; Grinblatt, M. Agnostic fundamental analysis works. *J. Financ. Econ.* **2018**, *128*, 125–147. [CrossRef]
4. Nti, I.K.; Adekoya, A.F.; Weyori, B.A. A systematic review of fundamental and technical analysis of stock market predictions. *Artif. Intell. Rev.* **2020**, *53*, 3007–3057. [CrossRef]
5. Zapranis, A.; Tsinaslanidis, P.E. Identifying and evaluating horizontal support and resistance levels: An empirical study on US stock markets. *Appl. Financ. Econ.* **2012**, *22*, 1571–1585. [CrossRef]
6. Gregory-Williams, J.; Williams, B.M. *Trading Chaos: Maximize Profits with Proven Technical Techniques*, 2nd ed.; John Wiley & Sons: New York, NY, USA, 2004.
7. Niederhoffer, V.; Kenner, L. *Practical Speculation*; John Wiley & Sons: New York, NY, USA, 2005.
8. Colby, R.W.; Meyers, T.A. *The Encyclopedia of Technical Market Indicators*; IRWIN Professional Publishing: Burr Ridge, IL, USA, 2012.
9. Chordia, T.; Goyal, A.; Saretto, A. *p-Hacking: Evidence from Two Million Trading Strategies*; No. 17-37; Swiss Finance Institute: Zurich, Switzerland, 2018.
10. Peters, E.E. *Chaos and Order in the Capital Markets: A New View of Cycles, Prices, and Market Volatility*, 2nd ed.; John Wiley & Sons: New York, NY, USA, 1996.
11. Zhang, M.H.; Cheng, Q.S. Gaussian mixture modelling to detect random walks in capital markets. *Math. Comput. Model.* **2003**, *38*, 503–508. [CrossRef]
12. de Wolff, T.; Cuevas, A.; Tobar, F. Gaussian process imputation of multiple financial series. In Proceedings of the ICASSP 2020-2020 IEEE International Conference on Acoustics, Speech and Signal Processing (ICASSP), Barcelona, Spain, 4–8 May 2020; pp. 8444–8448.
13. McGroarty, F.; Booth, A.; Gerding, E.; Chinthalapati, V.R. High frequency trading strategies, market fragility and price spikes: An agent based model perspective. *Ann. Oper. Res.* **2019**, *282*, 217–244. [CrossRef]
14. Tykocinski, O.; Israel, R.; Pittman, T.S. Inaction inertia in the stock market. *J. Appl. Soc. Psychol.* **2004**, *34*, 1166–1175. [CrossRef]
15. Schmitt, N.; Westerhoff, F. Heterogeneity, spontaneous coordination and extreme events within large-scale and small-scale agent-based financial market models. *J. Evol. Econ.* **2017**, *27*, 1041–1070. [CrossRef]
16. King, T.; Koutmos, D. Herding and feedback trading in cryptocurrency markets. *Ann. Oper. Res.* **2021**, *300*, 79–96. [CrossRef]
17. Downey, A.B. *Think Bayes: Bayesian Essentials with R*, 2nd ed.; Springer: New York, NY, USA, 2014.
18. Stone, J.V. *Bayes' Rule: A Tutorial Introduction to Bayesian Analysis*; Sebtel Press: Sheffield, UK, 2013; p. 174.
19. Nyberg, S.J. *The Bayesian Way: Introductory Statistics for Economists and Engineers*; John Wiley & Sons: Hoboken, NJ, USA, 2018.
20. Kalman, R.E. A New Approach to Linear Filtering and Prediction Problems. *J. Basic Eng.* **1960**, *82*, 35–45. [CrossRef]
21. Guanrong, C. Chaos theory and applications: A new trend. *Chaos Theory Appl.* **2021**, *3*, 1–2.
22. Gardini, L.; Grebogi, C.; Lenci, S. Chaos theory and applications: A retrospective on lessons learned and missed or new opportunities. *Nonlinear Dyn.* **2020**, *102*, 643–644. [CrossRef]
23. Davies, B. *Exploring Chaos: Theory and Experiment (Studies in Nonlinearity)*; CRC Press: Boca Raton, FL, USA, 2018.
24. Feldman, D. *Chaos and Dynamical Systems*; Princeton University Press: Princeton, NJ, USA, 2019.
25. Faggini, M.; Parziale, A. The failure of economic theory. Lessons from Chaos Theory. *Mod. Econ.* **2012**, *3*, 16802. [CrossRef]
26. Glenn, E.J. *Chaos Theory: The Essentials for Military Applications*; The Newport Papers; CreateSpace Independent Publishing Platform: Scotts Valley, CA, USA, 2012.
27. Rosenthal, C.; Jones, N. *Chaos Engineering. System Resiliency in Practice*; O'Reilly Publishing: Newton, MA, USA, 2020.
28. Skiadas, C.H.; Skiadas, C. *Handbook of Applications of Chaos Theory*; Chapman and Hall: London, UK; CRC: London, UK, 2016.
29. Sundbye, L. *Discrete Dynamical Systems, Chaos Theory and Fractals*; CreateSpace Independent Publishing Platform: Scotts Valley, CA, USA, 2018.
30. Musaev, A.A.; Makshanov, A.V.; Grigoriev, D.A. Forecasting Multivariate Chaotic Processes with Precedent Analysis. *Computation* **2021**, *9*, 110. [CrossRef]
31. Gardner, E.S., Jr. Exponential smoothing: The state of the art—Part II. *Int. J. Forecast.* **2006**, *22*, 637–666. [CrossRef]
32. Musaev, A.A.; Grigoriev, D.A. Analyzing, Modeling and Utilizing Observation Series Correlation in Capital Markets. *Computation* **2021**, *9*, 88. [CrossRef]
33. Musaev, A.A.; Borovinskaya, E.E. Evolutionary Optimization of Case-Based Forecasting Algorithms in Chaotic Environments. *Symmetry* **2021**, *13*, 301. [CrossRef]
34. Maronna, R.A.; Martin, R.D.; Yohai, V.J.; Salibián-Barrera, M. *Robust Statistics: Theory and Methods (with R)*; John Wiley & Sons: Hoboken, NJ, USA, 2019.
35. Musaev, A.; Grigoriev, D. Multi-expert Systems: Fundamental Concepts and Application Examples. *J. Theor. Appl. Inf. Technol.* **2022**, *100*, 336–348.

36. Pejić Bach, M.; Krstić, Ž.; Seljan, S.; Turulja, L. Text mining for big data analysis in financial sector: A literature review. *Sustainability* **2019**, *11*, 1277. [CrossRef]
37. Musaev, A.; Grigoriev, D. Numerical Studies of Statistical Management Decisions in Conditions of Stochastic Chaos. *Mathematics* **2022**, *10*, 226. [CrossRef]

mathematics

MDPI

Article

A Study on Dynamics of CD4+ T-Cells under the Effect of HIV-1 Infection Based on a Mathematical Fractal-Fractional Model via the Adams-Bashforth Scheme and Newton Polynomials

Hashem Najafi [1,†], Sina Etemad [2,†], Nichaphat Patanarapeelert [3,†], Joshua Kiddy K. Asamoah [4,†], Shahram Rezapour [2,5,*,†] and Thanin Sitthiwirattham [6,*,†]

1 Department of Mathematics, College of Sciences, Shiraz University, Shiraz 7187919556, Iran; hnajafi@shirazu.ac.ir
2 Department of Mathematics, Azarbaijan Shahid Madani University, Tabriz 3751-71379, Iran; sina.etemad@azaruniv.ac.ir
3 Department of Mathematics, Faculty of Applied Science, King Mongkut's University of Technology North Bangkok, Bangkok 10800, Thailand; nichaphat.p@sci.kmutnb.ac.th
4 Department of Mathematics, Kwame Nkrumah University of Science and Technology, Kumasi, Ghana; jkkasamoah@knust.edu.gh
5 Department of Medical Research, China Medical University Hospital, China Medical University, Taichung 40402, Taiwan
6 Mathematics Department, Faculty of Science and Technology, Suan Dusit University, Bangkok 10300, Thailand
* Correspondence: sh.rezapour@azaruniv.ac.ir (S.R.); thanin_sit@dusit.ac.th (T.S.)
† These authors contributed equally to this work.

Citation: Najafi, H.; Etemad, S.; Patanarapeelert, N.; Asamoah, J.K.K.; Rezapour, S.; Sitthiwirattham, T. A Study on Dynamics of CD4+ T-Cells under the Effect of HIV-1 Infection Based on a Mathematical Fractal-Fractional Model via the Adams-Bashforth Scheme and Newton Polynomials. *Mathematics* **2022**, *10*, 1366. https://doi.org/10.3390/math10091366

Academic Editors: Mihaela Neamţu, Eva Kaslik and Anca Rădulescu

Received: 19 February 2022
Accepted: 5 April 2022
Published: 19 April 2022

Publisher's Note: MDPI stays neutral with regard to jurisdictional claims in published maps and institutional affiliations.

Copyright: © 2022 by the authors. Licensee MDPI, Basel, Switzerland. This article is an open access article distributed under the terms and conditions of the Creative Commons Attribution (CC BY) license (https://creativecommons.org/licenses/by/4.0/).

Abstract: In recent decades, AIDS has been one of the main challenges facing the medical community around the world. Due to the large human deaths of this disease, researchers have tried to study the dynamic behaviors of the infectious factor of this disease in the form of mathematical models in addition to clinical trials. In this paper, we study a new mathematical model in which the dynamics of CD4$^+$ T-cells under the effect of HIV-1 infection are investigated in the context of a generalized fractal-fractional structure for the first time. The kernel of these new fractal-fractional operators is of the generalized Mittag-Leffler type. From an analytical point of view, we first derive some results on the existence theory and then the uniqueness criterion. After that, the stability of the given fractal-fractional system is reviewed under four different cases. Next, from a numerical point of view, we obtain two numerical algorithms for approximating the solutions of the system via the Adams-Bashforth method and Newton polynomials method. We simulate our results via these two algorithms and compare both of them. The numerical results reveal some stability and a situation of lacking a visible order in the early days of the disease dynamics when one uses the Newton polynomial.

Keywords: existence; fractal-fractional derivative; HIV-1 infection; Newton polynomial; Adams-Bashforth

MSC: 34A08; 65P99; 49J15

1. Introduction

According to medical definitions and clinical findings and virology, human immunodeficiency virus (HIV) is a type of retrovirus that leads to acquired immunodeficiency syndrome (AIDS) in humans [1]. In fact, CD4$^+$ T-lymphocytes are the largest number of white blood cells in the human immune system that are attacked by HIV viruses, which attack CD4$^+$ T-cells and infect them, reducing their number and efficiency. They disrupt cells. Therefore, this process reduces the resistance of the immune system in the human body and weakens it [1]. Although the HIV virus infects other cells, it causes the most damage to T cells by causing the degradation and destruction of CD4$^+$ T-cells. As a result, the affected person's body gradually becomes sensitive to various types of infections and contaminants and completely loses its immunity. Usually, according to laboratory results, a healthy person has white blood cells with a normal $CD4^+$ T-lymphocyte count of 800 to $1200/mm^3$. The decrease in $CD4^+$ T-lymphocyte count may be due to thymus insufficiency or bone defects. One way to determine whether a person is infected with the virus is recognized through very small number of $CD4^+$ T-lymphocytes (less than $200/mm^3$).

Currently, AIDS is one of the most unknown and dangerous diseases of our time. According to UNAIDS 2017, which is published annually, about 36.7 million people worldwide are living with HIV, and nearly 1.8 million people have recently been infected with HIV, and almost one million people worldwide are affected by various complication, and died of AIDS in 2016. There are even several countries, especially in Africa and some less developed countries, where up to thirty-five percent of the population between the ages of fifteen and fifty are infected with the HIV virus [2]. Despite significant advances in the control of the disease, no vaccine for HIV has yet been found.

In recent decades, many attempts have been made to design, analyze, simulate and solve mathematical dynamic models, which include the basic rules in the analysis of control and prevention of the spread of various diseases and infections. In this regard, biology and engineering are among the most widely used fields for various types of mathematical modeling, which are particularly popular among researchers. For such a purpose, derivation operators and differential equations play an important role. To demonstrate such high efficiency, we can refer to a sample of articles in which various types of mathematical models are observed, including time delying model of COVID-19 [3], anthrax model for animals [4], Hepatitis C model [5], memristor chaotic model of circuit [6], Lorenz-Stenflo hyperchaotic model [7], dynamics of environmental persistence of infections [8], Langevin model [9], Mump virus [10], Zika virus [11], mosaic disease [12], Computer viruses [13], thermostat control model [14,15], pantograph model [16,17], canine distemper virus [18], Lassa fever [19], hybrid model of p-Laplacian operators [20], co-dynamics of COVID-19 and diabetes [21], checmical modeling of cyclohexane [22,23], Navier systems [24], etc.

Designing mathematical models to analyze the dynamics of HIV infection is also known as a valuable and efficient measure in this regard [25–27]. Usually in almost all of these dynamic mathematical models, we can see the association between HIV viruses and non-infected $CD4^+$ T-cells and the effects of a variety of drug therapies and clinical therapies on reducing and controlling infected cells. Perlson [28] designed and presented a simple model for primary HIV infection in 1989. This model, which is related to HIV infection, is important, and many other models have been developed later, which are inspired by this model. Four years later, Perlson et al. generalized the model and examined some of the behaviors and characteristics of the new model [29]. They considered the model by defining four different state functions: virus population, non-infected $CD4^+$ T-cells, productive infected $CD4^+$ T-cells, and latent infected $CD4^+$ T-cells.

In recent years, some other mathematicians developed the basic integer order models of HIV infection in the context of the fractional order systems. In [30], Ding et al. introduced a fractional version of the infection of HIV for $CD4^+$ T-cells and analyzed the non-negative solutions of this system. Arafa et al. [31] used the generalized Euler technique for his model to obtain solutions and investigated the impact of changes of viral particles on the blood. After that, Bulut et al. [32] combined the homotopy and sumudu techniques to staudy the dynamics of their fractional model of the infection of HIV on $CD4^+$ T-cells. Lichae et al. [33] extended their fractional HIV model based on the effect of drug therapy and solved three-compartmental model of $CD4^+$ with the help of LADM (Laplace Adomian decomposition method). In 2020, Nazir et al. [34] investigated not only existenece theory, but also stability criteria for the Caputo-Fabrizo HIV model of $CD4^+$ T-cells. Wang et al. [35] studied the time periodic reaction-diffusion equations for modeling 2-LTR dynamics in HIV-infected patients.

In 2017, Atangana [36] presented a generalized kind of operators entitled fractal-fractional operators. This definition used the existing notions of fractional calculus and fractal calculus with together. These operators are the convolution of the power-law, exponential law and generalized Mittag–Leffler law with fractal derivatives. There exist two components for such fractal-fractional operators: the fractional order and fractal dimension (order). Fractal-fractional differential equations transfer the order and dimension of every dynamical system into a rational order one. In fact, we are able to generalize each standard differential equation to the generalized systems having arbitrary order and dimension of derivatives. The main goal of such a combination is to analyze a vast range of nonlocal BVPs or IVPs that possess fractal behaviors. In this direction, a limited researchers obtained some results in which we see that the generalized fractal-fractional operators give accurate and more exact simulations for describing mathematical models of real-world phenomena. Some of new works in this regard are [37–41].

Due to the novelty and efficacy of these new fractal-fractional operators, in this paper, we aim to design a mathematical model of CD4^{+} T-cells under the effect of HIV-1 infection in which derivatives are fractal-fractional operators in the sense of Atangana-Baleanu. It is notable that in 2021, Ahmad et al. [42] studied the dynamics of HIV primary infection in the context of a model designed by the fractal-fractional operators. The main contribution and novelty of our work in comparison to their paper [42] is that we analyze all qualitative behaviors of such a fractal-fractional system. In other words, we first review the existence and uniqueness of solutions and further, we complete our study by giving new results about the stability (Ulam-Hyers-Rassias) of solutions. Also, for the first time, in this paper, we derive numerical schemes for the fractal-fractional CD4^{+}-HIV-1 model with the help of the Newton polynomials and by applying some real data, we compare our results with the Adams-Bashforth simulations. In this direction, we can see some dynamical behaviors of the solutions in our simulations.

The arrangement of the paper is as follows: we introduce our fractal-fractional model in Section 2 and describe its parameters and coefficients. The existence results are given in Section 3 and further, in Section 4, we investigate the uniqueness. The stability criterion are implemented in Section 5. In the sequel, we derive two numerical schemes. In other words, in Section 6, the Adams-Bashforth method are done and then, we derive another algorithm in Section 7 by using the Newton polynomials. We present some simulations and discussions about both numerical methods in Section 8. We end the paper by giving conclusions in Section 9.

2. The Structure of the Model for CD4^{+} T-Cells and HIV-1

In 2006, Wang and Li [1] formulated an integer-order classical mathematical structure of dynamics of CD4^{+} T-cells under the HIV-1 infection in three-compartmental model as

$$\begin{cases} \mathbb{T}'(s) = \theta - q\mathbb{V}\mathbb{T} - \rho\mathbb{T} + \zeta\mathbb{U}, \\ \mathbb{U}'(s) = q\mathbb{V}\mathbb{T} - (\zeta + \kappa)\mathbb{U}, \\ \mathbb{V}'(s) = \kappa N\mathbb{U} - \vartheta\mathbb{V}, \end{cases} \tag{1}$$

via the initial values $\mathbb{T}(0) = \mathbb{T}_0$, $\mathbb{U}(0) = \mathbb{U}_0$, and $\mathbb{V}(0) = \mathbb{V}_0$, and also the state functions $\mathbb{T}(s)$, $\mathbb{U}(s)$, $\mathbb{V}(s)$ are an amount of susceptible CD4^{+} T-cells, an amount of infectious CD4^{+} T-cells, and the free particles of the infection of the HIV virus in the blood at the time $s \in \mathbb{J} := [0, S], (S > 0)$, respectively. Moreover, the parameter N stands for the average number of infected particles by an existing infected cell, ϑ is the natural rate of death for the virus, ζ is the return rate of infected cells to susceptible compartment, κ is the rate of death for infected T-cells, q is the rate of infection T-cells, ρ is the natural rate of death, and θ shows the supply rate for new T-cells. They considered all parameters as positive values and $\mathbb{T}_0, \mathbb{U}_0, \mathbb{V}_0 \geq 0$.

To upgrade and improve the exact results, inspired by the standard model (1), we present a mathematical fractal-fractional model on dynamics of CD4^{+} T-cells under the effect of HIV-1 infection via the generalized Mittag-Leffler-type kernel (fractal-fractional CD4^{+}-HIV-1-model) as

$$\begin{cases} {}^{\mathbf{FFML}}\mathcal{D}_{0,s}^{(\delta,\sigma)}\mathbb{T}(s) = \theta - q\mathbb{V}(s)\mathbb{T}(s) - \rho\mathbb{T}(s) + \zeta\mathbb{U}(s), \\ {}^{\mathbf{FFML}}\mathcal{D}_{0,s}^{(\delta,\sigma)}\mathbb{U}(s) = q\mathbb{V}(s)\mathbb{T}(s) - (\zeta+\kappa)\mathbb{U}(s), \\ {}^{\mathbf{FFML}}\mathcal{D}_{0,s}^{(\delta,\sigma)}\mathbb{V}(s) = \kappa N\mathbb{U}(s) - \vartheta\mathbb{V}(s), \end{cases} \tag{2}$$

subject to

$$\mathbb{T}(0) = \mathbb{T}_0 \geq 0, \quad \mathbb{U}(0) = \mathbb{U}_0 \geq 0, \quad \mathbb{V}(0) = \mathbb{V}_0 \geq 0,$$

where all assumptions and parameters are similar to above classical model (1). Also, ${}^{\mathbf{FFML}}\mathcal{D}_{0,s}^{(\delta,\sigma)}$ is the (δ,σ)-fractal-fractional derivative with the fractional order $\delta \in (0,1]$ and the fractal order $\sigma \in (0,1]$ via the Mittag-Leffler-type kernel.

In other words, let a continuous map $\Psi : (a,b) \to [0,\infty)$ be fractal differentiable of dimension σ. In this case, the (δ,σ)-fractal-fractional derivative of Ψ of the generalized Mittag-Leffler-type kernel of order δ in the Riemann-Liouville sense is defined as

$${}^{\mathbf{FFML}}\mathcal{D}_{a,s}^{(\delta,\sigma)}\Psi(s) = \frac{\mathcal{AB}(\delta)}{1-\delta}\frac{\mathrm{d}}{\mathrm{d}s^{\sigma}}\int_a^s \mathbb{E}_\delta\left[-\frac{\delta}{1-\delta}(s-\mathfrak{w})^\delta\right]\Psi(\mathfrak{w})\,\mathrm{d}\mathfrak{w}, \quad 0<\delta,\sigma\leq 1, \tag{3}$$

where

$$\frac{\mathrm{d}\Psi(\mathfrak{w})}{\mathrm{d}\mathfrak{w}^\sigma} = \lim_{s\to\mathfrak{w}}\frac{\Psi(s)-\Psi(\mathfrak{w})}{s^\sigma-\mathfrak{w}^\sigma},$$

is the fractal derivative and $\mathcal{AB}(\delta) = 1-\delta+\frac{\delta}{\Gamma(\delta)}$ and $\mathcal{AB}(0) = \mathcal{AB}(1) = 1$ [36].

Moreover, accordingly, for such a function Ψ, the (δ,σ)-fractal-fractional integral via the Mittag-Leffler-type kernel is given by

$${}^{\mathbf{FFML}}\mathcal{I}_{a,s}^{(\delta,\sigma)}\Psi(s) = \frac{\delta\sigma}{\mathcal{AB}(\delta)\Gamma(\delta)}\int_a^s \mathfrak{w}^{\sigma-1}(s-\mathfrak{w})^{\delta-1}\Psi(\mathfrak{w})\,\mathrm{d}\mathfrak{w} + \frac{(1-\delta)\sigma s^{\sigma-1}}{\mathcal{AB}(\delta)}\Psi(s), \tag{4}$$

if it exists, where $\delta,\sigma > 0$ [36].

3. Existence Property

In this section, the existence property is investigated based on fixed point theory. For the qualitative analysis, make the Banach space $\mathbb{X} = \mathbb{M}^3$, where $\mathbb{M} = C(\mathbb{J},\mathbb{R})$ with

$$\|\mathbb{K}\|_{\mathbb{X}} = \|(\mathbb{T},\mathbb{U},\mathbb{V})\|_{\mathbb{X}} = \max\{|W(s)| : \ s\in\mathbb{J}\},$$

for $|W| := |\mathbb{T}| + |\mathbb{U}| + |\mathbb{V}|$. We reformulate the R.H.S. of the fractal-fractional CD4$^+$-HIV-1-model (2) as:

$$\begin{cases} \mathbb{Q}_1(s,\mathbb{T}(s),\mathbb{U}(s),\mathbb{V}(s)) = \theta - q\mathbb{V}(s)\mathbb{T}(s) - \rho\mathbb{T}(s) + \zeta\mathbb{U}(s), \\ \mathbb{Q}_2(s,\mathbb{T}(s),\mathbb{U}(s),\mathbb{V}(s)) = q\mathbb{V}(s)\mathbb{T}(s) - (\zeta+\kappa)\mathbb{U}(s), \\ \mathbb{Q}_3(s,\mathbb{T}(s),\mathbb{U}(s),\mathbb{V}(s)) = \kappa N\mathbb{U}(s) - \vartheta\mathbb{V}(s). \end{cases} \tag{5}$$

In this case, the fractal-fractional CD4$^+$-HIV-1-model (2) is transformed into the following system

$$\begin{cases} {}^{\mathbf{ABR}}\mathcal{D}_{0,s}^{\delta}\mathbb{T}(s) = \sigma s^{\sigma-1}\mathbb{Q}_1(s,\mathbb{T}(s),\mathbb{U}(s),\mathbb{V}(s)), \\ {}^{\mathbf{ABR}}\mathcal{D}_{0,s}^{\delta}\mathbb{U}(s) = \sigma s^{\sigma-1}\mathbb{Q}_2(s,\mathbb{T}(s),\mathbb{U}(s),\mathbb{V}(s)), \\ {}^{\mathbf{ABR}}\mathcal{D}_{0,s}^{\delta}\mathbb{V}(s) = \sigma s^{\sigma-1}\mathbb{Q}_3(s,\mathbb{T}(s),\mathbb{U}(s),\mathbb{V}(s)). \end{cases} \tag{6}$$

In view of (6), we rewrite the developed tripled-system with the compact IVP which takes the form

$$\begin{cases} {}^{\mathbf{ABR}}\mathcal{D}^{\delta}_{0,s}\mathbb{K}(s) = \sigma s^{\sigma-1}\mathbb{Q}(s,\mathbb{K}(s)), \\ \mathbb{K}(0) = \mathbb{K}_0, \end{cases} \tag{7}$$

where

$$\mathbb{K}(s) = (\mathbb{T}(s),\mathbb{U}(s),\mathbb{V}(s))^T, \qquad \mathbb{K}_0 = (\mathbb{T}_0,\mathbb{U}_0,\mathbb{V}_0)^T, \quad \delta,\sigma \in (0,1], \tag{8}$$

and

$$\mathbb{Q}(s,\mathbb{K}(s)) = \begin{cases} \mathbb{Q}_1(s,\mathbb{T}(s),\mathbb{U}(s),\mathbb{V}(s)), \\ \mathbb{Q}_2(s,\mathbb{T}(s),\mathbb{U}(s),\mathbb{V}(s)), \\ \mathbb{Q}_3(s,\mathbb{T}(s),\mathbb{U}(s),\mathbb{V}(s)), \quad s \in \mathbb{J}. \end{cases} \tag{9}$$

By the definition and by (7), we have

$$\frac{\mathcal{AB}(\delta)}{1-\delta}\frac{\mathrm{d}}{\mathrm{d}s}\int_0^s \mathbb{E}_\delta\big[-\frac{\delta}{1-\delta}(s-\mathfrak{w})^\delta\big]\mathbb{K}(\mathfrak{w})\,\mathrm{d}\mathfrak{w} = \sigma s^{\sigma-1}\mathbb{Q}(s,\mathbb{K}(s)). \tag{10}$$

In the sequel, operating the fractal-fractional Atangana-Baleanu integral on (10), we get

$$\mathbb{K}(s) = \mathbb{K}(0) + \frac{\delta\sigma}{\mathcal{AB}(\delta)\Gamma(\delta)}\int_0^s \mathfrak{w}^{\sigma-1}(s-\mathfrak{w})^{\delta-1}\mathbb{Q}(\mathfrak{w},\mathbb{K}(\mathfrak{w}))\,\mathrm{d}\mathfrak{w} + \frac{(1-\delta)\sigma s^{\sigma-1}}{\mathcal{AB}(\delta)}\mathbb{Q}(s,\mathbb{K}(s)). \tag{11}$$

Due to the above compact form of the fractal-fractional integral equation, the extended representation of it is illustrated as

$$\begin{cases} \mathbb{T}(s) = \mathbb{T}_0 + \dfrac{(1-\delta)\sigma s^{\sigma-1}}{\mathcal{AB}(\delta)}\mathbb{Q}_1(s,\mathbb{T}(s),\mathbb{U}(s),\mathbb{V}(s)) \\ \qquad + \dfrac{\delta\sigma}{\mathcal{AB}(\delta)\Gamma(\delta)}\displaystyle\int_0^s \mathfrak{w}^{\sigma-1}(s-\mathfrak{w})^{\delta-1}\mathbb{Q}_1(\mathfrak{w},\mathbb{T}(\mathfrak{w}),\mathbb{U}(\mathfrak{w}),\mathbb{V}(\mathfrak{w}))\,\mathrm{d}\mathfrak{w}, \\ \mathbb{U}(s) = \mathbb{U}_0 + \dfrac{(1-\delta)\sigma s^{\sigma-1}}{\mathcal{AB}(\delta)}\mathbb{Q}_2(s,\mathbb{T}(s),\mathbb{U}(s),\mathbb{V}(s)) \\ \qquad + \dfrac{\delta\sigma}{\mathcal{AB}(\delta)\Gamma(\delta)}\displaystyle\int_0^s \mathfrak{w}^{\sigma-1}(s-\mathfrak{w})^{\delta-1}\mathbb{Q}_2(\mathfrak{w},\mathbb{T}(\mathfrak{w}),\mathbb{U}(\mathfrak{w}),\mathbb{V}(\mathfrak{w}))\,\mathrm{d}\mathfrak{w}, \\ \mathbb{V}(s) = \mathbb{V}_0 + \dfrac{(1-\delta)\sigma s^{\sigma-1}}{\mathcal{AB}(\delta)}\mathbb{Q}_3(s,\mathbb{T}(s),\mathbb{U}(s),\mathbb{V}(s)) \\ \qquad + \dfrac{\delta\sigma}{\mathcal{AB}(\delta)\Gamma(\delta)}\displaystyle\int_0^s \mathfrak{w}^{\sigma-1}(s-\mathfrak{w})^{\delta-1}\mathbb{Q}_3(\mathfrak{w},\mathbb{T}(\mathfrak{w}),\mathbb{U}(\mathfrak{w}),\mathbb{V}(\mathfrak{w}))\,\mathrm{d}\mathfrak{w}. \end{cases} \tag{12}$$

We consider a self-map to derive a fixed-point problem, by defining $F : \mathbb{X} \to \mathbb{X}$ as

$$\begin{aligned} F(\mathbb{K}(s)) = \mathbb{K}(0) &+ \frac{(1-\delta)\sigma s^{\sigma-1}}{\mathcal{AB}(\delta)}\mathbb{Q}(s,\mathbb{K}(s)) \\ &+ \frac{\delta\sigma}{\mathcal{AB}(\delta)\Gamma(\delta)}\int_0^s \mathfrak{w}^{\sigma-1}(s-\mathfrak{w})^{\delta-1}\mathbb{Q}(\mathfrak{w},\mathbb{K}(\mathfrak{w}))\,\mathrm{d}\mathfrak{w}. \end{aligned} \tag{13}$$

The Leray-Schauder theorem is utilized to prove the existence property in relation to the fractal-fractional CD4$^+$-HIV-1-model (2).

Theorem 1. *[43] Let $\mathbb{X}$ be a Banach space, $\mathbb{E} \subset \mathbb{X}$ a convex closed bounded set, $\mathbb{O} \subset \mathbb{E}$ an open set, and $0 \in \mathbb{O}$. Then for the continuous and compact map $F : \bar{\mathbb{O}} \to \mathbb{E}$, either:*
(HY1)$\exists\, y \in \bar{\mathbb{O}}$ s.t. $y = F(y)$, or
(HY2)$\exists\, y \in \partial\mathbb{O}$ and $0 < \mu < 1$ s.t. $y = \mu F(y)$.

Theorem 2. *Let $\mathbb{Q} \in C(\mathbb{J} \times \mathbb{X}, \mathbb{X})$. Assume:*

(P1) $\exists\, \varphi \in L^1(\mathbb{J}, \mathbb{R}^+)$ and $\exists\, A \in C([0,\infty),(0,\infty))$ (A is non-decreasing) s.t. $\forall s \in \mathbb{J}$ and $\mathbb{K} \in \mathbb{X}$,

$$\big|\mathbb{Q}(s, \mathbb{K}(s))\big| \leq \varphi(s) A(|\mathbb{K}(s)|);$$

(P2) $\exists\, \omega > 0$ s.t.

$$\frac{\omega}{\mathbb{K}_0 + \Big[\frac{(1-\delta)\sigma S^{\sigma-1}}{\mathcal{AB}(\delta)} + \frac{\delta\sigma S^{\delta+\sigma-1}\Gamma(\sigma)}{\mathcal{AB}(\delta)\Gamma(\delta+\sigma)}\Big]\varphi_0^* A(\omega)} > 1, \tag{14}$$

with $\varphi_0^* = \sup_{s\in\mathbb{J}} |\varphi(s)|$.

Then there is a solution for the fractal-fractional problem (7) and accordingly, for the fractal-fractional CD4$^+$-HIV-1-model (2) on $\mathbb{J}$.

Proof. First, consider $F : \mathbb{X} \to \mathbb{X}$ which is defined by (13) and assume

$$N_r = \Big\{\mathbb{K} \in \mathbb{X} : \|\mathbb{K}\|_{\mathbb{X}} \leq r\Big\},$$

for some $r > 0$. Clearly, as $\mathbb{Q}$ is continuous, thus F is so. From (P1), we get

$$\begin{aligned}
\big|F(\mathbb{K}(s))\big| &\leq \big|\mathbb{K}(0)\big| + \frac{(1-\delta)\sigma s^{\sigma-1}}{\mathcal{AB}(\delta)}\big|\mathbb{Q}(s, \mathbb{K}(s))\big| \\
&\quad + \frac{\delta\sigma}{\mathcal{AB}(\delta)\Gamma(\delta)}\int_0^s \mathfrak{w}^{\sigma-1}(s-\mathfrak{w})^{\delta-1}\big|\mathbb{Q}(\mathfrak{w}, \mathbb{K}(\mathfrak{w}))\big|\,\mathrm{d}\mathfrak{w} \\
&\leq \mathbb{K}_0 + \frac{(1-\delta)\sigma s^{\sigma-1}}{\mathcal{AB}(\delta)}\varphi(s) A(|\mathbb{K}(s)|) \\
&\quad + \frac{\delta\sigma}{\mathcal{AB}(\delta)\Gamma(\delta)}\int_0^s \mathfrak{w}^{\sigma-1}(s-\mathfrak{w})^{\delta-1}\varphi(\mathfrak{w}) A(|\mathbb{K}(\mathfrak{w})|)\,\mathrm{d}\mathfrak{w} \\
&\leq \mathbb{K}_0 + \frac{(1-\delta)\sigma S^{\sigma-1}}{\mathcal{AB}(\delta)}\varphi_0^* A(r) + \frac{\delta\sigma S^{\delta+\sigma-1} B(\delta,\sigma)}{\mathcal{AB}(\delta)\Gamma(\delta)}\varphi_0^* A(r) \\
&= \mathbb{K}_0 + \frac{(1-\delta)\sigma S^{\sigma-1}}{\mathcal{AB}(\delta)}\varphi_0^* A(r) + \frac{\delta\sigma S^{\delta+\sigma-1}\Gamma(\sigma)}{\mathcal{AB}(\delta)\Gamma(\delta+\sigma)}\varphi_0^* A(r),
\end{aligned}$$

for $\mathbb{K} \in N_r$. Hence

$$\|F\mathbb{K}\|_{\mathbb{X}} \leq \mathbb{K}_0 + \Big[\frac{(1-\delta)\sigma S^{\sigma-1}}{\mathcal{AB}(\delta)} + \frac{\delta\sigma S^{\delta+\sigma-1}\Gamma(\sigma)}{\mathcal{AB}(\delta)\Gamma(\delta+\sigma)}\Big]\varphi_0^* A(r) < \infty. \tag{15}$$

Thus F is uniformly bounded on $\mathbb{X}$. Now, take $s, v \in [0, S]$ s.t. $s < v$ and $\mathbb{K} \in N_r$. By denoting

$$\sup_{(s,\mathbb{K})\in\mathbb{J}\times N_r} |\mathbb{Q}(s, \mathbb{K}(s))| = \mathbb{Q}^* < \infty,$$

we estimate

$$\begin{aligned}
\big|F(\mathbb{K}(v)) - F(\mathbb{K}(s))\big| &\leq \Bigg| \frac{(1-\delta)\sigma v^{\sigma-1}}{\mathcal{AB}(\delta)}\mathbb{Q}(v,\mathbb{K}(v)) - \frac{(1-\delta)\sigma s^{\sigma-1}}{\mathcal{AB}(\delta)}\mathbb{Q}(s,\mathbb{K}(s)) \\
&+ \frac{\delta\sigma}{\mathcal{AB}(\delta)\Gamma(\delta)}\int_0^v \mathfrak{w}^{\sigma-1}(v-\mathfrak{w})^{\delta-1}\mathbb{Q}(\mathfrak{w},\mathbb{K}(\mathfrak{w}))\,\mathrm{d}\mathfrak{w} \\
&- \frac{\delta\sigma}{\mathcal{AB}(\delta)\Gamma(\delta)}\int_0^s \mathfrak{w}^{\sigma-1}(s-\mathfrak{w})^{\delta-1}\mathbb{Q}(\mathfrak{w},\mathbb{K}(\mathfrak{w}))\,\mathrm{d}\mathfrak{w}\Bigg| \\
&\leq \frac{(1-\delta)\sigma\mathbb{Q}^*}{\mathcal{AB}(\delta)}(v^{\sigma-1}-s^{\sigma-1}) \\
&+ \frac{\delta\sigma\mathbb{Q}^*}{\mathcal{AB}(\delta)\Gamma(\delta)}\left|\int_0^v \mathfrak{w}^{\sigma-1}(v-\mathfrak{w})^{\delta-1}\,\mathrm{d}\mathfrak{w} - \int_0^s \mathfrak{w}^{\sigma-1}(s-\mathfrak{w})^{\delta-1}\,\mathrm{d}\mathfrak{w}\right| \\
&\leq \frac{(1-\delta)\sigma\mathbb{Q}^*}{\mathcal{AB}(\delta)}(v^{\sigma-1}-s^{\sigma-1}) + \frac{\delta\sigma\mathbb{Q}^* B(\delta,\sigma)}{\mathcal{AB}(\delta)\Gamma(\delta)}\left[v^{\delta+\sigma-1}-s^{\delta+\sigma-1}\right] \\
&= \frac{(1-\delta)\sigma\mathbb{Q}^*}{\mathcal{AB}(\delta)}(v^{\sigma-1}-s^{\sigma-1}) + \frac{\delta\sigma\mathbb{Q}^*\Gamma(\sigma)}{\mathcal{AB}(\delta)\Gamma(\delta+\sigma)}\left[v^{\delta+\sigma-1}-s^{\delta+\sigma-1}\right].
\end{aligned} \tag{16}$$

We see that (16) approaches to 0 independent of $\mathbb{K}$, as $v \to s$. Consequently

$$\|F(\mathbb{K}(v)) - F(\mathbb{K}(s))\|_{\mathbb{X}} \to 0,$$

when $v \to s$. This gives the equicontinuity of F, and accordingly the compactness of F on N_r by the Arzelá–Ascoli thoerem. Since all conditions of Theorem 1 are fulfilled on F, we have one of (HY1) or (HY2). From (P2), we set

$$\Phi := \Big\{\mathbb{K} \in \mathbb{X} : \ \|\mathbb{K}\|_{\mathbb{X}} < \omega\Big\},$$

for some $\omega > 0$ s.t.

$$\mathbb{K}_0 + \left[\frac{(1-\delta)\sigma S^{\sigma-1}}{\mathcal{AB}(\delta)} + \frac{\delta\sigma S^{\delta+\sigma-1}\Gamma(\sigma)}{\mathcal{AB}(\delta)\Gamma(\delta+\sigma)}\right]\varphi_0^* A(\omega) < \omega.$$

From (P1) and (15), we have

$$\|F\mathbb{K}\|_{\mathbb{X}} \leq \mathbb{K}_0 + \left[\frac{(1-\delta)\sigma S^{\sigma-1}}{\mathcal{AB}(\delta)} + \frac{\delta\sigma S^{\delta+\sigma-1}\Gamma(\sigma)}{\mathcal{AB}(\delta)\Gamma(\delta+\sigma)}\right]\varphi_0^* A(\|\mathbb{K}\|_{\mathbb{X}}). \tag{17}$$

Suppose that there are $\mathbb{K} \in \partial\Phi$ and $0 < \mu < 1$ s.t. $\mathbb{K} = \mu F(\mathbb{K})$. Then by (17), we write

$$\begin{aligned}
\omega = \|\mathbb{K}\|_{\mathbb{X}} = \mu\|F\mathbb{K}\|_{\mathbb{X}} &< \mathbb{K}_0 + \left[\frac{(1-\delta)\sigma S^{\sigma-1}}{\mathcal{AB}(\delta)} + \frac{\delta\sigma S^{\delta+\sigma-1}\Gamma(\sigma)}{\mathcal{AB}(\delta)\Gamma(\delta+\sigma)}\right]\varphi_0^* A(\|\mathbb{K}\|_{\mathbb{X}}) \\
&< \mathbb{K}_0 + \left[\frac{(1-\delta)\sigma S^{\sigma-1}}{\mathcal{AB}(\delta)} + \frac{\delta\sigma S^{\delta+\sigma-1}\Gamma(\sigma)}{\mathcal{AB}(\delta)\Gamma(\delta+\sigma)}\right]\varphi_0^* A(\omega) < \omega,
\end{aligned}$$

which cannot be held. Therefore, (HY2) does not hold and F possesses a fixed–point in Φ by Theorem 1. This confirms the existence of a solution to the fractal-fractional $CD4^+$-HIV-1-model (2). □

4. Uniqueness Property

Here, on the fractal-fractional CD4$^+$-HIV-1-model (2), we investigate the Lipschitz property in the first step and further, the uniqueness property.

Lemma 1. *Consider* $\mathbb{T}, \mathbb{U}, \mathbb{V}, \mathbb{T}^*, \mathbb{U}^*, \mathbb{V}^* \in \mathbb{M} := C(\mathbb{J}, \mathbb{R})$, *and let*

(C1) $\|\mathbb{T}\| \leq \beta_1$, $\|\mathbb{U}\| \leq \beta_2$, $\|\mathbb{V}\| \leq \beta_3$ *for some constants* $\beta_1, \beta_2, \beta_3 > 0$.

Then the kernels $\mathbb{Q}_1, \mathbb{Q}_2, \mathbb{Q}_3$ *defined in* (5) *are fulfilled the Lipschitz property with constants* $\alpha_1, \alpha_2, \alpha_3 > 0$ *w.r.t. the relevant components, where*

$$\alpha_1 = \rho + q\beta_3, \quad \alpha_2 = \zeta + \kappa, \quad \alpha_3 = \vartheta. \tag{18}$$

Proof. For $\mathbb{Q}_1$, we take $\mathbb{T}, \mathbb{T}^* \in \mathbb{M} := C(\mathbb{J}, \mathbb{R})$ arbitrarily, and we have

$$\begin{aligned}
\|\mathbb{Q}_1(s, \mathbb{T}(s), \mathbb{U}(s), \mathbb{V}(s)) &- \mathbb{Q}_1(s, \mathbb{T}^*(s), \mathbb{U}(s), \mathbb{V}(s))\| \\
&= \left\| \left(\theta - q\mathbb{V}(s)\mathbb{T}(s) - \rho\mathbb{T}(s) + \zeta\mathbb{U}(s)\right) \right. \\
&\quad \left. - \left(\theta - q\mathbb{V}(s)\mathbb{T}^*(s) - \rho\mathbb{T}^*(s) + \zeta\mathbb{U}(s)\right) \right\| \\
&\leq \left[\rho + q\|\mathbb{V}(s)\|\right] \|\mathbb{T}(s) - \mathbb{T}^*(s)\| \\
&\leq \left[\rho + q\beta_3\right] \|\mathbb{T}(s) - \mathbb{T}^*(s)\| \\
&= \alpha_1 \|\mathbb{T}(s) - \mathbb{T}^*(s)\|.
\end{aligned}$$

From the above, we find out that $\mathbb{Q}_1$ is Lipschitz w.r.t. $\mathbb{T}$ under the constant $\alpha_1 > 0$. For $\mathbb{Q}_2$, we choose two arbitrary elements $\mathbb{U}, \mathbb{U}^* \in \mathbb{M} := C(\mathbb{J}, \mathbb{R})$, and estimate

$$\begin{aligned}
\|\mathbb{Q}_2(s, \mathbb{T}(s), \mathbb{U}(s), \mathbb{V}(s)) &- \mathbb{Q}_2(s, \mathbb{T}(s), \mathbb{U}^*(s), \mathbb{V}(s))\| \\
&= \left\| \left(q\mathbb{V}(s)\mathbb{T}(s) - (\zeta + \kappa)\mathbb{U}(s)\right) - \left(q\mathbb{V}(s)\mathbb{T}(s) - (\zeta + \kappa)\mathbb{U}^*(s)\right) \right\| \\
&\leq [\zeta + \kappa] \|\mathbb{U}(s) - \mathbb{U}^*(s)\| \\
&= \alpha_2 \|\mathbb{U}(s) - \mathbb{U}^*(s)\|.
\end{aligned}$$

The above inequality means that $\mathbb{Q}_2$ is Lipschitz w.r.t. $\mathbb{U}$ under the constant $\alpha_2 > 0$. Finally, for both arbitrary elemets $\mathbb{V}, \mathbb{V}^* \in \mathbb{M} := C(\mathbb{J}, \mathbb{R})$, we have

$$\begin{aligned}
\|\mathbb{Q}_3(s, \mathbb{T}(s), \mathbb{U}(s), \mathbb{V}(s)) &- \mathbb{Q}_3(s, \mathbb{T}(s), \mathbb{U}(s), \mathbb{V}^*(s))\| \\
&= \left\| \left(\kappa N\mathbb{U}(s) - \vartheta\mathbb{V}(s)\right) - \left(\kappa N\mathbb{U}(s) - \vartheta\mathbb{V}^*(s)\right) \right\| \\
&\leq \vartheta \|\mathbb{V}(s) - \mathbb{V}^*(s)\| \\
&= \alpha_3 \|\mathbb{V}(s) - \mathbb{V}^*(s)\|.
\end{aligned}$$

This confirms that $\mathbb{Q}_3$ is Lipschitz w.r.t. $\mathbb{V}$ under the constant $\alpha_3 > 0$. Therefore kernel functions $\mathbb{Q}_1, \mathbb{Q}_2, \mathbb{Q}_3$ are Lipschitze w.r.t. under the constants $\alpha_1, \alpha_2, \alpha_3 > 0$, respectively. □

Now, by invoking the above lemma, we are able to prove the uniqueness property for solutions of the fractal-fractional system (2).

Theorem 3. *Consider (C1) to be held. Then the fractal-fractional* CD4$^+$*-HIV-1-model* (2) *possesses exactly one solution if*

$$\Big[\frac{(1-\delta)\sigma S^{\sigma-1}}{\mathcal{AB}(\delta)}+\frac{\delta\sigma S^{\delta+\sigma-1}\Gamma(\sigma)}{\mathcal{AB}(\delta)\Gamma(\delta+\sigma)}\Big]\alpha_\jmath<1, \tag{19}$$

for $\jmath\in\{1,2,3\}$, *and the Lipschitz constants* $\alpha_\jmath>0$ *introduced by* (18).

Proof. To prove the desired result, we consider the contrary of the conclusion of theorem. That is, consider the existence of another solution for the fractal-fractional CD4$^+$-HIV-1-model (2), namely $\big(\mathbb{T}^*(s),\mathbb{U}^*(s),\mathbb{V}^*(s)\big)$ under the initial conditions

$$\big(\mathbb{T}^*(0)=\mathbb{T}_0,\mathbb{U}^*(0)=\mathbb{U}_0,\mathbb{V}^*(0)=\mathbb{V}_0\big).$$

From (12), we have

$$\begin{aligned}\mathbb{T}^*(s)&=\mathbb{T}_0+\frac{(1-\delta)\sigma s^{\sigma-1}}{\mathcal{AB}(\delta)}\mathbb{Q}_1(s,\mathbb{T}^*(s),\mathbb{U}^*(s),\mathbb{V}^*(s))\\&\quad+\frac{\delta\sigma}{\mathcal{AB}(\delta)\Gamma(\delta)}\int_0^s\mathfrak{w}^{\sigma-1}(s-\mathfrak{w})^{\delta-1}\mathbb{Q}_1(\mathfrak{w},\mathbb{T}^*(\mathfrak{w}),\mathbb{U}^*(\mathfrak{w}),\mathbb{V}^*(\mathfrak{w}))\,d\mathfrak{w},\end{aligned}$$

$$\begin{aligned}\mathbb{U}^*(s)&=\mathbb{U}_0+\frac{(1-\delta)\sigma s^{\sigma-1}}{\mathcal{AB}(\delta)}\mathbb{Q}_2(s,\mathbb{T}^*(s),\mathbb{U}^*(s),\mathbb{V}^*(s))\\&\quad+\frac{\delta\sigma}{\mathcal{AB}(\delta)\Gamma(\delta)}\int_0^s\mathfrak{w}^{\sigma-1}(s-\mathfrak{w})^{\delta-1}\mathbb{Q}_2(\mathfrak{w},\mathbb{T}^*(\mathfrak{w}),\mathbb{U}^*(\mathfrak{w}),\mathbb{V}^*(\mathfrak{w}))\,d\mathfrak{w},\end{aligned}$$

and

$$\begin{aligned}\mathbb{V}^*(s)&=\mathbb{V}_0+\frac{(1-\delta)\sigma s^{\sigma-1}}{\mathcal{AB}(\delta)}\mathbb{Q}_3(s,\mathbb{T}^*(s),\mathbb{U}^*(s),\mathbb{V}^*(s))\\&\quad+\frac{\delta\sigma}{\mathcal{AB}(\delta)\Gamma(\delta)}\int_0^s\mathfrak{w}^{\sigma-1}(s-\mathfrak{w})^{\delta-1}\mathbb{Q}_3(\mathfrak{w},\mathbb{T}^*(\mathfrak{w}),\mathbb{U}^*(\mathfrak{w}),\mathbb{V}^*(\mathfrak{w}))\,d\mathfrak{w}.\end{aligned}$$

In this case, we estimate

$$\begin{aligned}|\mathbb{T}(s)-\mathbb{T}^*(s)|&\le\frac{(1-\delta)\sigma s^{\sigma-1}}{\mathcal{AB}(\delta)}\Big|\mathbb{Q}_1(s,\mathbb{T}(s),\mathbb{U}(s),\mathbb{V}(s))-\mathbb{Q}_1(s,\mathbb{T}^*(s),\mathbb{U}^*(s),\mathbb{V}^*(s))\Big|\\&\quad+\frac{\delta\sigma}{\mathcal{AB}(\delta)\Gamma(\delta)}\int_0^s\mathfrak{w}^{\sigma-1}(s-\mathfrak{w})^{\delta-1}\\&\quad\times\Big|\mathbb{Q}_1(\mathfrak{w},\mathbb{T}(\mathfrak{w}),\mathbb{U}(\mathfrak{w}),\mathbb{V}(\mathfrak{w}))-\mathbb{Q}_1(\mathfrak{w},\mathbb{T}^*(\mathfrak{w}),\mathbb{U}^*(\mathfrak{w}),\mathbb{V}^*(\mathfrak{w}))\Big|\,d\mathfrak{w}\\&\le\frac{(1-\delta)\sigma s^{\sigma-1}}{\mathcal{AB}(\delta)}\alpha_1\|\mathbb{T}-\mathbb{T}^*\|+\frac{\delta\sigma}{\mathcal{AB}(\delta)\Gamma(\delta)}\int_0^s\mathfrak{w}^{\sigma-1}(s-\mathfrak{w})^{\delta-1}\alpha_1\|\mathbb{T}-\mathbb{T}^*\|\,d\mathfrak{w}\\&\le\Big[\frac{(1-\delta)\sigma S^{\sigma-1}}{\mathcal{AB}(\delta)}+\frac{\delta\sigma S^{\delta+\sigma-1}\Gamma(\sigma)}{\mathcal{AB}(\delta)\Gamma(\delta+\sigma)}\Big]\alpha_1\|\mathbb{T}-\mathbb{T}^*\|,\end{aligned}$$

and so

$$\left(1-\Big[\frac{(1-\delta)\sigma S^{\sigma-1}}{\mathcal{AB}(\delta)}+\frac{\delta\sigma S^{\delta+\sigma-1}\Gamma(\sigma)}{\mathcal{AB}(\delta)\Gamma(\delta+\sigma)}\Big]\alpha_1\right)\|\mathbb{T}-\mathbb{T}^*\|\le 0.$$

From (19), we know that the above inequality holds if $\|\mathbb{T}-\mathbb{T}^*\| = 0$, or $\mathbb{T} = \mathbb{T}^*$. In the similar manner, from

$$\|\mathbb{U}-\mathbb{U}^*\| \le \Big[\frac{(1-\delta)\sigma S^{\sigma-1}}{\mathcal{AB}(\delta)} + \frac{\delta\sigma S^{\delta+\sigma-1}\Gamma(\sigma)}{\mathcal{AB}(\delta)\Gamma(\delta+\sigma)}\Big]\alpha_2\|\mathbb{U}-\mathbb{U}^*\|,$$

we obtain

$$\left(1-\Big[\frac{(1-\delta)\sigma S^{\sigma-1}}{\mathcal{AB}(\delta)} + \frac{\delta\sigma S^{\delta+\sigma-1}\Gamma(\sigma)}{\mathcal{AB}(\delta)\Gamma(\delta+\sigma)}\Big]\alpha_2\right)\|\mathbb{U}-\mathbb{U}^*\| \le 0,$$

and this gives $\|\mathbb{U}-\mathbb{U}^*\| = 0$ or $\mathbb{U} = \mathbb{U}^*$. Further,

$$\|\mathbb{V}-\mathbb{V}^*\| \le \Big[\frac{(1-\delta)\sigma S^{\sigma-1}}{\mathcal{AB}(\delta)} + \frac{\delta\sigma S^{\delta+\sigma-1}\Gamma(\sigma)}{\mathcal{AB}(\delta)\Gamma(\delta+\sigma)}\Big]\alpha_3\|\mathbb{V}-\mathbb{V}^*\|,$$

yields

$$\left(1-\Big[\frac{(1-\delta)\sigma S^{\sigma-1}}{\mathcal{AB}(\delta)} + \frac{\delta\sigma S^{\delta+\sigma-1}\Gamma(\sigma)}{\mathcal{AB}(\delta)\Gamma(\delta+\sigma)}\Big]\alpha_3\right)\|\mathbb{V}-\mathbb{V}^*\| \le 0.$$

Hence $\mathbb{V} = \mathbb{V}^*$. In consequence,

$$(\mathbb{T}(s),\mathbb{U}(s),\mathbb{V}(s)) = (\mathbb{T}^*(s),\mathbb{U}^*(s),\mathbb{V}^*(s)).$$

The latter equality confirms that the fractal-fractional CD4$^+$-HIV-1-model (2) possesses a unique solution. □

5. Ulam-Hyers-Rassias Stability

In this part, the stability result of solutions in the context of four types of the Ulam–Hyers, Ulam–Hyers–Rassias and their generalizations are proved on the tripled system of the fractal-fractional CD4$^+$-HIV-1-model (2).

Definition 1. *The fractal-fractional* CD4$^+$*-HIV-1-model* (2) *is Ulam–Hyers stable if* $\exists\, a_{\mathbb{Q}_1}, a_{\mathbb{Q}_2}, a_{\mathbb{Q}_3} \in \mathbb{R}^+$ *such that* $\forall\, r_j > 0$, $j = 1,2,3$, *and* $\forall\, (\mathbb{T}^*,\mathbb{U}^*,\mathbb{V}^*) \in \mathbb{X}$ *satisfying*

$$\begin{cases} \left|{}^{\mathbf{FFML}}\mathcal{D}_{0,s}^{(\delta,\sigma)}\mathbb{T}^*(s) - \mathbb{Q}_1(s,\mathbb{T}^*(s),\mathbb{U}^*(s),\mathbb{V}^*(s))\right| < r_1, \\ \left|{}^{\mathbf{FFML}}\mathcal{D}_{0,s}^{(\delta,\sigma)}\mathbb{U}^*(s) - \mathbb{Q}_2(s,\mathbb{T}^*(s),\mathbb{U}^*(s),\mathbb{V}^*(s))\right| < r_2, \\ \left|{}^{\mathbf{FFML}}\mathcal{D}_{0,s}^{(\delta,\sigma)}\mathbb{V}^*(s) - \mathbb{Q}_3(s,\mathbb{T}^*(s),\mathbb{U}^*(s),\mathbb{V}^*(s))\right| < r_3, \end{cases} \tag{20}$$

$\exists\, (\mathbb{T},\mathbb{U},\mathbb{V}) \in \mathbb{X}$ *satisfying the fractal-fractional* CD4$^+$*-HIV-1-model* (2) *s.t.*

$$\begin{cases} \left|\mathbb{T}^*(s) - \mathbb{T}(s)\right| \le a_{\mathbb{Q}_1} r_1, \\ \left|\mathbb{U}^*(s) - \mathbb{U}(s)\right| \le a_{\mathbb{Q}_2} r_2, \\ \left|\mathbb{V}^*(s) - \mathbb{V}(s)\right| \le a_{\mathbb{Q}_3} r_3. \end{cases}$$

Definition 2. *The fractal-fractional* CD4$^+$*-HIV-1-model* (2) *is generalized Ulam–Hyers stable if* $\exists\, a_{\mathbb{Q}_j} \in C(\mathbb{R}^+,\mathbb{R}^+)$, $(j \in \{1,2,3\})$ *with* $a_{\mathbb{Q}_j}(0) = 0$ *s.t.* $\forall\, r_j > 0$ *and* $\forall\, (\mathbb{T}^*,\mathbb{U}^*,\mathbb{V}^*) \in \mathbb{X}$

fulfilling (20), *there is* $(\mathbb{T},\mathbb{U},\mathbb{V})\in\mathbb{X}$ *as a solution of the given fractal-fractional* CD4$^+$*-HIV-1-model* (2) *s.t.*

$$\begin{cases}|\mathbb{T}^*(s)-\mathbb{T}(s)|\le a_{\mathbb{Q}_1}(r_1),\\|\mathbb{U}^*(s)-\mathbb{U}(s)|\le a_{\mathbb{Q}_2}(r_2),\\|\mathbb{V}^*(s)-\mathbb{V}(s)|\le a_{\mathbb{Q}_3}(r_3).\end{cases}$$

Remark 1. $(\mathbb{T}^*,\mathbb{U}^*,\mathbb{V}^*)\in\mathbb{X}$ *is a solution for* (20) *iff* $\exists z_1,z_2,z_3\in C([0,S],\mathbb{R})$ *(each of them depend on* $\mathbb{T}^*,\mathbb{U}^*,\mathbb{V}^*$*, respectively) s.t.* $\forall s\in\mathbb{J}$,

(*i*) $|z_j(s)|<r_j$;
(*ii*) *We have*

$$\begin{cases}{}^{\mathbf{FFML}}\mathcal{D}_{0,s}^{(\delta,\sigma)}\mathbb{T}^*(s)=\mathbb{Q}_1(s,\mathbb{T}^*(s),\mathbb{U}^*(s),\mathbb{V}^*(s))+z_1(s),\\{}^{\mathbf{FFML}}\mathcal{D}_{0,s}^{(\delta,\sigma)}\mathbb{U}^*(s)=\mathbb{Q}_2(s,\mathbb{T}^*(s),\mathbb{U}^*(s),\mathbb{V}^*(s))+z_2(s),\\{}^{\mathbf{FFML}}\mathcal{D}_{0,s}^{(\delta,\sigma)}\mathbb{V}^*(s)=\mathbb{Q}_3(s,\mathbb{T}^*(s),\mathbb{U}^*(s),\mathbb{V}^*(s))+z_3(s).\end{cases}$$

Definition 3. *The fractal-fractional* CD4$^+$*-HIV-1-model* (2) *is Ulam–Hyers–Rassias stable w.r.t. functions* $\hbar_j$, $(j\in\{1,2,3\})$ *if* $\exists\,0<a_{(\mathbb{Q}_j,\hbar_j)}\in\mathbb{R}$ *s.t.* $\forall r_j>0$ *and* $\forall(\mathbb{T}^*,\mathbb{U}^*,\mathbb{V}^*)\in\mathbb{X}$ *fulfilling*

$$\begin{cases}\left|{}^{\mathbf{FFML}}\mathcal{D}_{0,s}^{(\delta,\sigma)}\mathbb{T}^*(s)-\mathbb{Q}_1(s,\mathbb{T}^*(s),\mathbb{U}^*(s),\mathbb{V}^*(s))\right|<r_1\hbar_1(s),\\\left|{}^{\mathbf{FFML}}\mathcal{D}_{0,s}^{(\delta,\sigma)}\mathbb{U}^*(s)-\mathbb{Q}_2(s,\mathbb{T}^*(s),\mathbb{U}^*(s),\mathbb{V}^*(s))\right|<r_2\hbar_2(s),\\\left|{}^{\mathbf{FFML}}\mathcal{D}_{0,s}^{(\delta,\sigma)}\mathbb{V}^*(s)-\mathbb{Q}_3(s,\mathbb{T}^*(s),\mathbb{U}^*(s),\mathbb{V}^*(s))\right|<r_3\hbar_3(s),\end{cases}\tag{21}$$

$\exists(\mathbb{T},\mathbb{U},\mathbb{V})\in\mathbb{X}$ *as a solution of the fractal-fractional* CD4$^+$*-HIV-1-model* (2) *s.t.*

$$\begin{cases}|\mathbb{T}^*(s)-\mathbb{T}(s)|\le r_1a_{(\mathbb{Q}_1,\hbar_1)}\hbar_1(s),\quad\forall s\in\mathbb{J},\\|\mathbb{U}^*(s)-\mathbb{U}(s)|\le r_2a_{(\mathbb{Q}_2,\hbar_2)}\hbar_2(s),\quad\forall s\in\mathbb{J},\\|\mathbb{V}^*(s)-\mathbb{V}(s)|\le r_3a_{(\mathbb{Q}_3,\hbar_3)}\hbar_3(s),\quad\forall s\in\mathbb{J},\end{cases}$$

in which $\hbar_1,\hbar_2,\hbar_3\in C([0,S],\mathbb{R}^+)$.

Definition 4. *The fractal-fractional* CD4$^+$*-HIV-1-model* (2) *is generalized Ulam–Hyers–Rassias stable w.r.t. functions* $\hbar_j$ *if* $\exists\,0<a_{(\mathbb{Q}_j,\hbar_j)}\in\mathbb{R}$ *s.t.* $\forall(\mathbb{T}^*,\mathbb{U}^*,\mathbb{V}^*)\in\mathbb{X}$ *satisfying*

$$\begin{cases}\left|{}^{\mathbf{FFML}}\mathcal{D}_{0,s}^{(\delta,\sigma)}\mathbb{T}^*(s)-\mathbb{Q}_1(s,\mathbb{T}^*(s),\mathbb{U}^*(s),\mathbb{V}^*(s))\right|<\hbar_1(s),\\\left|{}^{\mathbf{FFML}}\mathcal{D}_{0,s}^{(\delta,\sigma)}\mathbb{U}^*(s)-\mathbb{Q}_2(s,\mathbb{T}^*(s),\mathbb{U}^*(s),\mathbb{V}^*(s))\right|<\hbar_2(s),\\\left|{}^{\mathbf{FFML}}\mathcal{D}_{0,s}^{(\delta,\sigma)}\mathbb{V}^*(s)-\mathbb{Q}_3(s,\mathbb{T}^*(s),\mathbb{U}^*(s),\mathbb{V}^*(s))\right|<\hbar_3(s),\end{cases}$$

$\exists(\mathbb{T},\mathbb{U},\mathbb{V})\in\mathbb{X}$ *as a solution of the fractal-fractional* CD4$^+$*-HIV-1-model* (2) *s.t.*

$$\begin{cases}|\mathbb{T}^*(s)-\mathbb{T}(s)|\le a_{(\mathbb{Q}_1,\hbar_1)}\hbar_1(s),\\|\mathbb{U}^*(s)-\mathbb{U}(s)|\le a_{(\mathbb{Q}_2,\hbar_2)}\hbar_2(s),\\|\mathbb{V}^*(s)-\mathbb{V}(s)|\le a_{(\mathbb{Q}_3,\hbar_3)}\hbar_3(s).\end{cases}$$

If we take $\hbar_j(s) = 1$, in this case Definition 3 yields the Ulam-Hyers criterion for the stability of solutions.

Remark 2. *$(\mathbb{T}^*, \mathbb{U}^*, \mathbb{V}^*) \in \mathbb{X}$ is a solution for (21) iff $\exists\, z_1, z_2, z_3 \in C([0, S], \mathbb{R})$ (each of them depend on $\mathbb{T}^*, \mathbb{U}^*, \mathbb{V}^*$, respectively) s.t. $\forall\, s \in \mathbb{J}$,*

(i) $|z_j(s)| < r_j \hbar_j(s)$;

(ii) *We have*

$$\begin{cases} {}^{\mathbf{FFML}}\mathcal{D}_{0,s}^{(\delta,\sigma)}\mathbb{T}^*(s) = \mathbb{Q}_1\big(s, \mathbb{T}^*(s), \mathbb{U}^*(s), \mathbb{V}^*(s)\big) + z_1(s), \\ {}^{\mathbf{FFML}}\mathcal{D}_{0,s}^{(\delta,\sigma)}\mathbb{U}^*(s) = \mathbb{Q}_2\big(s, \mathbb{T}^*(s), \mathbb{U}^*(s), \mathbb{V}^*(s)\big) + z_2(s), \\ {}^{\mathbf{FFML}}\mathcal{D}_{0,s}^{(\delta,\sigma)}\mathbb{V}^*(s) = \mathbb{Q}_3\big(s, \mathbb{T}^*(s), \mathbb{U}^*(s), \mathbb{V}^*(s)\big) + z_3(s). \end{cases}$$

The following lemmas are useful for our main theorems.

Lemma 2. *For each $r_1, r_2, r_3 > 0$, suppose that $(\mathbb{T}^*, \mathbb{U}^*, \mathbb{V}^*) \in \mathbb{X}$ is considered as a solution of (20). Then the functions $\mathbb{T}^*, \mathbb{U}^*, \mathbb{V}^* \in \mathbb{M}$ fulfill the inequalities*

$$\bigg|\mathbb{T}^*(s) - \Big(\mathbb{T}_0 + \frac{(1-\delta)\sigma s^{\sigma-1}}{\mathcal{AB}(\delta)}\mathbb{Q}_1(s, \mathbb{T}^*(s), \mathbb{U}^*(s), \mathbb{V}^*(s)) + \frac{\delta\sigma}{\mathcal{AB}(\delta)\Gamma(\delta)}\int_0^s \mathfrak{w}^{\sigma-1}(s-\mathfrak{w})^{\delta-1}$$
$$\times\, \mathbb{Q}_1(\mathfrak{w}, \mathbb{T}^*(\mathfrak{w}), \mathbb{U}^*(\mathfrak{w}), \mathbb{V}^*(\mathfrak{w}))\, d\mathfrak{w}\Big)\bigg| \leq \Big[\frac{(1-\delta)\sigma S^{\sigma-1}}{\mathcal{AB}(\delta)} + \frac{\delta\sigma S^{\delta+\sigma-1}\Gamma(\sigma)}{\mathcal{AB}(\delta)\Gamma(\delta+\sigma)}\Big] r_1, \tag{22}$$

and

$$\bigg|\mathbb{U}^*(s) - \Big(\mathbb{U}_0 + \frac{(1-\delta)\sigma s^{\sigma-1}}{\mathcal{AB}(\delta)}\mathbb{Q}_2(s, \mathbb{T}^*(s), \mathbb{U}^*(s), \mathbb{V}^*(s)) + \frac{\delta\sigma}{\mathcal{AB}(\delta)\Gamma(\delta)}\int_0^s \mathfrak{w}^{\sigma-1}(s-\mathfrak{w})^{\delta-1}$$
$$\times\, \mathbb{Q}_2(\mathfrak{w}, \mathbb{T}^*(\mathfrak{w}), \mathbb{U}^*(\mathfrak{w}), \mathbb{V}^*(\mathfrak{w}))\, d\mathfrak{w}\Big)\bigg| \leq \Big[\frac{(1-\delta)\sigma S^{\sigma-1}}{\mathcal{AB}(\delta)} + \frac{\delta\sigma S^{\delta+\sigma-1}\Gamma(\sigma)}{\mathcal{AB}(\delta)\Gamma(\delta+\sigma)}\Big] r_2, \tag{23}$$

and

$$\bigg|\mathbb{V}^*(s) - \Big(\mathbb{V}_0 + \frac{(1-\delta)\sigma s^{\sigma-1}}{\mathcal{AB}(\delta)}\mathbb{Q}_3(s, \mathbb{T}^*(s), \mathbb{U}^*(s), \mathbb{V}^*(s)) + \frac{\delta\sigma}{\mathcal{AB}(\delta)\Gamma(\delta)}\int_0^s \mathfrak{w}^{\sigma-1}(s-\mathfrak{w})^{\delta-1}$$
$$\times\, \mathbb{Q}_3(\mathfrak{w}, \mathbb{T}^*(\mathfrak{w}), \mathbb{U}^*(\mathfrak{w}), \mathbb{V}^*(\mathfrak{w}))\, d\mathfrak{w}\Big)\bigg| \leq \Big[\frac{(1-\delta)\sigma S^{\sigma-1}}{\mathcal{AB}(\delta)} + \frac{\delta\sigma S^{\delta+\sigma-1}\Gamma(\sigma)}{\mathcal{AB}(\delta)\Gamma(\delta+\sigma)}\Big] r_3. \tag{24}$$

Proof. Let $r_1 > 0$ be arbitrary. Since $\mathbb{T}^* \in \mathbb{M}$ satisfies

$$\Big|{}^{\mathbf{FFML}}\mathcal{D}_{0,s}^{(\delta,\sigma)}\mathbb{T}^*(s) - \mathbb{Q}_1\big(s, \mathbb{T}^*(s), \mathbb{U}^*(s), \mathbb{V}^*(s)\big)\Big| < r_1,$$

so, by Remark 1, we are allowed to take a function $z_1(s)$ s.t.

$${}^{\mathbf{FFML}}\mathcal{D}_{0,s}^{(\delta,\sigma)}\mathbb{T}^*(s) = \mathbb{Q}_1\big(s, \mathbb{T}^*(s), \mathbb{U}^*(s), \mathbb{V}^*(s)\big) + z_1(s),$$

and $|z_1(s)| \leq r_1$. Clearly,

$$\mathbb{T}^*(s) = \mathbb{T}_0 + \frac{(1-\delta)\sigma s^{\sigma-1}}{\mathcal{AB}(\delta)}\Big[\mathbb{Q}_1(s, \mathbb{T}^*(s), \mathbb{U}^*(s), \mathbb{V}^*(s)) + z_1(s)\Big]$$

$$+ \frac{\delta\sigma}{\mathcal{AB}(\delta)\Gamma(\delta)}\int_0^s \mathfrak{w}^{\sigma-1}(s-\mathfrak{w})^{\delta-1}\Big[\mathbb{Q}_1(\mathfrak{w}, \mathbb{T}^*(\mathfrak{w}), \mathbb{U}^*(\mathfrak{w}), \mathbb{V}^*(\mathfrak{w})) + z_1(\mathfrak{w})\Big]\,d\mathfrak{w}.$$

In this case, we estimate

$$\Bigg|\mathbb{T}^*(s) - \Big(\mathbb{T}_0 + \frac{(1-\delta)\sigma s^{\sigma-1}}{\mathcal{AB}(\delta)}\mathbb{Q}_1(s, \mathbb{T}^*(s), \mathbb{U}^*(s), \mathbb{V}^*(s))$$

$$+ \frac{\delta\sigma}{\mathcal{AB}(\delta)\Gamma(\delta)}\int_0^s \mathfrak{w}^{\sigma-1}(s-\mathfrak{w})^{\delta-1}\mathbb{Q}_1(\mathfrak{w}, \mathbb{T}^*(\mathfrak{w}), \mathbb{U}^*(\mathfrak{w}), \mathbb{V}^*(\mathfrak{w}))\,d\mathfrak{w}\Big)\Bigg|$$

$$\leq \frac{(1-\delta)\sigma s^{\sigma-1}}{\mathcal{AB}(\delta)}|z_1(s)| + \frac{\delta\sigma}{\mathcal{AB}(\delta)\Gamma(\delta)}\int_0^s \mathfrak{w}^{\sigma-1}(s-\mathfrak{w})^{\delta-1}\, z_1(\mathfrak{w})|\,d\mathfrak{w}$$

$$\leq \frac{(1-\delta)\sigma S^{\sigma-1}}{\mathcal{AB}(\delta)}r_1 + \frac{\delta\sigma S^{\delta+\sigma-1}\Gamma(\sigma)}{\mathcal{AB}(\delta)\Gamma(\delta+\sigma)}r_1$$

$$= \Big[\frac{(1-\delta)\sigma S^{\sigma-1}}{\mathcal{AB}(\delta)} + \frac{\delta\sigma S^{\delta+\sigma-1}\Gamma(\sigma)}{\mathcal{AB}(\delta)\Gamma(\delta+\sigma)}\Big]r_1.$$

This states that the inequality (22) is fulfilled. Similarly, we can obtain the inequalities (23) and (24). □

To prove the next result, we regard the following:
(C2) $\exists$ increasing mappings $\hbar_\jmath \in C([0,S], \mathbb{R}^+)$, $(\jmath \in \{1,2,3\})$ and $\exists\,\Delta_{\hbar_\jmath} > 0$ provided that

$$^{\mathbf{FFML}}\mathcal{I}_{0,s}^{(\delta,\sigma)}\hbar_\jmath(s) < \Delta_{\hbar_\jmath}\hbar_\jmath(s),\ (\jmath \in \{1,2,3\}), \forall s \in \mathbb{J}. \tag{25}$$

Lemma 3. *Let (C2) to be held. For each $r_1, r_2, r_3 > 0$, suppose that $(\mathbb{T}^*, \mathbb{U}^*, \mathbb{V}^*) \in \mathbb{X}$ is considered as a solution of (21). Then the functions $\mathbb{T}^*, \mathbb{U}^*, \mathbb{V}^* \in \mathbb{M}$ fulfill the inequalities*

$$\Big|\mathbb{T}^*(s) - \Big(\mathbb{T}_0 + {}^{\mathbf{FFML}}\mathcal{I}_{0,s}^{(\delta,\sigma)}\mathbb{Q}_1(s, \mathbb{T}^*(s), \mathbb{U}^*(s), \mathbb{V}^*(s))\Big)\Big| \leq r_1\Delta_{\hbar_1}\hbar_1(s), \tag{26}$$

$$\Big|\mathbb{U}^*(s) - \Big(\mathbb{U}_0 + {}^{\mathbf{FFML}}\mathcal{I}_{0,s}^{(\delta,\sigma)}\mathbb{Q}_2(s, \mathbb{T}^*(s), \mathbb{U}^*(s), \mathbb{V}^*(s))\Big)\Big| \leq r_2\Delta_{\hbar_2}\hbar_2(s), \tag{27}$$

$$\Big|\mathbb{V}^*(s) - \Big(\mathbb{V}_0 + {}^{\mathbf{FFML}}\mathcal{I}_{0,s}^{(\delta,\sigma)}\mathbb{Q}_3(s, \mathbb{T}^*(s), \mathbb{U}^*(s), \mathbb{V}^*(s))\Big)\Big| \leq r_3\Delta_{\hbar_3}\hbar_3(s). \tag{28}$$

Proof. Let $r_1 > 0$. Since $\mathbb{T}^* \in \mathbb{M}$ satisfies

$$\Big|{}^{\mathbf{FFML}}\mathcal{D}_{0,s}^{(\delta,\sigma)}\mathbb{T}^*(s) - \mathbb{Q}_1(s, \mathbb{T}^*(s), \mathbb{U}^*(s), \mathbb{V}^*(s))\Big| < r_1\hbar_1(s),$$

thus, from Remark 2, we are allowed to take a function $z_1(s)$ s.t.

$$^{\mathbf{FFML}}\mathcal{D}_{0,s}^{(\delta,\sigma)}\mathbb{T}^*(s) = \mathbb{Q}_1(s, \mathbb{T}^*(s), \mathbb{U}^*(s), \mathbb{V}^*(s)) + z_1(s),$$

and $|z_1(s)| \leq r_1\hbar_1(s)$. Evidently,

$$\mathbb{T}^*(s) = \mathbb{T}_0 + \frac{(1-\delta)\sigma s^{\sigma-1}}{\mathcal{AB}(\delta)}\Big[\mathbb{Q}_1(s,\mathbb{T}^*(s),\mathbb{U}^*(s),\mathbb{V}^*(s)) + z_1(s)\Big]$$

$$+ \frac{\delta\sigma}{\mathcal{AB}(\delta)\Gamma(\delta)}\int_0^s \mathfrak{w}^{\sigma-1}(s-\mathfrak{w})^{\delta-1}\Big[\mathbb{Q}_1(\mathfrak{w},\mathbb{T}^*(\mathfrak{w}),\mathbb{U}^*(\mathfrak{w}),\mathbb{V}^*(\mathfrak{w})) + z_1(\mathfrak{w})\Big]\,d\mathfrak{w}.$$

Then, we estimate

$$\Big|\mathbb{T}^*(s) - \Big(\mathbb{T}_0 + {}^{\mathbf{FFML}}\mathcal{I}_{0,s}^{(\delta,\sigma)}\mathbb{Q}_1(s,\mathbb{T}^*(s),\mathbb{U}^*(s),\mathbb{V}^*(s))\Big)\Big| =$$

$$\Bigg|\mathbb{T}^*(s) - \Big(\mathbb{T}_0 + \frac{(1-\delta)\sigma s^{\sigma-1}}{\mathcal{AB}(\delta)}\mathbb{Q}_1(s,\mathbb{T}^*(s),\mathbb{U}^*(s),\mathbb{V}^*(s))$$

$$+ \frac{\delta\sigma}{\mathcal{AB}(\delta)\Gamma(\delta)}\int_0^s \mathfrak{w}^{\sigma-1}(s-\mathfrak{w})^{\delta-1}\mathbb{Q}_1(\mathfrak{w},\mathbb{T}^*(\mathfrak{w}),\mathbb{U}^*(\mathfrak{w}),\mathbb{V}^*(\mathfrak{w}))\,d\mathfrak{w}\Big)\Bigg|$$

$$\leq \frac{(1-\delta)\sigma s^{\sigma-1}}{\mathcal{AB}(\delta)}|z_1(s)| + \frac{\delta\sigma}{\mathcal{AB}(\delta)\Gamma(\delta)}\int_0^s \mathfrak{w}^{\sigma-1}(s-\mathfrak{w})^{\delta-1}|z_1(\mathfrak{w})|\,d\mathfrak{w}$$

$$= {}^{\mathbf{FFML}}\mathcal{I}_{0,s}^{(\delta,\sigma)}|z_1(s)|$$

$$\leq {}^{\mathbf{FFML}}\mathcal{I}_{0,s}^{(\delta,\sigma)}r_1\hbar_1(s)$$

$$\leq r_1\Delta_{\hbar_1}\hbar_1(s).$$

Similarly, we can obtain the remaining inequalities. □

The Ulam–Hyers stability is checked about the fractal-fractional CD4$^+$-HIV-1-model (2).

Theorem 4. *Let (C1) be fulfilled. Then the fractal-fractional* CD4$^+$*-HIV-1-model* (2) *is Ulam–Hyers stable on* $\mathbb{J} := [0,S]$ *and also is generalized Ulam–Hyers stable s.t.*

$$\Big[\frac{(1-\delta)\sigma S^{\sigma-1}}{\mathcal{AB}(\delta)} + \frac{\delta\sigma S^{\delta+\sigma-1}\Gamma(\sigma)}{\mathcal{AB}(\delta)\Gamma(\delta+\sigma)}\Big]\alpha_\jmath < 1, \quad \jmath \in \{1,2,3\},$$

in which $\alpha_1,\alpha_2,\alpha_3 > 0$ *are given by* (18).

Proof. Let $r_1 > 0$ and $\mathbb{T}^* \in \mathbb{M}$ be an arbitrary solution of (20). Also, from Theorem 3, we assume $\mathbb{T} \in \mathbb{M}$ as a unique solution of the fractal-fractional CD4$^+$-HIV-1-model (2). Then $\mathbb{T}(s)$ is defined as

$$\mathbb{T}(s) = \mathbb{T}_0 + \frac{(1-\delta)\sigma s^{\sigma-1}}{\mathcal{AB}(\delta)}\mathbb{Q}_1(s,\mathbb{T}(s),\mathbb{U}(s),\mathbb{V}(s))$$

$$+ \frac{\delta\sigma}{\mathcal{AB}(\delta)\Gamma(\delta)}\int_0^s \mathfrak{w}^{\sigma-1}(s-\mathfrak{w})^{\delta-1}\mathbb{Q}_1(\mathfrak{w},\mathbb{T}(\mathfrak{w}),\mathbb{U}(\mathfrak{w}),\mathbb{V}(\mathfrak{w}))\,d\mathfrak{w}.$$

Therefore, by Lemma 2 and with the help of the triangle inequality, we estimate

$$|\mathbb{T}^*(s) - \mathbb{T}(s)| \leq \Big|\mathbb{T}^*(s) - \mathbb{T}_0 - \frac{(1-\delta)\sigma s^{\sigma-1}}{\mathcal{AB}(\delta)}\mathbb{Q}_1(s,\mathbb{T}(s),\mathbb{U}(s),\mathbb{V}(s))$$

$$
\begin{aligned}
&- \frac{\delta\sigma}{\mathcal{AB}(\delta)\Gamma(\delta)} \int_0^s \mathfrak{w}^{\sigma-1}(s-\mathfrak{w})^{\delta-1}\mathbb{Q}_1(\mathfrak{w},\mathbb{T}(\mathfrak{w}),\mathbb{U}(\mathfrak{w}),\mathbb{V}(\mathfrak{w}))\,d\mathfrak{w}\Big| \\
&\leq \Bigg|\mathbb{T}^*(s) - \Big(\mathbb{T}_0 + \frac{(1-\delta)\sigma s^{\sigma-1}}{\mathcal{AB}(\delta)}\mathbb{Q}_1(s,\mathbb{T}^*(s),\mathbb{U}^*(s),\mathbb{V}^*(s)) \\
&+ \frac{\delta\sigma}{\mathcal{AB}(\delta)\Gamma(\delta)} \int_0^s \mathfrak{w}^{\sigma-1}(s-\mathfrak{w})^{\delta-1}\mathbb{Q}_1(\mathfrak{w},\mathbb{T}^*(\mathfrak{w}),\mathbb{U}^*(\mathfrak{w}),\mathbb{V}^*(\mathfrak{w}))\,d\mathfrak{w}\Big)\Bigg| \\
&+ \frac{(1-\delta)\sigma s^{\sigma-1}}{\mathcal{AB}(\delta)}|\mathbb{Q}_1(s,\mathbb{T}^*(s),\mathbb{U}^*(s),\mathbb{V}^*(s)) - \mathbb{Q}_1(s,\mathbb{T}(s),\mathbb{U}(s),\mathbb{V}(s))| \\
&+ \frac{\delta\sigma}{\mathcal{AB}(\delta)\Gamma(\delta)} \int_0^s \mathfrak{w}^{\sigma-1}(s-\mathfrak{w})^{\delta-1}\big|\mathbb{Q}_1(\mathfrak{w},\mathbb{T}^*(\mathfrak{w}),\mathbb{U}^*(\mathfrak{w}),\mathbb{V}^*(\mathfrak{w})) - \mathbb{Q}_1(\mathfrak{w},\mathbb{T}(\mathfrak{w}),\mathbb{U}(\mathfrak{w}),\mathbb{V}(\mathfrak{w}))\big|\,d\mathfrak{w} \\
&\leq \Big[\frac{(1-\delta)\sigma S^{\sigma-1}}{\mathcal{AB}(\delta)} + \frac{\delta\sigma S^{\delta+\sigma-1}\Gamma(\sigma)}{\mathcal{AB}(\delta)\Gamma(\delta+\sigma)}\Big]r_1 + \frac{(1-\delta)\sigma S^{\sigma-1}}{\mathcal{AB}(\delta)}\alpha_1\|\mathbb{T}^* - \mathbb{T}\| \\
&+ \frac{\delta\sigma S^{\delta+\sigma-1}\Gamma(\sigma)}{\mathcal{AB}(\delta)\Gamma(\delta+\sigma)}\alpha_1\|\mathbb{T}^* - \mathbb{T}\| \\
&\leq \Big[\frac{(1-\delta)\sigma S^{\sigma-1}}{\mathcal{AB}(\delta)} + \frac{\delta\sigma S^{\delta+\sigma-1}\Gamma(\sigma)}{\mathcal{AB}(\delta)\Gamma(\delta+\sigma)}\Big]r_1 + \Big[\frac{(1-\delta)\sigma S^{\sigma-1}}{\mathcal{AB}(\delta)} + \frac{\delta\sigma S^{\delta+\sigma-1}\Gamma(\sigma)}{\mathcal{AB}(\delta)\Gamma(\delta+\sigma)}\Big]\alpha_1\|\mathbb{T} - \mathbb{T}^*\|.
\end{aligned}
$$

Hence,

$$
\|\mathbb{T}^* - \mathbb{T}\| \leq \frac{\Big[\frac{(1-\delta)\sigma S^{\sigma-1}}{\mathcal{AB}(\delta)} + \frac{\delta\sigma S^{\delta+\sigma-1}\Gamma(\sigma)}{\mathcal{AB}(\delta)\Gamma(\delta+\sigma)}\Big]r_1}{1 - \Big[\frac{(1-\delta)\sigma S^{\sigma-1}}{\mathcal{AB}(\delta)} + \frac{\delta\sigma S^{\delta+\sigma-1}\Gamma(\sigma)}{\mathcal{AB}(\delta)\Gamma(\delta+\sigma)}\Big]\alpha_1}.
$$

We set $a_{\mathbb{Q}_1} = \dfrac{\Big[\frac{(1-\delta)\sigma S^{\sigma-1}}{\mathcal{AB}(\delta)} + \frac{\delta\sigma S^{\delta+\sigma-1}\Gamma(\sigma)}{\mathcal{AB}(\delta)\Gamma(\delta+\sigma)}\Big]}{1 - \Big[\frac{(1-\delta)\sigma S^{\sigma-1}}{\mathcal{AB}(\delta)} + \frac{\delta\sigma S^{\delta+\sigma-1}\Gamma(\sigma)}{\mathcal{AB}(\delta)\Gamma(\delta+\sigma)}\Big]\alpha_1}$. In this case, $\|\mathbb{T}^* - \mathbb{T}\| \leq a_{\mathbb{Q}_1}r_1$.

Similarly, we obtain

$$
\begin{aligned}
\|\mathbb{U}^* - \mathbb{U}\| &\leq a_{\mathbb{Q}_2}r_2, \\
\|\mathbb{V}^* - \mathbb{V}\| &\leq a_{\mathbb{Q}_3}r_3,
\end{aligned}
$$

where

$$
a_{\mathbb{Q}_\jmath} = \frac{\Big[\frac{(1-\delta)\sigma S^{\sigma-1}}{\mathcal{AB}(\delta)} + \frac{\delta\sigma S^{\delta+\sigma-1}\Gamma(\sigma)}{\mathcal{AB}(\delta)\Gamma(\delta+\sigma)}\Big]}{1 - \Big[\frac{(1-\delta)\sigma S^{\sigma-1}}{\mathcal{AB}(\delta)} + \frac{\delta\sigma S^{\delta+\sigma-1}\Gamma(\sigma)}{\mathcal{AB}(\delta)\Gamma(\delta+\sigma)}\Big]\alpha_\jmath}, \quad (\jmath \in \{2,3\}).
$$

Hence, the Ulam–Hyers stability of the fractal-fractional CD4$^+$-HIV-1-model (2) is fulfilled. On the other hand, if we take

$$
a_{\mathbb{Q}_\jmath}(r_\jmath) = \frac{\Big[\frac{(1-\delta)\sigma S^{\sigma-1}}{\mathcal{AB}(\delta)} + \frac{\delta\sigma S^{\delta+\sigma-1}\Gamma(\sigma)}{\mathcal{AB}(\delta)\Gamma(\delta+\sigma)}\Big]r_\jmath}{1 - \Big[\frac{(1-\delta)\sigma S^{\sigma-1}}{\mathcal{AB}(\delta)} + \frac{\delta\sigma S^{\delta+\sigma-1}\Gamma(\sigma)}{\mathcal{AB}(\delta)\Gamma(\delta+\sigma)}\Big]\alpha_\jmath}, \quad (\jmath \in \{1,2,3\}),
$$

then $a_{\mathbb{Q}_\jmath}(0) = 0$, and the generalized Ulam–Hyers stability is simply proved. □

The Ulam–Hyers-Rassias stability is checked about the fractal-fractional CD4$^+$-HIV-1-model (2) in the next theorem.

Theorem 5. *The hypotheses (C1) and (C2) are considered to be held. Then the given fractal-fractional* CD4$^+$*-HIV-1-model* (2) *is the Ulam–Hyers–Rassias and generalized Ulam–Hyers–Rassias stable.*

Proof. Let $r_1 > 0$, and $\mathbb{T}^* \in \mathbb{M}$ satisfying (21). By Theorem 3, let $\mathbb{T} \in \mathbb{M}$ be the unique solution of the given fractal-fractional CD4$^+$-HIV-1-model (2). Then $\mathbb{T}(s)$ becomes

$$\begin{aligned}\mathbb{T}(s) = \mathbb{T}_0 + \frac{(1-\delta)\sigma s^{\sigma-1}}{\mathcal{AB}(\delta)}\mathbb{Q}_1(s,\mathbb{T}(s),\mathbb{U}(s),\mathbb{V}(s)) \\ + \frac{\delta\sigma}{\mathcal{AB}(\delta)\Gamma(\delta)}\int_0^s \mathfrak{w}^{\sigma-1}(s-\mathfrak{w})^{\delta-1}\mathbb{Q}_1(\mathfrak{w},\mathbb{T}(\mathfrak{w}),\mathbb{U}(\mathfrak{w}),\mathbb{V}(\mathfrak{w}))\,d\mathfrak{w}.\end{aligned}$$

Therefore, by Lemma 3 and with the help of the triangle inequality, we estimate

$$\begin{aligned}\left|\mathbb{T}^*(s) - \mathbb{T}(s)\right| \le & \Big|\mathbb{T}^*(s) - \mathbb{T}_0 - \frac{(1-\delta)\sigma s^{\sigma-1}}{\mathcal{AB}(\delta)}\mathbb{Q}_1(s,\mathbb{T}(s),\mathbb{U}(s),\mathbb{V}(s)) \\ & - \frac{\delta\sigma}{\mathcal{AB}(\delta)\Gamma(\delta)}\int_0^s \mathfrak{w}^{\sigma-1}(s-\mathfrak{w})^{\delta-1}\mathbb{Q}_1(\mathfrak{w},\mathbb{T}(\mathfrak{w}),\mathbb{U}(\mathfrak{w}),\mathbb{V}(\mathfrak{w}))\,d\mathfrak{w}\Big| \\ \le & \Bigg|\mathbb{T}^*(s) - \Big(\mathbb{T}_0 + \frac{(1-\delta)\sigma s^{\sigma-1}}{\mathcal{AB}(\delta)}\mathbb{Q}_1(s,\mathbb{T}^*(s),\mathbb{U}^*(s),\mathbb{V}^*(s)) \\ & + \frac{\delta\sigma}{\mathcal{AB}(\delta)\Gamma(\delta)}\int_0^s \mathfrak{w}^{\sigma-1}(s-\mathfrak{w})^{\delta-1}\mathbb{Q}_1(\mathfrak{w},\mathbb{T}^*(\mathfrak{w}),\mathbb{U}^*(\mathfrak{w}),\mathbb{V}^*(\mathfrak{w}))\,d\mathfrak{w}\Big)\Bigg| \\ & + \frac{(1-\delta)\sigma s^{\sigma-1}}{\mathcal{AB}(\delta)}\left|\mathbb{Q}_1(s,\mathbb{T}^*(s),\mathbb{U}^*(s),\mathbb{V}^*(s)) - \mathbb{Q}_1(s,\mathbb{T}(s),\mathbb{U}(s),\mathbb{V}(s))\right| \\ & + \frac{\delta\sigma}{\mathcal{AB}(\delta)\Gamma(\delta)}\int_0^s \mathfrak{w}^{\sigma-1}(s-\mathfrak{w})^{\delta-1}\left|\mathbb{Q}_1(\mathfrak{w},\mathbb{T}^*(\mathfrak{w}),\mathbb{U}^*(\mathfrak{w}),\mathbb{V}^*(\mathfrak{w})) - \mathbb{Q}_1(\mathfrak{w},\mathbb{T}(\mathfrak{w}),\mathbb{U}(\mathfrak{w}),\mathbb{V}(\mathfrak{w}))\right|d\mathfrak{w} \\ \le & \left|\mathbb{T}^*(s) - \left(\mathbb{T}_0 + {}^{\mathbf{FFML}}\mathcal{I}_{0,s}^{(\delta,\sigma)}\mathbb{Q}_1(s,\mathbb{T}^*(s),\mathbb{U}^*(s),\mathbb{V}^*(s))\right)\right| \\ & + \frac{(1-\delta)\sigma s^{\sigma-1}}{\mathcal{AB}(\delta)}\left|\mathbb{Q}_1(s,\mathbb{T}^*(s),\mathbb{U}^*(s),\mathbb{V}^*(s)) - \mathbb{Q}_1(s,\mathbb{T}(s),\mathbb{U}(s),\mathbb{V}(s))\right| \\ & + \frac{\delta\sigma}{\mathcal{AB}(\delta)\Gamma(\delta)}\int_0^s \mathfrak{w}^{\sigma-1}(s-\mathfrak{w})^{\delta-1}\left|\mathbb{Q}_1(\mathfrak{w},\mathbb{T}^*(\mathfrak{w}),\mathbb{U}^*(\mathfrak{w}),\mathbb{V}^*(\mathfrak{w})) - \mathbb{Q}_1(\mathfrak{w},\mathbb{T}(\mathfrak{w}),\mathbb{U}(\mathfrak{w}),\mathbb{V}(\mathfrak{w}))\right|d\mathfrak{w} \\ \le & r_1\Delta_{\hbar_1}\hbar_1(s) + \frac{(1-\delta)\sigma S^{\sigma-1}}{\mathcal{AB}(\delta)}\alpha_1\|\mathbb{T}^* - \mathbb{T}\| + \frac{\delta\sigma S^{\delta+\sigma-1}\Gamma(\sigma)}{\mathcal{AB}(\delta)\Gamma(\delta+\sigma)}\alpha_1\|\mathbb{T}^* - \mathbb{T}\| \\ \le & r_1\Delta_{\hbar_1}\hbar_1(s) + \left[\frac{(1-\delta)\sigma S^{\sigma-1}}{\mathcal{AB}(\delta)} + \frac{\delta\sigma S^{\delta+\sigma-1}\Gamma(\sigma)}{\mathcal{AB}(\delta)\Gamma(\delta+\sigma)}\right]\alpha_1\|\mathbb{T} - \mathbb{T}^*\|.\end{aligned}$$

Accordingly, it gives

$$\|\mathbb{T}^* - \mathbb{T}\| \leq \frac{r_1 \Delta_{\hbar_1} \hbar_1(s)}{1 - \left[\frac{(1-\delta)\sigma S^{\sigma-1}}{\mathcal{AB}(\delta)} + \frac{\delta\sigma S^{\delta+\sigma-1}\Gamma(\sigma)}{\mathcal{AB}(\delta)\Gamma(\delta+\sigma)}\right]\alpha_1}.$$

Set

$$a_{(\mathbb{Q}_1,\hbar_1)} = \frac{\Delta_{\hbar_1}}{1 - \left[\frac{(1-\delta)\sigma S^{\sigma-1}}{\mathcal{AB}(\delta)} + \frac{\delta\sigma S^{\delta+\sigma-1}\Gamma(\sigma)}{\mathcal{AB}(\delta)\Gamma(\delta+\sigma)}\right]\alpha_1}.$$

Then $\|\mathbb{T}^* - \mathbb{T}\| \leq r_1 a_{(\mathbb{Q}_1,\hbar_1)} \hbar_1(s)$. Similarly,

$$\|\mathbb{U}^* - \mathbb{U}\| \leq r_2 a_{(\mathbb{Q}_2,\hbar_2)} \hbar_2(s),$$

$$\|\mathbb{V}^* - \mathbb{V}\| \leq r_3 a_{(\mathbb{Q}_3,\hbar_3)} \hbar_3(s),$$

where

$$a_{(\mathbb{Q}_\jmath,\hbar_\jmath)} = \frac{\Delta_{\hbar_\jmath}}{1 - \left[\frac{(1-\delta)\sigma S^{\sigma-1}}{\mathcal{AB}(\delta)} + \frac{\delta\sigma S^{\delta+\sigma-1}\Gamma(\sigma)}{\mathcal{AB}(\delta)\Gamma(\delta+\sigma)}\right]\alpha_\jmath}, \quad (\jmath \in \{2,3\}).$$

In consequence, the fractal-fractional CD4$^+$-HIV-1-model (2) is stable in the context of the Ulam–Hyers–Rassias criterion. Along with this, by defining $r_\jmath = 1$, $(\jmath \in \{1,2,3\})$, the mentioned fractal-fractional CD4$^+$-HIV-1-model (2) is generalized Ulam–Hyers–Rassias stable. □

6. Numerical Scheme via Adams-Bashforth Method

In the present section of the paper, we aim to derive a numerical algorithm for the fractal-fractional CD4$^+$-HIV-1-model (2). To do this, we apply a technique based on two-step Lagrange polynomials entitled fractional Adams-Bashforth method. We re-define fractal-fractional integral equations (12) at s_{k+1}. In fact, we discretize these integral equations (12) for $s = s_{k+1}$ as follows

$$\begin{cases} \mathbb{T}(s_{k+1}) = \mathbb{T}_0 + \dfrac{(1-\delta)\sigma s_k^{\sigma-1}}{\mathcal{AB}(\delta)} \mathbb{Q}_1(s_k, \mathbb{T}(s_k), \mathbb{U}(s_k), \mathbb{V}(s_k)) \\ \qquad + \dfrac{\delta\sigma}{\mathcal{AB}(\delta)\Gamma(\delta)} \displaystyle\int_0^{s_{k+1}} \mathfrak{w}^{\sigma-1}(s_{k+1} - \mathfrak{w})^{\delta-1} \mathbb{Q}_1(\mathfrak{w}, \mathbb{T}(\mathfrak{w}), \mathbb{U}(\mathfrak{w}), \mathbb{V}(\mathfrak{w}))\, d\mathfrak{w}, \\ \mathbb{U}(s_{k+1}) = \mathbb{U}_0 + \dfrac{(1-\delta)\sigma s_k^{\sigma-1}}{\mathcal{AB}(\delta)} \mathbb{Q}_2(s_k, \mathbb{T}(s_k), \mathbb{U}(s_k), \mathbb{V}(s_k)) \\ \qquad + \dfrac{\delta\sigma}{\mathcal{AB}(\delta)\Gamma(\delta)} \displaystyle\int_0^{s_{k+1}} \mathfrak{w}^{\sigma-1}(s_{k+1} - \mathfrak{w})^{\delta-1} \mathbb{Q}_2(\mathfrak{w}, \mathbb{T}(\mathfrak{w}), \mathbb{U}(\mathfrak{w}), \mathbb{V}(\mathfrak{w}))\, d\mathfrak{w}, \\ \mathbb{V}(s_{k+1}) = \mathbb{V}_0 + \dfrac{(1-\delta)\sigma s_k^{\sigma-1}}{\mathcal{AB}(\delta)} \mathbb{Q}_3(s_k, \mathbb{T}(s_k), \mathbb{U}(s_k), \mathbb{V}(s_k)) \\ \qquad + \dfrac{\delta\sigma}{\mathcal{AB}(\delta)\Gamma(\delta)} \displaystyle\int_0^{s_{k+1}} \mathfrak{w}^{\sigma-1}(s_{k+1} - \mathfrak{w})^{\delta-1} \mathbb{Q}_3(\mathfrak{w}, \mathbb{T}(\mathfrak{w}), \mathbb{U}(\mathfrak{w}), \mathbb{V}(\mathfrak{w}))\, d\mathfrak{w}. \end{cases}$$

The approximation of above integrals are formulated by

$$\begin{cases} \mathbb{T}(s_{k+1}) = \mathbb{T}_0 + \dfrac{(1-\delta)\sigma s_k^{\sigma-1}}{\mathcal{AB}(\delta)}\mathbb{Q}_1(s_k, \mathbb{T}(s_k), \mathbb{U}(s_k), \mathbb{V}(s_k)) \\ \qquad + \dfrac{\delta\sigma}{\mathcal{AB}(\delta)\Gamma(\delta)}\displaystyle\sum_{\ell=1}^{k}\int_{s_\ell}^{s_{\ell+1}} \mathfrak{w}^{\sigma-1}(s_{k+1}-\mathfrak{w})^{\delta-1}\mathbb{Q}_1(\mathfrak{w}, \mathbb{T}(\mathfrak{w}), \mathbb{U}(\mathfrak{w}), \mathbb{V}(\mathfrak{w}))\, d\mathfrak{w}, \\ \mathbb{U}(s_{k+1}) = \mathbb{U}_0 + \dfrac{(1-\delta)\sigma s_k^{\sigma-1}}{\mathcal{AB}(\delta)}\mathbb{Q}_2(s_k, \mathbb{T}(s_k), \mathbb{U}(s_k), \mathbb{V}(s_k)) \\ \qquad + \dfrac{\delta\sigma}{\mathcal{AB}(\delta)\Gamma(\delta)}\displaystyle\sum_{\ell=1}^{k}\int_{s_\ell}^{s_{\ell+1}} \mathfrak{w}^{\sigma-1}(s_{k+1}-\mathfrak{w})^{\delta-1}\mathbb{Q}_2(\mathfrak{w}, \mathbb{T}(\mathfrak{w}), \mathbb{U}(\mathfrak{w}), \mathbb{V}(\mathfrak{w}))\, d\mathfrak{w}, \\ \mathbb{V}(s_{k+1}) = \mathbb{V}_0 + \dfrac{(1-\delta)\sigma s_k^{\sigma-1}}{\mathcal{AB}(\delta)}\mathbb{Q}_3(s_k, \mathbb{T}(s_k), \mathbb{U}(s_k), \mathbb{V}(s_k)) \\ \qquad + \dfrac{\delta\sigma}{\mathcal{AB}(\delta)\Gamma(\delta)}\displaystyle\sum_{\ell=1}^{k}\int_{s_\ell}^{s_{\ell+1}} \mathfrak{w}^{\sigma-1}(s_{k+1}-\mathfrak{w})^{\delta-1}\mathbb{Q}_3(\mathfrak{w}, \mathbb{T}(\mathfrak{w}), \mathbb{U}(\mathfrak{w}), \mathbb{V}(\mathfrak{w}))\, d\mathfrak{w}. \end{cases}$$

Next, we approximate three functions $\mathfrak{w}^{\sigma-1}\mathbb{Q}_\jmath(\mathfrak{w}, \mathbb{T}(\mathfrak{w}), \mathbb{U}(\mathfrak{w}), \mathbb{V}(\mathfrak{w}))$, $\jmath = 1,2,3$, on the interval $[s_\ell, s_{\ell+1}]$ by applying two-step Lagrange interpolation polynomials under the step size $\mathbf{h} = s_\ell - s_{\ell-1}$. By straighforward computations, we obtain algorithms which yield the numerical solutions to the fractal-fractional CD4$^+$-HIV-1-model (2) as

$$\mathbb{T}_{k+1} = \mathbb{T}_0 + \frac{(1-\delta)\sigma s_k^{\sigma-1}}{\mathcal{AB}(\delta)}\mathbb{Q}_1(s_k, \mathbb{T}_k, \mathbb{U}_k, \mathbb{V}_k) + \frac{\sigma\mathbf{h}^\delta}{\mathcal{AB}(\delta)\Gamma(\delta+2)}$$
$$\times\sum_{\ell=1}^{k}\Big[s_\ell^{\sigma-1}\mathbb{Q}_1(s_\ell, \mathbb{T}_\ell, \mathbb{U}_\ell, \mathbb{V}_\ell)\hat{\Upsilon}_1(k,\ell) - s_{\ell-1}^{\sigma-1}\mathbb{Q}_1(s_{\ell-1}, \mathbb{T}_{\ell-1}, \mathbb{U}_{\ell-1}, \mathbb{V}_{\ell-1})\hat{\Upsilon}_2(k,\ell)\Big], \tag{29}$$

$$\mathbb{U}_{k+1} = \mathbb{U}_0 + \frac{(1-\delta)\sigma s_k^{\sigma-1}}{\mathcal{AB}(\delta)}\mathbb{Q}_2(s_k, \mathbb{T}_k, \mathbb{U}_k, \mathbb{V}_k) + \frac{\sigma\mathbf{h}^\delta}{\mathcal{AB}(\delta)\Gamma(\delta+2)}$$
$$\times\sum_{\ell=1}^{k}\Big[s_\ell^{\sigma-1}\mathbb{Q}_2(s_\ell, \mathbb{T}_\ell, \mathbb{U}_\ell, \mathbb{V}_\ell)\hat{\Upsilon}_1(k,\ell) - s_{\ell-1}^{\sigma-1}\mathbb{Q}_2(s_{\ell-1}, \mathbb{T}_{\ell-1}, \mathbb{U}_{\ell-1}, \mathbb{V}_{\ell-1})\hat{\Upsilon}_2(k,\ell)\Big], \tag{30}$$

$$\mathbb{V}_{k+1} = \mathbb{V}_0 + \frac{(1-\delta)\sigma s_k^{\sigma-1}}{\mathcal{AB}(\delta)}\mathbb{Q}_3(s_k, \mathbb{T}_k, \mathbb{U}_k, \mathbb{V}_k) + \frac{\sigma\mathbf{h}^\delta}{\mathcal{AB}(\delta)\Gamma(\delta+2)}$$
$$\times\sum_{\ell=1}^{k}\Big[s_\ell^{\sigma-1}\mathbb{Q}_3(s_\ell, \mathbb{T}_\ell, \mathbb{U}_\ell, \mathbb{V}_\ell)\hat{\Upsilon}_1(k,\ell) - s_{\ell-1}^{\sigma-1}\mathbb{Q}_3(s_{\ell-1}, \mathbb{T}_{\ell-1}, \mathbb{U}_{\ell-1}, \mathbb{V}_{\ell-1})\hat{\Upsilon}_2(k,\ell)\Big], \tag{31}$$

where

$$\hat{\Upsilon}_1(k,\ell) = (k+1-\ell)^\delta(k-\ell+2+\delta) - (k-\ell)^\delta(k-\ell+2+2\delta),$$

and

$$\hat{\Upsilon}_2(k,\ell) = (k+1-\ell)^{\delta+1} - (k-\ell)^\delta(k-\ell+1+\delta).$$

7. Numerical Scheme via Newton Polynomials Method

In this section, we produce a new numerical scheme for solutions of our fractal-fractional CD4$^+$-HIV-1-model (2) which was introduced by Atangana and Araz in the book [44] in 2021. To do this, we again use the compact form of IVP (7) with the conditions (8) and (9). In this case, we have

$$\mathbb{K}(s) - \mathbb{K}(0) = \frac{\delta\sigma}{\mathcal{AB}(\delta)\Gamma(\delta)} \int_0^s \mathfrak{w}^{\sigma-1}(s-\mathfrak{w})^{\delta-1}\mathbb{Q}(\mathfrak{w}, \mathbb{K}(\mathfrak{w}))\, d\mathfrak{w} + \frac{(1-\delta)\sigma s^{\sigma-1}}{\mathcal{AB}(\delta)}\mathbb{Q}(s, \mathbb{K}(s)).$$

Set $\mathbb{Q}^*(s, \mathbb{K}(s)) = \sigma s^{\sigma-1}\mathbb{Q}(s, \mathbb{K}(s))$. Then

$$\mathbb{K}(s) - \mathbb{K}(0) = \frac{\delta}{\mathcal{AB}(\delta)\Gamma(\delta)} \int_0^s (s-\mathfrak{w})^{\delta-1}\mathbb{Q}^*(\mathfrak{w}, \mathbb{K}(\mathfrak{w}))\, d\mathfrak{w} + \frac{(1-\delta)}{\mathcal{AB}(\delta)}\mathbb{Q}^*(s, \mathbb{K}(s)).$$

By discretizing the above equation at $s = s_{k+1} = (k+1)\mathbf{h}$, we get

$$\mathbb{K}(s_{k+1}) - \mathbb{K}(0) = \frac{\delta}{\mathcal{AB}(\delta)\Gamma(\delta)} \int_0^{s_{k+1}} (s_{k+1}-\mathfrak{w})^{\delta-1}\mathbb{Q}^*(\mathfrak{w}, \mathbb{K}(\mathfrak{w}))\, d\mathfrak{w} + \frac{(1-\delta)}{\mathcal{AB}(\delta)}\mathbb{Q}^*(s_k, \mathbb{K}(s_k)).$$

If we approximate the above integral, then it becomes

$$\begin{aligned}\mathbb{K}(s_{k+1}) &= \mathbb{K}_0 + \frac{(1-\delta)}{\mathcal{AB}(\delta)}\mathbb{Q}^*(s_k, \mathbb{K}(s_k)) \\ &\quad + \frac{\delta}{\mathcal{AB}(\delta)\Gamma(\delta)} \sum_{\ell=2}^{k} \int_{s_\ell}^{s_{\ell+1}} (s_{k+1}-\mathfrak{w})^{\delta-1}\mathbb{Q}^*(\mathfrak{w}, \mathbb{K}(\mathfrak{w}))\, d\mathfrak{w}. \end{aligned} \tag{32}$$

In this step, the function $\mathbb{Q}^*(s, \mathbb{K}(s))$ is approximated by the Newton polynomial as

$$\begin{aligned} P_k^*(\mathfrak{w}) &= \mathbb{Q}^*(s_{k-2}, \mathbb{K}(s_{k-2})) + \frac{\mathbb{Q}^*(s_{k-1}, \mathbb{K}(s_{k-1})) - \mathbb{Q}^*(s_{k-2}, \mathbb{K}(s_{k-2}))}{\mathbf{h}}(\mathfrak{w} - s_{k-2}) \\ &\quad + \frac{\mathbb{Q}^*(s_k, \mathbb{K}(s_k)) - 2\mathbb{Q}^*(s_{k-1}, \mathbb{K}(s_{k-1})) + \mathbb{Q}^*(s_{k-2}, \mathbb{K}(s_{k-2}))}{2\mathbf{h}^2}(\mathfrak{w} - s_{k-2})(\mathfrak{w} - s_{k-1}). \end{aligned} \tag{33}$$

Substitute (33) into (32):

$$\begin{aligned} \mathbb{K}_{k+1} &= \mathbb{K}_0 + \frac{(1-\delta)}{\mathcal{AB}(\delta)}\mathbb{Q}^*(s_k, \mathbb{K}(s_k)) + \frac{\delta}{\mathcal{AB}(\delta)\Gamma(\delta)} \sum_{\ell=2}^{k} \int_{s_\ell}^{s_{\ell+1}} \Big[\mathbb{Q}^*(s_{\ell-2}, \mathbb{K}_{\ell-2}) \\ &\quad + \frac{\mathbb{Q}^*(s_{\ell-1}, \mathbb{K}_{\ell-1}) - \mathbb{Q}^*(s_{\ell-2}, \mathbb{K}_{\ell-2})}{\mathbf{h}}(\mathfrak{w} - s_{\ell-2}) \\ &\quad + \frac{\mathbb{Q}^*(s_\ell, \mathbb{K}_\ell) - 2\mathbb{Q}^*(s_{\ell-1}, \mathbb{K}_{\ell-1}) + \mathbb{Q}^*(s_{\ell-2}, \mathbb{K}_{\ell-2})}{2\mathbf{h}^2}(\mathfrak{w} - s_{\ell-2})(\mathfrak{w} - s_{\ell-1})\Big] \\ &\quad \times (s_{k+1} - \mathfrak{w})^{\delta-1}\, d\mathfrak{w}. \end{aligned}$$

We simplify the above relations, and we get

$$\begin{aligned} \mathbb{K}_{k+1} &= \mathbb{K}_0 + \frac{(1-\delta)}{\mathcal{AB}(\delta)}\mathbb{Q}^*(s_k, \mathbb{K}(s_k)) \\ &\quad + \frac{\delta}{\mathcal{AB}(\delta)\Gamma(\delta)} \sum_{\ell=2}^{k} \Big[\Big[\int_{s_\ell}^{s_{\ell+1}} \mathbb{Q}^*(s_{\ell-2}, \mathbb{K}_{\ell-2})(s_{k+1} - \mathfrak{w})^{\delta-1}\, d\mathfrak{w} \end{aligned}$$

$$+ \int_{s_\ell}^{s_{\ell+1}} \frac{\mathbb{Q}^*(s_{\ell-1}, \mathbb{K}_{\ell-1}) - \mathbb{Q}^*(s_{\ell-2}, \mathbb{K}_{\ell-2})}{\mathbf{h}} (\mathfrak{w} - s_{\ell-2})(s_{k+1} - \mathfrak{w})^{\delta-1} \, d\mathfrak{w}$$

$$+ \int_{s_\ell}^{s_{\ell+1}} \frac{\mathbb{Q}^*(s_\ell, \mathbb{K}_\ell) - 2\mathbb{Q}^*(s_{\ell-1}, \mathbb{K}_{\ell-1}) + \mathbb{Q}^*(s_{\ell-2}, \mathbb{K}_{\ell-2})}{2\mathbf{h}^2} (\mathfrak{w} - s_{\ell-2})(\mathfrak{w} - s_{\ell-1})$$

$$\times (s_{k+1} - \mathfrak{w})^{\delta-1} \, d\mathfrak{w} \Big].$$

In consequence,

$$\mathbb{K}_{k+1} = \mathbb{K}_0 + \frac{(1-\delta)}{\mathcal{AB}(\delta)} \mathbb{Q}^*(s_k, \mathbb{K}(s_k))$$

$$+ \frac{\delta}{\mathcal{AB}(\delta)\Gamma(\delta)} \sum_{\ell=2}^{k} \mathbb{Q}^*(s_{\ell-2}, \mathbb{K}_{\ell-2}) \int_{s_\ell}^{s_{\ell+1}} (s_{k+1} - \mathfrak{w})^{\delta-1} \, d\mathfrak{w}$$

$$+ \frac{\delta}{\mathcal{AB}(\delta)\Gamma(\delta)} \sum_{\ell=2}^{k} \frac{\mathbb{Q}^*(s_{\ell-1}, \mathbb{K}_{\ell-1}) - \mathbb{Q}^*(s_{\ell-2}, \mathbb{K}_{\ell-2})}{\mathbf{h}} \int_{s_\ell}^{s_{\ell+1}} (\mathfrak{w} - s_{\ell-2})(s_{k+1} - \mathfrak{w})^{\delta-1} \, d\mathfrak{w}$$

$$+ \frac{\delta}{\mathcal{AB}(\delta)\Gamma(\delta)} \sum_{\ell=2}^{k} \frac{\mathbb{Q}^*(s_\ell, \mathbb{K}_\ell) - 2\mathbb{Q}^*(s_{\ell-1}, \mathbb{K}_{\ell-1}) + \mathbb{Q}^*(s_{\ell-2}, \mathbb{K}_{\ell-2})}{2\mathbf{h}^2}$$

$$\times \int_{s_\ell}^{s_{\ell+1}} (\mathfrak{w} - s_{\ell-2})(\mathfrak{w} - s_{\ell-1}) \times (s_{k+1} - \mathfrak{w})^{\delta-1} \, d\mathfrak{w} \Big]. \quad (34)$$

On the other hand, we compute above three integrals separately, and we get

$$\int_{s_\ell}^{s_{\ell+1}} (s_{k+1} - \mathfrak{w})^{\delta-1} \, d\mathfrak{w} = \frac{\mathbf{h}^\delta}{\delta} \left[(k - \ell + 1)^\delta - (k - \ell)^\delta \right], \quad (35)$$

and

$$\int_{s_\ell}^{s_{\ell+1}} (\mathfrak{w} - s_{\ell-2})(s_{k+1} - \mathfrak{w})^{\delta-1} \, d\mathfrak{w} = \frac{\mathbf{h}^{\delta+1}}{\delta(\delta+1)} \left[(k - \ell + 1)^\delta (k - \ell + 3 + 2\delta) \right.$$

$$\left. - (k - \ell + 1)^\delta (k - \ell + 3 + 3\delta) \right], \quad (36)$$

and

$$\int_{s_\ell}^{s_{\ell+1}} (\mathfrak{w} - s_{\ell-2})(\mathfrak{w} - s_{\ell-1})(s_{k+1} - \mathfrak{w})^{\delta-1} \, d\mathfrak{w} = \frac{\mathbf{h}^{\delta+2}}{\delta(\delta+1)(\delta+2)} \Big((k - \ell + 1)^\delta \big[2(k - \ell)^2$$

$$+ (3\delta + 10)(k - \ell) + 2\delta^2 + 9\delta + 12 \big] - (k - \ell)^\delta \big[2(k - \ell)^2$$

$$+ (5\delta + 10)(k - \ell) + 6\delta^2 + 18\delta + 12 \big] \Big). \quad (37)$$

By putting (35)–(37) in (34), we obtain

$$\mathbb{K}_{k+1} = \mathbb{K}_0 + \frac{(1-\delta)}{\mathcal{AB}(\delta)} \mathbb{Q}^*(s_k, \mathbb{K}(s_k))$$

$$+ \frac{\delta \mathbf{h}^\delta}{\mathcal{AB}(\delta)\Gamma(\delta+1)} \sum_{\ell=2}^{k} \mathbb{Q}^*(s_{\ell-2}, \mathbb{K}_{\ell-2}) \left[(k - \ell + 1)^\delta - (k - \ell)^\delta \right]$$

$$+\frac{\delta \mathbf{h}^{\delta}}{\mathcal{AB}(\delta)\Gamma(\delta+2)}\sum_{\ell=2}^{k}\Big[\mathbb{Q}^*(s_{\ell-1},\mathbb{K}_{\ell-1})-\mathbb{Q}^*(s_{\ell-2},\mathbb{K}_{\ell-2})\Big]$$

$$\times\left[(k-\ell+1)^{\delta}(k-\ell+3+2\delta)-(k-\ell+1)^{\delta}(k-\ell+3+3\delta)\right]$$

$$+\frac{\delta \mathbf{h}^{\delta}}{2\mathcal{AB}(\delta)\Gamma(\delta+3)}\sum_{\ell=2}^{k}\Big[\mathbb{Q}^*(s_{\ell},\mathbb{K}_{\ell})-2\mathbb{Q}^*(s_{\ell-1},\mathbb{K}_{\ell-1})+\mathbb{Q}^*(s_{\ell-2},\mathbb{K}_{\ell-2})\Big]$$

$$\times\Big[(k-\ell+1)^{\delta}[2(k-\ell)^2+(3\delta+10)(k-\ell)+2\delta^2+9\delta+12]-(k-\ell)^{\delta}[2(k-\ell)^2$$

$$+(5\delta+10)(k-\ell)+6\delta^2+18\delta+12]\Big]. \tag{38}$$

Lastly, we replace $\mathbb{Q}^*(s,\mathbb{K}(s))=\sigma s^{\sigma-1}\mathbb{Q}(s,\mathbb{K}(s))$ in (38), and we get

$$\begin{aligned}\mathbb{K}_{k+1}=\mathbb{K}_0&+\frac{(1-\delta)\sigma s_k^{\sigma-1}}{\mathcal{AB}(\delta)}\mathbb{Q}(s_k,\mathbb{K}(s_k))\\&+\frac{\delta\sigma \mathbf{h}^{\delta}}{\mathcal{AB}(\delta)\Gamma(\delta+1)}\sum_{\ell=2}^{k}s_{\ell-2}^{\sigma-1}\mathbb{Q}(s_{\ell-2},\mathbb{K}_{\ell-2})\hat{\Psi}_1(k,\ell,\delta)\\&+\frac{\delta\sigma \mathbf{h}^{\delta}}{\mathcal{AB}(\delta)\Gamma(\delta+2)}\sum_{\ell=2}^{k}\Big[s_{\ell-1}^{\sigma-1}\mathbb{Q}(s_{\ell-1},\mathbb{K}_{\ell-1})-s_{\ell-2}^{\sigma-1}\mathbb{Q}(s_{\ell-2},\mathbb{K}_{\ell-2})\Big]\hat{\Psi}_2(k,\ell,\delta)\\&+\frac{\delta\sigma \mathbf{h}^{\delta}}{2\mathcal{AB}(\delta)\Gamma(\delta+3)}\sum_{\ell=2}^{k}\Big[s_{\ell}^{\sigma-1}\mathbb{Q}(s_{\ell},\mathbb{K}_{\ell})-2s_{\ell-1}^{\sigma-1}\mathbb{Q}(s_{\ell-1},\mathbb{K}_{\ell-1})+s_{\ell-2}^{\sigma-1}\mathbb{Q}(s_{\ell-2},\mathbb{K}_{\ell-2})\Big]\hat{\Psi}_3(k,\ell,\delta),\end{aligned} \tag{39}$$

where

$$\begin{aligned}\hat{\Psi}_1(k,\ell,\delta)&=(k-\ell+1)^{\delta}-(k-\ell)^{\delta},\\\hat{\Psi}_2(k,\ell,\delta)&=(k-\ell+1)^{\delta}(k-\ell+3+2\delta)-(k-\ell+1)^{\delta}(k-\ell-3+3\delta),\\\hat{\Psi}_3(k,\ell,\delta)&=(k-\ell+1)^{\delta}[2(k-\ell)^2+(3\delta+10)(k-\ell)+2\delta^2+9\delta+12]\\&\quad-(k-\ell)^{\delta}[2(k-\ell)^2+(5\delta+10)(k-\ell)+6\delta^2+18\delta+12].\end{aligned} \tag{40}$$

Based on the numerical scheme obtained in (39), numerical solutions of the fractal-fractional CD4$^+$-HIV-1-model (2) are given by

$$\begin{aligned}\mathbb{T}_{k+1}=\mathbb{T}_0&+\frac{(1-\delta)\sigma s_k^{\sigma-1}}{\mathcal{AB}(\delta)}\mathbb{Q}_1(s_k,\mathbb{T}(s_k),\mathbb{U}(s_k),\mathbb{V}(s_k))\\&+\frac{\delta\sigma \mathbf{h}^{\delta}}{\mathcal{AB}(\delta)\Gamma(\delta+1)}\sum_{\ell=2}^{k}s_{\ell-2}^{\sigma-1}\mathbb{Q}_1(s_{\ell-2},\mathbb{T}_{\ell-2},\mathbb{U}_{\ell-2},\mathbb{V}_{\ell-2})\hat{\Psi}_1(k,\ell,\delta)\\&+\frac{\delta\sigma \mathbf{h}^{\delta}}{\mathcal{AB}(\delta)\Gamma(\delta+2)}\sum_{\ell=2}^{k}\Big[s_{\ell-1}^{\sigma-1}\mathbb{Q}_1(s_{\ell-1},\mathbb{T}_{\ell-1},\mathbb{U}_{\ell-1},\mathbb{V}_{\ell-1})-s_{\ell-2}^{\sigma-1}\mathbb{Q}_1(s_{\ell-2},\mathbb{T}_{\ell-2},\mathbb{U}_{\ell-2},\mathbb{V}_{\ell-2})\Big]\hat{\Psi}_2(k,\ell,\delta)\\&+\frac{\delta\sigma \mathbf{h}^{\delta}}{2\mathcal{AB}(\delta)\Gamma(\delta+3)}\sum_{\ell=2}^{k}\Big[s_{\ell}^{\sigma-1}\mathbb{Q}_1(s_{\ell},\mathbb{T}_{\ell},\mathbb{U}_{\ell},\mathbb{V}_{\ell})-2s_{\ell-1}^{\sigma-1}\mathbb{Q}_1(s_{\ell-1},\mathbb{T}_{\ell-1},\mathbb{U}_{\ell-1},\mathbb{V}_{\ell-1})\end{aligned}$$

$$+ s_{\ell-2}^{\sigma-1}\mathbb{Q}_1(s_{\ell-2}, \mathbb{T}_{\ell-2}, \mathbb{U}_{\ell-2}, \mathbb{V}_{\ell-2})\Big]\hat{\Upsilon}_3(k,\ell,\delta), \tag{41}$$

and

$$\begin{aligned}
\mathbb{U}_{k+1} &= \mathbb{U}_0 + \frac{(1-\delta)\sigma s_k^{\sigma-1}}{\mathcal{AB}(\delta)}\mathbb{Q}_2(s_k, \mathbb{T}(s_k), \mathbb{U}(s_k), \mathbb{V}(s_k)) \\
&+ \frac{\delta\sigma \mathbf{h}^{\delta}}{\mathcal{AB}(\delta)\Gamma(\delta+1)}\sum_{\ell=2}^{k} s_{\ell-2}^{\sigma-1}\mathbb{Q}_2(s_{\ell-2}, \mathbb{T}_{\ell-2}, \mathbb{U}_{\ell-2}, \mathbb{V}_{\ell-2})\hat{\Upsilon}_1(k,\ell,\delta) \\
&+ \frac{\delta\sigma \mathbf{h}^{\delta}}{\mathcal{AB}(\delta)\Gamma(\delta+2)}\sum_{\ell=2}^{k}\Big[s_{\ell-1}^{\sigma-1}\mathbb{Q}_2(s_{\ell-1}, \mathbb{T}_{\ell-1}, \mathbb{U}_{\ell-1}, \mathbb{V}_{\ell-1}) - s_{\ell-2}^{\sigma-1}\mathbb{Q}_2(s_{\ell-2}, \mathbb{T}_{\ell-2}, \mathbb{U}_{\ell-2}, \mathbb{V}_{\ell-2})\Big]\hat{\Upsilon}_2(k,\ell,\delta) \\
&+ \frac{\delta\sigma \mathbf{h}^{\delta}}{2\mathcal{AB}(\delta)\Gamma(\delta+3)}\sum_{\ell=2}^{k}\Big[s_{\ell}^{\sigma-1}\mathbb{Q}_2(s_{\ell}, \mathbb{T}_{\ell}, \mathbb{U}_{\ell}, \mathbb{V}_{\ell}) - 2s_{\ell-1}^{\sigma-1}\mathbb{Q}_2(s_{\ell-1}, \mathbb{T}_{\ell-1}, \mathbb{U}_{\ell-1}, \mathbb{V}_{\ell-1}) \\
&+ s_{\ell-2}^{\sigma-1}\mathbb{Q}_2(s_{\ell-2}, \mathbb{T}_{\ell-2}, \mathbb{U}_{\ell-2}, \mathbb{V}_{\ell-2})\Big]\hat{\Upsilon}_3(k,\ell,\delta),
\end{aligned} \tag{42}$$

and

$$\begin{aligned}
\mathbb{V}_{k+1} &= \mathbb{V}_0 + \frac{(1-\delta)\sigma s_k^{\sigma-1}}{\mathcal{AB}(\delta)}\mathbb{Q}_3(s_k, \mathbb{T}(s_k), \mathbb{U}(s_k), \mathbb{V}(s_k)) \\
&+ \frac{\delta\sigma \mathbf{h}^{\delta}}{\mathcal{AB}(\delta)\Gamma(\delta+1)}\sum_{\ell=2}^{k} s_{\ell-2}^{\sigma-1}\mathbb{Q}_3(s_{\ell-2}, \mathbb{T}_{\ell-2}, \mathbb{U}_{\ell-2}, \mathbb{V}_{\ell-2})\hat{\Upsilon}_1(k,\ell,\delta) \\
&+ \frac{\delta\sigma \mathbf{h}^{\delta}}{\mathcal{AB}(\delta)\Gamma(\delta+2)}\sum_{\ell=2}^{k}\Big[s_{\ell-1}^{\sigma-1}\mathbb{Q}_3(s_{\ell-1}, \mathbb{T}_{\ell-1}, \mathbb{U}_{\ell-1}, \mathbb{V}_{\ell-1}) - s_{\ell-2}^{\sigma-1}\mathbb{Q}_3(s_{\ell-2}, \mathbb{T}_{\ell-2}, \mathbb{U}_{\ell-2}, \mathbb{V}_{\ell-2})\Big]\hat{\Upsilon}_2(k,\ell,\delta) \\
&+ \frac{\delta\sigma \mathbf{h}^{\delta}}{2\mathcal{AB}(\delta)\Gamma(\delta+3)}\sum_{\ell=2}^{k}\Big[s_{\ell}^{\sigma-1}\mathbb{Q}_3(s_{\ell}, \mathbb{T}_{\ell}, \mathbb{U}_{\ell}, \mathbb{V}_{\ell}) - 2s_{\ell-1}^{\sigma-1}\mathbb{Q}_3(s_{\ell-1}, \mathbb{T}_{\ell-1}, \mathbb{U}_{\ell-1}, \mathbb{V}_{\ell-1}) \\
&+ s_{\ell-2}^{\sigma-1}\mathbb{Q}_3(s_{\ell-2}, \mathbb{T}_{\ell-2}, \mathbb{U}_{\ell-2}, \mathbb{V}_{\ell-2})\Big]\hat{\Upsilon}_3(k,\ell,\delta),
\end{aligned} \tag{43}$$

where the constants $\hat{\Upsilon}_j(k,\ell,\delta)$ are introduced in (40) for $j = 1,2,3$.

8. Simulations and Discussions

In this section, we simulate and discuss dynamical behaviors of the fractal-fractional $CD4^+$-HIV-1-model (2) via real data for parameters and initial values assumed by [45]. Based on the given data in this source, we let $\theta = 10$, $q = 0.000024$, $\rho = 0.01$, $\zeta = 0.2$, $\kappa = 0.16$, $\vartheta = 3.4$, and $N = 1000$. Moreover, the initial values for three state functions are $\mathbb{T}(0) = \mathbb{T}_0 = 1000, \mathbb{U}(0) = \mathbb{U}_0 = 0, \mathbb{V}(0) = \mathbb{V}_0 = 0.001$. It is assumed that all parameters are in days.

In the provided figures, we show the behaviors of three state functions $\mathbb{T}, \mathbb{U}, \mathbb{V}$ under the effect of different values for the fractal-fractional orders $\delta = \sigma = 1.00, 0.98, 0.96, 0.94, 0.92, 0.90$ simultaneously. We also compare the results in the graphs with respect to two numerical algorithms given by (29)–(31) and (41)–(43) as shown in Tables 1–3.

In these graphs, the quantity of the time s interprets the number of days. Note that Figures 1 and 2 demonstrate the dynamics of the $CD4^+$ T-cells via the Adams-Bashforth method (Section 6). Figure 1a indicates that when the fractal dimensions and fractional orders decrease, the number of susceptible CD4$^+$ T-cells steadily increases from the 42nd day to the 64th day, and steadily decreases from the 65th day to the 102nd day. Further it increases again from the 103rd day and finally converges to the integer-order at the 250th day to the end of the simulation period.

Figure 1b,c show that when the fractal dimensions and fractional orders move from the integer order, the peak of the amount of infectious CD4$^+$ T-cells and the free particles of the infection of the HIV-1 virus in the blood decrease respectively and also give slight differences in their asymptotic stabilities.

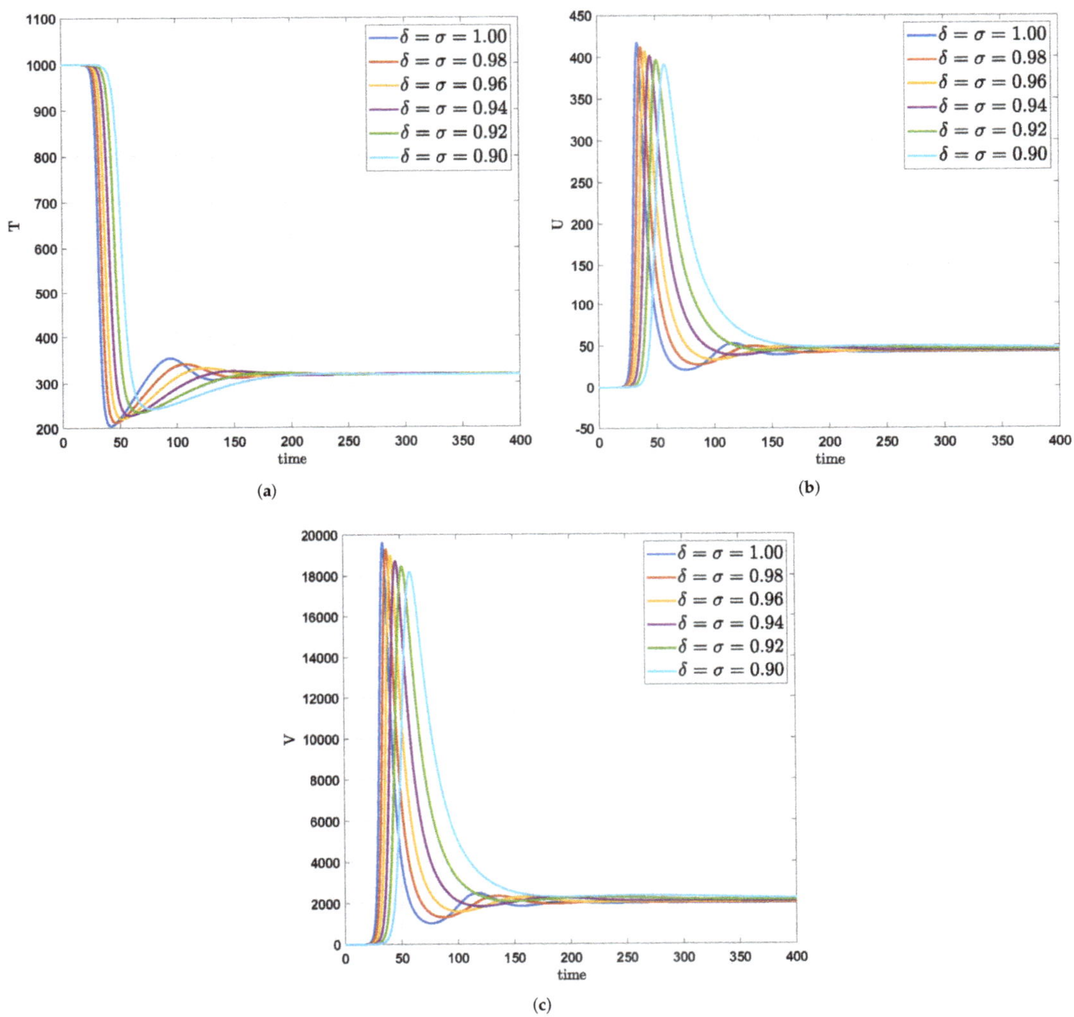

Figure 1. Numerical trajectories by varying both fractal order and fractional order $\delta = \sigma = 1.00$, 0.98, 0.96, 0.94, 0.92, 0.90 via the Adams-Bashforth method. (**a**) Amount of susceptible CD4$^+$ T-cells. (**b**) Amount of infectious CD4$^+$ T-cells. (**c**) Free particles of the infection of the HIV-1 virus in the blood.

In Figure 2a,c, we show a 0.01-variation in fractal-fractional order. In this case, the numerical trajectories show that a slight change in the fractal-fractional order produces a slight changes in the asymptotic behavior of the HIV-1 virus on CD4$^+$ T-cells.

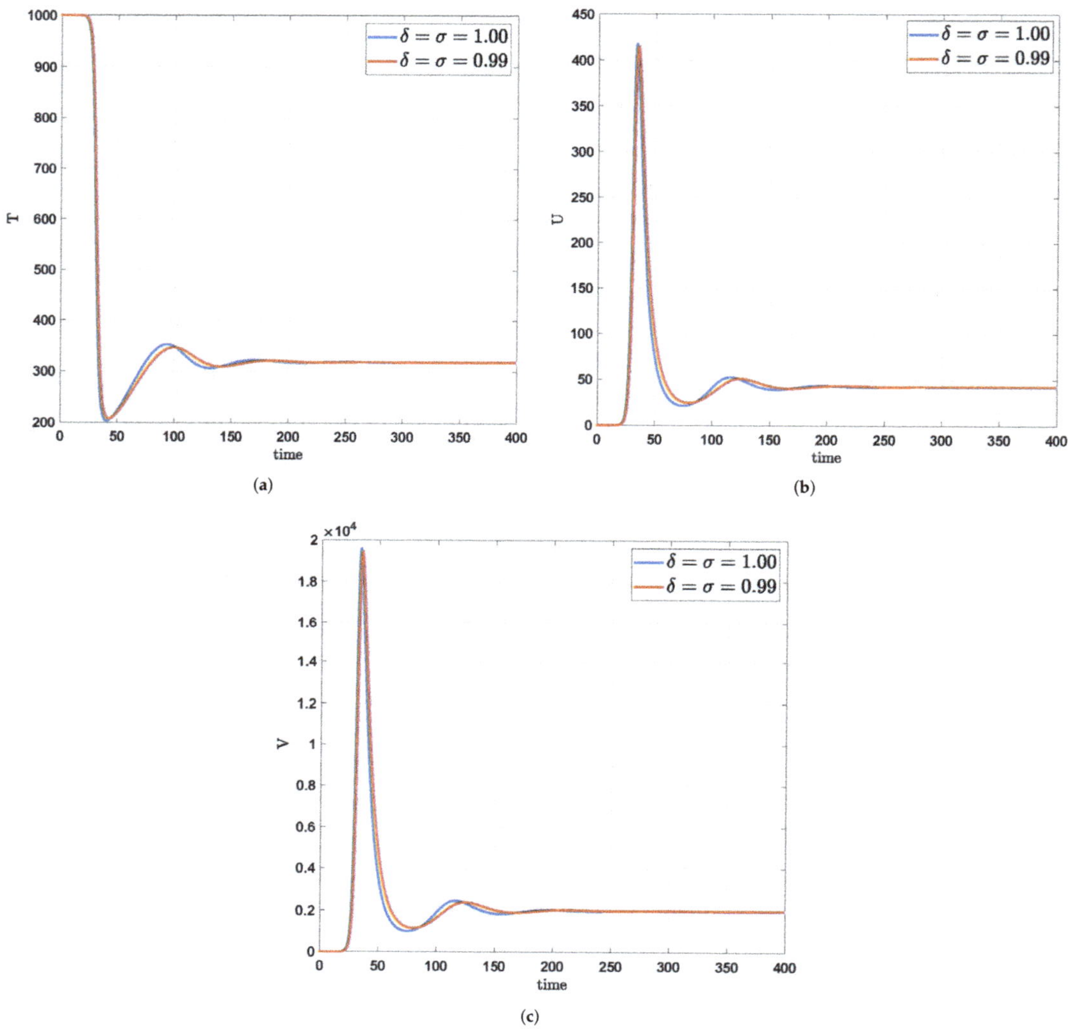

Figure 2. Numerical trajectories by varying both fractal order and fractional orders $\delta = \sigma = 1.00, 0.99$ via the Adams-Bashforth method. (**a**) A 0.01-variation in fractal-fractional orders of susceptible $CD4^+$ T-cells. (**b**) A 0.01-variation in fractal-fractional orders of infectious $CD4^+$ T-cells. (**c**) A 0.01-variation in fractal-fractional orders of free particles of the infection of the HIV-1 virus in the blood.

Figures 3 and 4 demonstrate the dynamics of the $CD4^+$ T-cells via the Newton polynomials method (Section 7). Figure 3a also indicates that when the fractal dimensions and fractional orders decrease, the number of susceptible $CD4^+$ T-cells steadily increases from the 42nd day to the 64th day, and steadily decreases from the 65th day to the 102nd day. Also it increases again from the 103rd day and finally converges to the integer-order at the 250th day to the end of the simulation period, but it shows a disorganized behavior from the 11th day to the 24th day, which indicates the early stages of HIV in the $CD4^+$ T-cells.

Figure 3b,c show that when the fractal dimensions and fractional orders move from the integer order, the peak of the amount of infectious $CD4^+$ T-cells and the free particles of the infection of the HIV-1 virus in the blood decrease, respectively and also give slight differences in their asymptotic stabilities, but again, they show a disorganized behavior from the 11th day to the 24th day.

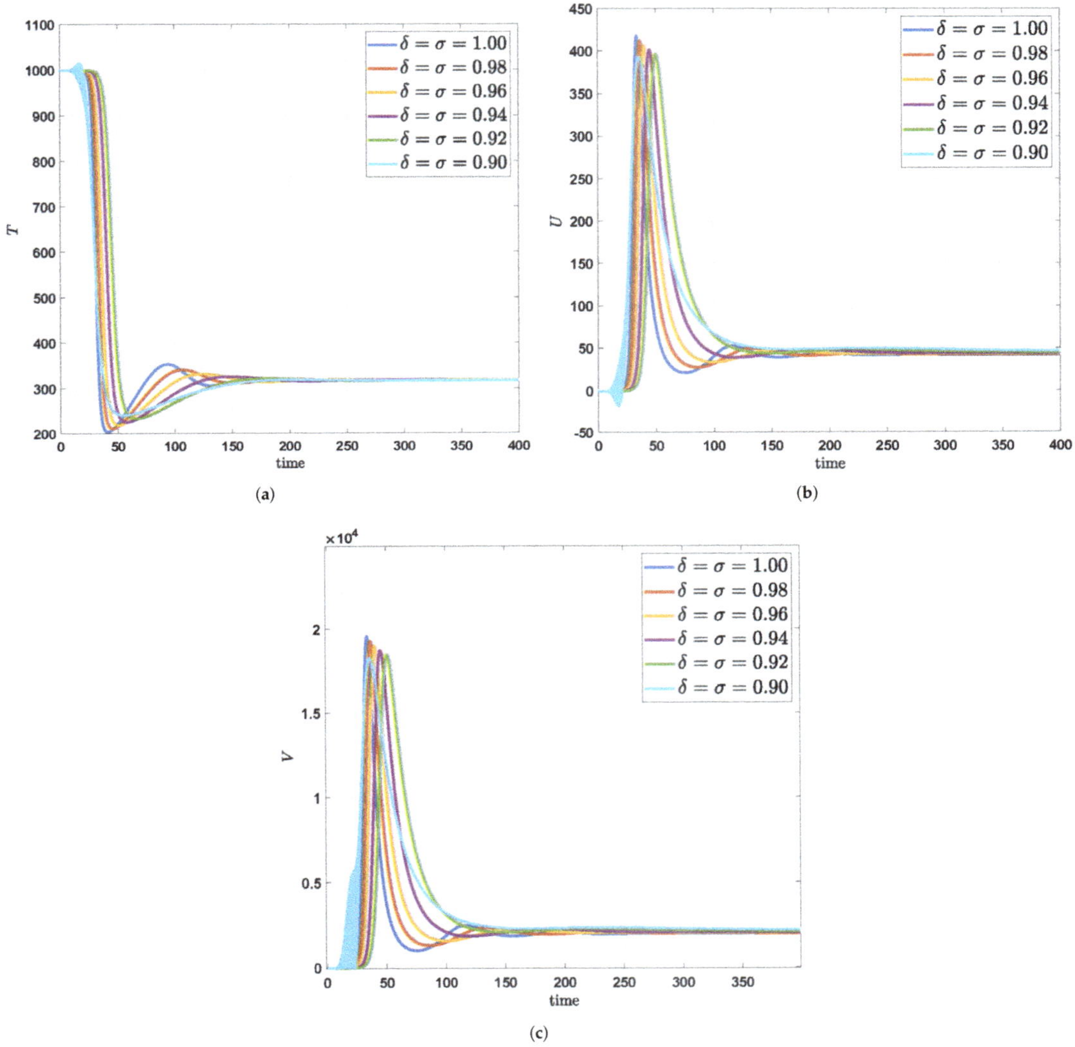

Figure 3. Numerical trajectories by varying both fractal order and fractional order $\delta = \sigma$ = 1.00, 0.98, 0.96, 0.94, 0.92, 0.90 via the Newton polynomials method. (**a**) Amount of susceptible CD4^{+} T-cells. (**b**) Amount of infectious CD4^{+} T-cells. (**c**) Free particles of the infection of the HIV-1 virus in the blood.

In Figure 4a,c, we show a 0.01-variation in fractal-fractional orders, and the numerical trajectories show that a slight change in the fractal-fractional orders produce slight changes in the asymptotic behavior of the HIV-1 virus on CD4^{+} T-cells.

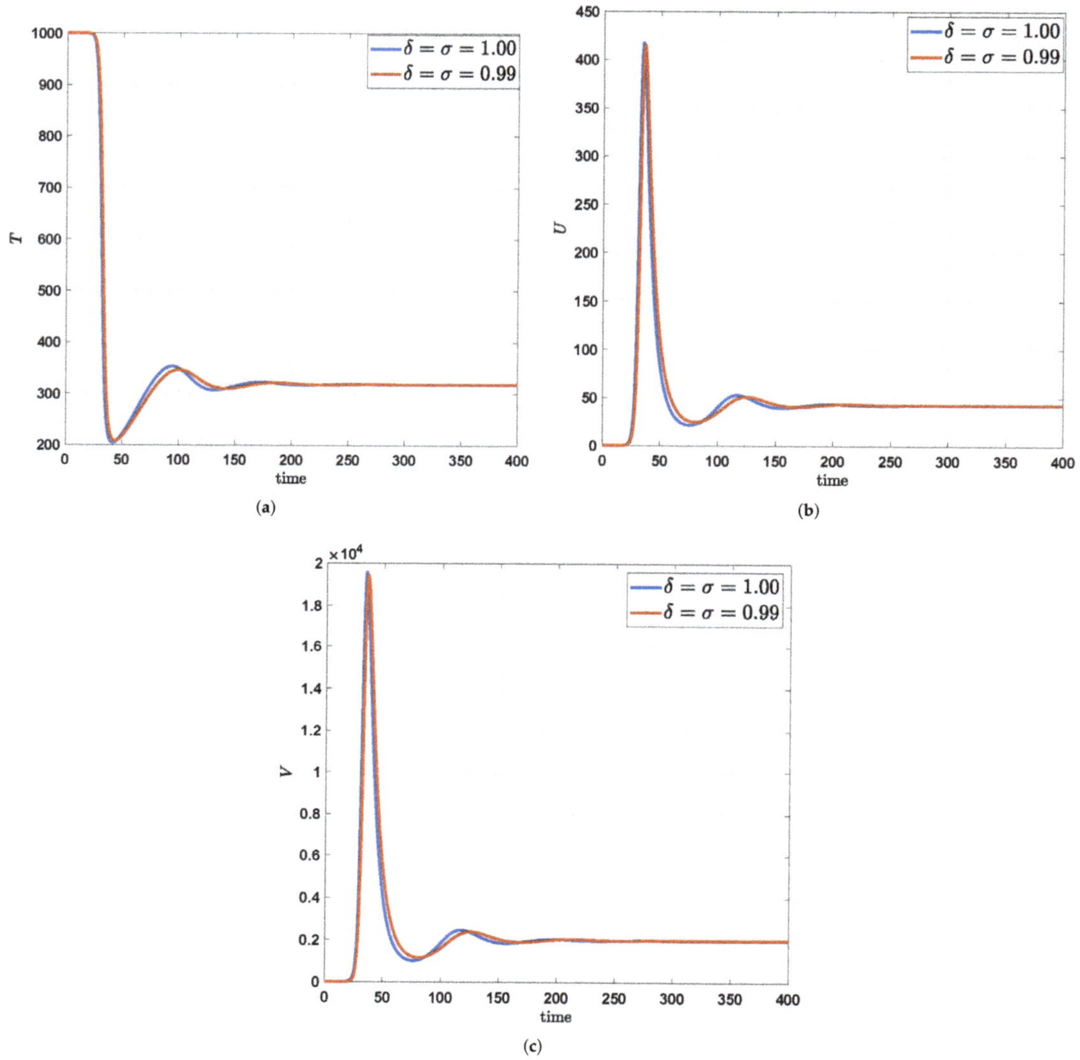

Figure 4. Numerical trajectories by varying both fractal order and fractional order $\delta = \sigma = 1.00, 0.99$ via the Newton polynomials method. (**a**) A 0.01-variation in fractal-fractional orders of susceptible $CD4^+$ T-cells. (**b**) A 0.01-variation in fractal-fractional orders of infectious $CD4^+$ T-cells. (**c**) A 0.01-variation in fractal-fractional orders of free particles of the infection of the HIV-1 virus in the blood.

In Figure 5, based on the Adams-Bashforth method, we see the numerical trajectories of three state functions by varying the average number of infected particles N for the values $N = 500, 600, 700, 800, 900, 1000$ under the fractal-fractional order $\delta = \sigma = 0.95$. Thus, in Figure 5a, it shows that the susceptible $CD4^+$ T-cells increase as the average number of infected particles by an existing infected cell reduces by 10%, and that of the peak of the amount of infectious $CD4^+$ T-cells and the free particles of the infection of the HIV-1 virus in the blood decrease, respectively in Figure 5b,c.

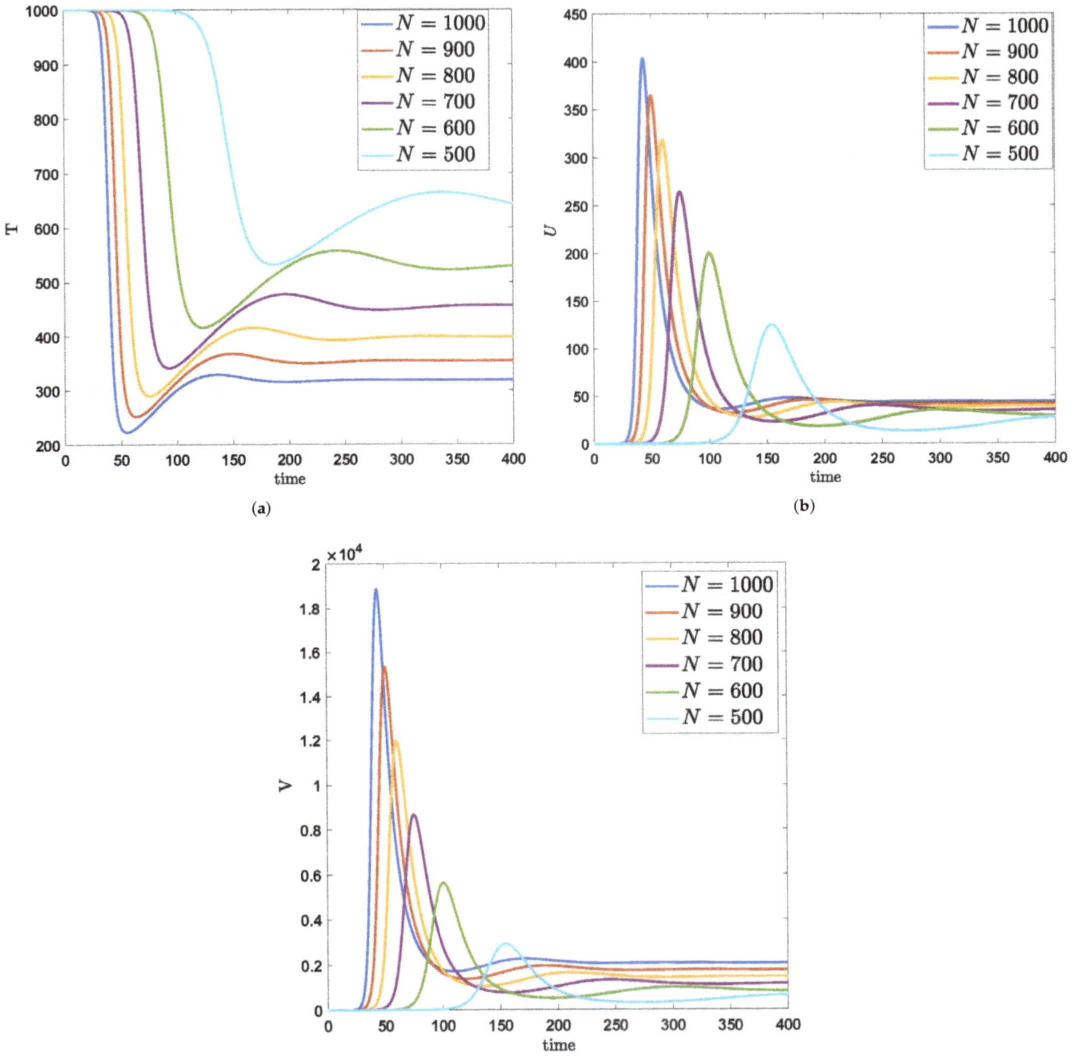

Figure 5. Numerical trajectories by varying the average number of infected particles N under the fractal order and fractional order $\delta = \sigma = 0.95$ via the Adams-Bashforth method. (**a**) The effects of N on $\mathbb{T}$. (**b**) The effects of N on $\mathbb{U}$. (**c**) The effects of N on $\mathbb{V}$.

In Figure 6, based on the Adams-Bashforth method, we see the numerical trajectories by varying the supply rate for new T-cells for $\theta = 10, 15, 20, 25, 30, 35$ under the fractal order and fractional order $\delta = \sigma = 0.95$. Thus, in Figure 6a, it shows that the susceptible CD4$^+$ T-cells increase for a period of time and gradually decrease as the number of the supply rate θ increases before converging at the 150th day.

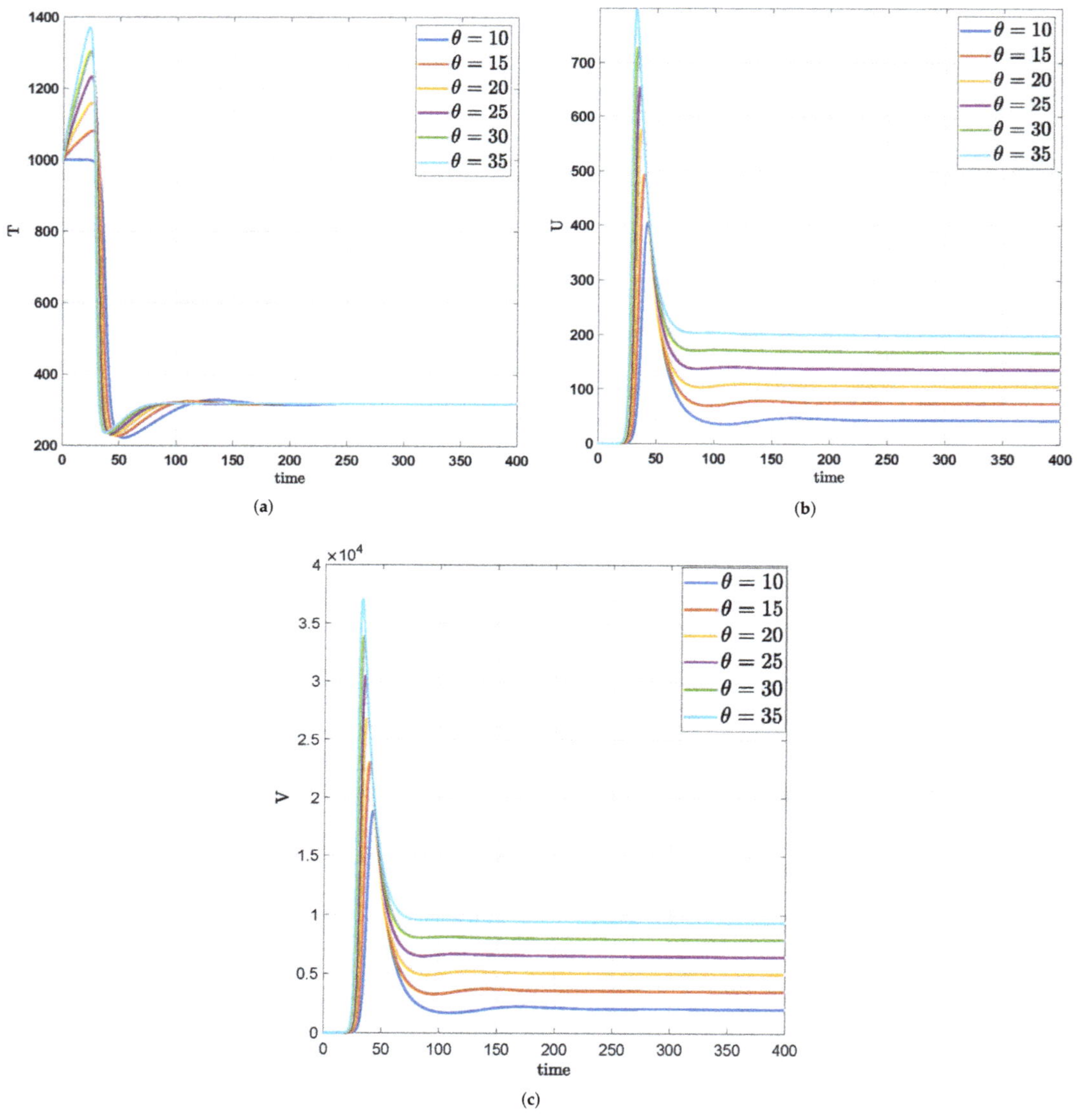

Figure 6. Numerical trajectories by varying the supply rate for new T-cells θ under the fractal order and fractional order $\delta = \sigma = 0.95$ via the Adams-Bashforth method. (**a**) The effects of θ on $\mathbb{T}$. (**b**) The effects of θ on $\mathbb{U}$. (**c**) The effects of θ on $\mathbb{V}$.

In Figure 6b,c, we noticed that the amount of infectious CD4$^+$ T-cells and the free particles of the infection of the HIV-1 virus in the blood increase, respectively. Figure 7a–c display the numerical trajectories when we compare the numerical solutions of the Adams-Bashforth method with the Newton polynomial method under the fractal dimension and fractional order $\delta = \sigma = 0.90$.

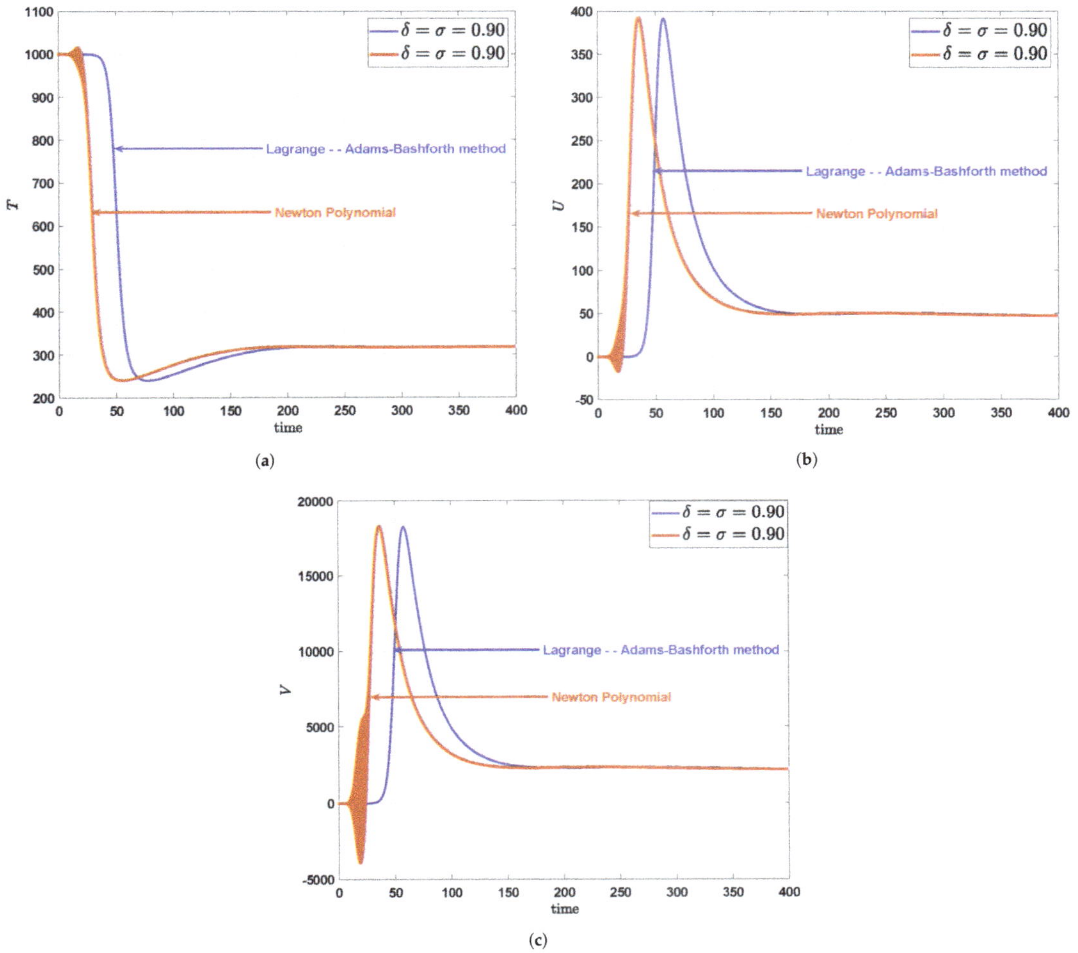

Figure 7. Numerical trajectories in the Comparison of the numerical solutions of the Adams-Bashforth method with Newton polynomials method under the fractal dimension and fractional order $\delta = \sigma = 0.90$. (**a**) Comparison of the graphs w.r.t. two different numerical algorithms on $\mathbb{T}$. (**b**) Comparison of the graphs w.r.t. two different numerical algorithms on $\mathbb{U}$. (**c**) Comparison of the graphs w.r.t. two different numerical algorithms on $\mathbb{V}$.

In Tables 1–3, we see some results of the two numerical schemes (the Adams-Bashforth method and Newton polynomials method) for all of three state functions under the fractal dimension and fractional order $\delta = \sigma = 0.95$ with step size $\mathbf{h} = 0.1$.

Table 1. Results of the two numerical schemes for the susceptible CD4$^+$ T-cells $\mathbb{T}$ (s) under the fractal order and fractional order $\delta = \sigma = 0.95$ with the step size $\mathbf{h} = 0.1$.

Time: (s)	0.0001	100	200	300	400
Adams-Bashforth method	1000	300.8396	315.4653	318.7946	318.8066
Newton polynomials method	1000	300.7881	315.4643	318.7951	318.8066

Table 2. Results of the two numerical schemes for the infectious CD4$^+$ T-cells $\mathbb{U}$ (s) under the fractal order and fractional order $\delta = \sigma = 0.95$ with the step size $\mathbf{h} = 0.1$.

Time: (s)	0.0001	100	200	300	400
Adams-Bashforth method	0	37.7693	46.0612	44.2611	43.7066
Newton polynomials method	0	37.7778	46.0645	44.2612	43.7066

Table 3. Results of the two numerical schemes for free particles of the infection of the HIV virus in the blood $\mathbb{V}$ (s) under the fractal order and fractional order $\delta = \sigma = 0.95$ with the step size $\mathbf{h} = 0.1$.

Time: (s)	0.0001	100	200	300	400
Adams-Bashforth method	0	1787.4317	2169.7278	2082.8420	2082.8420
Newton polynomials method	0	1787.8611	2169.8866	2082.8597	2082.8472

9. Conclusions

In this paper, we designed a fractal-fractional CD4$^+$-HIV-1-model and analyzed the dynamics of CD4$^+$ T-cells under the infection of HIV-1 virus. We considered three compartments for this model by defining three state functions $\mathbb{T}$, $\mathbb{U}$, and $\mathbb{V}$ for the amount of susceptible CD4$^+$ T-cells, amount of infectious CD4$^+$ T-cells, and the free particles of the infection of the HIV virus in the blood. We derived three fractal-fractional integral equations and proved that their kernels are Lipschitz. In this direction, we could prove the existence and uniqueness criteria for solutions of the fractal-fractional CD4$^+$-HIV-1-model. In the sequel, we investigated four stability results with the help of two auxiliary inequality. We extracted two algorithms via the Adams-Bashforth method and also via the Newton polynomials and simulated our real data in relation to the given fractal-fractional CD4$^+$-HIV-1-model. The numerical and graphical results showed that these two numerical algorithms give the same outcomes and differences are small. Also, we investigated the effect of frcatal dimensions and fractional orders on these simulations. Also, the effect of different values for the average number of infected particles and the supply rate of new T-cells were simulated in some graphs under the Adams-Bashforth method. This stude showed that we can predict the next behavior of the fractal-fractional CD4$^+$-HIV-1-model via the two mentioned numerical methods and their results are more accurate and identical. This shows the power of simulation of the frcatal-fractional models in comparison to the fractional models. In the next researches, we can develope our numerical methods on different fractal-fractional models of diseases.

Author Contributions: Conceptualization, H.N. and S.E.; Formal analysis, H.N., S.E., N.P., J.K.K.A., S.R. and T.S.; Funding acquisition, N.P. and T.S.; Methodology, H.N., S.E., N.P., J.K.K.A., S.R. and T.S.; Software, S.E. and J.K.K.A. All authors have read and agreed to the published version of the manuscript.

Funding: This research was funded by King Mongkut's University of Technology North Bangkok. Contract no. KMUTNB-62-KNOW-27.

Institutional Review Board Statement: Not applicable.

Informed Consent Statement: Not applicable.

Data Availability Statement: Data sharing not applicable to this article as no datasets were generated or analyzed during the current study.

Acknowledgments: The first author would like to thank Shiraz University. Also, the second and fifth authors would like to thank Azarbaijan Shahid Madani University. Also, we would like to thank dear reviewers for their constructive and helpful comments and remarks to improve the quality of the paper.

Conflicts of Interest: The authors declare no conflict of interest.

References

1. Wang, L.; Li, M.Y. Mathematical analysis of the global dynamics of a model for HIV infection of CD4$^+$ T cells. *Math. Biosci.* **2006**, *200*, 44–57. [CrossRef] [PubMed]
2. UNAIDS. Report on the Global AIDS Epidemic. 2016. Available online: http://www.unaids.org (accessed on 18 February 2022).
3. Kumar, P.; Erturk, V.S. The analysis of a time delay fractional COVID-19 model via Caputo type fractional derivative. *Math. Methods Appl. Sci.* **2022**. [CrossRef] [PubMed]
4. Rezapour, S.; Etemad, S.; Mohammadi, H. A mathematical analysis of a system of Caputo-Fabrizio fractional differential equationsfor the anthrax disease model in animals. *Adv. Differ. Equ.* **2020**, *2020*, 481. [CrossRef]
5. Alshehri, H.M.; Khan, A. A fractional order Hepatitis C mathematical model with Mittag-Leffler kernel. *J. Funct. Spaces* **2021**, *2021*, 2524027. [CrossRef]
6. Deressa, C.T.; Etemad, S.; Rezapour, S. On a new four-dimensional model of memristor-based chaotic circuit in the context of nonsingular Atangana-Baleanu-Caputo operators. *Adv. Differ. Equ.* **2021**, *2021*, 444. [CrossRef]
7. Deressa, C.T.; Etemad, S.; Kaabar, M.K.A.; Rezapour, S. Qualitative analysis of a hyperchaotic Lorenz-Stenflo mathematical model via the Caputo fractional operator. *J. Funct. Spaces* **2022**, *2022*, 4975104. [CrossRef]
8. Kumar, P.; Erturk, V.S. Environmental persistence influences infection dynamics for a butterfly pathogen via new generalised Caputo type fractional derivative. *Chaos Solitons Fractals* **2021**, *144*, 110672. [CrossRef]
9. Devi, A.; Kumar, A.; Abdeljawad, T.; Khan, A. Stability analysis of solutions and existence theory of fractional Lagevin equation. *Alex. Eng. J.* **2021**, *60*, 3641–3647. [CrossRef]
10. Mohammadi, H.; Kumar, S.; Rezapour, S.; Etemad, S. A theoretical study of the Caputo-Fabrizio fractional modeling for hearing loss due to Mumps virus with optimal control. *Chaos Solitons Fractals* **2021**, *144*, 110668. [CrossRef]
11. Begum, R.; Tunc, O.; Khan, H.; Gulzar, H.; Khan, A. A fractional order Zika virus model with Mittag-Leffler kernel. *Chaos Solitons Fractals* **2021**, *146*, 110898. [CrossRef]
12. Kumar, P.; Erturk, V.S.; Almusawa, H. Mathematical structure of mosaic disease using microbial biostimulants via Caputo and Atangana-Baleanu derivatives. *Results Phys.* **2021**, *24*, 104186. [CrossRef]
13. Zarin, R.; Khaliq, H.; Khan, A.; Khan, D.; Akgul, A.; Humphries, U.W. Deterministic and fractional modeling of a computer virus propagation. *Results Phys.* **2022**, *33*, 105130. [CrossRef]
14. Baleanu, D.; Etemad, S.; Rezapour, S. A hybrid Caputo fractional modeling for thermostat with hybrid boundary value conditions. *Bound. Value Probl.* **2020**, *2020*, 64. [CrossRef]
15. Thaiprayoon, C.; Sudsutad, W.; Alzabut, J.; Etemad, S.; Rezapour, S. On the qualitative analysis of the fractional boundary valueproblem describing thermostat control model via ψ-Hilfer fractional operator. *Adv. Differ. Equ.* **2021**, *2021*, 201. [CrossRef]
16. Alzabut, J.; Selvam, G.M.; El-Nabulsi, R.A.; Vignesh, D.; Samei, M.E. Asymptotic stability of nonlinear discrete fractional pantograph equations with non-local initial conditions. *Symmetry* **2021**, *13*, 473. [CrossRef]
17. Wongcharoen, A.; Ntouyas, S. K.; Tariboon, J. Nonlocal boundary value problems for Hilfer type pantograph fractional differentialequations and inclusions. *Adv. Differ. Equ.* **2020**, *2020*, 279. [CrossRef]
18. Kumar, P.; Erturk, V.S.; Yusuf, A.; Nisar, K.S.; Abdelwahab, S.F. A study on canine distemper virus (CDV) and rabies epidemics in the red fox population via fractional derivatives. *Results Phys.* **2021**, *25*, 104281. [CrossRef]
19. Asamoah, J.K.K.; Okyere, E.; Yankson, E.; Opoku, A.A.; Adom-Konadu, A.; Acheampong, E.; Arthur, Y.D. Non-fractional and fractional mathematical analysis and simulations for Q fever. *Chaos Solitons Fractals* **2022**, *156*, 111821. [CrossRef]
20. Khan, H.; Tunc, C.; Chen, W.; Khan, A. Existence theorems and Hyers-Ulam stability for a class of hybrid fractional differentialequations with p-Laplacial operator. *J. Appl. Anal. Comput.* **2018**, *8*, 1211–1226.
21. Omame, A.; Nwajeri, U.K.; Abbas, M.; Onyenegecha, C.P. A fractional order control model for Diabetes and COVID-19 co-dynamics with Mittag-Leffler function. *Alex. Eng. J.* **2022**, *61*, 7619–7635. [CrossRef]
22. Ali, W.; Turab, A.; Nieto, J.J. On the novel existence results of solutions for a class of fractional boundary value problems on the cyclohexane graph. *J. Inequalities Appl.* **2022**, *2022*, 5. [CrossRef]
23. Baleanu, D.; Etemad, S.; Mohammadi, H.; Rezapour, S. A novel modeling of boundary value problems on the glucose graph. *Commun. Nonlinear Sci. Numer. Simul.* **2021**, *100*, 105844. [CrossRef]
24. Rezapour, S.; Tellab, B.; Deressa, C.T.; Etemad, S.; Nonlaopon, K. H-U-type stability and numerical solutions for a nonlinear model of the coupled systems of Navier BVPs via the generalized differential transform method. *Fractal Fract.* **2021**, *5*, 166. [CrossRef]
25. Culshaw, R.V.; Ruan, S. A delay-differential equation model of HIV infection of CD4$^+$ T-cells. *Math. Biosci.* **2000**, *165*, 27–39. [CrossRef]
26. Cai, L.; Li, X.; Ghosh, M.; Guo, B. Stability analysis of an HIV/AIDS epidemic model with treatment. *J. Comput. Appl. Math.* **2009**, *229*, 313–323. [CrossRef]
27. Mohyud-Din, S.T.; Nazir, A.; Almohsin, B.; Ahmed, N.; Khan, U.; Waheed, A.; Hussain, T. On mathematical model of HIV CD4+ T-cells. *Alex. Eng. J.* **2021**, *60*, 995–1000. [CrossRef]
28. Perlson, A.S. Modeling the interaction of the immune system with HIV. *Lec. Notes Biomath.* **1989**, *83*, 350–370.
29. Perlson, A.S.; Kirschner, D.E.; Boer, R.D. Dynamics of HIV infection of $CD4^+$ T cells. *Math. Biosci.* **1993**, *114*, 81–125. [CrossRef]
30. Ding, Y.; Ye, H. A fractional-order differential equation model of HIV infection of CD4+ T-cells. *Math. Comput. Model.* **2009**, *50*, 386–392. [CrossRef]

31. Arafa, A.A.M.; Rida, S.Z.; Khalil, M. Fractional modeling dynamics of HIV and $CD4^+$ T-cells during primary infection. *Nonlinear Biomed. Phys.* **2012**, *6*, 1. [CrossRef]
32. Bulut, H.; Kumar, D.; Singh, J.; Swroop, R.; Baskonus, H.M. Analytic study for a fractional model of HIV infection of $CD4^+$T lymphocyte cells. *Math. Nat. Sci.* **2018**, *2*, 33–43. [CrossRef]
33. Lichae, B.H.; Biazar, J.; Ayati, Z. The fractional differential model of HIV-1 infection of $CD4^+$ T-cells with description of the effect of antiviral drug treatment. *Comput. Math. Methods Med.* **2019**, *2019*, 4059549. [CrossRef] [PubMed]
34. Nazir, G.; Shah, K.; Debbouche, A.; Khan, R.A. Study of HIV mathematical model under nonsingular kernel type derivative of fractional order. *Chaos Solitons Fractals* **2020**, *139*, 110095. [CrossRef]
35. Wang, W.; Wang, X.; Feng, Z. Time periodic reaction-diffusion equations for modeling 2-LTR dynamics in HIV-infected patients. *Nonlin. Anal. Real World Appl.* **2021**, *57*, 103184. [CrossRef]
36. Atangana, A. Fractal-fractional differentiation and integration: Connecting fractal calculus and fractional calculus to predict complex system. *Chaos Solitons Fractals* **2017**, *102*, 396–406. [CrossRef]
37. Shah, K.; Arfan, M.; Mahariq, I.; Ahmadian, A.; Salahshour, S.; Ferrara, M. Fractal-fractional mathematical model addressing the situation of Corona virus in Pakistan. *Results Phys.* **2020**, *19*, 103560. [CrossRef] [PubMed]
38. Gomez-Aguilar, J.F.; Cordova-Fraga, T.; Abdeljawad, T.; Khan, A.; Khan, H. Analysis of fractal-fractional Malaria transmission model. *Fractals* **2020**, *28*, 2040041. [CrossRef]
39. Ali, Z.; Rabiei, F.; Shah, K.; Khodadadi, T. Qualitative analysis of fractal-fractional order COVID-19 mathematical model with case study of Wuhan. *Alex. Eng. J.* **2021**, *60*, 477–489. [CrossRef]
40. Asamoah, J.K.K. Fractal-fractional model and numerical scheme based on Newton polynomial for Q fever disease under Atangana-Baleanu derivative. *Res. Phys.* **2022**, *34*, 105189. [CrossRef]
41. Khan, H.; Alam, K.; Gulzar, H.; Etemad, S.; Rezapour, S. A case study of fractal-fractional tuberculosis model in China: Existence and stability theories along with numerical simulations. *Math. Comput. Simul.* **2022**, *198*, 455–473. [CrossRef]
42. Ahmad, S.; Ullah, A.; Akgul, A.; De la Sen, M. Study of HIV disease and its association with immune cells under nonsingular and nonlocal fractal-fractional operator. *Complexity* **2021**, *2021*, 1904067. [CrossRef]
43. Granas, A.; Dugundji, J. *Fixed Point Theory*; Springer: New York, NY, USA, 2003.
44. Atangana, A.; Araz, S.I. *New Numerical Scheme with Newton Polynomial: Theory, Methods, and Applications*; Academic Press: Cambridge, MA, USA, 2021.
45. Baleanu, D.; Mohammadi, H.; Rezapour, S. Analysis of the model of HIV-1 infection of $CD4^+$ T-cell with a new approach of fractional derivative. *Adv. Differ. Equ.* **2020**, *2020*, 71. [CrossRef]

mathematics

MDPI

Article

Finite-Time Stability Analysis of Linear Differential Systems with Pure Delay

Ahmed M. Elshenhab [1,2,*], **Xingtao Wang** [1], **Omar Bazighifan** [3] and **Jan Awrejcewicz** [4,*]

1 School of Mathematics, Harbin Institute of Technology, Harbin 150001, China; xingtao@hit.edu.cn
2 Department of Mathematics, Faculty of Science, Mansoura University, Mansoura 35516, Egypt
3 Section of Mathematics, International Telematic University Uninettuno, CorsoVittorio Emanuele II, 39, 00186 Roma, Italy; o.bazighifan@gmail.com
4 Department of Automation, Biomechanics and Mechatronics, Lodz University of Technology, 1/15 Stefanowski St., 90-924 Lodz, Poland
* Correspondence: ahmedelshenhab@stu.hit.edu.cn (A.M.E.); jan.awrejcewicz@p.lodz.pl (J.A.)

Abstract: Nonhomogeneous systems governed by second-order linear differential equations with pure delay are considered. As an application, the exact solutions of these systems and their delayed matrix functions are used to obtain the finite-time stability results. Our results extend and improve some previous results by removing some restrictive conditions. Finally, an example is provided to illustrate our theoretical results.

Keywords: delayed matrix function; finite-time stability; delay differential equations

MSC: 34K20; 34K06

Citation: Elshenhab, A.M.; Wang, X.; Bazighifan, O.; Awrejcewicz, J. Finite-Time Stability Analysis of Linear Differential Systems with Pure Delay. *Mathematics* **2022**, *10*, 1359. https://doi.org/10.3390/math10091359

Academic Editors: Eva Kaslik, Mihaela Neamțu and Anca Rădulescu

Received: 6 March 2022
Accepted: 15 April 2022
Published: 19 April 2022

Publisher's Note: MDPI stays neutral with regard to jurisdictional claims in published maps and institutional affiliations.

Copyright: © 2022 by the authors. Licensee MDPI, Basel, Switzerland. This article is an open access article distributed under the terms and conditions of the Creative Commons Attribution (CC BY) license (https://creativecommons.org/licenses/by/4.0/).

1. Introduction

Numerous processes in mechanical and technological systems were described using delay differential equations. These systems are frequently utilized in the modeling of phenomena in technological and scientific problems. These models have applications in diffusion processes, forced oscillations, signal analysis, control theory, viscoelastic systems, modeling disease, biology, computer engineering, finance, and population dynamics. Time-delays are frequently associated with the economy, electric networks, physico-chemical processes, heredity in population growth, hydraulic networks, and other relevant industries. Generally, these mathematical models have a peculiarity, which is that the rate of change of these processes is determined by their history. On the other hand, in 2003, Khusainov and Shuklin [1] represented the solutions of linear delay differentiaL equations by constructing a new concept of a delayed exponential matrix function. In 2008, Khusainov et al. [2] adopted this approach to represent the solutions of an oscillating system with pure delay by establishing a delayed matrix sine and a delayed matrix cosine. This pioneering research yielded plenty of novel results on the exact solutions that were used in the stability analysis and control problems of time-delay systems; see for example [3–13] and the references therein.

Finite-time stability is a novel definition that involves a fixed finite-time interval and a prescribed constraint for the system, as opposed to the exponential/asymptotic stability definition, which is exposed to an infinite-time interval. In the literature, there has been a considerable interest in finite-time stability analysis of differential or fractional delay systems, and several methods for studying finite-time stability of differential or fractional delay systems have been developed; for example, fundamental matrix and the largest singular value of matrix coefficients [14], Lyapunov-like approach with Jensen's and Coppel's inequality [15], Grownwall's approach [16], method of steps [17], Hölder inequality [18], delayed Mittag–Leffler matrix function [19], Gronwall inequality [20], linear matrix inequality [21], the delayed matrix exponential and Jensen and Coppel inequalities [22], the delayed matrix exponential function and Gronwall integral inequalities [23], the delayed matrix cosine and sine [24], the explicit solution of the system [25], and conformable delayed matrix functions [26].

However, to the best of our knowledge, no study exists dealing with finite-time stability analysis of a system of second-order linear differential equations with pure delay of the form

$$\begin{aligned} y''(x) &= -By(x-h) + f(x), \quad \text{for } h > 0,\ x \in W := [0, L], \\ y(x) &\equiv \psi(x),\ y'(x) \equiv \psi'(x) \quad \text{for } -h \le x \le 0, \end{aligned} \tag{1}$$

where h is a delay, L is a pre-fixed positive number, $y(x) \in \mathbb{R}^n$, $\psi \in C^2([-h,0],\mathbb{R}^n)$, $B \in \mathbb{R}^{n\times n}$ is a constant nonzero matrix and $f \in C([0,\infty),\mathbb{R}^n)$ is a given function.

Very recently, Elshenhab and Wang [8] gave a new representation of solutions of (1) as follows:

$$\begin{aligned} y(x) = {} & \mathcal{H}_h(B(x-h))\psi(0) + \mathcal{M}_h(B(x-h))\psi'(0) \\ & - B\int_{-h}^{0} \mathcal{M}_h(B(x-2h-\varrho))\psi(\varrho)d\varrho \\ & + \int_{0}^{x} \mathcal{M}_h(B(x-h-\varrho))f(\varrho)d\varrho, \end{aligned} \tag{2}$$

and they also derived alternative formulas of solutions of (1) as follows:

$$\begin{aligned} y(x) = {} & \mathcal{H}_h(Bx)\psi(-h) + \mathcal{M}_h(Bx)\psi'(-h) \\ & + \int_{-h}^{0} \mathcal{M}_h(B(x-h-\varrho))\psi''(\varrho)d\varrho \\ & + \int_{0}^{x} \mathcal{M}_h(B(x-h-\varrho))f(\varrho)d\varrho, \end{aligned} \tag{3}$$

or

$$\begin{aligned} y(x) = {} & \mathcal{H}_h(Bx)\psi(-h) + \mathcal{M}_h(B(x-h))\psi'(0) \\ & + \int_{-h}^{0} \mathcal{H}_h(B(x-h-\varrho))\psi'(\varrho)d\varrho \\ & + \int_{0}^{x} \mathcal{M}_h(B(x-h-\varrho))f(\varrho)d\varrho, \end{aligned} \tag{4}$$

where $\mathcal{H}_h(Bx)$ and $\mathcal{M}_h(Bx)$ are called the delayed matrix functions formulated by

$$\mathcal{H}_h(Bx) := \begin{cases} \Theta, & -\infty < x < -h, \\ I, & -h \le x < 0, \\ I - B\frac{x^2}{2!}, & 0 \le x < h, \\ \vdots & \vdots \\ I - B\frac{x^2}{2!} + B^2\frac{(x-h)^4}{4!} & \\ + \cdots + (-1)^m B^m \frac{(x-(m-1)h)^{2m}}{(2m)!}, & (m-1)h \le x < mh, \end{cases} \tag{5}$$

$$\mathcal{M}_h(Bx) := \begin{cases} \Theta, & -\infty < x < -h, \\ I(x+h), & -h \le x < 0, \\ I(x+h) - B\frac{x^3}{3!}, & 0 \le x < h, \\ \vdots & \vdots \\ I(x+h) - B\frac{x^3}{3!} + B^2\frac{(x-h)^5}{5!} & \\ + \cdots + (-1)^m B^m \frac{(x-(m-1)h)^{2m+1}}{(2m+1)!}, & (m-1)h \le x < mh, \end{cases} \tag{6}$$

respectively, where $m = 0, 1, 2, \ldots$, the notation I is the $n \times n$ identity matrix and Θ is the $n \times n$ null matrix.

Motivated by [8,24], as an application, the explicit formulas of solutions of the system (1) and the delayed matrix functions are used to obtain finite-time stability results on $W = [0, L]$.

The rest of this paper is arranged as follows: In Section 2, we present some basic definitions and estimations of norms for the delayed matrix functions, which are used while discussing finite-time stability. In Section 3, as an application, the representation of the solutions of (1) are used to obtain finite-time stability results. Finally, we give an example to illustrate the main results.

2. Preliminaries

Throughout the paper, we denote the vector norm as $\|y\| = \sum_{i=1}^{n} |y_i|$ and the matrix norm as $\|B\| = \max_{1 \le j \le n} \sum_{i=1}^{n} |a_{ij}|$; y_i and a_{ij} are the elements of the vector y and the matrix B, respectively. Denote $C(W, \mathbb{R}^n)$ the Banach space of vector-valued continuous function from $W \to \mathbb{R}^n$ endowed with the norm $\|y\|_C = \max_{x \in W} \|y(x)\|$ for a norm $\|\cdot\|$ on $\mathbb{R}^n$. We introduce a space $C^1(W, \mathbb{R}^n) = \{y \in C(W, \mathbb{R}^n) : y' \in C(W, \mathbb{R}^n)\}$. Furthermore, we see $\|\psi\|_C = \max_{\varrho \in [-h, 0]} \|\psi(\varrho)\|$.

We recall some basic definitions used further in this paper.

Definition 1 ([27]). *The two-parameter Mittag–Leffler function is given by*

$$\mathbb{E}_{\alpha,\gamma}(z) = \sum_{k=0}^{\infty} \frac{z^k}{\Gamma(\alpha k + \gamma)}, \quad \alpha, \ \gamma > 0, \ z \in \mathbb{C}.$$

Especially, if $\gamma = 1$, then

$$\mathbb{E}_{\alpha,1}(z) = \mathbb{E}_{\alpha}(z) = \sum_{k=0}^{\infty} \frac{z^k}{\Gamma(\alpha k + 1)}, \quad \alpha > 0.$$

Definition 2 ([16]). *The system (1) is finite-time stable with respect to $\{0, W, h, \delta, \rho\}$, $\delta < \rho$ if and only if $\varrho < \delta$ implies $\|y(x)\| < \rho$ for all $x \in W$, where $\varrho = \max\{\|\psi\|_C, \|\psi'\|_C, \|\psi''\|_C\}$ and δ, ρ are real positive numbers.*

To conclude this section, we provide estimations of norms for the delayed matrix functions, which are used in discussing finite-time stability.

Lemma 1. *For any $x \in [(m-1)h, mh]$, $m = 1, 2, \ldots$, we have*

$$\|\mathcal{H}_h(Bx)\| \le \mathbb{E}_2\left(\|B\|x^2\right).$$

Proof. Using (5), we get

$$\begin{aligned}
\|\mathcal{H}_h(Bx)\| &\le 1 + \|B\|\frac{x^2}{2!} + \|B\|^2\frac{(x-h)^4}{4!} \\
&\quad + \cdots + \|B\|^m\frac{(x-(m-1)h)^{2m}}{(2m)!} \\
&\le 1 + \|B\|\frac{x^2}{2!} + \|B\|^2\frac{x^4}{4!} + \cdots + \|B\|^m\frac{x^{2m}}{(2m)!} \\
&\le \sum_{k=0}^{\infty} \frac{\left(\|B\|x^2\right)^k}{(2k)!} = \mathbb{E}_2\left(\|B\|x^2\right).
\end{aligned}$$

This completes the proof. □

Lemma 2. *For any $x \in [(m-1)h, mh]$, $m = 1, 2, \ldots$, we have*

$$\|\mathcal{M}_h(Bx)\| \leq (x+h)\mathbb{E}_{2,2}\Big(\|B\|(x+h)^2\Big).$$

Proof. Using (6), we get

$$\begin{aligned}\|\mathcal{M}_h(Bx)\| &\leq (x+h) + \|B\|\frac{x^3}{3!} + \|B\|^2\frac{(x-h)^5}{5!} \\ &+ \cdots + \|B\|^m\frac{(x-(m-1)h)^{2m+1}}{(2m+1)!} \\ &\leq (x+h) + \|B\|\frac{(x+h)^3}{3!} + \|B\|^2\frac{(x+h)^5}{5!} \\ &+ \cdots + \|B\|^m\frac{(x+h)^{2m+1}}{(2m+1)!} \\ &\leq \sum_{k=0}^{\infty}\frac{\left[\|B\|(x+h)^2\right]^k(x+h)}{(2k+1)!} = (x+h)\mathbb{E}_{2,2}\Big(\|B\|(x+h)^2\Big).\end{aligned}$$

This completes the proof. □

3. Main Results

In this section, we derived finite-time stability results of (1) by making use of the three possible formulas of solutions (2), (3) and (4), respectively.

Theorem 1. *The system (1) is finite-time stable with respect to $\{0, W, h, \delta, \rho\}$, if*

$$\mathbb{E}_2\Big(\|B\|(L-h)^2\Big) < \frac{\rho - \Big(\delta L + \frac{L^2}{2}(\delta\|B\| + \|f\|_C)\Big)\mathbb{E}_{2,2}(\|B\|L^2)}{\delta}. \tag{7}$$

Proof. By using Definition 2 and (2), we have $\varrho < \delta$ and

$$\begin{aligned}\|y(x)\| &\leq \|\mathcal{H}_h(B(x-h))\|\|\psi(0)\| + \|\mathcal{M}_h(B(x-h))\|\|\psi'(0)\| \\ &+ \|B\|\left\|\int_{-h}^{0}\mathcal{M}_h(B(x-2h-\varrho))\psi(\varrho)d\varrho\right\| \\ &+ \left\|\int_0^x\mathcal{M}_h(B(x-h-\varrho))f(\varrho)d\varrho\right\| \\ &\leq \|\mathcal{H}_{h,\alpha}(B(x-h))\|\|\psi(0)\| + \|\mathcal{M}_h(B(x-h))\|\|\psi'(0)\| \\ &+ \|B\|\int_{-h}^{0}\|\mathcal{M}_h(B(x-2h-\varrho))\|\|\psi(\varrho)\|d\varrho \\ &+ \int_0^x\|\mathcal{M}_h(B(x-h-\varrho))\|\|f(\varrho)\|d\varrho \\ &\leq \delta\|\mathcal{H}_{h,\alpha}(B(x-h)^\alpha)\| + \delta\|\mathcal{M}_{h,\alpha}(B(x-h)^\alpha)\| \\ &+ \delta\|B\|\int_{-h}^{0}\|\mathcal{M}_h(B(x-2h-\varrho))\|d\varrho \\ &+ \|f\|_C\int_0^x\|\mathcal{M}_h(B(x-h-\varrho))\|d\varrho.\end{aligned} \tag{8}$$

From Lemma 2, we have

$$\begin{aligned}\|\mathcal{M}_h(B(x-2h-\varrho))\| &\leq (x-h-\varrho)\mathbb{E}_{2,2}\Big(\|B\|(x-h-\varrho)^2\Big) \\ &\leq (x-h-\varrho)\mathbb{E}_{2,2}\Big(\|B\|x^2\Big),\end{aligned} \tag{9}$$

for $-h \leq \varrho \leq 0$, $x \in W$, and since $\mathbb{E}_{2,2}(\|B\|x^2)$ is increasing function when $x \geq 0$. From (9), we get

$$\int_{-h}^{0} \|\mathcal{M}_h(B(x-2h-\varrho)^{\alpha})\| d\varrho \leq \frac{x^2}{2}\mathbb{E}_{2,2}\Big(\|B\|x^2\Big), \tag{10}$$

and

$$\begin{aligned}\int_{0}^{x} \|\mathcal{M}_h(B(x-h-\varrho)^{\alpha})\| d\varrho &\leq \mathbb{E}_{2,2}\Big(\|B\|x^2\Big) \int_{0}^{x} (x-\varrho) d\varrho \\ &= \frac{x^2}{2}\mathbb{E}_{2,2}\Big(\|B\|x^2\Big). \end{aligned} \tag{11}$$

From (8), (10) and (11), we have

$$\begin{aligned}\|y(x)\| \leq{}& \delta\mathbb{E}_2\Big(\|B\|(x-h)^2\Big) + \delta x\mathbb{E}_{2,2}\Big(\|B\|x^2\Big) \\ &+ \frac{x^2}{2}(\delta\|B\| + \|f\|_C)\mathbb{E}_{2,2}\Big(\|B\|x^2\Big), \end{aligned} \tag{12}$$

for all $x \in W$. Combining (7) with (12), we get $\|y(x)\| < \rho$ for all $x \in W$. This ends the proof. □

Theorem 2. *The system (1) is finite-time stable with respect to $\{0, W, h, \delta, \rho\}$, if*

$$\mathbb{E}_2\Big(\|B\|L^2\Big) < \frac{\rho - \frac{\delta(L+h)(L+h+2)}{2}\mathbb{E}_{2,2}\Big(\|B\|(L+h)^2\Big) - \frac{L^2\|f\|_C}{2}\mathbb{E}_{2,2}(\|B\|L^2)}{\delta}. \tag{13}$$

Proof. By using Definition 2 and (3), we have $\varrho < \delta$ and

$$\begin{aligned}\|y(x)\| \leq{}& \|\mathcal{H}_h(Bx)\|\|\psi(-h)\| + \|\mathcal{M}_h(Bx)\|\|\psi'(-h)\| \\ &+ \left\|\int_{-h}^{0} \mathcal{M}_h(B(x-h-\varrho))\psi''(\varrho)d\varrho\right\| \\ &+ \left\|\int_{0}^{x} \mathcal{M}_h(B(x-h-\varrho))f(\varrho)d\varrho\right\| \\ \leq{}& \delta\|\mathcal{H}_h(Bx)\| + \delta\|\mathcal{M}_h(Bx)\| + \delta\int_{-h}^{0}\|\mathcal{M}_h(B(x-h-\varrho))\|d\varrho \\ &+ \|f\|_C\int_{0}^{x}\|\mathcal{M}_h(B(x-h-\varrho))\|d\varrho. \end{aligned} \tag{14}$$

From Lemma 2, we have

$$\begin{aligned}\int_{-h}^{0}\|\mathcal{M}_h(B(x-h-\varrho))\|d\varrho &\leq \mathbb{E}_{2,2}\Big(\|B\|(x+h)^2\Big)\int_{-h}^{0}(x-\varrho)d\varrho \\ &\leq \frac{(x+h)^2}{2}\mathbb{E}_{2,2}\Big(\|B\|(x+h)^2\Big). \end{aligned} \tag{15}$$

From (11), (14) and (15), we get

$$\begin{aligned}\|y(x)\| \leq{}& \delta\mathbb{E}_2\Big(\|B\|x^2\Big) + \delta(x+h)\mathbb{E}_{2,2}\Big(\|B\|(x+h)^2\Big) \\ &+ \frac{\delta(x+h)^2}{2}\mathbb{E}_{2,2}\Big(\|B\|(x+h)^2\Big) + \frac{\|f\|_C}{2}x^2\mathbb{E}_{2,2}\Big(\|B\,|x^2\Big) \end{aligned} \tag{16}$$

for all $x \in W$. Combining (13) with (16), we have $\|y(x)\| < \rho$ for all $x \in W$. This ends the proof. □

Theorem 3. *The system (1) is finite-time stable with respect to* $\{0, W, h, \delta, \rho\}$, *if*

$$\mathbb{E}_2\Big(\|B\|L^2\Big) < \frac{\rho - L\Big(\delta + \frac{\|f\|_C}{2}L\Big)\mathbb{E}_{2,2}(\|B\|L^2)}{\delta(1+h)}. \tag{17}$$

Proof. By using Definition 2 and (4), we have $\eta < \delta$ and

$$\begin{aligned}\|y(x)\| &\le \|\mathcal{H}_h(Bx)\|\|\psi(-h)\| + \|\mathcal{M}_h(B(x-h))\|\|\psi'(0)\| \\ &+ \left\|\int_{-h}^{0} \mathcal{H}_h(B(x-h-\varrho))\psi'(\varrho)d\varrho\right\| \\ &+ \left\|\int_{0}^{x} \mathcal{M}_h(B(x-h-\varrho))f(\varrho)d\varrho\right\| \\ &\le \delta\|\mathcal{H}_h(Bx)\| + \delta\|\mathcal{M}_h(B(x-h))\| \\ &+ \delta\int_{-h}^{0}\|\mathcal{H}_h(B(x-h-\varrho))\|d\varrho \\ &+ \|f\|_C\int_{0}^{x}\|\mathcal{M}_h(B(x-h-\varrho))\|d\varrho.\end{aligned} \tag{18}$$

From Lemma 1, we have

$$\int_{-h}^{0}\|\mathcal{H}_h(B(x-h-\varrho))\|d\varrho \le h\mathbb{E}_2\Big(\|B\|x^2\Big). \tag{19}$$

From (11), (18) and (19), we get

$$\begin{aligned}\|y(x)\| &\le \delta\mathbb{E}_2\Big(\|B\|x^2\Big) + \delta x\mathbb{E}_{2,2}\Big(\|B\|x^2\Big) \\ &+ \delta h\mathbb{E}_2\Big(\|B\|x^2\Big) + \frac{\|f\|_C}{2}x^2\mathbb{E}_{2,2}\Big(\|B\|x^2\Big).\end{aligned} \tag{20}$$

for all $x \in W$. Combining (17) with (20), we have $\|y(x)\| < \rho$ for all $x \in W$. This ends the proof. □

Remark 1. *We see that by dropping the nonsingularity criterion on a matrix coefficient B and making the matrix B an arbitrary, not necessarily squared matrix* B^2, *our results in Theorems 1–3 improve and extend the corresponding results in Theorems 3.1–3.3 in [24].*

4. An Example

Consider the delay differential equations

$$\begin{aligned} y''(x) &= -By(x-0.5) + f(x), \ \ x \in [0,1], \\ \psi(x) = (0.1x^2, 0.2x)^T, \ \psi'(x) &= (0.2x, 0.2)^T, \ \psi''(x) = (0.2, 0)^T, \ \ -0.5 \le x \le 0,\end{aligned} \tag{21}$$

where

$$h = 0.5, \ \ B = \begin{pmatrix} 2 & 0 \\ 0 & 0 \end{pmatrix}, \ \ f(x) = \begin{pmatrix} 1 \\ 2 \end{pmatrix}.$$

From (3), for all $0 \le x \le 1$, and through a basic calculation, we can obtain

$$\begin{aligned} y(x) &= \begin{pmatrix} 0.025\mathcal{H}_{0.5}(2x) \\ -0.1 \end{pmatrix} + \begin{pmatrix} -0.1\mathcal{M}_{0.5}(2x) \\ 0.2(x+0.5) \end{pmatrix} \\ &+ \begin{pmatrix} 0.2\int_{-0.5}^{0}\mathcal{M}_{0.5}(2(x-0.5-\varrho))d\varrho \\ 0 \end{pmatrix} \\ &+ \begin{pmatrix} \int_0^x \mathcal{M}_{0.5}(2(x-0.5-\varrho))d\varrho \\ 2\int_0^x (x-\varrho)d\varrho \end{pmatrix} = \begin{pmatrix} y_1(x) \\ y_2(x) \end{pmatrix},\end{aligned}$$

which implies that

$$\begin{aligned} y_1(x) &= 0.025\mathcal{H}_{0.5}(2x) - 0.1\mathcal{M}_{0.5}(2x) \\ &+ 0.2\int_{-0.5}^{0} \mathcal{M}_{0.5}(2(x-0.5-\varrho))d\varrho \\ &+ \int_{0}^{x} \mathcal{M}_{0.5}(2(x-0.5-\varrho))d\varrho, \end{aligned}$$

and

$$y_2(x) = x^2 + \frac{1}{5}x,$$

where

$$\mathcal{H}_{0.5}(2x) = \begin{cases} 1, & -0.5 \le x < 0, \\ 1 - x^2, & 0 \le x < 0.5, \\ 1 - x^2 + \frac{1}{6}(x-0.5)^4, & 0.5 \le x < 1, \end{cases}$$

and

$$\mathcal{M}_{0.5}(2x) = \begin{cases} (x+0.5), & -0.5 \le x < 0, \\ (x+0.5) - \frac{1}{3}x^3, & 0 \le x < 0.5, \\ (x+0.5) - \frac{1}{3}x^3 + \frac{1}{30}(x-0.5)^5, & 0.5 \le x < 1. \end{cases}$$

Thus, the explicit solutions of (21) are

$$\begin{aligned} y_1(x) &= 0.025\mathcal{H}_{0.5}(2x) - 0.1\mathcal{M}_{0.5}(2x) \\ &+ 0.2\int_{-0.5}^{x-0.5} \mathcal{M}_{0.5}(2(x-0.5-\varrho))d\varrho \\ &+ 0.2\int_{x-0.5}^{0} \mathcal{M}_{0.5}(2(x-0.5-\varrho))d\varrho \\ &+ \int_{0}^{x} \mathcal{M}_{0.5}(2(x-0.5-\varrho))d\varrho, \end{aligned}$$

$$y_2(x) = x^2 + \frac{1}{5}x,$$

where $0 \le x \le 0.5$, which implies that

$$y_1(x) = -\frac{1}{60}x^4 + \frac{1}{30}x^3 + \frac{19}{40}x^2,$$

$$y_2(x) = x^2 + \frac{1}{5}x,$$

and

$$\begin{aligned} y_1(x) &= 0.025\mathcal{H}_{0.5}(2x) - 0.1\mathcal{M}_{0.5}(2x) \\ &+ 0.2\int_{-0.5}^{x-1} \mathcal{M}_{0.5}(2(x-0.5-\varrho))d\varrho \\ &+ 0.2\int_{x-1}^{0} \mathcal{M}_{0.5}(2(x-0.5-\varrho))d\varrho \\ &+ \int_{0}^{x-0.5} \mathcal{M}_{0.5}(2(x-0.5-\varrho))d\varrho \\ &+ \int_{x-0.5}^{x} \mathcal{M}_{0.5}(2(x-0.5-\varrho))d\varrho, \end{aligned}$$

$$y_2(x) = x^2 + \frac{1}{5}x,$$

where $0.5 \leq x \leq 1$, which implies that

$$
\begin{aligned}
y_1(x) = {} & \frac{1}{900}(x-0.5)^6 - \frac{1}{300}(x-0.5)^5 - \frac{1}{16}(x-0.5)^4 \\
& + \frac{19}{40}x^2 + \frac{1}{30}x^3 - \frac{1}{60}x^4,
\end{aligned}
$$

$$y_2(x) = x^2 + \frac{1}{5}x.$$

By calculating we get $\eta = \max\{\|\psi\|_C, \|\psi'\|_C, \|\psi''\|_C\} = 0.3$, $\|B\| = 2$, $\|f\|_C = 3$, $\mathbb{E}_2(2L^2) = 2.1782$, $\mathbb{E}_{2,2}\left(2(L+0.5)^2\right) = 1.938$, $\mathbb{E}_{2,2}(2L^2) = 1.3683$, then we choose $\delta = 0.31 > 0.3 = \eta$. Figure 1 shows the state $y(x)$ and the norm $\|y(x)\|$ of (21). Now, Theorems 1–3 imply that $\|y(x)\| \leq 3.29158$, $\|y(x)\| \leq 4.3047395$ and $\|y(x)\| \leq 3.489486$, respectively, we simply take $\rho = 3.3, 4.31, 3.49$, respectively. Table 1 shows the data.

We can see $\|y(x)\| < \rho$ for all $x \in W$ and (21) is finite-time stable under Theorems 1–3. Concerning the definition of finite-time stable, we need to determine a specific threshold ρ. By checking the value of ρ in Theorems 1–3, we find that in this example the result of Theorem 1 is the optimal.

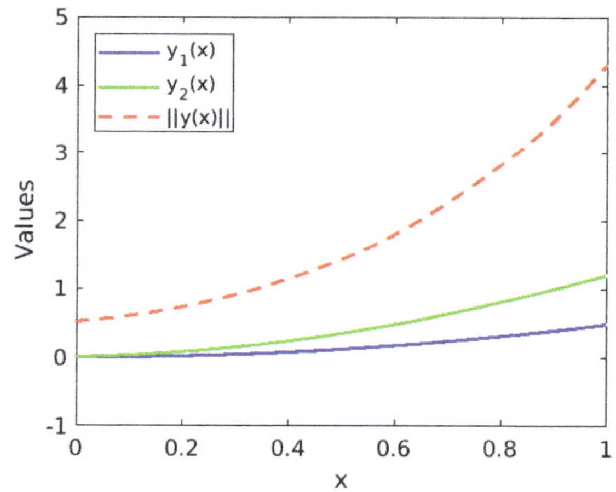

Figure 1. The state $y(x)$ and $||y(x)||$ of (21).

Table 1. Finite-time stability results of (21) and fixed the time $L = 1$.

Theorem	L	$\|B\|$	δ	$\|y(x)\|$	ρ	h	Finite-Time Stability
1	1	2	0.31	$\leq$3.29158	3.3 (optimal)	0.5	Yes
2	1	2	0.31	$\leq$4.30474	4.31	0.5	Yes
3	1	2	0.31	$\leq$3.48949	3.49	0.5	Yes

Remark 2. *We note that Theorems 3.1–3.3 in [24] cannot be applied to (21) because the matrix B is singular, and an arbitrary, not necessarily squared matrix* B^2.

5. Conclusions

In this work, by making use of three possible formulas of solutions of nonhomogeneous systems governed by second-order linear delay differential equations, and estimations of norms for the delayed matrix functions, we derived finite-time stability results of these systems. Finally, we provided an example to demonstrate the effectiveness of the obtained results. The results are applicable to all singular, non-singular and arbitrary matrices, not necessarily squared. Consequently, our results improve and extend upon the existing results in [24].

One possible direction in which to extend the results of this paper is toward that of fractional differential and conformable fractional differential systems of order $\alpha \in (1,2]$.

Author Contributions: Conceptualization, A.M.E., X.W., O.B. and J.A.; Formal analysis, A.M.E., X.W., O.B. and J.A.; Investigation, A.M.E., X.W. and O.B.; Methodology, A.M.E., X.W. and J.A.; Project administration, A.M.E.; Resources, A.M.E., O.B. and J.A.; Software, A.M.E.; Supervision, X.W.; Validation, A.M.E. and X.W.; Visualization, A.M.E. and J.A.; Writing—original draft, A.M.E., X.W., O.B. and J.A.; Writing—review & editing, A.M.E., X.W., O.B. and J.A. All authors have read and agreed to the published version of the manuscript.

Funding: This research received no external funding.

Acknowledgments: The authors sincerely appreciate the editor and anonymous referees for their careful reading and helpful comments to improve this paper.

Conflicts of Interest: The authors declare no conflict of interest.

References

1. Khusainov, D.Y.; Shuklin, G.V. Linear autonomous time-delay system with permutation matrices solving. *Stud. Univ. Zilina Math. Ser.* **2003**, *17*, 101–108.
2. Khusainov, D.Y.; Diblík, J.; Růžičková, M.; Lukáčová, J. Representation of a solution of the Cauchy problem for an oscillating system with pure delay. *Nonlinear Oscil.* **2008**, *11*, 276–285. [CrossRef]
3. Diblík, J.; Fečkan, M.; Pospíšil, M. Representation of a solution of the Cauchy problem for an oscillating system with multiple delays and pairwise permutable matrices. *Abstr. Appl. Anal.* **2013**, *2013*, 931493. [CrossRef]
4. Diblík, J.; Fečkan, M.; Pospíšil, M. On the new control functions for linear discrete delay systems. *SIAM J. Control Optim.* **2014**, *52*, 1745–1760. [CrossRef]
5. Diblík, J.; Khusainov, D.Y.; Baštinec, J.; Sirenko, A.S. Exponential stability of linear discrete systems with constant coefficients and single delay. *Appl. Math. Lett.* **2016**, *51*, 68–73. [CrossRef]
6. Diblík, J.; Mencáková, K. Representation of solutions to delayed linear discrete systems with constant coefficients and with second-order differences. *Appl. Math. Lett.* **2020**, *105*, 106309. [CrossRef]
7. Elshenhab, A.M.; Wang, X.T. Representation of solutions for linear fractional systems with pure delay and multiple delays. *Math. Meth. Appl. Sci.* **2021**, *44*, 12835–12850. [CrossRef]
8. Elshenhab, A.M.; Wang, X.T. Representation of solutions of linear differential systems with pure delay and multiple delays with linear parts given by non-permutable matrices. *Appl. Math. Comput.* **2021**, *410*, 126443. [CrossRef]
9. Elshenhab, A.M.; Wang, X.T. Representation of solutions of delayed linear discrete systems with permutable or nonpermutable matrices and second-order differences. *RACSAM Rev. R. Acad. Cienc. Exactas Fís. Nat. Ser. A Mat.* **2022**, *116*, 58. [CrossRef]
10. Li, M.; Wang, J.R. Exploring delayed Mittag–Leffler type matrix functions to study finite time stability of fractional delay differential equations. *Appl. Math. Comput.* **2018**, *324*, 254–265. [CrossRef]
11. Liu, L.; Dong, Q.; Li, G. Exact solutions and Hyers–Ulam stability for fractional oscillation equations with pure delay. *Appl. Math. Lett.* **2021**, *112*, 106666. [CrossRef]
12. Nawaz, M.; Jiang, W.; Sheng, J. The controllability of nonlinear fractional differential system with pure delay. *Adv. Differ. Equ.* **2020**, *2020*, 30. [CrossRef]
13. Elshenhab, A.M.; Wang, X.T. Controllability and Hyers–Ulam stability of differential systems with pure delay *Mathematics* **2022**, *10*, 1248. [CrossRef]
14. Lazarević, M.P.; Debeljković, D.; Nenadić, Z. Finite-time stability of delayed systems. *IMA J. Math. Control Inf* **2000**, *17*, 101–109. [CrossRef]
15. Debeljković, D.L.; Stojanovic, S.B.; Jovanović, A.M. Finite-time stability of continuous time delay systems: Lyapunov-like approach with Jensen's and Coppel's inequality. *Acta Polytech. Hung.* **2013**, *10*, 135–150.
16. Lazarević, M.P.; Spasić, A.M. Finite-time stability analysis of fractional order time-delay system: Grownwall's approach. *Math. Comput. Model.* **2009**, *49*, 475–481. [CrossRef]
17. Du, F.; Jia, B. Finite-time stability of nonlinear fractional order systems with a constant delay. *J. Nonlinear Model. Anal.* **2020**, *2*, 1–13.
18. Du, F.; Jia, B. Finite-time stability of a class of nonlinear fractional delay difference systems. *Appl. Math. Lett.* **2019**, *98*, 233–239. [CrossRef]
19. Li, M.; Wang, J.R. Finite time stability of fractional delay differential equations. *Appl. Math. Lett.* **2017**, *64*, 170–176. [CrossRef]
20. Phat, V.N.; Thanh, N.T. New criteria for finite-time stability of nonlinear fractional-order delay systems: A Gronwall inequality approach. *Appl. Math. Lett.* **2018**, *83*, 169–175. [CrossRef]

21. Thanh, N.T.; Phat, V.N. Improved approach for finite-time stability of nonlinear fractional-order systems with interval time-varying delay. *IEEE Trans Circuits Syst. II Exp. Briefs* **2019**, *66*, 1356–1360. [CrossRef]
22. Luo, Z.; Wang, J. Finite time stability analysis of systems based on delayed exponential matrix. *J. Appl. Math. Comput.* **2017**, *55*, 335–351. [CrossRef]
23. Luo, Z.; Wei, W.; Wang, J. Finite time stability of semilinear delay differential equations. *Nonlinear Dyn.* **2017**, *89*, 713–722. [CrossRef]
24. Liang, C.; Wei, W.; Wang, J. Stability of delay differential equations via delayed matrix sine and cosine of polynomial degrees. *Adv. Differ. Equ.* **2017**, *2017*, 1–17. [CrossRef]
25. Cao, X.; Wang, J. Finite-time stability of a class of oscillating systems with two delays. *Math. Meth. Appl. Sci.* **2018**, *41*, 4943–4954. [CrossRef]
26. Elshenhab, A.M.; Wang, X.T.; Mofarreh, F.; Bazighifan, O. Exact solutions and finite time stability of linear conformable fractional systems with pure delay. *CMES* 2022, *in press*.
27. Kilbas, A.A.; Srivastava, H.M.; Trujillo, J.J. *Theory and Applications of Fractional Differential Equations*; Elsevier Science BV: Amsterdam, The Netherkands, 2006.

mathematics

Article

Statistical Analysis of Current Financial Instrument Quotes in the Conditions of Market Chaos

Alexander Musaev [1], Andrey Makshanov [2] and Dmitry Grigoriev [3,*]

[1] Saint-Petersburg State Institute of Technology, Technical University, St. Petersburg Institute for Informatics and Automation of the Russian Academy of Sciences, 190013 St. Petersburg, Russia; amusaev@technolog.edu.ru
[2] Department of Computing Systems and Computer Science, Admiral Makarov State University of Maritime and Inland Shipping, 198035 St. Petersburg, Russia; andrey.makshanov@mail.ru
[3] Center of Econometrics and Business Analytics (CEBA), St. Petersburg State University, 199034 St. Petersburg, Russia
* Correspondence: d.a.grigoriev@spbu.ru

Abstract: In this paper, the problem of estimating the current value of financial instruments using multidimensional statistical analysis is considered. The research considers various approaches to constructing regression computational schemes using quotes of financial instruments correlated to the data as regressors. An essential feature of the problem is the chaotic nature of its observation series, which is due to the instability of the probabilistic structure of the initial data. These conditions invalidate the constraints under which traditional statistical estimates remain non-biased and effective. Violation of experiment repeatability requirements obstructs the use of the conventional data averaging approach. In this case, numeric experiments become the main method for investigating the efficiency of forecasting and analysis algorithms of observation series. The empirical approach does not provide guaranteed results. However, it can be used to build sufficiently effective rational strategies for managing trading operations.

Keywords: stochastic chaos; multidimensional statistical analysis; multi-regression estimation; sliding observation window; asset management

Citation: Musaev, A.; Makshanov, A.; Grigoriev, D. Statistical Analysis of Current Financial Instrument Quotes in the Conditions of Market Chaos. *Mathematics* **2022**, *10*, 587. https://doi.org/10.3390/math10040587

Academic Editors: Manuel Alberto M. Ferreira and David Barilla

Received: 17 December 2021
Accepted: 11 February 2022
Published: 14 February 2022

Publisher's Note: MDPI stays neutral with regard to jurisdictional claims in published maps and institutional affiliations.

Copyright: © 2022 by the authors. Licensee MDPI, Basel, Switzerland. This article is an open access article distributed under the terms and conditions of the Creative Commons Attribution (CC BY) license (https://creativecommons.org/licenses/by/4.0/).

1. Introduction

The principal problem of managing any objects in conditions of non-stationarity, non-uniformity, and/or chaotic dynamics is the lack of repeatability, which obstructs the use of conventional statistical research techniques. In this case, statistical extrapolation, which is the basis of automatic generation of management decisions, turns out to be ineffective, or even inapplicable [1–3]. Hence, there is a need to search for regularizing effects that stabilize (at least locally) the prediction process in a given range of changes in the values of characteristic parameters.

This research considers a multidimensional time series of current readings of a number of financial instruments that evaluate, each from its own viewpoint, a certain market asset. The current market value of an asset is monitored based on the readings of the corresponding instrument. These readings can be adjusted, taking into account other similar instruments. In principle, all readings have a common base and are mutually correlated. The hypothesis is that the system of correlations between instruments has a significant degree of inertia [4–6] and is approximately preserved with a small time lag.

Within the framework of this article, correlations between quotations of instruments are used as a regularizing mechanism. Similarly to other probabilistic characteristics of observation series, the parameters of the relationship change over time, which corresponds to the non-stationarity of the observed processes. However, the variability in correlations is significantly lower than that of the initial quotation changes. Moreover, with the growth of the observation window used to estimate the relationship parameters, the correlation characteristics stabilize near asymptotically stable values.

Hence, an assumption arises about the possibility of using these relationships to build multi-regression indicators that reflect market estimates of the current value of financial instruments. The mismatch of the market estimate of an instrument with its current value allows us to obtain an idea of the further dynamics of its quotation, which, as a rule, approaches equalizing the value of the instrument in relation to the "fair" market price. This, in turn, creates the basis for building a proactive management strategy based on this discrepancy.

It would be irrational to evaluate the quality of forecasting in conditions of market chaos using conventional indicators of the effectiveness of statistical estimates due to the above-mentioned features of the considered observation series. Rather, the effectiveness of such estimates should be assessed via terminal performance indicators of management strategies based on them. The corresponding numerical studies are provided in this paper.

This article is structured as follows. Section 2 contains the main conceptual research contributions of this work. It presents quotation data on several financial instruments, which demonstrate that they are highly correlated. Further on, it contains a detailed description of the proposed problem, a basic asset management strategy that is based on corrected estimates of a selected indicator. Next, the task specifics that decrease the effectiveness of traditional statistical methods are considered as well as the approaches to improving the estimate quality via selecting the indicator kernel and removing some of the outliers. Section 3 contains the resulting scheme for correcting estimates using an evolutionary algorithm. Finally, Sections 4 and 5 are dedicated to a discussion and our conclusions, respectively.

2. Analyzed Data and Methodology

2.1. Data Model

Papers [7–10] justify describing quotation observation series on the basis of the Wald theorem by an additive model

$$y_k = x_k + v_k, k = 1, \dots, n, \tag{1}$$

where x_k, $k = 1, \dots, n$ is a smoothed system component used for constructing management strategies and v_k, $k = 1, \dots, n$ is noise. Similarly, instead of y_k, first finite differences of their logarithms could be used (i.e., the ARCH or GARCH models) [11,12], but this complicates returning to an estimation of the initial y_k. Let us highlight two important features of the presented observation model:

- the system component is comprised of an oscillatory non-periodic observation series typical for non-linear dynamics processes; and
- the noise v_k, $k = 1, \dots, n$ is a non-stationary random process roughly described by the Gaussian model with fluctuating parameters [9–15].

As illustrations justifying the choice of the model, Figure 1 shows examples of quotations of financial assets on 10-day observation intervals.

To isolate the system component, an exponential filter

$$x_k = \alpha y_k + (1 - \alpha) y_{(k-1)} = x_{(k-1)} + \alpha(y_k - x_{(k-1)}), k = 2, \dots, n \tag{2}$$

with a smoothing coefficient $\alpha = 0.01$–0.3 was used.

The observed process y_k, $k = 1, \dots, n$ does not conform to the efficient market hypothesis [7] and, as shown in [16,17], has almost no inertia. The latter statement leads to a complete failure of management strategies based on mechanistic prolongation of the detected trends. At the same time, useful patterns can also be found in the system of correlations of trading asset quotations, as they are represented by multivariate observation series. In particular, increasing the observation interval makes it possible to detect stable correlations between various instruments of the currency and other markets [18,19].

As an example illustrating the last statement, Figure 2 shows time-synchronized charts of quotations of five different currency instruments on an 85-day observation area.

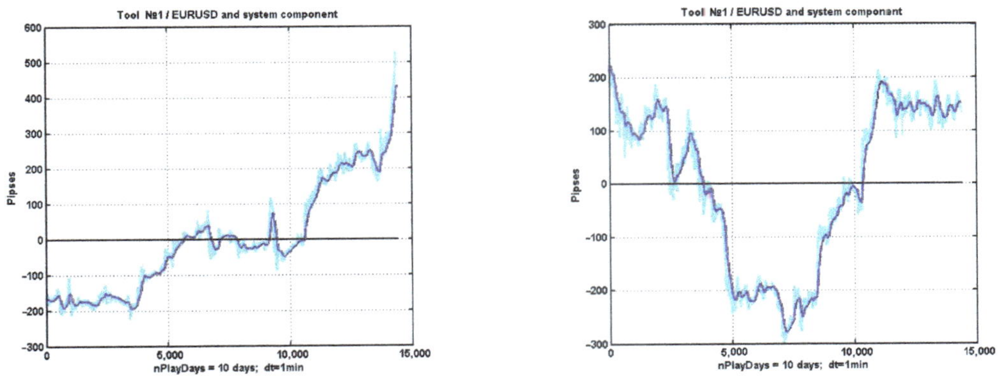

Figure 1. Two examples of observing the quotation of the EURUSD currency pair at two 10-day intervals with the system component identified by means of exponential filtering.

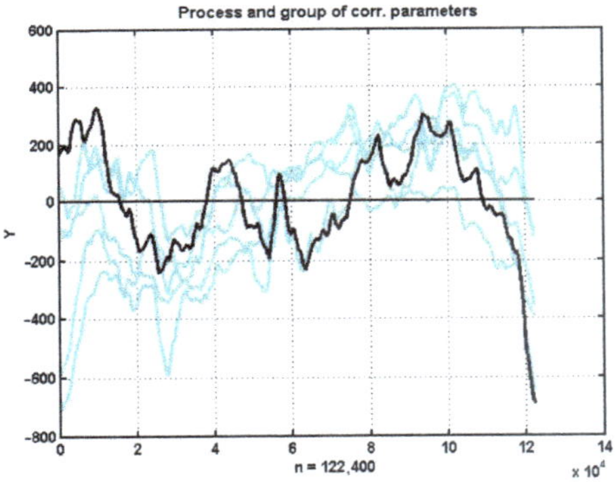

Figure 2. The process of quote change in the EURUSD currency pair and its five most correlated currency instruments' quotes.

Figure 3 shows a tonal matrix reflecting the values of estimated pairwise correlations in the same observation interval for 16 currency pairs presented in Table 1.

Figure 4 shows a similar tonal matrix for the pairwise correlations of EURUSD and its five most correlated currency pairs (#11: GBPUSD, #14: CHFJPY, #16: NZDJPY, #2: EURJPY, and #12: AUDJPY). These exact instruments are plotted in Figure 2.

Table 1. Financial instruments and their respective numbers.

Financial Instruments					
1	2	3	4	5	6
EURUSD	EURJPY	EURGBP	EURCHF	EURCAD	USDCAD
7	8	9	10	11	12
USDCHF	USDJPY	GBPCHF	GBPJPY	GBPUSD	GBPUSD
13	14	15	16	17	18
AUDUSD	CHFJPY	NZDUSD	NZDJPY	FTSE	DJ

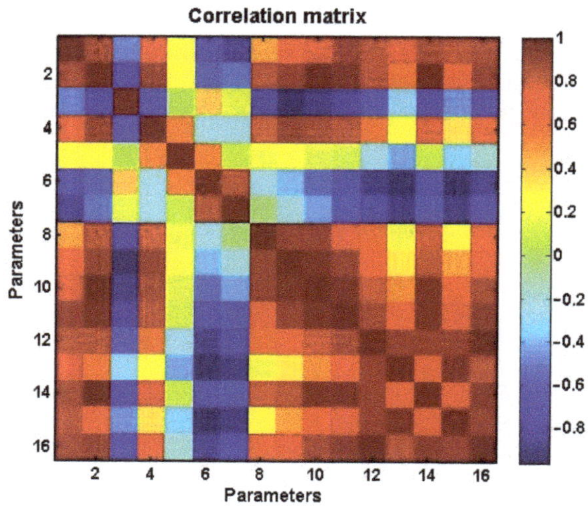

Figure 3. Tonal representation of the correlation matrix of 16 financial instruments.

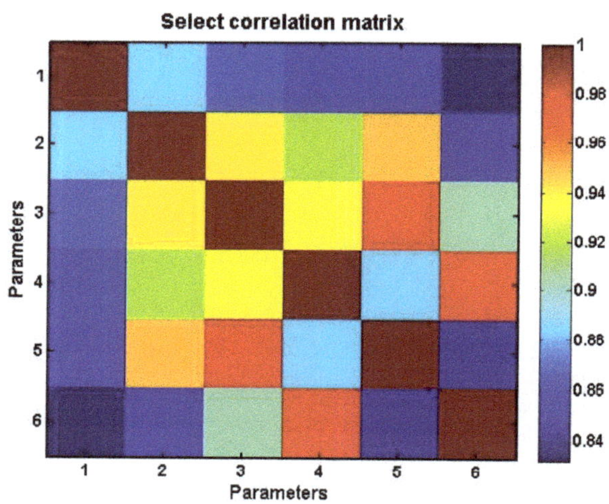

Figure 4. Tonal representation of the EURUSD correlation matrix and its five financial instruments with the highest correlations.

The presence of ordered relationships with lower variability than the quotes of working assets themselves allows us to construct mutual statistical dependencies, which, in turn, make it possible to adjust the current value of a financial instrument using its correlated instruments. Hence, there is a real possibility of constructing management strategies based on a statistical indicator, the value of which at any given time will be determined by the difference between the current quotation of a currency instrument and its estimate based on a group correlated with it. The most natural way of constructing such a forecast is multivariate linear regression (multi-regression) based on the modifications of the least squares method (LSM) [20,21].

It is obvious that the difference between the estimate and the current quotation of an asset can be used to manage assets with a variety of criteria that determine the time of opening and closing positions during the trading process. At the same time, in conditions of non-linearity and non-stationarity, analytical methods for comparing these criteria cannot be implemented. Hence arises the problem of numerical research on analytic and management techniques based on multi-regression (MR) estimation of a currency instrument, which is considered in this article.

2.2. Problem Statement

We consider a number of observations of the value of a non-stationary random process described by model (1), the system component of which is an oscillatory non-periodic process classified as deterministic chaos. On the same observation interval, an m-dimensional non-stationary random vector process is present, the elements of which are correlated with (1) and correspond to the same model of stochastic dynamics

$$\begin{gathered} Y_{Rk} = X_{Rk} + V_{Rk}, \\ \text{where } Y_{Rk} = (y_{R1}, y_{R2}, \ldots, y_{Rm})_k, X_{Rk} = (x_{R1}, x_{R2}, \ldots, x_{Rm})_k, \\ V_{Rk} = (v_{R1}, v_{R2}, \ldots, v_{Rm}), k = 1, \ldots, n. \end{gathered} \tag{3}$$

As a result of numerical experiments, it was found [18] that in general, among a set of financial instruments, it is possible to choose a vector subset (3) that has a significant correlation with the working (i.e., used in financial transactions) instrument y_k, $k = 1, \ldots, n$. In the future, the elements Y_R will be used as regressors in the traditional MR model

$$\hat{y}_k = C_k Y_{Rk} + v_k, k = 1, \ldots, n \tag{4}$$

The most natural way to estimate the value of the transfer coefficient C_k, $k = 1, \ldots, n$, as already noted, is via the use of a conventional computational scheme based on the LSM:

$$\begin{gathered} \hat{C}_k = (Y_{R[1:k-1,m]}{}^T Y_{R[1:k-1,m]})^{-1} Y_{R[1:k-1,m]}{}^T Y_{[1:k-1,1]}, k = 2, \ldots, n \\ \text{where } Y_{R[1:k-1,m]} = (Y_{R,1}, \ldots, Y_{R,k-1}), Y_{[1:k-1,1]} = (y_1, \ldots, y_{k-1}), k = 2, \ldots, m \end{gathered} \tag{5}$$

All data accumulated during the observation from 1 to $k - 1$ are initially used as a training sample.

The analysis of the condition of a working tool consists in assessing the significance of the discrepancy between its current value y_k, $k = 1, \ldots, n$ and its regression estimate (4), reflecting the opinion of the market, represented by a set of regressor instruments, about its real value. If the regressors are representative in terms of their ability to reflect significant market variations, then the difference

$$d_k = \hat{y}_k - y_k, k = 1, \ldots, n \tag{6}$$

will reflect the degree of under- or overpricing of the selected working instrument. This directly implies recommendations for a management strategy: the opening of a position should be carried out in a direction that compensates for the resulting fluctuation discrepancy (6).

2.3. A Simple Management Strategy Based on Multiregression Estimation

Despite the validity and constructiveness of the considered forecasting and management technique, the solution of this problem faces a number of significant issues due to the specifics of the data model (1). Let us consider this question in more detail.

The simplest asset management strategy based on computational regression estimates can be constructed upon criterion K: $|d_k| > d^*$, $k = 1, \ldots, n$. If $d_k > d^*$, this means that $\hat{y}_k > y + d^*$, i.e., the financial instrument is underpriced, and its price can be expected to go up. Vice versa, $d_k < -d^*$ indicates an overprice, and therefore the instrument's price should be expected to go down.

The conventional statistical approach involves determining the critical value d^* based on its distribution tables (or the distribution of an associated statistic) and a given level of confidence. In the conditions of non-stationary dynamics with a chaotic systemic component, such an approach is unfeasible. The critical value has to be selected based on preliminary analysis of retrospective information drawn from a large observation interval.

As an example, let us consider the change in (6) for USDCHF on a 100-day observation interval with a 5-min step. The respective plot, its smoothed version, and the decision levels are presented in Figure 5.

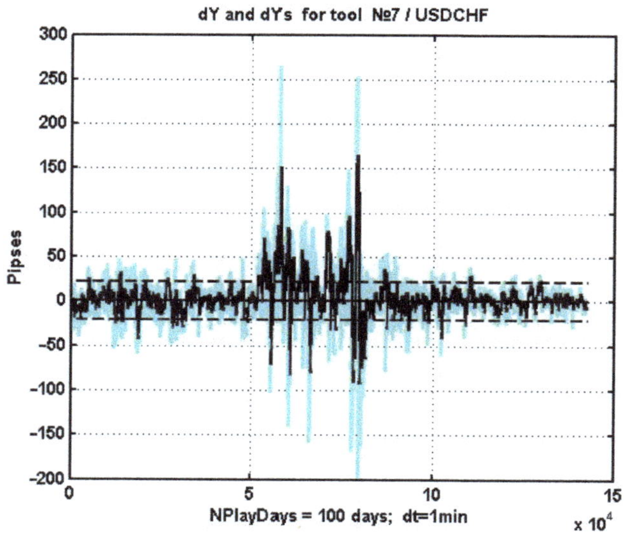

Figure 5. Process d_k, its smoothed version, and the levels of SD.

The threshold value d* for making a management decision is the standard deviation (SD) of d_k estimated on the selected observation interval. The corresponding estimate of the SD of the difference process was s = 24.7 and s = 21.5 for the smoothed process.

Figure 6 contains a histogram of d_k, k = 1, ..., n, which demonstrates the weak convergence of the given difference's distribution to the Gaussian law. This picture is static; considering it dynamics-wise, it can be seen that all the moments of the distribution change over time, and therefore the process is purely non-stationary.

Next, let us consider the combined plots of the initial process and the criterion statistics (an oscillator) d_k, k = 1, ..., n (Figure 7). For easier visualization, a relatively small observation interval (5 days) is considered, and the oscillator and its critical value are enlarged by 1.5 times.

According to the described management strategy, if the smoothed oscillator surpasses the threshold value, i.e., $d_k > d^*$, a long position should be opened. Alternatively, one can open a short position if $d_k < -d^*$. Positions can be closed along with a reverse crossing of the threshold d*, or by the conventional methods of setting Take Profit and Stop Loss levels.

Figure 7 shows that there is no stable trend for either underprice or overprice in this particular example. One can assume that, on a sideways ("flat") trend, the multi-regression oscillator will work quite effectively with correctly selected values of the management strategy's parameters. However, if there is a strong trend caused by factors external to the market, the corresponding trend will prevail, majoring the influence of the discrepancy between the current price of the asset in relation to its market value.

Thus, in order to construct an effective management strategy based on the discrepancy between the asset price and its market estimate in conditions of non-linearity and non-stationarity, it is necessary to form a set of regressors most associated with the working instrument for the current short observation interval. This approach involves structural adaptation of the MR model (4), which in turn requires a more careful study of the features of the estimation task in the conditions of non-stationary dynamics of market parameters.

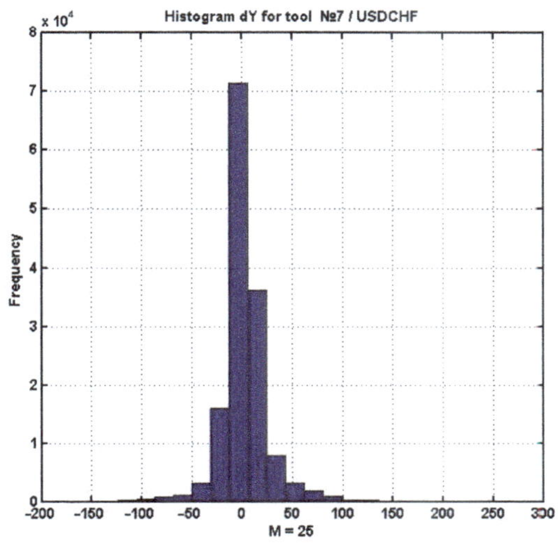

Figure 6. Histogram of smoothed $d_{k.}$

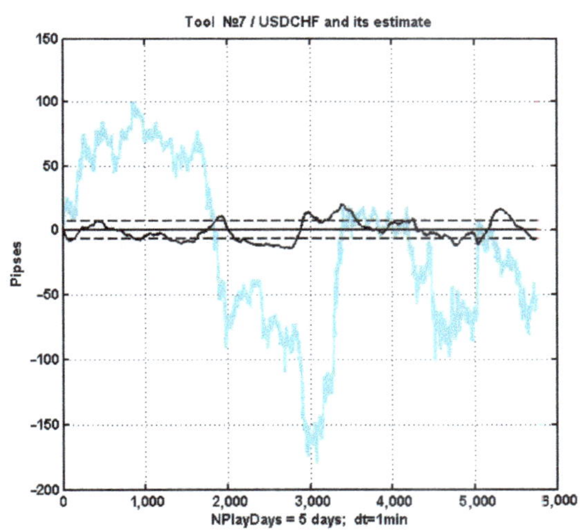

Figure 7. Changes in the quotation of a currency instrument and a decision statistic (an oscillator) on a five-day interval.

2.4. Specifics of Financial Instrument Value Analysis in the Conditions of Non-Stationary and Non-Linear Dynamics

The traditional MR estimation meets the conditions of non-bias, consistency, and asymptotic efficiency only when a number of constraints are met. Let us describe some of them:

1. The system component of a series of observations (1) x_k, $k = 1, \ldots, n$ is an unknown deterministic process, which in some cases can be expressed analytically;
2. Regressors are not mutually correlated, i.e.,

$$cov(y_{Ri}, y_{rj}) = 0, \forall i \neq j, i, j = 1, \ldots, m; \text{ and}$$

3. The noise component of the model (1) v_k, k = 1, . . . , n is stationary, centered relative to the system component, and uncorrelated to the regressor's random process

$$E\{v_k\} = 0, cov\{v_i,v_j\}_k = 0; \forall i \neq j;$$
$$E\{v_k, y_{Ri,k}\} = 0, i = 1, \ldots ,m, k = 1, \ldots , n.$$

The simplest methods of data analysis based on algorithms of statistical hypothesis testing show that real observation series obtained via monitoring values of market assets do not fully meet the constraints listed above. This means that the price estimates of the working instrument obtained via statistical analysis are not only not optimal, but also, as a rule, are biased. At the same time, it is not possible to construct analytical methods for calculating a market estimate of a financial instrument for a model of type (1). Thus, the main method of analyzing the quality of evaluation, as has already been noted, is computational studies.

One of the known ways to reduce the bias of statistical estimates for non-stationary processes is to limit the length of the series of observations via a sliding observation window $Y(\Delta_k)$, where $\Delta_k = (k - L, k)$, $k = L + 1, \ldots, n$, and L is the window size. At the same time, the most up-to-date data should be processed, reflecting the current situation to a greater extent. The multipliers in the estimation algorithm using the LSM (5) are given by expressions

$$Y_{R[k-L:k-1,m]} = (Y_{R,k-L}, \ldots , Y_{R,k-1}), Y_{[k-L:k-1,1]} = (y_{k-L}, \ldots , y_{k-1}), k = 1, \ldots , m$$

The size of the observation window is usually chosen to minimize the mean square of the error L*: $d^2 = s^2 + b^2 = \min(L)$, where s is the estimate of the SD, and b is the estimated bias. However, for a model of the form (1) with a chaotic system component, it is not possible to obtain stable estimates of SD and bias values. In this regard, the choice of the observation window is carried out empirically, by comparing the estimation results on the training observation window preceding the current time.

Another important feature of the statistical analysis of financial instruments is the presence of significant variations in the correlation matrix of the used regressors. Figure 8 shows changes in the estimates of the pairwise correlation between two correlated currency instruments with an increasing sliding observation window of size L = 10, 25, 50, and 75 counts, respectively.

It is not difficult to see that as the sample size increases, the estimates become more stable. However, for a current assessment of the value of a financial instrument, the best composition of regressors Y_{Rk}, k = 1, . . . , n in the computational scheme (4) will be different at each time moment. This justifies the use of structural adaptation of the MR model with a step-by-step selection of financial instruments used as regressors.

Another important feature of the considered problem is the separation of the system component from the initial process (1). In real time, this procedure is carried out by sequential filtering methods of type (2). It is quite obvious that an increase in the memory size of the filter or the weight characteristics of already smoothed observations leads to a decrease in the noise level v_k, k = 1, . . . , n. At the same time, this approach inevitably leads to delays in the reaction of current estimates to significant changes in the relative dynamics of quotations of financial instruments. In other words, there is a contradiction between the quality of smoothing random fluctuations in observations and the growth of the estimation bias caused by dynamic errors. An illustration of this problem is given by the example of the allocation of the system component of the quotes of the EURUSD currency pair using the exponential filter (2) for the filter coefficients $\alpha = 0.1$ (Figure 9a) and $\alpha = 0.01$ (Figure 9b).

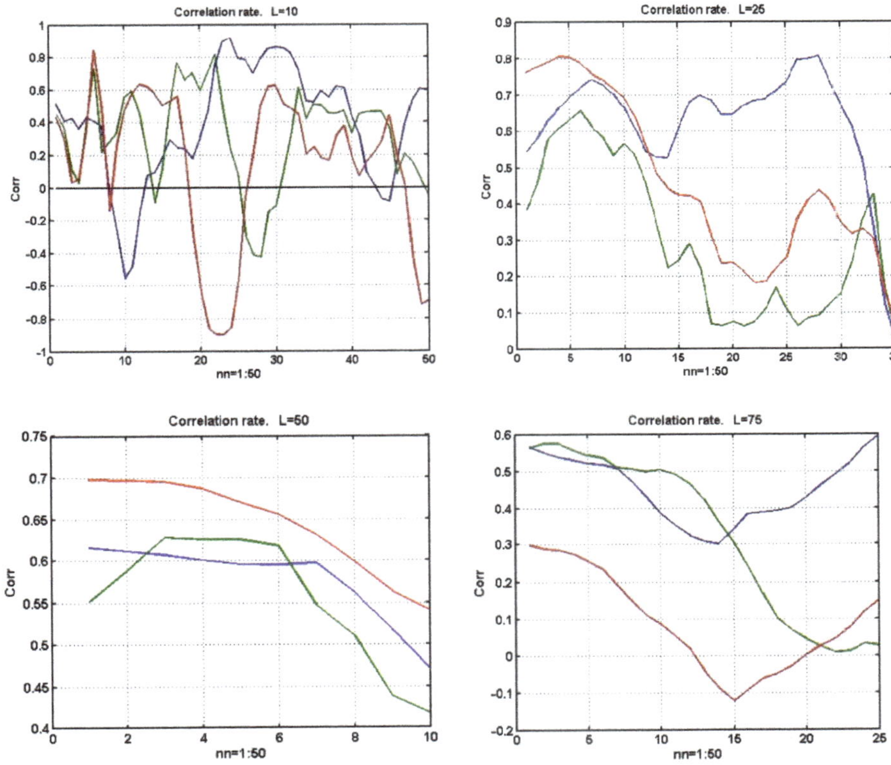

Figure 8. The change in the estimated pairwise correlation between two currency instruments on sliding observation windows of size L = 10, 25, 50, and 75 counts.

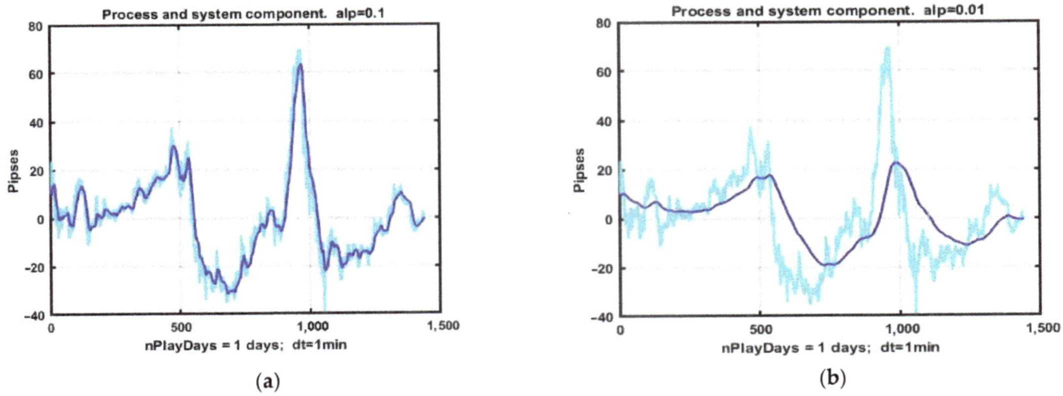

Figure 9. System component selection using an exponential filter with coefficients $\alpha = 0.1$ (**a**) and $\alpha = 0.01$ (**b**).

Thus, the quality of recovery of the system component in data model (1) is determined by the compromise between the values of statistical and dynamic estimation errors. Furthermore, the shift in the balance between them depends either on the filter coefficient α (for a filter of type (2)) or on the size of the filter memory [22]. In the conditions of chaotic dynamics described by model (1), the choice of this parameter also has no analytically sound recommendations and is based on empirical fitting to the results of retrospective analysis on previous observation segments that serve as a dataset. Double exponential filtration (back and forth) with α decreased to minimize bias could be a reasonable variant.

Based on the studies presented above, it can be concluded that the nature of the considered series of observations is deeply inconsistent with the known limitations of statistical analysis, which makes it possible to obtain effective estimates. Hence, an incorrect task of constructing management strategies arises in conditions when the quality of assessments of the state of market parameters is very uncertain. Traditional approaches to obtaining stable results in such situations are most often based on the principles of adaptation of estimates. The question of how this technique can be useful in conditions of chaotic dynamics remains open. Some aspects of this problem are touched upon in this article.

3. Results

As an example of structural adaptation of the estimation model, consider the problem of choosing a group of regressors for a currency asset that consists of USDCHF.

Figure 10 shows the dynamics of the quotation over ten days against the background of changes in the quotes of the five currency pairs most correlated with it at the specified observation interval. The corresponding group consists of currency pairs with numbers 1, 9, 8, 10, and 16 (see Table 1).

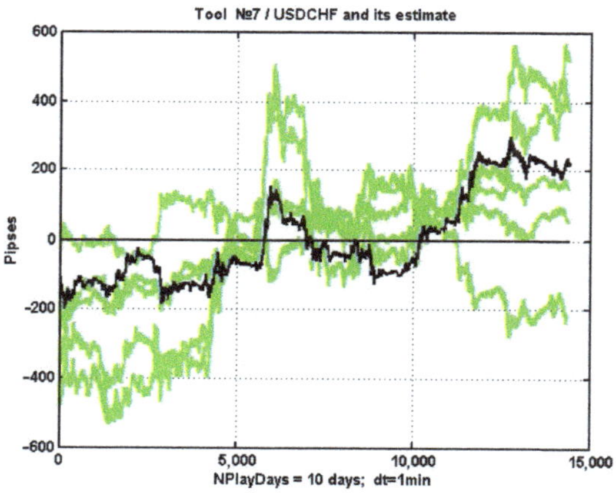

Figure 10. Changes in the USDCHF quotation and the group of its five most correlated currency pairs.

Note that one of the elements behaves counterphase-like with respect to the considered process. This is due to the degree of the relationship being evaluated by the absolute value. A currency pair with a strong negative correlation also carries a large amount of information about the behavior of the associated instrument. In this case, the corresponding regression coefficient before this term will be negative.

Within the listed constraints, the correlation matrix of the market (i.e., all 16 currency pairs) was recalculated. From the resulting matrix, a row corresponding to the number of the working asset was selected and sorted in descending order of the absolute values of pairwise correlations. The second to (m + 1)th observations of the obtained variation series determine the group of regressors with the largest absolute correlation value. The corresponding results obtained for 24 non-overlapping 10-h observation intervals are presented in Table 2.

It can be seen that, during the first seven observation intervals, the optimal group of regressors <1 9 8 10 16> did not change. At steps 8–10, the composition of the group was preserved as well, but the first and the ninth regressors swapped places. Further evolution of the group's composition is clear from the data given in the table. The general conclusion is that the composition changes quite slowly and the 10-day adaptation interval is acceptable for producing regression estimates with given regressors.

Table 2. Lists of regressors in descending order of correlation on 10-h non-overlapping observation intervals.

Intervals	Regressors				
1–7	1	9	8	10	16
8–10	9	1	10	8	16
11	9	10	1	8	16
12–14	9	8	10	16	1
15–17	9	8	10	16	1
18–19	8	9	10	16	4
20–21	8	9	10	4	16
22–24	9	8	10	4	16

Figure 11 shows a centered plot of the quotation dynamics of the USDCHF currency instrument (bottom) and a smoothed plot of d_k (top) on a three-day observation interval. The same figure shows an example of the implementation of a simple management strategy based on an adaptive MR estimate. If the difference d_k between the estimate and the current value of a currency instrument turns out to be greater (in absolute value) than the threshold value d*, a recommendation is made to open a position in the appropriate direction. The asterisk marks the state of the quote at the time of opening a long position. The diamond corresponds to the position closing at d_k intersecting back the threshold value d*.

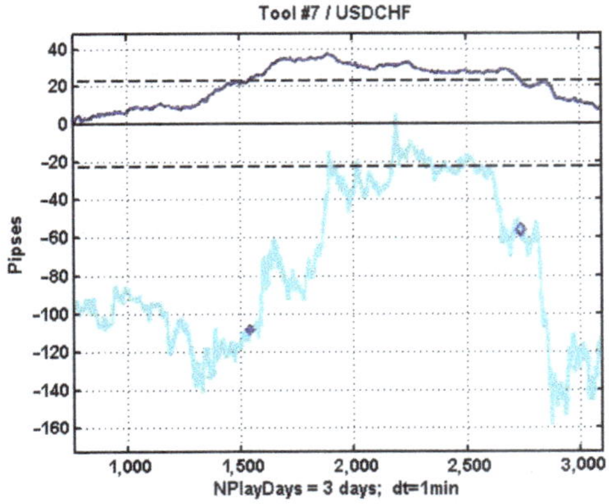

Figure 11. An example implementation of the simplest control strategy with an adaptive MR oscillator.

The use of structural adaptation in solving the asset management problem with MR estimation increases the frequency of profitable decisions by 5–10%, but does not solve the problem of stability of management in chaotic environments in general.

The reason for this is clear: moving in the oscillator's direction is determined by its estimate of the prices of financial instruments used as regressors. At the same time, there are external factors that lead to the appearance of dynamic trends. The sum of this movements produces the final form of the dynamics, the direction of which is determined by a vector sum of heterogeneous and hardly forecasted influencing factors.

The absence of an analytical representation of the initial chaotic process does not allow us to obtain an accurate assessment of the potential capabilities of the chosen management strategy. The most effective way to obtain such an estimate is numerical analysis based on random search for optimal parameters of the management strategy. One of the implementations of random search is evolutionary modeling, which has found wide application in numerical optimization problems [23–28].

Let us consider the problem of estimating the potential characteristics of the adaptive MR oscillator discussed above based on the evolutionary optimization method. In accordance with the strategy described in the article, a position is opened when the smoothed value d_k, k = 1, . . . , n of the threshold value d* is crossed. The position is closed when the process $d_{k+\tau}$ crosses the level $d^*_{k+\tau}$ back or the zero level $d_{k+\tau} = 0$, where τ is the time of the crossing.

The list of optimized strategy parameters G = [nW, α, d*] (in terms of evolutionary modeling, genome G) includes the size of the sliding observation window nW, which is used to produce a regression estimate, an exponential smoothing coefficient α, and the decision level d*.

At the initial stage, by introducing small (within the limits of the corresponding parameters) variations in all parameters, a group of ancestral genomes (AG) of size N_a is formed. Further, in a loop over the number of generations N_g, a new generation is created, consisting of an already existing group of ancestral genomes and a newly generated group of descendant genomes (DG). Descendant genomes are constructed from ancestral genomes in three main ways, including:

1. Small single changes made to one of the AG parameters. The parameter in question is selected by a random draw. If changes are to be made sequentially to each parameter, then each AG receives m_g modifications, where m_g is the size of the genome. In this case, there are $N^{(1)}{}_d = N_a m_g$ descendants with a given type of modification, and only one parameter (gene) is modified in each of them. In this case, $m_g = 3$; therefore, if $N_a = 4$ best variants (ancestors) are preserved in each generation, $N^{(1)}{}_d = 12$ versions of the first type of DG will be obtained.
2. Small group changes. They are carried out similarly to the previous case, but are made to all parameters at once instead. Thus, there are $N^{(2)}{}_d = 4$ more versions of DGs with slow changes in all genes.
3. Strong single mutation or parametric mutation. The AG and the gene number are selected via a random draw. With the probability of parametric mutation, P_{pm} produces $N^{(3)}{}_d$ descendants, in each of which one gene in the range $|\Delta| > 3\sigma$ is modified.

As an example, we used a program with $N_{gc} = 9$ generation changes in the same time interval of 10 days. As a starting genome, we used a vector $G_0 = [nW_0, \alpha_0, d^*_0] = [5, 0.01, 0.6]$. During genome modification, we used rough estimates of the SD of the three listed parameters: SD(G) = [3, 0.02, 0.5].

To compare the potential effectiveness of management strategies based on the MR assessment of the condition of the used asset, we considered two options:

1. A fixed group of five regressors is selected before the start of trading operations maximizing the correlation with the working instrument (asset). The correlation matrix for all 16 financial instruments is evaluated based on observations of their quotes during the 15 days preceding the start of trading.
2. Sequential adjustment of a group of five regressors. The adjustment is carried out based on the maximum correlation with the working instrument (asset) with an interval of 10 h. The correlation matrix is evaluated based on the results of observations of their quotes on a sliding observation window of 15 days.

Since the evaluation of gain was carried out via random search, we can only talk about some approximation of the optimal solution, which theoretically could be obtained via brute-force search in the values of the management strategy parameters.

Figure 12a shows an example of the implementation of the best variant of management strategy parameters obtained via evolutionary parametric optimization over nine generations of the corresponding programs. The description of the plots is similar to the description of the processes in Figure 11. This plot corresponds to the first option, i.e., non-adaptive with respect to the choice of the regressor set. Figure 12b shows the dependence of the gain growth on the generation number for the non-adaptive strategy.

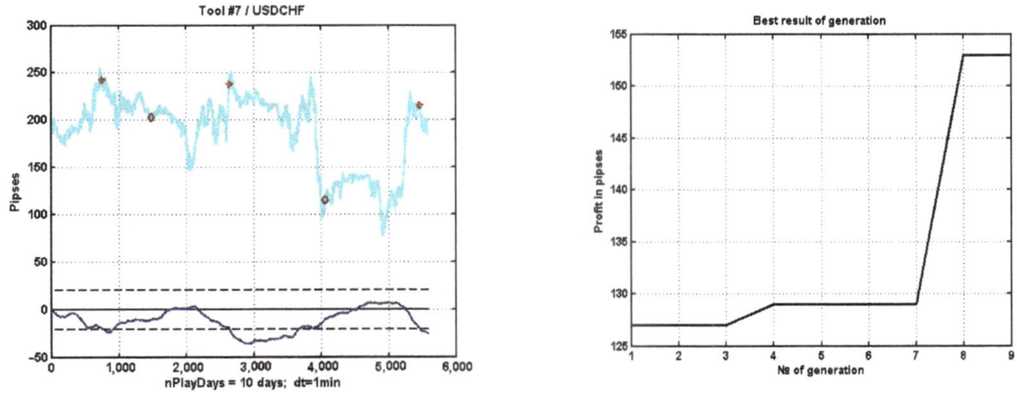

(a) (b)

Figure 12. (**a**) An example of a suboptimal management strategy using a non-adaptive MR oscillator. (**b**) The dependence of the gain increase on the generation number for this strategy.

Similar plots of the implementation of a suboptimal strategy and the dependence of the gain on the generation number for the second option are shown, respectively, in Figure 13a,b. A comparison of the two examples shows that adaptation during the choice of a group of regressors improves the quality of the estimate and, as a result, increases the potential gain by about 10–15%. Of course, a single example does not give an objective picture of the gain. To generalize the result, we averaged one hundred 10-day observation segments of quotations. Note that averaging over implementations in this case is not equivalent to averaging over one segment of a length equal to the sum of individual implementations. This is due to the fact that the process of price dynamics is not ergodic. Therefore, the task was reproduced for both averaging schemes and, ultimately, showed that the gain from adaptation varies in a wide range of 5% to 15%.

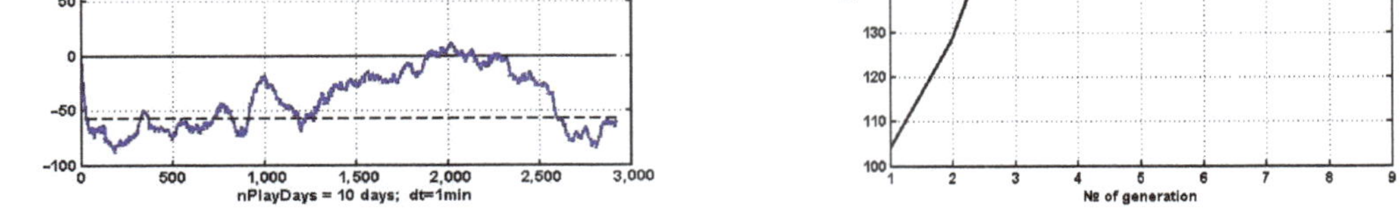

(a) (b)

Figure 13. (**a**) An example of a suboptimal control strategy using an adaptive MR oscillator. (**b**) The dependence of the gain growth on the generation number for this strategy.

4. Discussion

The research presented in this article is focused on establishing the fundamental possibility of effective proactive management in multidimensional non-stationary, nonlinear, and/or chaotic environments. As a basic hypothesis, the assumption is put forward that the market seeks to eliminate the mismatch between the current value of a financial instrument and its valuation, formed by the "opinion" of the market about its "fair price". The traditional MR evaluation of the instrument was used as such a price in the conducted studies.

In general, the provided numerical studies confirm the hypothesis stated above, but its implementation for an effective management strategy encountered a number of significant difficulties. In particular, the nature of real market processes leads to a significant decrease in the accuracy of the generated estimates of the market value and, consequently, to a decrease in the effectiveness of the asset management process.

The conventional approach to increasing estimate stability for stochastic processes is based on adapting the estimation model to variations in the statistical structure of observation series. However, the effectiveness of adaptation in the tasks of constructing management strategies for chaotic dynamics can be ambiguous. This is due to the fact that real processes, due to their non-stationarity and non-ergodicity, do not allow us to close the adaptation circuit fast enough to have time to track changes in the structure of the observed dynamic process. Therefore, we propose to use the correlation structure of the initial data as a regularizing factor based on processes with relatively slow changes. In particular, as shown in this paper, the adaptation of the estimating model to variations in the correlation structure of multidimensional quotation dynamics can improve the quality of asset value recovery, and, as a result, raise the level of the potential efficiency of the MR oscillator.

5. Conclusions

Selecting a single working component and obtaining its current adjustment based on auxiliary components are problems that arise naturally during the analysis of multidimensional time series that estimate the current values of an indicator from different points of view using different sources of information. In a number of formulations, it turns out that the system of correlations between components has a known inertia, which allows it to be used with some time lag to adjust the current values of the working component. One such formulation of the problem of estimating the current value of a financial instrument is considered in this paper.

Further improvement of estimation quality involves the use of self-organizing algorithms of data analysis and management. In this paper, we investigated the construction of such algorithms using evolutionary modeling, in which the best versions of the observation model are formed by random changes in both the model's structure and its parameters, with further selection of the most effective solutions in the process of changing generations of models.

The use of self-organization in the task of MR data analysis made it possible to obtain a fundamental confirmation of the viability of the proposed asset management method. However, a stable result with permanent profit was not achieved. This is due to the choice of regressors used for estimating the instrument value being limited to currency pairs available at the electronic Forex exchange. However, other parameters may have a greater influence on a particular financial instrument in a given time interval. These parameters could describe dynamic processes in the stock or commodity market or in the market infrastructure associated with political, military, environmental, and other factors.

Another direction of development of this approach is based on the robustification of estimation algorithms [29,30], i.e., reducing the sensitivity of estimates to variations in the statistical structure of the data. In principle, this approach could serve as an alternative to adaptation technologies that are insufficiently effective due to the inertia-free and chaotic nature of the initial series of observations. However, the price of increased stability is a decrease in accuracy. Thus, this issue requires separate research.

Author Contributions: conceptualization and methodology, A.M. (Andrey Makshanov); validation and writing, A.M. (Alexander Musaev); review and editing, investigation, programming, visualization, administration, scientific discussions, supervision, and funding acquisition, D.G. All authors have read and agreed to the published version of the manuscript.

Funding: The research of Alexander Musaev described in this paper was partially supported by the Russian Foundation for Basic Research (grant 20-08-01046, state research FFZF-2022-0004). Dmitry Grigoriev's research for this paper was supported by The Endowment Fund of St Petersburg University.

Institutional Review Board Statement: Not applicable.

Informed Consent Statement: Not applicable.

Data Availability Statement: The data presented in this article can be found at Finam.ru (https://www.finam.ru/, accessed on 9 January 2022).

Acknowledgments: The authors are grateful to the participants in the Center for Econometrics and Business Analytics (ceba-lab.org, accessed on 9 January 2022, CEBA) seminar series for their helpful comments and suggestions.

Conflicts of Interest: The authors declare no conflict of interest.

References

1. Panayiotis, T. Oil volatility index and Chinese stock markets during financial crisis: A time-varying perspective. *J. Chin. Econ. Foreign Trade Stud.* **2021**, *14*, 187–201.
2. Lee, R. Quantum Trader—A Multiagent-Based Quantum Financial Forecast and Trading System. In *Quantum Finance*; Springer: Singapore, 2020; pp. 375–398.
3. Lee, R. COSMOS trader–Chaotic Neuro-oscillatory multiagent financial prediction and trading system. *J. Financ. Data Sci.* **2019**, *5*, 61–82. [CrossRef]
4. Marti, G.; Nielsen, F.; Binkowski, M.; Donnat, P. A review of two decades of correlations, hierarchies, networks and clustering in financial markets. *arXiv* **2017**, arXiv:1703.00485.
5. Moews, B.; Herrmann, J.M.; Ibikunle, G. Lagged correlation-based deep learning for directional trend change prediction in financial time series. *Expert Syst. Appl.* **2019**, *120*, 197–206. [CrossRef]
6. Wątorek, M.; Drożdż, S.; Oświęcimka, P.; Stanuszek, M. Multifractal cross-correlations between the world oil and other financial markets in 2012–2017. *Energy Econ.* **2019**, *81*, 874–885. [CrossRef]
7. Peters, E.E. *Chaos and Order in the Capital Markets: A New View of Cycles, Prices, and Market Volatility*, 2nd ed.; John Wiley & Sons: New York, NY, USA, 1996; 288p.
8. Gregory-Williams, J.; Williams, B.M. *Trading Chaos: Maximize Profits with Proven Technical Techniques*, 2nd ed.; John Wiley & Sons: New York, NY, USA, 2004; 251p.
9. Sage, A.P.; White, C.C.; Siouris, G.M. Optimum systems control. *IEEE Trans. Syst. Man Cybern.* **1979**, *9*, 102–103. [CrossRef]
10. Ahlberg, J.H.; Nilson, E.N.; Walsh, J.L. *The Theory of Splines and Their Applications: Mathematics in Science and Engineering: A Series of Monographs and Textbooks*; Elsevier: Amsterdam, The Netherlands, 2016; Volume 38.
11. Zumbach, G.O. On the Short Term Stability of Financial ARCH Price Processes. Available online: https://arxiv.org/abs/2107.06758 (accessed on 16 December 2021).
12. Xing, D.Z.; Li, H.F.; Li, J.C.; Long, C. Forecasting price of financial market crash via a new nonlinear potential GARCH model. *Phys. A Stat. Mech. Its Appl.* **2021**, *566*, 125649. [CrossRef]
13. Prigogine, I.; Stengers, I. Order Out of Chaos: Man's New Dialogue with Nature. In *Bantam New Age Books*; Flamingo: New York, NY, USA, 1985; ISBN 978-0-00-654115-8.
14. Smith, L. *Chaos: A Very Short Introduction*; Oxford University Press: Oxford, UK, 2007; 180p.
15. Kautz, R. *Chaos: The Science of Predictable Random Motion*; Oxford University Press: Oxford, UK, 2011; ISBN 978-0-19-959458-0.
16. Yusupov, R.M.; Musaev, A.A.; Grigoriev, D.A. Evaluation of Statistical Forecast Method Efficiency in the Conditions of Dynamic Chaos. In Proceedings of the 2021 IV International Conference on Control in Technical Systems (CTS), Saint-Petersburg, Russia, 21–23 September 2021; pp. 178–180. [CrossRef]
17. Musaev, A.; Grigoriev, D. Numerical Studies of Statistical Management Decisions in Conditions of Stochastic Chaos. *Mathematics* **2022**, *10*, 226. [CrossRef]
18. Musaev, A.; Makshanov, A.; Grigoriev, D. Forecasting Multivariate Chaotic Processes with Precedent Analysis. *Computation* **2021**, *9*, 110. [CrossRef]
19. Musaev, A.; Grigoriev, D. Analyzing, Modeling, and Utilizing Observation Series Correlation in Capital Markets. *Computation* **2021**, *9*, 88. [CrossRef]
20. Axelsson, O. A generalized conjugate gradient, least square method. *Numer. Math.* **1937**, *51*, 209–227. [CrossRef]
21. Bolch, B.W.; Huang, C.J. *Multivariate Statistical Methods for Business and Economics*; Prentice-Hall: Hoboken, NJ, USA, 1974; 317p.
22. Kim, P.S. Selective finite memory structure filtering using the chi-square test statistic for temporarily uncertain systems. *Appl. Sci.* **2019**, *9*, 4257. [CrossRef]
23. Fogel, L.J.; Owens, A.J.; Walsh, M.J. *Artificial Intelligence through Simulated Evolution*, 1st ed.; John Wiley & Sons: New York, NY, USA, 1966; 231p.

24. Song, W.; Choi, L.C.; Park, S.C.; Ding, X.F. Fuzzy evolutionary optimization modeling and its applications to unsupervised categorization and extractive summarization. *Expert Syst. Appl.* **2011**, *38*, 9112–9121. [CrossRef]
25. Al-Ahmad, B.; Al-Zoubi, A.M.; Abu Khurma, R.; Aljarah, I. An Evolutionary Fake News Detection Method for COVID-19 Pandemic Information. *Symmetry* **2021**, *13*, 1091. [CrossRef]
26. Maier, H.R.; Razavi, S.; Kapelan, Z.; Matott, L.S.; Kasprzyk, J.; Tolson, B.A. Introductory overview: Optimization using evolutionary algorithms and other metaheuristics. *Environ. Model. Softw.* **2019**, *114*, 195–213. [CrossRef]
27. Fernandez, A.; Herrera, F.; Cordon, O.; del Jesus, M.J.; Marcelloni, F. Evolutionary fuzzy systems for explainable artificial intelligence: Why, when, what for, and where to? *IEEE Comput. Intell. Mag.* **2019**, *14*, 69–81. [CrossRef]
28. Mukhopadhyay, A.; Maulik, U.; Bandyopadhyay, S.; Coello, C.A.C. A Survey of Multiobjective Evolutionary Algorithms for Data Mining. *IEEE Trans. Evol. Comput.* **2014**, *18*, 4–35. [CrossRef]
29. Huber, P.J. Robust Statistics. In *Wiley Series in Probability and Statistics*; John Wiley & Sons: New York, NY, USA, 1981; ISBN 978-0-471-41805-4.
30. Holopainen, M.; Sarlin, P. Toward robust early-warning models: A horse race, ensembles and model uncertainty. *Quant. Financ.* **2017**, *17*, 1933–1963. [CrossRef]

mathematics MDPI

Article

The Least Squares Homotopy Perturbation Method for Systems of Differential Equations with Application to a Blood Flow Model

Mădălina Sofia Paşca [1,2,*,†], Olivia Bundău [1,†], Adina Juratoni [1,†] and Bogdan Căruntu [1,†]

1 Department of Mathematics, Politehnica University Timişoara, 300006 Timişoara, Romania; olivia.bundau@upt.ro (O.B.); adina.juratoni@upt.ro (A.J.); bogdan.caruntu@upt.ro (B.C.)
2 Department of Mathematics, West University of Timişoara, 300223 Timişoara, Romania
* Correspondence: madalina.pasca@upt.ro or madalina.pasca79@e-uvt.ro
† These authors contributed equally to this work.

Abstract: In this paper, least squares homotopy perturbation is presented as a straightforward and accurate method to compute approximate analytical solutions for systems of ordinary differential equations. The method is employed to solve a problem related to a laminar flow of a viscous fluid in a semi-porous channel, which may be used to model the blood flow through a blood vessel, taking into account the effects of a magnetic field. The numerical computations show that the method is both easy to use and very accurate compared to the other methods previously used to solve the given problem.

Keywords: least squares homotopy perturbation method; system of nonlinear differential equations; approximate analytical solutions; non-Newtonian fluid; magnetohydrodynamics

Citation: Paşca, M.S.; Bundău, O.; Juratoni, A.; Căruntu, B. The Least Squares Homotopy Perturbation Method for Systems of Differential Equations with Application to a Blood Flow Model. *Mathematics* **2022**, *10*, 546. https://doi.org/10.3390/math10040546

Academic Editor: Marco Pedroni

Received: 2 November 2021
Accepted: 6 February 2022
Published: 10 February 2022

Publisher's Note: MDPI stays neutral with regard to jurisdictional claims in published maps and institutional affiliations.

Copyright: © 2022 by the authors. Licensee MDPI, Basel, Switzerland. This article is an open access article distributed under the terms and conditions of the Creative Commons Attribution (CC BY) license (https://creativecommons.org/licenses/by/4.0/).

1. Introduction

The least squares homotopy perturbation method was introduced in 2017 by Bota and Caruntu in [1], and its main feature is an accelerated convergence compared to the regular homotopy perturbation method. In the few years since its introduction, the method (or slightly modified versions of it) was used by several researchers with very good results in finding approximate solutions for various types of problems, among which, we mention:

- Boundary value problems for ordinary differential equations [2,3].
- Fractional partial differential equations [4–6].
- Fractional order integro-differential equations [7].
- Systems of fractional partial differential equations [8].

In the present paper, we employ the least squares homotopy perturbation method to compute approximate analytical solutions for boundary problems consisting of systems of nonlinear ordinary differential equations of the type:

$$\begin{aligned} \mathcal{L}_i(U_1(y),U_2(y),\ldots,U_n(y))+\mathcal{N}_i(U_1(y),U_2(y),\ldots,U_n(y))-f_i(y)=0, \qquad i=\overline{1,n} \\ \mathcal{B}_j(U_i(y))=0, \qquad j=\overline{1,k} \end{aligned} \tag{1}$$

where $U_i(y)$ are the unknown functions, $\mathcal{L}_i$ are linear operators, $\mathcal{N}_i$ are nonlinear operators, $f(t)_i$ are given functions, y denotes the variable, and $\mathcal{B}_j$ are boundary operators.

2. The Least Squares Homotopy Perturbation Method

In this section, the least squares homotopy perturbation method (LSHPM) is presented. Since the numerical application considered in the following section only contains two equations, we introduce LSHPM for the case of a system consisting of two equations. Obviously, LSHPM can be easily generalized for systems consisting of as many equations as needed. We should also note that LSHPM works as well, if instead of $\mathcal{B}_j(U_i(y))=0$, we have other types of conditions, such as $\mathcal{B}_j(U_i(y), V_i(y)) = 0$, initial-type conditions, or any combinations of the above.

Thus we consider the system:

$$\begin{aligned} \mathcal{L}_1(U(y), V(y)) + \mathcal{N}_1(U(y), V(y)) - f_1(y) = 0, \\ \mathcal{L}_2(U(y), V(y)) + \mathcal{N}_2(U(y), V(y)) - f_2(y) = 0, \\ \mathcal{B}_i(U(y)) = 0,\ \mathcal{B}_j(V(y)) = 0, \quad i = \overline{1, n1} \quad j = \overline{1, n2} \end{aligned} \tag{2}$$

where $U(y)$ and $V(y)$ are the unknown functions, $\mathcal{L}_1, \mathcal{L}_1$ are linear operators, $\mathcal{N}_1, \mathcal{N}_2$ are nonlinear operators, and $\mathcal{B}_i$,$\mathcal{B}_j$ are boundary operators.

Let $\tilde{U}(y)$ and $\tilde{V}(y)$ be approximate solutions of the system (2). The error obtained by replacing the exact solutions $U(y), V(y)$ of the system (2) with the approximate ones $\tilde{U}(y); \tilde{V}(y)$ is given by the remainders:

$$\begin{aligned} R_1(y, \tilde{U}) = \mathcal{L}_1(\tilde{U}(y)) + \mathcal{N}_1(\tilde{U}(y)) - f_1(y), \ y \in \mathbb{R} \\ R_2(y, \tilde{V}) = \mathcal{L}_2(\tilde{V}(y)) + \mathcal{N}_2(\tilde{V}(y)) - f_2(y), \ y \in \mathbb{R} \end{aligned} \tag{3}$$

Following the homotopy perturbation method [9–11], the first step in applying LSHPM is to attach to the system (3) the family of equations:

$$\begin{aligned} (1-p)[\mathcal{L}_1(\Phi_1(y,p)) - f_1(y)] + p\,[\mathcal{L}_1(\Phi_1(y,p)) + \mathcal{N}_1(\Phi_1(y,p)) - f_1(y)] = 0 \\ (1-p)[\mathcal{L}_2(\Phi_2(y,p)) - f_2(y)] + p\,[\mathcal{L}_2(\Phi_2(y,p)) + \mathcal{N}_2(\Phi_2(y,p)) - f_2(y)] = 0 \end{aligned} \tag{4}$$

where $p \in [0,1]$ is an embedding parameter and $\Phi_i(y,p)$ with $i = \overline{1,2}$ are unknown functions.

When p increases from 0 to 1, the solutions $\Phi_i(y,p)$ of system (4) vary from $\Phi_1(y,0) = U_0(y)$ and $\Phi_2(y,0) = V_0(y)$ to the solutions $\Phi_1(y,1) = U(y)$ and $\Phi_2(y,1) = V(y)$ of the system (2). The functions $U_0(y)$ and $V_0(y)$ are the solutions of the system:

$$\begin{aligned} \mathcal{L}_1(U_0(y)) - f_1(y) = 0 \\ \mathcal{L}_2(V_0(y)) - f_2(y) = 0 \\ \mathcal{B}_i(U(y)) = 0,\ \mathcal{B}_j(V(y)) = 0, \quad i = \overline{1, n1} \quad j = \overline{1, n2} \end{aligned} \tag{5}$$

We consider the following expansions of $\Phi_i(y,p)$:

$$\begin{aligned} \Phi_1(y,p) = U_0(y) + \sum_{m\geq 1} U_m(y)\, p^m \\ \Phi_2(y,p) = V_0(y) + \sum_{m\geq 1} V_m(y)\, p^m \end{aligned} \tag{6}$$

Substituting the relations (6) in (4), collecting the same powers of p and equating the coefficients of the powers of p, we obtain:

$$\begin{aligned} \mathcal{L}_1(U_m(y)) = -\mathcal{N}_1^{m-1}(U_0(y), U_1(y), \ldots, U_{m-1}(y)) \\ \mathcal{L}_2(V_m(y)) = -\mathcal{N}_1^{m-1}(V_0(y), V_1(y), \ldots, V_{m-1}(y)) \\ \mathcal{B}_i(U_m(y)) = 0,\ \mathcal{B}_j(V_m(y)) = 0, \quad i = \overline{1, n1} \quad j = \overline{1, n2} \end{aligned} \tag{7}$$

where $\mathcal{N}_i^j$, $j \geq 0$ are the coefficients of p^j in the nonlinear operator $\mathcal{N}_i$:

$$\begin{aligned} \mathcal{N}_1(U(y)) = \mathcal{N}_1^0(U_0(y)) + p\mathcal{N}_1^1(U_0(y), U_1(y)) + p^2\mathcal{N}_1^2(U_0(y), U_1(y), U_2(y)) + \ldots \\ \mathcal{N}_2(V(y)) = \mathcal{N}_1^0(V_0(y)) + p\mathcal{N}_1^1(V_0(y), V_1(y)) + p^2\mathcal{N}_1^2(V_0(y), V_1(y), V_2(y)) + \ldots \end{aligned} \tag{8}$$

Now we can denote by

$$\begin{aligned} f_{1m} = U_0 + U_1 + \ldots + U_m, \\ f_{2m} = V_0 + V_1 + \ldots + V_m \end{aligned} \tag{9}$$

where $U_m, m \geq 1$, and $V_m, m \geq 1$, are obtained from the linear Equation (7).

For $m = 0, 1, 2, \ldots$ let us consider the set S_{im} containing the functions

$$\varphi_{im0}, \varphi_{im1}, \varphi_{im2}, \ldots, \varphi_{imn_m}, \tag{10}$$

chosen as linearly independent functions in the vector space of the continuous functions on the real interval I, such that $S_{im-1} \subseteq S_{im}$ and f_{im} is a real linear combination of these functions where $i = \overline{1,2}$.

Using the functions given by (10), we define some types of approximate solutions of the system (2).

Definition 1. *A sequence of functions $\{s_{im}(y)\}_{m\in\mathbb{N}}$ of the form*

$$s_{im}(y) = \sum_{k=0}^{n_m} \alpha_{im}^k \varphi_{imk}\,, \quad m \in \mathbb{N},\ \alpha_m^k \in \mathbb{R}, i = \overline{1,2} \tag{11}$$

are called HP-sequences of system (2).

Functions of the HP-sequences are called HP-functions of system (2).

The HP-sequences $\{s_{im}(y)\}_{m\in\mathbb{N}}$ with the property

$$\lim_{m\to\infty} R_i(y, s_{1m}(y), s_{2m}(y)) = 0, \quad i = \overline{1,2}$$

is called convergent to the solution of the system (2).

Definition 2. *The HP-functions $\tilde{U}$ and $\tilde{V}$ satisfying the conditions*

$$\begin{array}{ll} |R_1(y, \tilde{U}, \tilde{V})| < \epsilon, & \mathcal{B}_i(\tilde{U}) = 0 \\ |R_2(y, \tilde{U}, \tilde{V})| < \epsilon, & \mathcal{B}_j(\tilde{V}) = 0 \end{array} \tag{12}$$

are called ε-approximate HP-solutions of the system (2).

Definition 3. *HP-function $\tilde{U}$ and $\tilde{V}$ satisfying the conditions*

$$\int_I R_1^2(y, \tilde{U}, \tilde{V})dy \leq \delta, \quad \mathcal{B}_i(\tilde{y}) = 0 \tag{13}$$

$$\int_I R_2^2(y, \tilde{v})dy \leq \delta, \quad \mathcal{B}_j(\tilde{y}) = 0 \tag{14}$$

are called weak δ-approximate HP-solutions of the system (2) on the real interval I.

Remark 1. *It is easy to see that any ε-approximate HP-solution of the system (2) is also a weak approximate HP-solution. It follows that the set of weak approximate HP-solutions of the system (2) also contains the approximate HP-solutions of the system.*

The following theorem states the existence of weak approximate HP-solutions of the system (2) and furnishes the way to construct them.

Theorem 1. *The system (2) admits a sequence of weak approximate HP-solutions.*

Proof. The first step of the proof is to construct the HP-sequences $\{s_{im}(t)\}_{m\in\mathbb{N}}$, $i = \overline{1,2}$.

Let us consider the approximate HP-solutions of the type:

$\tilde{U} = \sum\limits_{k=0}^{n_m} c_m^k \varphi_{1mk}$, where $m = 0, 1, \ldots$ and

$\tilde{V} = \sum\limits_{k=0}^{n_m} d_m^k \varphi_{2mk}$, where $m = 0, 1, \ldots$.

In the following, the unknown constants c_m^k and d_m^k $k \in \{0, 1, \ldots, k_m\}$, will be determined.

Substituting the approximate solutions $\tilde{U}$ and $\tilde{V}$ in the system (2), one gets the expression:

$$\begin{aligned} \mathcal{R}_1(y, c_m^k) &:= R_1(y, \tilde{U}) \\ \mathcal{R}_2(y, d_m^k) &:= R_2(y, \tilde{V}). \end{aligned} \tag{15}$$

Attaching to the system (2) the following real functionals

$$\begin{aligned} J_1(c_m^k) &= \int_I \mathcal{R}_1^2(y, c_m^k)dy \\ J_2(d_m^k) &= \int_I \mathcal{R}_2^2(y, d_m^k)dy \end{aligned} \tag{16}$$

and imposing the boundary conditions, we can determine $l \in \mathbb{N}$, $l \leq m$, such that $c_0^m, c_1^m, \ldots, c_l^m$ and $d_0^m, d_1^m, \ldots, d_l^m$ are computed as functions of $c_{l+1}^m, c_{l+2}^m, \ldots, c_n^m$ respectively $d_{l+1}^m, d_{l+2}^m, \ldots, d_n^m$.

Replacing $c_0^m, c_1^m, \ldots, c_l^m$ and $d_0^m, d_1^m, \ldots, d_l^m$ in (16), the values of $\tilde{c}_{l+1}^m, \tilde{c}_{l+2}^m, \ldots, \tilde{c}_n^m$ respectively $\tilde{d}_{l+1}^m, \tilde{d}_{l+2}^m, \ldots, \tilde{d}_n^m$ are computed as the values, which give the minimum of the functional (16).

Using again the boundary conditions, the values of $\tilde{c}_0^m, \tilde{c}_1^m, \ldots, \tilde{c}_l^m$ as functions of $\tilde{c}_{l+1}^m$, $\tilde{c}_{l+2}^m, \ldots, \tilde{c}_n^m$ are determined and the values of $\tilde{d}_0^m, \tilde{d}_1^m, \ldots, \tilde{d}_l^m$ as functions of $\tilde{d}_{l+1}^m, \tilde{d}_{l+2}^m, \ldots, \tilde{d}_n^m$ are determined.

Using the constants $\tilde{c}_0^m, \ldots, \tilde{c}_n^m$ and $\tilde{d}_0^m, \ldots, \tilde{d}_n^m$ thus determined, the following HP-functions

$$\begin{aligned} s_{1m}(t) &= \sum_{k=0}^{n_m} \tilde{c}_m^k \varphi_{mk} \\ s_{2m}(t) &= \sum_{k=0}^{n_m} \tilde{d}_m^k \varphi_{mk} \end{aligned} \tag{17}$$

are constructed.

The second step of the proof is to show that the above HP-functions $s_{im}(y)$ are weak approximate solutions of the system (2).

Based on the way the HP-functions $s_{im}(y)$ are computed and taking into account that f_im given by (9) are HP-functions for system (2), it follows:

$$0 \leq \int_I R_i^2(y, s_{im}(y))dy \leq \int_I R_i^2(y, f_{im}(y))dy\,, \ \forall m \in \mathbb{N}, i = \overline{1,2}.$$

Therefore,

$$0 \leq \lim_{m \to \infty} \int_I R_i^2(y, s_{im}(y))dy \leq \lim_{m \to \infty} \int_I R_i^2(y, f_{im}(y))dy, i = \overline{1,2}.$$

Since f_{im} are convergent to the solution of the system (2), we obtain:

$$\lim_{m \to \infty} \int_I R_i^2(y, s_{im}(y))dy = 0.$$

It follows that for all $\epsilon > 0$ there exists $m_0 \in \mathbb{N}$ such that for all $m \in \mathbb{N}$, $m > m_0$, the sequence $s_{im}(y)$ is a weak ϵ-approximate HP-solution of the system (2). □

Remark 2. *The proof of the above theorem give us a way to determine a weak approximate HP-solution of the system (2), $\tilde{U}, \tilde{V}$. Moreover, taking into account the Remark 1, if $|R_1(y, \tilde{U})| < \epsilon$ and $|R(y, \tilde{V})| < \epsilon$ then $\tilde{U}$ and $\tilde{V}$ are also ϵ-approximate HP-solutions of the considered system.*

3. Numerical Application

The application presented in this section is the one included in the paper by Basiri Parsa, Rashidi et al. [12], where the authors employed the well-known homotopy analysis method and differential transform method to find approximate analytical solutions for the following boundary value problem:

$$\begin{gathered} UV' - VU' = \frac{1}{Re}(U'' - Ha^2 U) \\ V^{IV} = Ha^2 V'' + Re(V'V'' - VV''') \\ U(0) = 1,\ V(0) = 0,\ V'(0) = 0,\ U(1) = 0,\ V(1) = 1,\ V'(1) = 0 \end{gathered} \tag{18}$$

These equations model a laminar magnetohydrodynamic flow of a non-Newtonian viscous fluid in a semi-porous channel under the influence of an axial uniform static magnetic field. U and V are the mean axial and normal velocity components, respectively, Ha is the Hartmann number, and Re is the Reynolds number.

In order to find analytical solutions for this type of problems, various approximation methods were employed over the years, with various rates of success, methods among which we mention: the homotopy perturbation method [10], the variational iteration method [11], the Adomian decomposition method [13], and the optimal homotopy asymptotic method [14]. While these methods (and many others) have been successfully employed, due to the nature of the equations, the computations involved are usually very difficult.

We remark that the system (18) may be used to study the influence of a magnetic field on the blood flow through a blood vessel.Numerous models have been established for the study of the hydrodynamic blood flow through the vessels, for example in [15]. Here, the authors analyze the flow of blood in tubes with reduced diameters, while in [16], the authors engage themselves in the study of the blood flow in small arched tubes, which are modeled. Blood flow has been analyzed trough the effect of the magnetic field as an great electrically conductive fluid. Knowing that blood is a ferrofluid, it can be concluded that there is the possibility of controlling the blood pressure and its flow behavior by using a fitting magnetic field. In [17], the authors came up with a mathematical representation of the blood flow in a blood vessel of reduced dimensions, in the presence of a magnetic field. Moreover, in [18], the authors investigated the apparatus of interaction between the red blood cells and an external magnetic field. The results will show the capacity of a magnetic field to modulate the blood flow. Other research on the magnetic properties of the blood are based on [12,19–27].

Many mathematical models show parts of the human circulatory system (for example [28–31]), most of the time, the blood flow is modeled by using differential equations, mostly nonlinear ones. However, it is usually nearly impossible to find exact solutions for these types of equations. Such cases require approximation methods for calculating almost exact solutions, because these approximated solutions may provide important information about the phenomenon.

In the following, we apply the least squares homotopy perturbation method to compute approximate polynomial solutions for the system (18) for two cases with particular significant values of the Hartmann number Ha and of the Reynolds number Re, and we compare our solutions with previous ones obtained in the literature.

3.1. The Case $Re = 1$ and $Ha = 0$

The case $Re = 1$ and $Ha = 0$ corresponds to a non-conducting blood flow. In [12], Basiri Parsa et al. computed approximate solutions of the system (18) by using the homotopy analysis method (HAM) and the differential transform method (DTM), and in [32], Caruntu et al. computed approximate solutions by using the polynomial least squares method (PLSM).

In this case, employing LSHPM for the system (18), we compute the approximate solutions as follows:

The linear operators are:

$$\begin{gathered} \mathcal{L}_1(\Phi_1(y,p)) = -\frac{1}{Re}U'' \\ \mathcal{L}_2(\Phi_2(y,p)) = V^{IV} \end{gathered} \tag{19}$$

and the nonlinear operators are:

$$\begin{aligned} \mathcal{N}_1(\Phi_1(t,p)) &= \frac{Ha^2}{Re}U + UV' - U'V \\ \mathcal{N}_2(\Phi_2(t,p)) &= -Ha^2V'' - Re(V'V'' - VV'''). \end{aligned} \tag{20}$$

We obtain the HPM approximations:

- First-term approximations:

$$\begin{aligned} U_0(y) &= 1 - y \\ V_0(y) &= 3y^2 - y^3 \end{aligned} \tag{21}$$

- Second-term approximations:

$$\begin{aligned} U_1(y) &= \frac{y^5}{5} - \frac{3y^4}{4} + y^3 - \frac{29y}{20} + 1 \\ V_1(y) &= \frac{2y^7}{35} - \frac{y^6}{5} + \frac{3y^5}{10} - \frac{167y^3}{70} + \frac{113y^2}{35} \end{aligned} \tag{22}$$

- Third-term approximations:

$$\begin{aligned} U_2(y) &= \frac{2y^9}{315} - \frac{19y^8}{560} + \frac{y^7}{20} - \frac{y^6}{20} + \frac{23y^5}{70} - \frac{1643y^4}{1680} + \frac{113y^3}{105} - \frac{1763y}{1260} + 1 \\ V_2(y) &= \frac{4y^{11}}{5775} - \frac{2y^{10}}{525} + \frac{y^9}{210} - \frac{3y^8}{560} + \frac{97y^7}{1225} - \frac{533y^6}{2100} + \frac{121y^5}{350} - \frac{774469y^3}{323400} + \frac{2087479y^2}{646800} \end{aligned} \tag{23}$$

For the second-term approximation $U_0 + U_1$, the set S_{1m} consist of the functions $\{y, y^3, y^4, y^5\}$ and for $V_0 + V_1$ the set is $S_{2m} = \{y^2, y^3, y^5, y^6, y^7\}$.

We will compute the approximate solutions $\tilde{U}(y) = c_0 + c_1 y + c_2 y^3 + c_3 y^4 + c_4 y^5$ and $\tilde{V}(y) = d_0 y^2 + d_1 y^3 + d_2 y^5 + d_3 y^6 + d_4 y^7$.

From the initial conditions: $\tilde{U}(0) = 1, \tilde{U}(1) = 0$ and $\tilde{V}(0) = 0, \tilde{V}(1) = 1, \tilde{V}'(0) = 0, \tilde{V}'(1) = 0$ we obtain: $c_0 = 1$ and $c_1 = -1 - c_2 - c_3 - c_4$ respectively $-d_0 = 2d_2 + 3d_3 + 4d_4 + 3$ and $d_1 = -3d_2 - 4d_3 - 5d_4 - 2$.

Replacing these expressions of c_0, c_1, d_0, and d_1 in the corresponding remainders (15) are:

$$\begin{aligned} \mathcal{R}_1(y, \tilde{U}) &= \mathcal{R}(y, c_2, c_3, c_4) \\ \mathcal{R}_2(y, \tilde{V}) &= \mathcal{R}(y, d_2, d_3, d_4) \end{aligned} \tag{24}$$

Next, we compute the corresponding functionals (16) (too long to be included here):

$$\begin{aligned} J_1(c_2, c_3, c_4) &= \int_0^1 \mathcal{R}_1^2(y, c_2, c_3, c_4)dy \\ J_2(d_2, d_3, d_4) &= \int_0^1 \mathcal{R}_2^2(y, d_2, d_3, d_4)dy \end{aligned} \tag{25}$$

and we compute the minimum of this functionals, determining the coefficients c_j and d_j, $j = \overline{2,4}$, thus finding the approximate solutions of the system (18) by LSHPM.

In a similar way, we compute the third-term approximations by LSHPM, and the solutions are:

- Second-term approximations:
  $\tilde{U}(y) = 0.25139431098009168490y^5 - 0.89052033909540654567y^4$
  $+ 1.0440071928069991650y^3 - 1.4048811646916843042y + 1$
  $\tilde{V}(y) = 0.05831712746233677986 3y^7 - 0.21889392791262717538y^6$
  $+ 0.32522212582352969927y^5 - 2.3916763031317642956y^3 + 3.2270309777585249919y^2$
- Third-term approximations:
  $\tilde{U}(y) = -0.012783174216369037776y^9 + 0.079312017411340841618y^8$
  $- 0.21264602372589301173y^7 + 0.26096763633169609359y^6$

$$+ 0.13236850157788926145y^5 - 0.90665763559833919190y^4 + 1.06533237499444804 12y^3$$
$$- 1.40589369677477299 64y + 1$$
$$\tilde{V}(y) = 0.00077252197269867863129y^{11} - 0.0033512835587865196224y^{10}$$
$$+ 0.0028365610269174046158y^9 - 0.0031087019658224288029y^8$$
$$+ 0.079915383004597381009y^7 - 0.25695066296630124619y^6$$
$$+ 0.34706280997410686745y^5 - 2.3943088377575857329y^3 + 3.2271322102701755958y^2$$

The comparison of these LSHPM solutions with the previous approximate solutions computed in [12] using HAM and DTM, and in [32] using PLSM, is presented in Figures 1 and 2. Since no exact solutions are available, the comparison is done by means of computing the error relative to a fourth-order Runge–Kutta method numerical solution (i.e., the absolute errors are computed as the difference between our approximate solutions and the numerical solutions).

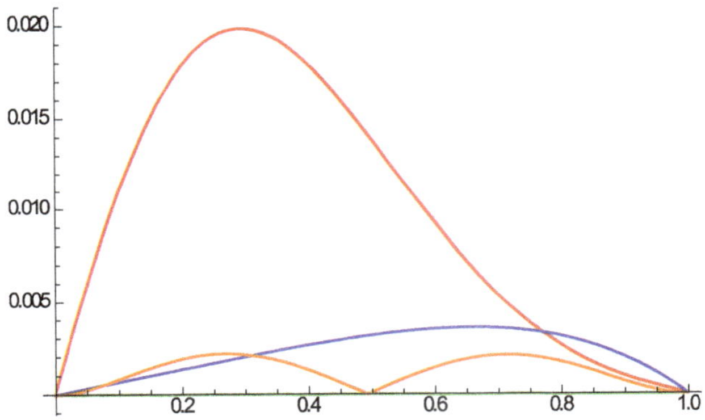

Figure 1. Comparison of absolute errors corresponding to the approximation from [12] U_{DTM} (red curve), [10] U_{HPM} 3 terms (blue curve), [32] U_{PLSM} (orange curve), and our LSHPM approximation U_{LSHPM} (green curve).

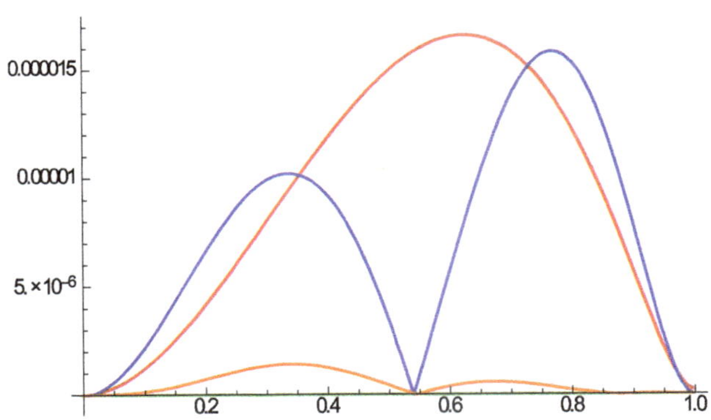

Figure 2. Comparison of absolute errors corresponding to the approximation from [12] V_{DTM} (red curve), [10] V_{HPM} 3 terms (blue curve), [32] V_{PLSM} (orange curve), and our LSHPM approximation V_{LSHPM} (green curve).

The comparison is further illustrated by Tables 1 and 2, which includes the results obtained in [12], by means of the HAM and DTM, the results obtained in [32] by PLSM, and the results computed by classical HPM and by LSHPM. The comparison lead to the same conclusion: the approximate solutions obtained by LSHPM are more precise.

Table 1. Comparison of the absolute errors of the approximate solutions U in case $Re = 1$ and $Ha = 0$.

y	U_{HAM}	U_{DTM}	U_{PLSM}	$U_{HPM2term.}$	$U_{HPM3t.}$	$U_{LSHPM2t.}$	$U_{LSHPM3t.}$
0.1	9.04×10^{-3}	1.08×10^{-2}	8.06×10^{-4}	4.45×10^{-3}	6.77×10^{-4}	8.61×10^{-5}	3.68×10^{-6}
0.2	1.77×10^{-2}	1.77×10^{-2}	1.90×10^{-3}	9.08×10^{-3}	1.35×10^{-3}	8.00×10^{-5}	1.56×10^{-6}
0.3	2.71×10^{-2}	1.98×10^{-2}	2.14×10^{-3}	1.37×10^{-2}	2.01×10^{-3}	4.18×10^{-6}	3.41×10^{-6}
0.4	3.45×10^{-2}	1.79×10^{-2}	1.34×10^{-3}	1.78×10^{-2}	2.64×10^{-3}	9.71×10^{-5}	9.99×10^{-8}
0.5	3.73×10^{-2}	1.39×10^{-2}	8.14×10^{-5}	2.10×10^{-2}	3.16×10^{-3}	1.29×10^{-4}	2.42×10^{-6}
0.6	3.46×10^{-2}	9.33×10^{-3}	1.43×10^{-3}	2.24×10^{-2}	3.50×10^{-3}	7.69×10^{-5}	1.22×10^{-6}
0.7	2.77×10^{-2}	5.39×10^{-3}	2.09×10^{-3}	2.15×10^{-2}	3.54×10^{-3}	2.59×10^{-5}	9.10×10^{-7}
0.8	1.87×10^{-2}	2.63×10^{-3}	1.74×10^{-3}	1.78×10^{-2}	3.12×10^{-3}	9.63×10^{-5}	7.61×10^{-7}
0.9	9.27×10^{-3}	9.33×10^{-4}	6.99×10^{-4}	1.07×10^{-2}	2.02×10^{-3}	6.71×10^{-5}	4.05×10^{-7}

Table 2. Comparison of the absolute errors of the approximate solutions V in case $Re = 1$ and $Ha = 0$.

y	V_{HAM}	V_{DTM}	V_{PLSM}	$V_{HPM2term.}$	$V_{HPM3t.}$	$V_{LSHPM2t.}$	$V_{LSHPM3t.}$
0.1	2.21×10^{-4}	1.19×10^{-6}	1.48×10^{-7}	2.25×10^{-5}	2.13×10^{-6}	1.41×10^{-6}	2.20×10^{-8}
0.2	5.75×10^{-4}	4.10×10^{-6}	7.17×10^{-7}	1.14×10^{-4}	6.55×10^{-6}	1.21×10^{-5}	9.78×10^{-9}
0.3	6.47×10^{-2}	7.91×10^{-6}	1.34×10^{-6}	2.83×10^{-4}	9.89×10^{-6}	3.20×10^{-5}	4.69×10^{-9}
0.4	3.03×10^{-4}	1.17×10^{-5}	1.26×10^{-6}	4.95×10^{-4}	9.25×10^{-6}	5.07×10^{-5}	1.40×10^{-8}
0.5	2.63×10^{-4}	1.49×10^{-5}	4.45×10^{-7}	6.84×10^{-4}	3.56×10^{-6}	5.66×10^{-5}	1.53×10^{-8}
0.6	7.02×10^{-4}	1.65×10^{-5}	3.80×10^{-7}	7.75×10^{-4}	5.54×10^{-6}	4.58×10^{-5}	1.84×10^{-8}
0.7	7.62×10^{-4}	1.59×10^{-5}	5.63×10^{-7}	7.12×10^{-4}	1.37×10^{-5}	2.53×10^{-5}	1.67×10^{-8}
0.8	4.73×10^{-4}	1.23×10^{-5}	2.24×10^{-7}	4.87×10^{-4}	1.54×10^{-5}	7.48×10^{-6}	1.46×10^{-8}
0.9	1.25×10^{-4}	5.88×10^{-6}	2.11×10^{-8}	1.80×10^{-4}	8.05×10^{-6}	2.77×10^{-7}	1.41×10^{-8}

3.2. The Case $Re = 1$ and $Ha = 1$

In the case $Re = 1$ and $Ha = 1$, the influence of the magnetic field on the blood flow is non-negligible and the flow is weakly magnetic. The computations by LSHPM are similar to the ones in the previous case.

The approximations terms by HPM are:

- First-term approximations:

$$\begin{aligned} U_0(y) &= 1 - y \\ V_0(y) &= 3y^2 - 2y^3 \end{aligned} \tag{26}$$

- Second-term approximations:

$$\begin{aligned} U_1(y) &= \frac{y^5}{5} - \frac{3y^4}{4} + \frac{5y^3}{6} + \frac{y^2}{2} - \frac{107y}{60} + 1 \\ V_1(y) &= \frac{2y^7}{35} - \frac{y^6}{5} + \frac{y^5}{5} + \frac{y^4}{4} - \frac{181y^3}{70} + \frac{459y^2}{140} \end{aligned} \tag{27}$$

- Third-term approximations:

$$\begin{aligned} U_2(y) &= \frac{2y^9}{315} - \frac{19y^8}{560} + \frac{9y^7}{140} - \frac{2y^6}{15} + \frac{2129y^5}{4200} - \frac{451y^4}{420} + \frac{401y^3}{504} + \frac{y^2}{2} - \frac{41129y}{25200} + 1 \\ V_2(y) &= \frac{4y^{11}}{5775} - \frac{2y^{10}}{525} + \frac{y^9}{140} - \frac{9y^8}{560} + \frac{1507y^7}{14700} - \frac{1129y^6}{4200} + \frac{317y^5}{1400} + \frac{153y^4}{560} \\ &\quad - \frac{838379y^3}{323400} + \frac{352623y^2}{107800} \end{aligned} \tag{28}$$

The corresponding solutions obtained by using LSHPM are:

- Second-term approximations:

$\tilde{U}(y) = 0.24252310472551437387y^5 - 0.79607170987751279666y^4$
$+ 0.70039529434560076978y^3 + 0.51680281524640722276y^2$
$- 1.6636495044400095698y + 1$
$\tilde{V}(y) = 0.058199385875154810585y^7 - 0.20176414373359385855y^6$
$+ 0.17947791112252870094y^5 + 0.28462933749944678084y^4 - 2.5916327628078782832y^3$
$+ 3.2710902720443418494y^2$

- Third-term approximations:
  $\tilde{U}(y) = -0.039570918805227895983y^9 + 0.18039801278862870509y^8$
  $- 0.33738809541609613996y^7 + 0.25556139173308396128y^6 + 0.28898499486097221037y^5$
  $- 0.98687905924276726841y^4 + 0.80192640217575596712y^3 + 0.50063075107844706510y^2$
  $- 1.6636634791727966046y + 1$
  $\tilde{V}(y) = 0.00070249628967946626576y^{11} - 0.0034005002946346048030y^{10}$
  $+ 0.0063398061612668565765y^9 - 0.017261143110682215908y^8$
  $+ 0.10727677845680360313y^7 - 0.27266199870132195652y^6 + 0.22694957760601929024y^5$
  $+ 0.27260900187074289280y^4 - 2.5917328827530869045y^3 + 3.2711788644752135727y^2$

Again, the comparisons included in Tables 3 and 4 allow us the reach the same conclusion as in the previous case, namely that the approximations obtained by LSHPM are more accurate than the previous ones by other methods.

We mention the fact that we computed approximations for a wide range of values of the parameters *Re* and *Ha*, and the above conclusions remained valid for all of the computed solutions.

Table 3. Comparison of the absolute errors of the approximate solutions for U for the case $Re = 1$ and $Ha = 1$.

y	U_{HAM}	U_{DTM}	U_{PLSM}	$U_{HPM2term.}$	$U_{HPM3t.}$	$U_{LSHPM2t.}$	$U_{LSHPM3t.}$
0.1	1.19×10^{-2}	3.18×10^{-2}	8.22×10^{-5}	1.19×10^{-2}	3.13×10^{-3}	8.03×10^{-5}	4.19×10^{-7}
0.2	1.59×10^{-2}	5.15×10^{-2}	1.17×10^{-4}	2.33×10^{-2}	6.14×10^{-3}	1.14×10^{-4}	1.14×10^{-6}
0.3	1.25×10^{-2}	5.95×10^{-2}	3.15×10^{-5}	3.35×10^{-2}	8.85×10^{-3}	2.72×10^{-5}	1.41×10^{-7}
0.4	5.92×10^{-3}	5.84×10^{-2}	9.22×10^{-5}	4.14×10^{-2}	1.10×10^{-2}	9.64×10^{-5}	1.32×10^{-6}
0.5	1.62×10^{-4}	5.14×10^{-2}	1.48×10^{-4}	4.60×10^{-2}	1.25×10^{-2}	1.52×10^{-4}	3.46×10^{-7}
0.6	2.79×10^{-3}	4.11×10^{-2}	9.29×10^{-5}	4.65×10^{-2}	1.29×10^{-2}	9.57×10^{-5}	1.10×10^{-6}
0.7	3.20×10^{-3}	2.99×10^{-2}	3.03×10^{-5}	4.24×10^{-2}	1.22×10^{-2}	2.86×10^{-5}	3.63×10^{-7}
0.8	2.34×10^{-3}	1.92×10^{-2}	1.16×10^{-4}	3.33×10^{-2}	9.21×10^{-3}	1.15×10^{-4}	7.95×10^{-7}
0.9	1.21×10^{-3}	9.05×10^{-3}	8.04×10^{-5}	1.92×10^{-2}	5.92×10^{-3}	8.01×10^{-5}	1.05×10^{-7}

Table 4. Comparison of the absolute errors of the approximate solutions for V for the case $Re = 1$ and $Ha = 1$.

y	V_{HAM}	V_{DTM}	V_{PLSM}	$V_{HPM2term.}$	$V_{HPM3t.}$	$V_{LSHPM2t.}$	$V_{LSHPM3t.}$
0.1	4.90×10^{-4}	2.70×10^{-4}	3.06×10^{-8}	7.74×10^{-5}	1.53×10^{-6}	1.14×10^{-8}	1.91×10^{-8}
0.2	1.15×10^{-3}	5.83×10^{-4}	8.27×10^{-8}	3.03×10^{-4}	8.01×10^{-6}	5.24×10^{-6}	7.27×10^{-9}
0.3	1.19×10^{-3}	5.51×10^{-4}	1.37×10^{-7}	6.22×10^{-4}	2.06×10^{-6}	1.87×10^{-5}	4.18×10^{-10}
0.4	4.36×10^{-4}	1.45×10^{-4}	1.04×10^{-7}	9.38×10^{-4}	3.79×10^{-6}	3.38×10^{-5}	3.16×10^{-9}
0.5	7.25×10^{-4}	4.47×10^{-4}	1.51×10^{-8}	1.14×10^{-3}	5.51×10^{-6}	4.05×10^{-5}	1.35×10^{-9}
0.6	1.66×10^{-3}	9.52×10^{-4}	9.11×10^{-8}	1.16×10^{-3}	6.56×10^{-6}	3.41×10^{-5}	3.91×10^{-10}
0.7	1.90×10^{-3}	1.12×10^{-3}	6.89×10^{-8}	9.70×10^{-4}	6.30×10^{-5}	1.91×10^{-5}	1.24×10^{-9}
0.8	1.35×10^{-3}	8.72×10^{-4}	1.55×10^{-8}	6.05×10^{-4}	4.48×10^{-5}	5.55×10^{-6}	7.28×10^{-10}
0.9	4.73×10^{-4}	3.42×10^{-4}	6.71×10^{-10}	2.04×10^{-4}	1.71×10^{-6}	1.06×10^{-7}	3.35×10^{-9}

4. Discussion of the Results

As we mention in the previous section, we computed approximations for a wide range of values of the parameters *Re* and *Ha*, and the conclusions of our study are in very good agreement with previous ones included in [12,33–35] and other studies. This is to be expected because, even though our approximations are much more precise than the ones included in the previous studies, the overall phenomena described by the solutions are obviously the same. In the following, we will briefly summarize the results of the study.

The first results are related to the influence of the Reynolds number on the velocities U and V, when the value of the Hartmann number is fixed. We were able to draw similar conclusions for both cases studied in [12] (the non-conducting case $Ha = 0$ and the weakly magnetic flow case $Ha = 1$). The consequences of any increase of the Reynolds number are a modest increase to the $V(y)$ component of the velocity of the blood flow, and a major decrease of the $U(y)$ component. The effect of this phenomenon is a major deceleration of the blood velocity in the x-direction. For the case $Ha = 0$, these conclusions are illustrated by the Figures 3 and 4, while for the case $Ha = 1$ (and, actually, for any other value of Ha on its nominal interval $[0, 2]$), the figures look very similar.

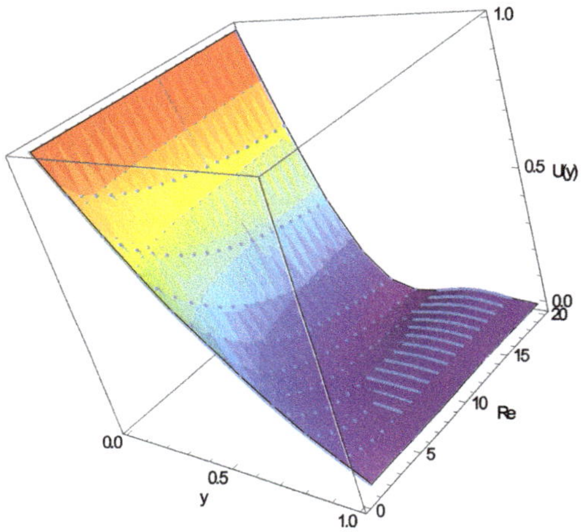

Figure 3. The effect of the increase of Re on U for the case $Ha = 0$—three dimensional plot.

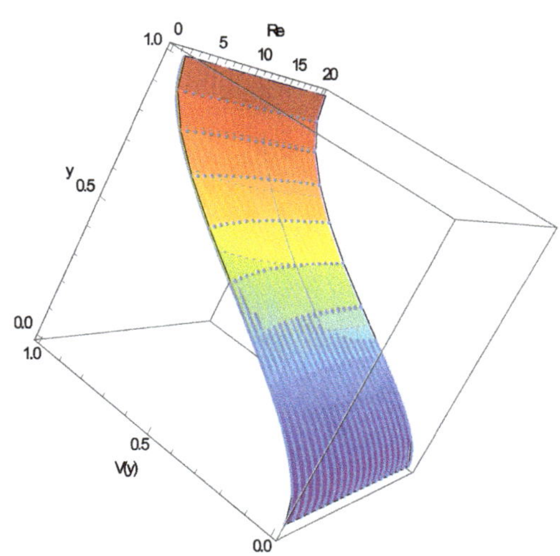

Figure 4. The effect of the increase of Re on V for the case $Ha = 0$—three dimensional plot.

The next study item is the influence of the Hartmann number on the velocities U and V for fixed values of the Reynolds number. Because the magnetic field is applied in the y-direction only, there is no visible influence of the increase of the Ha $V(y)$ component of the blood velocity while there is a decrease of the $U(y)$ component, as expected. These conclusions are illustrated for $Re = 1$ in Figures 5 and 6 (again, we mention that for any values of the Reynolds number on its nominal interval [1, 20], the figures are similar).

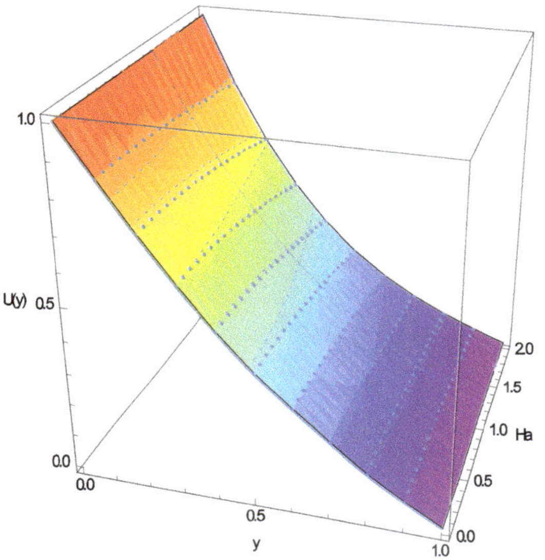

Figure 5. The effect of the increase of Ha on U for the case $Re = 1$—three dimensional plot.

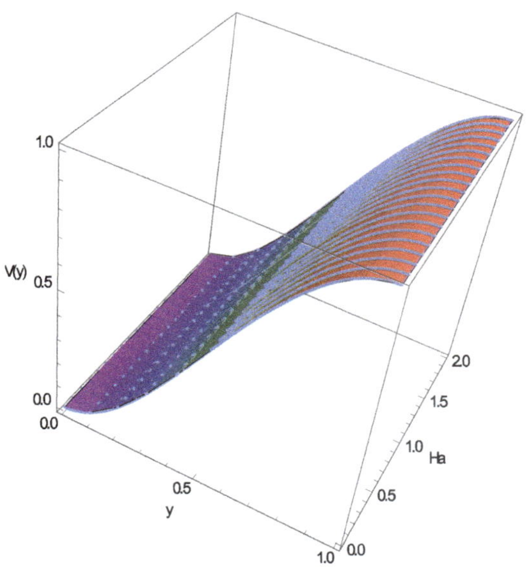

Figure 6. The effect of the increase of Ha on V for the case $Re = 1$—three dimensional plot.

In the last part of the study, we investigated the merged impact of Re and Ha on $U(y)$ and $V(y)$, impact highlighted in the Figures 7 and 8 for $y = 0.2$ (red surface), $y = 0.4$ (blue surface), $y = 0.6$ (yellow surface) and $y = 0.8$ (green surface).

Figure 7 is a good synthesis of the research done on the impact of Re and Ha on a replicated blood flow in a semi-porous channel, as it is obvious that the increase in both

Re and *Ha* conduct to the reduction of $U(y)$. At the same time, Figure 8 gives new insight regarding the interplay between the impact of *Re* and *Ha* on the blood flow velocity in the y-direction, particularly the fact that the decelerative consequence of the increase of *Ha* greatly depends on *Re*. This effect is greater for reduced values of *Re* and, as a result, in situations where, due to practical discussions, a strong suction at the upper wall (defined by a large value of *Re*) cannot be achieved, a decrease of blood flow velocity can be achieved by boosting the intensity of the magnetic field applied. Even if it is practically possible, an increase of the suction at the upper wall is apparently the preferable method for reducing the flow, since the effect of the increase of Re seems to be considerably greater than the effect of the increase of Ha. Furthermore, if the value of the Reynolds number is large, the consequences of the increase of the Hartmann number is small.

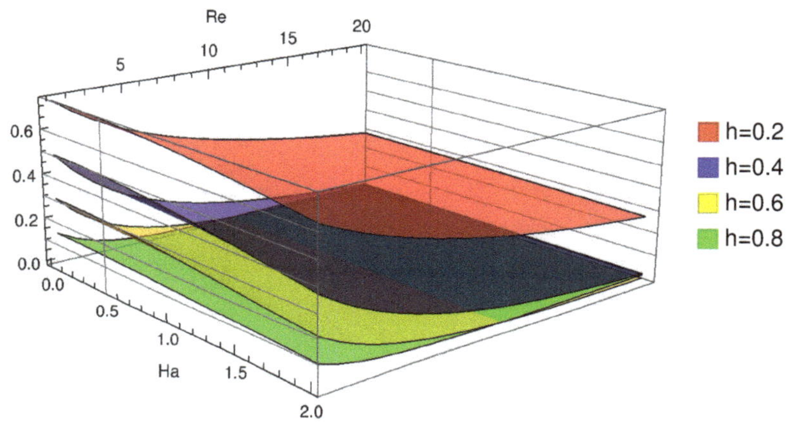

Figure 7. The combined influence of *Re* and *Ha* on U. The red surface corresponds to $y = 0.2$, the blue surface to $y = 0.4$, the yellow surface to $y = 0.6$ and the green one to $y = 0.8$.

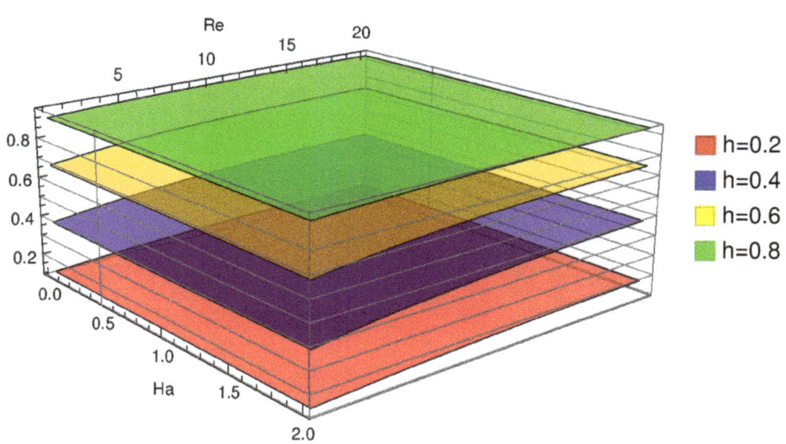

Figure 8. The combined influence of *Re* and *Ha* on V. The red surface corresponds to $y = 0.2$, the blue surface to $y = 0.4$, the yellow surface to $y = 0.6$, and the green one to $y = 0.8$.

5. Conclusions

The least squares homotopy perturbation method is introduced as a straightforward and very accurate method to compute analytically approximate solutions for systems of ordinary differential equations.

The method was employed for a system of equations modeling the blood flow through a blood vessel under the action of a magnetic field. The comparison with approximate solutions computed by using well-known methods, such as the homotopy analysis method, the differential transform method and the homotopy perturbation method, clearly illustrate the precision of our method.

Author Contributions: Conceptualization, M.S.P. and B.C.; Data curation, M.S.P., O.B., A.J. and B.C.; Formal analysis, B.C.; Funding acquisition, M.S.P., O.B., A.J. and B.C.; Investigation, M.S.P., O.B., A.J. and B.C.; Methodology, B.C.; Software, M.S.P. and B.C.; Supervision, B.C.; Validation, M.S.P. and B.C.; Visualization, M.S.P., A.J. and B.C.; Writing—original draft, M.S.P., O.B., A.J. and B.C.; Writing—review and editing, M.S.P. and B.C. All authors have read and agreed to the published version of the manuscript.

Funding: This research received no external funding.

Institutional Review Board Statement: Not applicable.

Informed Consent Statement: Not applicable.

Conflicts of Interest: The authors declare no conflict of interest.

References

1. Bota, C.; Caruntu, B. Approximate analytical solutions of nonlinear differential equations using the Least Squares Homotopy Perturbation Method. *J. Math. Anal. Appl.* **2017**, *448*, 401–408. [CrossRef]
2. Bota, C.; Caruntu, B.; Lazureanu, C. The Least Squares Homotopy Perturbation Method for boundary value problems. *Appl. Comput. Math.* **2017**, *16*, 39–47.
3. Qayyum, M.; Oscar, I. Least Square Homotopy Perturbation Method for Ordinary Differential Equations. *J. Math.* **2021**, *2021*, 7059194. [CrossRef]
4. Thabet, H.; Kendre, S. Modified least squares homotopy perturbation method for solving fractional partial differential equations. *Malaya J. Mat.* **2018**, *6*, 420–427. [CrossRef]
5. Kumar, R.; Koundal, R.; Shehzad, S.A. Generalized least square homotopy perturbation solution of fractional telegraph equations. *Comput. Appl. Math.* **2019**, *38*, 184. [CrossRef]
6. Zhang, J.; Wei, Z.; Li, L.; Zhou, C. Least-Squares Residual Power Series Method for the Time-Fractional Differential Equations. *Complexity* **2019**, *2019*, 6159024. [CrossRef]
7. Das, P.; Rana, S. Theoretical prospects of fractional order weakly singular Volterra Integro differential equations and their approximations with convergence analysis. *Math. Methods Appl. Sci.* **2021**, *44*, 9419–9440. [CrossRef]
8. Kumar, R.; Koundal, R.; Shehzad, S.A. Modified homotopy perturbation approach for the system of fractional partial differential equations: A utility of fractional Wronskian. *Math. Methods Appl. Sci.* **2022**, *45*, 809–826. [CrossRef]
9. He, J.H. Homotopy perturbation technique. *Comput. Methods Appl. Mech. Eng.* **1999**, *178*, 257–262. [CrossRef]
10. He, J.H. Application of homotopy perturbation method to nonlinear wave equations. *Chaos Solitons Fractals* **2005**, *26*, 695–700. [CrossRef]
11. He, J.H. Variational iteration method, a kind of nonlinear analytical technique. Some examples. *Int. J. Non-Linear Mech.* **1999**, *34*, 699–708. [CrossRef]
12. Parsa, A.B.; Rashidi, M.M.; Beg, O.A.; Sardi, S.M. Semi-computational simulation of magneto-hemodynamic flow in a semi-porous channel using optimal homotopy and differential transform methods. *Comput. Biol. Med.* **2013**, *43*, 1142–1153. [CrossRef] [PubMed]
13. Adomian, G.A. Review of the decomposition method in applied mathematics. *J. Math. Anal. Appl.* **1998**, *135*, 501–544. [CrossRef]
14. Marinca, V.; Herisanu, N.; Bota, C. An optimal homotopy asymptotic method applied to the steady flow of a fourth-grade fluid past a porous plate. *Appl. Math. Lett.* **2009**, *22*, 245–251. [CrossRef]
15. Srivastava, V.P. A Theoretical Model for Blood Flow in Small Vessels. *Int. J. Appl. Appl. Math.* **2007**, *2*, 51–65.
16. Wang, C.Y.; Bassingthwaighte, J.B. Blood Flow in Small Curved Tubes. *J. Biomech. Eng.* **2003**, *125*, 910–913. [CrossRef] [PubMed]
17. Bali, R.; Awasthi, U. Mathematical model of Blood Flow in Small Blood Vessel in the Presence of Magnetic Field. *Appl. Math.* **2001**, *2*, 264–269. [CrossRef]
18. Yamamoto, T.; Nagayama, Y.; Tamura, M. A blood-oxygenation-dependent increase in blood viscosity due to a static magnetic field. *Phys. Med. Biol.* **2004**, *49*, 3267–3267. [CrossRef] [PubMed]
19. Alshare, A.; Tashtoush, B.; El-Khalil, H.H. Computational Modeling of Non-Newtonian Blood Flow Through Stenosed Arteries in the Presence of Magnetic Field. *J. Biomech. Eng.* **2013**, *135*, 1145–1153. [CrossRef]
20. Weng, H.C. Hydrodynamic Modeling of Targeted Magnetic-Particle Delivery in a Blood Vessel. *J. Biomech. Eng.* **2013**, *135*, 034504. [CrossRef] [PubMed]
21. Mekheimer, K.S. Peristaltic flow of blood under effect of a magnetic field in a non-uniform channels. *Appl. Math. Comput.* **2004**, *153*, 763–777. [CrossRef]

22. Tenforde, T.S. Magnetically induced electric fields and currents in the circulatory system. *Prog. Biophys. Mol. Biol.* **2005**, *87*, 279–288. [CrossRef] [PubMed]
23. Bhatti, M.M.; Zeeshan, A.; Bashir, F.; Sait, S.M.; Ellahi, R. Sinusoidal motion of small particles through a Darcy-Brinkman-Forchheimer microchannel filled with non-Newtonian fluid under electro-osmotic forces. *J. Taibah Univ. Sci.* **2021**, *15*, 514–529. [CrossRef]
24. Rostami, S.; Ellahi, R.; Oztop, H.F.; Goldanlou, A.S. A study on the effect of magnetic field and the sinusoidal boundary condition on free convective heat transfer of non-Newtonian power-law fluid in a square enclosure with two constant-temperature obstacles using lattice Boltzmann method. *J. Therm. Anal. Calorim.* **2021**, *144*, 2557–2573. [CrossRef]
25. Khan, A.S.; Xu, H.Y.; Khan, W. Magnetohydrodynamic Hybrid Nanofluid Flow Past an Exponentially Stretching Sheet with Slip Conditions. *Mathematics* **2021**, *9*, 3291–3305. [CrossRef]
26. Rehman, A.; Salleh, Z. Influence of Marangoni Convection on Magnetohydrodynamic Viscous Dissipation and Heat Transfer on Hybrid Nanofluids in a Rotating System among Two Surfaces. *Mathematics* **2021**, *9*, 2242. [CrossRef]
27. Ali, B.; Naqvi, R.A.; Haider, A.; Hussain, D.; Hussain, S. Finite Element Study of MHD Impacts on the Rotating Flow of Casson Nanofluid with the Double Diffusion Cattaneo—Christov Heat Flux Model. *Mathematics* **2020**, *8*, 1555. [CrossRef]
28. Ryu, J.; Hu, X.; Shadden, S.C. A Coupled Lumped-Parameter and Distributed Network Model for Cerebral Pulse-Wave Hemodynamics. *J. Biomech. Eng.* **2015**, *137*, 101009. [CrossRef]
29. Srinivasacharya, D.; Rao, G.M. Mathematical model for blood flow through a bifurcated artery using couple stress fluid. *Math. Biosci.* **2016**, *278*, 37–47. [CrossRef]
30. Sinha, A. MHD flow and heat transfer of a third order fluid in a porous channel with stretching wall: Application to hemodynamics. *Alex. Eng. J.* **2015**, *54*, 1243–1252. [CrossRef]
31. Zaman, A.; Ali, N.; Sajid, M. Numerical simulation of pulsatile flow of blood in a porous-saturated overlapping stenosed artery. *Math. Comput. Simul.* **2017**, *134*, 1–16. [CrossRef]
32. Caruntu, B.; Bota, C.; Bundau, O. Analytical simulation of magneto-hemodynamic flow in a semi-porous channel using the Polynomial Least Squares Method. *ITM Web Conf.* **2019**, *29*, 1–13. [CrossRef]
33. Skalak, F.M.; Wang, C.Y. On the non-unique solutions of laminar flow through a porous tube or channel. *SIAM J. Appl. Math.* **1978**, *34*, 535–544. [CrossRef]
34. Quaile, J.P.; Levy, E.K. Laminar flow in a porous tube with suction. *Int. J. Heat Mass Transf.* **1975**, *97*, 223–243. [CrossRef]
35. Evans, E.A.; Skalak, R. Mechanics and Thermodynamics of Biomembranes. In *Elsevier Biomedical*; Hue, L., Van de Werve, G., Eds.; Elsevier Biomedical: Amsterdam, NY, USA, 1981; 464p.

Σ mathematics

MDPI

Article

Lie Geometric Methods in the Study of Driftless Control Affine Systems with Holonomic Distribution and Economic Applications

Liviu Popescu [1,*], Daniel Militaru [2] and Gabriel Tică [1]

1 Department of Statistics and Economic Informatics, University of Craiova, 200585 Craiova, Romania; tica.leonard.q5k@student.ucv.ro
2 Independent Researcher, 200585 Craiova, Romania; militarunicolaedaniel@yahoo.com
* Correspondence: liviu.popescu@edu.ucv.ro

Abstract: In the present paper, two optimal control problems are studied using Lie geometric methods and applying the Pontryagin Maximum Principle at the level of a new working space, called Lie algebroid. It is proved that the framework of a Lie algebroid is more suitable than the cotangent bundle in order to find the optimal solutions of some driftless control affine systems with holonomic distributions. Finally, an economic application is given.

Keywords: control affine systems; controllability; optimal control; Hamilton–Jacobi–Bellman equations; Lie geometric methods; holonomic distribution; economic applications

Citation: Popescu, L.; Militaru, D.; Tică, G. Lie Geometric Methods in the Study of Driftless Control Affine Systems with Holonomic Distribution and Economic Applications. *Mathematics* **2022**, *10*, 545. https://doi.org/10.3390/math10040545

Academic Editors: Mihaela Neamţu, Eva Kaslik and Anca Rădulescu

Received: 30 December 2021
Accepted: 7 February 2022
Published: 10 February 2022

Publisher's Note: MDPI stays neutral with regard to jurisdictional claims in published maps and institutional affiliations.

Copyright: © 2022 by the authors. Licensee MDPI, Basel, Switzerland. This article is an open access article distributed under the terms and conditions of the Creative Commons Attribution (CC BY) license (https://creativecommons.org/licenses/by/4.0/).

1. Introduction

In the last decades, Lie geometric methods have been applied successfully in different domains of research such as dynamical systems or optimal control theory. In this paper, some Lagrangian systems with some external holonomic constraints are studied. These types of nonlinear systems have many applications in different areas of optimal control theory, cybernetics or mathematical economics. In [1], an introduction to optimal control problems in life sciences and economics is presented, while in [2], some applications of the control theory of economic growth are given. The book [3] presents a modern and thorough exposition of the fundamental mathematical formalism used to study optimal control theory, i.e., continuous time dynamic economic processes and to interpret dynamic economic behavior. In the book [4], the notions of deterministic optimal control systems governed by ordinary differential equations are studied. These models cover the problems of economic growth, exploitation of (non)renewable resources, pollution control, behavior of firms or differential games. The monograph [5] presents some optimal control models with management science applications. The book [6] covers the main results of optimal control theory, in particular necessary and sufficient optimality conditions. In the paper [7], an optimal control problem regarding a production–inventory system with customer impatience is studied, and optimizing a production–inventory system under a cost target is investigated in [8]. In addition, a new approach to maximizing the profit in a stock-dependent demand inventory model is presented in [9]. Otherwise, there are a multitude of papers that study the optimization of production and storage costs with various restrictions (see form instance, [10–15]).

The notion of the Lie algebroid was introduced in differential geometry in the early 1950s, but it can also be found in physics or algebra under a wide variety of names. Using the geometry of Lie algebroids, in [16] a generalized theory of Lagrangian mechanics is developed and the equations of motions are obtained using the Poisson structure on the dual of a Lie algebroid and Legendre transformation associated with a regular Lagrangian. Later, in [17–19], the same equations of motion are obtained using the symplectic formalism for Lagrangians and Hamiltonians, similarly with the J. Klein formalism for the classical Lagrangian mechanics. The first step in studying the mechanical control systems on Lie algebroids seems to be performed in [20], where the problems of accessibility and controllability are also approached. The Pontryagin Maximum Principle on Lie algebroids is presented in [21] and later extended on almost Lie algebroids in [22]. Some aspects regarding the abnormality problem in control theory on Lie algebroids are presented in [23]. The link between optimal control and connection theory on Lie algebroids can be found in [24–26]. Lie geometric methods in control theory have been applied in many papers. In [27], the connection between Lie theory and control is presented. The book [28] deals with the issue of being able to steer the system from any point of departure to any desired destination and also studies the optimal control problems and the question of finding the best possible trajectory. In addition, some facts and methods of control theory treated from the geometric point of view are presented in [29]. The geometry of Lie algebroids is used in the study of distributional systems (driftless control afine system) in the papers [12,14,30–36].

One of the most well-known and useful methods in the geometric approach is the analysis of the solution for the optimal control problem, as given by Pontryagin's Maximum Principle. It is known that a curve given by $c(t) = (x(t), u(t))$ is an optimal trajectory if there exists a lifting of $x(t)$ to the dual space $(x(t), p(t))$, which satisfies the Hamilton–Jacobi–Bellman equations. On the other hand, finding the optimal solution to the control system remains an extremely difficult problem for several reasons. Firstly, we need to integrate a Hamiltonian system, which is generally difficult to achieve, depending on the shape of the dynamic equations and the Hamiltonian function. Thus, if the Lagrangian has a complicated expression, we cannot guarantee that the Hamiltonian can be calculated without any dependence on the control. The best situation happens for systems with quadratic cost, or the so called linear quadratic problems. Secondly, there are some more special solutions, the so-called abnormal ones, which should be studied and which do not depend on the shape of the Hamiltonian function. Finally, even if all the solutions are found, the problem of selecting the optimal solutions from them remains extremely difficult. For these reasons, we believe that it is important to find new methods and workspaces that could simplify the study. The optimal trajectories of a driftless control affine system with holonomic distribution can be regarded as the geodesics in the geometry of Lie algebroids [37]. In fact, sub-Riemannian problems are distributional systems with quadratic cost and nonholonomic distribution (bracket generating), see [38,39]. If the distribution is holonomic, then the system is not controllable, and the distribution determines a foliation with the property that any curve is contained in a single leaf of the foliation and the restriction to each leaf of the foliation is bracket-generating. In many cases, it is not possible to find the exact solution to the optimal control problem. Thus, using the geometry of the space, we can find information about their local or global behavior. Moreover, if the geodesic curves belong to a manifold with a constant positive curvature, then the geodesics focus, and contrarily, the negative curvature spreads the geodesics.

In this paper, we solve two optimal control problems and prove that the framework of a Lie algebroid is more suitable than the cotangent space in the study of some driftless control affine systems with holonomic distributions. The paper is organized as follows. In the second section, the known results about Lie geometric methods in optimal control theory for control affine systems are presented, including the controllability issues in the case of holonomic and nonholonomic distributions. In addition, only the necessary notions about Lie algebroids and their prolongation over the vector bundle projection are given, and the geometric viewpoint of the optimal control systems on this space is presented. Moreover, the relation between the Hamiltonian $\mathcal{H}$ on dual the Lie algebroid E^* and the Hamiltonian H on the cotangent space T^*M is given. Our strategy is to apply the Pontryagin Maximum Principle at the level of a Lie algebroid built in the case of control affine systems with holonomic distribution. The last two sections contain the novelty of the paper. In the third section, we give an application of driftless control affine system with positive homogeneous cost, which is more general than the quadratic cost and show that the Hamilton–Jacobi–Bellman equations, provided by the Pontryagin Maximum Principle in the cotangent bundle, lead to a very complicated system of differential equations. Moreover, it is very difficult to find the Hamiltonian function without dependence on control variables. For these reasons, we will use a different approach considering the framework of a Lie algebroid, which simplifies the study. We also prove that the distribution generated by the vector fields is holonomic, and it determines a foliation in three-dimensional space. Using the Lie algebra of the vectors of distribution we study the controllability of the system and find the surfaces that generate the foliation. These aspects are not investigated in other previous papers of the authors. In the last part of the third section, we find the complete solution of the problem using the framework of Lie algebroids. The fourth section deals with the study of an economic problem of inventory and production using the mathematical model of optimal control and Lie geometric methods for controllability issues. We prove that the system is not controllable, meaning that we cannot produce any final quantity of products. The problem has a solution (it is controllable) if and only if there is a certain relationship between the final quantities of products. The mathematical models and final results are completely new and different from other previous works. The optimal solution is obtained using the Pontryagin Maximum Principle on a Lie algebroid. This approach simplifies the study and shows the connection between the geometry of Lie algebroids and optimal control for distributional systems.

2. Lie Geometric Methods in Optimal Control

Let M be a smooth n-dimensional manifold and a continuous-time control system given by differential equations, in the following form:

$$\frac{dx^i}{dt} = f^i(x,u),$$

where $x \in M$ are the state of the system, and $u \in U \subset R^m$ represents the controls. Let x_0 and x_1 be two points of M. We consider an optimal control problem, which consists of finding the optimal trajectories of the control system that connects x_0 and x_1 and minimizing the functional costs:

$$\min \int_0^T L(x(t),u(t))dt, \quad x(0) = x_0,\ x(T) = x_1,$$

where L is the *Lagrangian* function (energy, cost, time, distance, etc.). We have to remark that the time T can be fixed or free. Fixing the initial point x_0 and letting the final point x_1 vary in some domain, we obtain a family of optimal control problems. Similarly, we can fix x_1 and let x_0 vary. The theory of control deals with some systems whose evolution can be influenced by some external agents. It is known that one of the most important methods for studying the optimal solutions in control theory is Pontryagin's Maximum Principle. It generates the differential equations of first order, which are only necessary for the optimal solutions. For each optimal trajectory, $c(t) = (x(t),u(t))$, it generates a lift on the cotangent space $(x(t),p(t))$, satisfying the Hamilton–Jacobi–Bellman equations. The Hamiltonian function is given by the relation:

$$H(x,p,u) = \langle p, f(x,u)\rangle - L(x,u), \quad p \in T^*M,$$

and the maximization condition with respect to the control variables u, given by:

$$H(x(t),p(t),u(t)) = \max_v H(x(t),p(t),v),$$

which leads to $\frac{\partial H}{\partial u} = 0$, and the extreme trajectories satisfy the equations:

$$\dot{x} = \frac{\partial H}{\partial p}, \quad \dot{p} = -\frac{\partial H}{\partial x}. \tag{1}$$

2.1. Control Affine Systems

Let us consider a control affine system in the following form [40]:

$$\dot{x} = X_0(x) + \sum_{i=1}^{m} u_i X_i(x), \tag{2}$$

where $x = (x_1, ..., x_n)$ are local coordinates on a smooth n-dimensional manifold M, $u(t) = (u_1(t), ..., u_m(t)) \in U \subset \mathbb{R}^m$, $m \leq n$ are control variables, and $X_0, X_1...X_m$ are smooth vector fields on M. In addition, X_0 is usually called the drift vector field, describing the dynamics of the system in the absence of controls, and the vector fields X_i, $i = \overline{1,m}$ are called the input vector fields.

Definition 1. *A control system is named controllable if for any two points x_0 and x_1 on M there exists a finite time T and an admissible control $u : [0, T] \to U$, such that for x satisfying $x(0) = x_0$, we have $x(T) = x_1$.*

In other words, the control system is controllable if for any two points x_0, x_1 there exists a trajectory of the system (2) that connects x_0 to x_1. Controllability is the ability to steer a system from a given initial state to any final state, in finite time, using the available controls. The reachable set $\mathcal{R}$ of a point $x_0 \in M$ characterizes the states $x \in M$ that can be reached from a given initial state x_0 in positive time, by choosing various controls and switching from one to another from time to time. A system is controllable if $\mathcal{R}(x) = M$, $\forall x \in M$. Controllability does not depend on the quality of the trajectory between two states of the system and neither the amount nor the effort made for this.

Definition 2. *A distribution Δ on the smooth manifold M is a map which assigns to each point in M a subspace of the tangent bundle at this point:*

$$x \to \Delta(x) \subset T_x M.$$

The distribution Δ is named locally finitely generated if there is a set of vector fields $\{X_i\}_{i=\overline{1,m}}$, called local generators, which spans Δ, that is $\Delta(x) = span\{X_1(x), ..., X_m(x)\}$. In addition, the distribution Δ has dimension k if $\dim \Delta(x) = k$, for all points x in M. We recall that the Lie bracket of two vector fields is given by the relation:

$$[f, g](x) = \frac{\partial g}{\partial x}(x) f(x) - \frac{\partial f}{\partial x}(x) g(x),$$

($\frac{\partial g}{\partial x}$ is the Jacobian matrix of g). A distribution Δ on M is called involutive if $\forall\, x \in M$, then:

$$f(x), g(x) \in \Delta(x) \Rightarrow [f, g](x) \in \Delta(x).$$

Moreover, if the involutive distribution is generated by the vector fields $\{X_i\}_{i=\overline{1,m}}$, then we have:

$$[X_i, X_j](x) = \sum_{k=1}^{m} L_{ij}^k(x) X_k(x).$$

In other words, every Lie bracket can be expressed as a linear combination of the vector fields from distribution, and therefore, it already belongs to Δ. A foliation $\{S_\alpha\}_{\alpha \in A}$ of manifold M is a partition of $M = \cup S_\alpha$ into disjoint connected (immersed) submanifolds S_α, called leaves.

Definition 3. *A distribution Δ of constant dimension on M is called integrable (or holonomic) if there exists a foliation $\{S_\alpha\}_{\alpha\in A}$ on M whose tangent space is just Δ, i.e., $T_x S = \Delta(x)$, where S is the leaf passing through x.*

Theorem 1. *(Frobenius) If Δ is a distribution with a constant dimension, then Δ is integrable if and only if Δ is involutive.*

Definition 4. *The distribution $\Delta = span\{X_1, ..., X_m\}$ of the manifold M is called bracket-generating if the iterated Lie brackets*

$$X_i, [X_i, X_j], [X_i, [X_j, X_k]], \cdots, \quad 1 \le i, j, k \le m,$$

span the tangent space TM of M at every point.

Using the Lie brackets of vector fields, we construct the flag of subsheaves:

$$\Delta \subset \Delta^2 \subset \cdots \subset \Delta^r \subset \cdots \subset TM$$

$$\Delta^2 = \Delta + [\Delta, \Delta], ..., \Delta^{r+1} = \Delta^r + [\Delta, \Delta^r]$$

where

$$[\Delta, \Delta^r] = span\{[X, Y] : X \in \Delta,\ Y \in \Delta^r\}.$$

If there exists an $r \ge 2$ such that $\Delta^r = TM$, we say that Δ is a bracket-generating distribution, and r is called the step of the distribution Δ. In this case, the distribution Δ is not integrable and is called nonholonomic. This condition is also known as a *strong Hörmander condition* or a *Lie algebra rank condition*. If $r = 2$, the distribution is called a *strong bracket-generating* distribution. Next, we consider the *driftless control affine system* ($X_0 = 0$) or distributional systems in the following form:

$$\dot{x} = \sum_{i=1}^{m} u_i X_i(x). \tag{3}$$

The vector fields X_i, $i = \overline{1, m}$, generate a distribution Δ on the manifold M (assumed to be connected) such that the rank of Δ is assumed to be constant. For x_0 and x_1, two points on M, we consider an optimal control problem that consists of finding those trajectories of the distributional system which connect x_0 and x_1 and minimizing the cost:

$$\min_{u(\cdot)} \int_0^T \mathcal{F}(u(t))dt, \tag{4}$$

where $\mathcal{F}$ is a positive homogeneous function of Δ. We will characterize the controllability using the properties of vector fields, which generate the distribution Δ.

Theorem 2. *(Chow–Rashevsky) If the distribution $\Delta = span\{X_1, ..., X_m\}$ is bracket-generating (nonholonomic), then the driftless control affine system is controllable.*

If Δ is not bracket-generating and is integrable (holonomic), then the system (3) is not controllable, and Δ determines a foliation on M, with the property that any curve is contained in a single leaf of the foliation, and the restriction of Δ to each leaf of the foliation is bracket-generating. We will study in this paper the case of holonomic distributions.

If we assume that the distribution $\Delta = span\{X_1, X_2, ..., X_m\}$ is holonomic with constant rank, which means that $[X_i, X_j] \in \Delta$ for every $i, j = \overline{1, m}$, $i \ne j$, then from the Frobenius theorem, it results that the distribution Δ is integrable, it determines a foliation on M and two points can be joined if and only if they are situated on the same leaf.

Next, we will present some notions about Lie algebroids, which are useful in the study of driftless control affine systems.

2.2. Lie Algebroids

Let M be a real, C^∞-differentiable, n-dimensional manifold and T_xM its tangent space at $x \in M$. The tangent bundle of M is denoted (TM, π_M, M), where $TM = \cup T_xM$ and π_M is the canonical projection map $\pi_M : TM \to M$ taking a tangent vector $X(x) \in T_xM \subset TM$ to the base point $x \in M$. A vector bundle is a triple (E, π, M) where E and M are manifolds, called the total space and the base space and the map $\pi : E \to M$ is a surjective submersion. Using [41], we have:

Definition 5. *A Lie algebroid over a manifold M is given by a triple $(E, [\cdot,\cdot]_E, \sigma)$, where (E, π, M) is a vector bundle of rank m over M, which satisfies the following conditions:*
a) $C^\infty(M)$-module of sections $\Gamma(E)$ is equipped with a Lie algebra structure $[\cdot,\cdot]_E$;
b) $\sigma : E \to TM$ is a bundle map, called the anchor, which induces a Lie algebra homomorphism from the Lie algebra of sections $(\Gamma(E), [\cdot,\cdot]_E)$ to the Lie algebra of vector fields $(\mathcal{X}(M), [\cdot,\cdot])$, satisfying the Leibniz rule:

$$[s_1, fs_2]_E = f[s_1, s_2]_E + (\sigma(s_1)f)s_2, \ \forall s_1, s_2 \in \Gamma(E), \ f \in C^\infty(M).$$

In addition, we have the following relations:
1° $[\cdot,\cdot]_E$ is a $\mathbb{R}$-bilinear operation,
2° $[\cdot,\cdot]_E$ is skew-symmetric, i.e., $[s_1, s_2]_E = -[s_2, s_1]_E, \quad \forall s_1, s_2 \in \Gamma(E)$,

3° $[\cdot,\cdot]_E$ verifies the Jacobi identity:

$$[s_1, [s_2, s_3]_E]_E + [s_2, [s_3, s_1]_E]_E + [s_3, [s_1, s_2]_E]_E = 0,$$

and σ, being a Lie algebra homomorphism, then satisfies the relation:

$$\sigma[s_1, s_2]_E = [\sigma(s_1), \sigma(s_2)].$$

For a function f on M, then $df(x) \in E^*_x$ is given by $\langle df(x), a\rangle = \sigma(a)f$ for $\forall a \in E_x$. If $\omega \in \bigwedge^k(E^*)$, then the *exterior derivative* $d^E\omega \in \bigwedge^{k+1}(E^*)$ is given by the formula:

$$\begin{aligned} d^E\omega(s_1, ..., s_{k+1}) &= \sum_{i=1}^{k+1} (-1)^{i+1}\sigma(s_i)\omega(s_1, ..., \hat{s_i}, ..., s_{k+1}) \\ &+ \sum_{1\le i<j\le k+1} (-1)^{i+j}\omega([s_i, s_j]_E, s_1, ..., \hat{s_i}, ..., \hat{s_j}, ...s_{k+1}), \end{aligned}$$

where $s_i \in \Gamma(E)$, $i = \overline{1, k+1}$, and the hat over an argument means the absence of the argument. In addition, it results that $(d^E)^2 = 0$. If we consider the local coordinates (x^i) on $U \subset M$ and a local basis $\{s_\alpha\}$ of the sections of the bundle $\pi^{-1}(U) \to U$, then these generate local coordinates (x^i, y^α) on E. The local functions $\sigma^i_\alpha(x)$, $L^\gamma_{\alpha\beta}(x)$ on M are given by the following relations:

$$\sigma(s_\alpha) = \sigma^i_\alpha \frac{\partial}{\partial x^i}, \quad [s_\alpha, s_\beta]_E = L^\gamma_{\alpha\beta} s_\gamma, \quad i = \overline{1, n}, \quad \alpha, \beta, \gamma = \overline{1, m},$$

and are called the structure functions of Lie algebroids. Some examples of Lie algebroids are as follows:

Example 1. *The tangent bundle $E = TM$ itself, with identity mapping as anchor. With respect to the usual coordinates $(x, \dot{x})$, the structure functions are $L^i_{jk} = 0$, $\sigma^i_j = \delta^i_j$, but if we were to change to another basis for the vector fields, the structure functions would become nonzero.*

Example 2. *Any integrable subbundle of tangent bundle TM is a Lie algebroid with the inclusion as anchor and the induced Lie bracket.*

2.3. The Prolongation of a Lie Algebroid

Let us consider $\tau : E^* \to M$ as the dual bundle of $\pi : E \to M$ and $(E, [\cdot,\cdot]_E, \sigma)$ a Lie algebroid structure over M. A Lie algebroid structure over E^* can be constructed by taking the prolongation of $(E, [\cdot,\cdot]_E, \sigma)$ over $\tau : E^* \to M$ (see [42–44]). This structure is given by:

(i) The associated vector bundle is $(\mathcal{T}E^*, \tau_1, E^*)$, where $\mathcal{T}E^* = \cup \mathcal{T}_{u^*}E^*$, $u^* \in E^*$,

$$\mathcal{T}_{u^*}E^* = \{(u_x, v_{u^*}) \in E_x \times T_{u^*}E^* | \sigma(u_x) = T_{u^*}\tau(v_{u^*}), \tau(u^*) = x \in M\},$$

and the projection $\tau_1 : \mathcal{T}E^* \to E^*$, $\tau_1(u_x, v_{u^*}) = u^*$.

(ii) The Lie algebra structure $[\cdot,\cdot]_{\mathcal{T}E^*}$ on $\Gamma(\mathcal{T}E^*)$ is defined as follows: If $\rho_1, \rho_2 \in \Gamma(\mathcal{T}E^*)$ are such that $\rho_i(u^*) = (X_i(\tau(u^*)), U_i(u^*))$, where $X_i \in \Gamma(E), U_i \in \chi(E^*)$ and $\sigma(X_i(\tau(u^*)) = T_{u^*}\tau(U_i(u^*))$, $i = 1, 2$, then

$$[\rho_1, \rho_2]_{\mathcal{T}E^*}(u^*) = ([X_1, X_2]_{\mathcal{T}E^*}(\tau(u^*)), [U_1, U_2]_{\mathcal{T}E^*}(u^*)).$$

(iii) The anchor map is the projection $\sigma^1 : \mathcal{T}E^* \to TE^*$, $\sigma^1(u, v) = v$.

We remark that if $\mathcal{T}\tau : \mathcal{T}E^* \to E$, $\mathcal{T}\tau(u, v) = u$ then $(V\mathcal{T}E^*, \tau_{1|V\mathcal{T}E^*}, E^*)$ with $V\mathcal{T}E^* = Ker\mathcal{T}\tau$ is a sub-bundle of $(\mathcal{T}E^*, \tau_1, E^*)$, called the vertical sub-bundle. If (x^i, μ_α) are local coordinates on E^* at u^* and s_α is a local basis of sections of $\pi : E \to M$, then a local basis of $\Gamma(\mathcal{T}E^*)$ is $\{\mathcal{X}_\alpha, \mathcal{P}^\alpha\}$, where:

$$\mathcal{X}_\alpha(u^*) = \left(s_\alpha(\tau(u^*)), \sigma^i_\alpha \frac{\partial}{\partial x^i}|_{u^*}\right), \quad \mathcal{P}^\alpha(u^*) = \left(0, \frac{\partial}{\partial \mu_\alpha}|_{u^*}\right).$$

The structure functions of $\mathcal{T}E^*$ are given by the following formulas:

$$\sigma^1(\mathcal{X}_\alpha) = \sigma^i_\alpha \frac{\partial}{\partial x^i}, \quad \sigma^1(\mathcal{P}^\alpha) = \frac{\partial}{\partial \mu_\alpha},$$

$$[\mathcal{X}_\alpha, \mathcal{X}_\beta]_{\mathcal{T}E^*} = L^\gamma_{\alpha\beta}\mathcal{X}_\gamma, \quad [\mathcal{X}_\alpha, \mathcal{P}^\alpha]_{\mathcal{T}E^*} = 0, \quad [\mathcal{P}^\alpha, \mathcal{P}^\beta]_{\mathcal{T}E^*} = 0,$$

and therefore:

$$d^E x^i = \sigma^i_\alpha \mathcal{X}^\alpha, \; d^E \mu_\alpha = \mathcal{P}_\alpha, \; d^E \mathcal{X}^\gamma = -\frac{1}{2} L^\gamma_{\alpha\beta} \mathcal{X}^\alpha \wedge \mathcal{X}^\beta, \quad d^E \mathcal{P}_\alpha = 0,$$

where $\{\mathcal{X}^\alpha, \mathcal{P}_\alpha\}$ is the dual basis of $\{\mathcal{X}_\alpha, \mathcal{P}^\alpha\}$. In local coordinates, the Liouville section is given by:

$$\theta_E = \mu_\alpha \mathcal{X}^\alpha.$$

The canonical symplectic section ω_E is defined by:

$$\omega_E = -d^E \theta_E,$$

and it results in a nondegenerate 2-section and $d^E \omega_E = 0$. We obtain:

$$\omega_E = \mathcal{X}^\alpha \wedge \mathcal{P}_\alpha + \frac{1}{2} \mu_\alpha L^\alpha_{\beta\gamma} \mathcal{X}^\beta \wedge \mathcal{X}^\gamma.$$

By a control system on the Lie algebroid $(E, [\cdot,\cdot]_E, \sigma)$ (see [21]) with the control space given by $\tau : A \to M$, we understand a section ρ of E along τ. A trajectory of the system ρ is an integral curve of the vector field $\sigma(\rho)$. Considering the cost function $\mathcal{L} \in C^\infty(A)$, we must to minimize the integral of $\mathcal{L}$ over the family of those system trajectories which satisfy certain constraints. The Hamiltonian function $\mathcal{H} \in C^\infty(E^* \times_M A)$ is given by:

$$\mathcal{H}(\mu, u) = \langle \mu, \rho(u) \rangle - \mathcal{L}(u),$$

and the associated Hamiltonian control system ρ_H is given by the symplectic equation on Lie algebroid:

$$i_{\rho_H}\omega_E = d^E\mathcal{H}.$$

In local coordinates, the solution of the previous equation reads as:

$$\rho_H = \frac{\partial \mathcal{H}}{\partial \mu_\alpha}\mathcal{X}_\alpha - \left(\sigma^i_\alpha \frac{\partial \mathcal{H}}{\partial x^i} + \mu_\gamma L^\gamma_{\alpha\beta}\frac{\partial \mathcal{H}}{\partial \mu_\beta}\right)\mathcal{P}^\alpha,$$

on the subset where

$$\frac{\partial \mathcal{H}}{\partial u^A} = 0.$$

The critical trajectories are given by [21]:

$$\frac{\partial \mathcal{H}}{\partial u^A} = 0, \quad \frac{dx^i}{dt} = \sigma^i_\alpha \frac{\partial \mathcal{H}}{\partial \mu_\alpha}, \quad \frac{d\mu_\alpha}{dt} = -\sigma^i_\alpha \frac{\partial \mathcal{H}}{\partial x^i} - \mu_\gamma L^\gamma_{\alpha\beta}\frac{\partial \mathcal{H}}{\partial \mu_\beta}. \tag{5}$$

We have to remark that it can be associated to any positive homogeneous cost $\mathcal{L} : E \to \mathbb{R}$ on Lie algebroids E. A cost L on $Im\sigma \subset TM$ is defined in the following form:

$$L(v) = \{\mathcal{L}(u) | u \in E_x, \quad \sigma(u) = v\},$$

where $v \in (Im\sigma)_x \subset T_xM$, $x \in M$. From [35], we have:

Theorem 3. *The relation between the Hamiltonian function H on the cotangent bundle T^*M and the Hamiltonian function $\mathcal{H}$ on the dual Lie algebroid E^* has the form:*

$$H(p) = \mathcal{H}(\sigma^\star(p)), \quad \mu = \sigma^\star(p), \quad p \in T^*_xM, \quad \mu \in E^*_x. \tag{6}$$

Proof. From the Fenchel–Legendre dual of Lagrangian L, we obtain the Hamiltonian H given by:

$$\begin{aligned} H(p) &= \sup_v\{\langle p, v\rangle - L(v)\} = \sup_v\{\langle p, v\rangle - \mathcal{L}(u); \sigma(u) = v\} \\ &= \sup_u\{\langle p, \sigma(u)\rangle - \mathcal{L}(u)\} = \sup_u\{\langle \sigma^\star(p), u\rangle - \mathcal{L}(u)\} = \mathcal{H}(\sigma^\star(p)), \end{aligned}$$

Furthermore, it results in the following:

$$H(p) = \mathcal{H}(\mu), \quad \mu = \sigma^\star(p), \quad p \in T^*_xM, \quad \mu \in E^*_x,$$

or in local coordinates:

$$\mu_\alpha = \sigma^{*i}_\alpha p_i, \tag{7}$$

where the Hamiltonian $H(p)$ is degenerate on $Ker\sigma^\star \subset T^*M$. □

3. Application to Optimal Control

Let us consider the following driftless control affine system (distributional system) with positive homogeneous cost (Randers type metric):

$$\begin{cases} \dot{x}^1 = u_2 \\ \dot{x}^2 = u_1 + u_2 x^2 \\ \dot{x}^3 = u_1 + u_2 x^3 \end{cases} \tag{8}$$

$$\min_{u(\cdot)} \int_0^T \left(\sqrt{u_1^2+u_2^2}+\varepsilon u_1\right)dt, \quad 0 \le \varepsilon < 1.$$

We are looking for the optimal trajectories of the system, starting from the initial point $(0,1,0)^t$, which are parameterized by constant speed, that is, $\sqrt{u_1^2+u_2^2}+\varepsilon u_1 = const.$ (minimum time problem), and have free endpoints. The control system can be written in the following form:

$$\dot{x} = u_1 X_1 + u_2 X_2, \quad x = \begin{pmatrix} x^1 \\ x^2 \\ x^3 \end{pmatrix} \in R^3,\ X_1 = \begin{pmatrix} 0 \\ 1 \\ 1 \end{pmatrix},\ X_2 = \begin{pmatrix} 1 \\ x^2 \\ x^3 \end{pmatrix} \tag{9}$$

$$\min_{u(\cdot)} \int_0^T \mathcal{F}(u(t))dt,\ \mathcal{F}(u) = \sqrt{u_1^2+u_2^2}+\varepsilon u_1, \quad 0 \le \varepsilon < 1.$$

The vector fields of distributional system are given by:

$$X_1 = \frac{\partial}{\partial x^2} + \frac{\partial}{\partial x^3}, \quad X_2 = \frac{\partial}{\partial x^1} + x^2\frac{\partial}{\partial x^2} + x^3\frac{\partial}{\partial x^3}.$$

The Lie bracket of the vector fields is:

$$[X_1, X_2] = \left[\frac{\partial}{\partial x^2} + \frac{\partial}{\partial x^3}, \frac{\partial}{\partial x^1} + x^2\frac{\partial}{\partial x^2} + x^3\frac{\partial}{\partial x^3}\right] = X_1,$$

and it results that the associated distribution $\Delta = span\{X_1, X_2\}$ is holonomic and has the constant rank 2. Moreover, from the system (8), we obtain:

$$\dot{x}^2 - \dot{x}^3 = \dot{x}^1(x^2 - x^3),$$

which yields the following by integration:

$$\ln\left|x^2 - x^3\right| = x^1 + c, \tag{10}$$

where c is a constant, and it results that Δ determines a foliation on $\mathbb{R}^3$, given by the surfaces (10). Furthermore, the coordinate point $(0,1,0)^t$ is on the surface given by $\ln|x^2 - x^3| = x^1$, and the optimal trajectories of the system belong to the same surface. In order to solve this optimal control problem, the Pontryagin Maximum Principle in the cotangent bundle can be used. The Lagrangian has the form $\mathcal{L} = \frac{1}{2}\mathcal{F}^2$ (every minimizer parametrized by arclength, or constant speed $\mathcal{F} = 1$, is also a minimizer of the so-called energy cost $\mathcal{L} = \frac{1}{2}\mathcal{F}^2$; see [34] for more details), and we obtain the Hamiltonian as follows:

$$H(u,x,p) = p_i\dot{x}^i - \mathcal{L} = p_1u_2 + p_2(u_1 + u_2x^2) + p_3(u_1 + u_2x^3) - \frac{1}{2}\left(\sqrt{u_1^2+u_2^2}+\varepsilon u_1\right)^2.$$

The Hamilton–Jacobi–Bellman equations on the cotangent bundle given by $\frac{\partial H}{\partial u_i} = 0$, $\frac{dx^i}{dt} = \frac{\partial H}{\partial p_i}, \frac{dp_i}{dt} = -\frac{\partial H}{\partial x^i}$ lead to the following system:

$$\begin{cases} p_2 + p_3 - \left(\sqrt{u_1^2+u_2^2}+\varepsilon u_1\right)\left(\varepsilon + \frac{u_1}{\sqrt{u_1^2+u_2^2}}\right) = 0 \\ p_1 + p_2x^2 + p_3x^3 - \left(\sqrt{u_1^2+u_2^2}+\varepsilon u_1\right)\frac{u_2}{\sqrt{u_1^2+u_2^2}} = 0 \end{cases} \tag{11}$$

and a very complicated system of implicit differential equations. From (11), it is difficult to find the Hamiltonian H without dependence on the control variables. For this reason, a different approach will be used, involving the framework of a Lie algebroid.

In order to use the Pontryagin Maximum Principle in the framework of Lie algebroids, we will consider $E = \Delta$ (holonomic distribution with constant rank). The anchor $\sigma : E \to TM$ is the inclusion, and $[\cdot,\cdot]_E$ the induced Lie bracket. In the case of the previous control system (9), the anchor σ has the local components given by:

$$\sigma^i_\alpha = \begin{pmatrix} 0 & 1 \\ 1 & x^2 \\ 1 & x^3 \end{pmatrix},$$

and we consider the Lagrangian function given by:

$$\mathcal{L} = \frac{1}{2}\left(\sqrt{u_1^2 + u_2^2} + \varepsilon u_1\right)^2.$$

Using [45], we can find the Hamiltonian on E^* in the following form:

$$\mathcal{H}(\mu) = \frac{1}{2}\left(\sqrt{\frac{(\mu_1)^2}{(1-\varepsilon^2)^2} + \frac{(\mu_2)^2}{1-\varepsilon^2}} - \frac{\varepsilon\mu_1}{1-\varepsilon^2}\right)^2. \tag{12}$$

Using the relations (6) and (7), we can calculate the Hamiltonian H on T^*M given by $H(x,p) = \mathcal{H}(\mu)$, $\mu = \sigma^\star(p)$, where:

$$\begin{pmatrix} \mu_1 \\ \mu_2 \end{pmatrix} = \begin{pmatrix} 0 & 1 & 1 \\ 1 & x^2 & x^3 \end{pmatrix}\begin{pmatrix} p_1 \\ p_2 \\ p_3 \end{pmatrix}.$$

We obtain that $\mu_1 = p_2 + p_3$, $\mu_2 = p_1 + p_2x^2 + p_3x^3$, and it results in the Hamiltonian in the cotangent bundle:

$$H(x,p) = \frac{1}{2}\left(\sqrt{\frac{(p_2+p_3)^2}{(1-\varepsilon^2)^2} + \frac{(p_1+p_2x^2+p_3x^3)^2}{1-\varepsilon^2}} - \frac{\varepsilon(p_2+p_3)}{1-\varepsilon^2}\right)^2. \tag{13}$$

Unfortunately, the Equations (1) on T^*M with $H(x,p)$ from (13) lead to a complicated system of implicit differential equations. We will use the framework of a Lie algebroid. From the relation $[X_\alpha, X_\beta] = L^\gamma_{\alpha\beta}X_\gamma$, we obtain the non-zero components $L^1_{12} = 1$, $L^1_{21} = -1$, while from (5), we deduce the following:

$$\begin{aligned} \dot{x}^1 &= \frac{\partial\mathcal{H}}{\partial\mu_2},\ \dot{x}^2 = \frac{\partial\mathcal{H}}{\partial\mu_1} + x^2\frac{\partial\mathcal{H}}{\partial\mu_2},\ \dot{x}^3 = \frac{\partial\mathcal{H}}{\partial\mu_1} + x^3\frac{\partial\mathcal{H}}{\partial\mu_2}, \\ \dot{\mu}_1 &= -\mu_1\frac{\partial\mathcal{H}}{\partial\mu_2},\ \dot{\mu}_2 = \mu_1\frac{\partial\mathcal{H}}{\partial\mu_1}, \end{aligned}$$

where

$$\frac{\partial\mathcal{H}}{\partial\mu_1} = \frac{(1+\varepsilon^2)\mu_1}{(1-\varepsilon^2)^2} - \frac{\varepsilon\sqrt{\frac{(\mu_1)^2}{(1-\varepsilon^2)^2} + \frac{(\mu_2)^2}{1-\varepsilon^2}}}{1-\varepsilon^2} - \frac{\varepsilon\mu_1^2}{(1-\varepsilon^2)^3\sqrt{\frac{(\mu_1)^2}{(1-\varepsilon^2)^2} + \frac{(\mu_2)^2}{1-\varepsilon^2}}},$$

$$\frac{\partial\mathcal{H}}{\partial\mu_2} = \frac{\mu_2}{1-\varepsilon^2} - \frac{\varepsilon\mu_1\mu_2}{(1-\varepsilon^2)^2\sqrt{\frac{(\mu_1)^2}{(1-\varepsilon^2)^2} + \frac{(\mu_2)^2}{1-\varepsilon^2}}}.$$

We consider the following change of variables:

$$\mu_1(t) = (1-\varepsilon^2)r(t)\,\mathrm{sech}\theta(t), \quad \mu_2(t) = \sqrt{1-\varepsilon^2}r(t)\tanh\theta(t). \tag{14}$$

where

$$\sinh\theta = \frac{e^{\theta} - e^{-\theta}}{2}, \ \cosh\theta = \frac{e^{\theta} + e^{-\theta}}{2}, \ \tanh\theta = \frac{\sinh\theta}{\cosh\theta}, \ \text{sech}\vartheta = \frac{1}{\cosh\theta}.$$

In these conditions, we have:

$$\sqrt{\frac{(\mu_1)^2}{(1-\varepsilon^2)^2} + \frac{(\mu_2)^2}{1-\varepsilon^2}} = |r|,$$

and the differential equations:

$$\dot{\mu}_1 = -\mu_1 \frac{\partial \mathcal{H}}{\partial \mu_2},$$

with the relation (14) yielding:

$$\sqrt{1-\varepsilon^2}\left(\frac{\dot{r}}{r} - \dot{\theta}\tanh\theta\right) = r(-\tanh\theta + \varepsilon\text{sech}\theta\tanh\theta). \tag{15}$$

In addition, from the equation

$$\dot{\mu}_2 = \mu_1 \frac{\partial \mathcal{H}}{\partial \mu_1},$$

and (14), we obtain:

$$\sqrt{1-\varepsilon^2}\left(\frac{\dot{r}}{r}\tanh\theta + \dot{\theta}\,\text{sech}^2\theta\right) = r((1+\varepsilon)^2\text{sech}^2\theta - \varepsilon\text{sech}\theta - \varepsilon\text{sech}^3\theta). \tag{16}$$

Now, reducing $\dot{\theta}$ and $\frac{\dot{r}}{r}$ from the Equations (15) and (16), we obtain:

$$\sqrt{1-\varepsilon^2}\dot{r} = r^2\varepsilon\text{sech}\theta\tanh\theta(\varepsilon\text{sech}\theta - 1),$$

and

$$\sqrt{1-\varepsilon^2}\dot{\theta} = r(\varepsilon\text{sech}\theta - 1)^2.$$

The last two equations lead to:

$$\frac{\dot{r}}{\dot{\theta}} = \frac{r\varepsilon\text{sech}\theta\tanh\theta}{\varepsilon\text{sech}\theta - 1},$$

and to:

$$\frac{dr}{r} = \frac{\varepsilon\text{sech}\theta\tanh\theta}{\varepsilon\text{sech}\theta - 1}d\theta,$$

respectively, with the solution given by:

$$\ln|r| = -\ln(\varepsilon\text{sech}\theta - 1) - \ln c.$$

which leads to:

$$|r| = \frac{1}{c(\varepsilon\text{sech}\theta - 1)}.$$

Since the optimal trajectories are parameterized by arclength, the conclusion corresponds to the 1/2 level of the Hamiltonian, and we obtain:

$$\mathcal{H} = \frac{r^2}{2}(1 - \varepsilon\text{sech}\theta)^2 = \frac{1}{2c^2}.$$

Now, $c = \pm 1$ and

$$r = \pm \frac{1}{\varepsilon \text{sech}\theta - 1}.$$

The equation

$$\dot{\mu}_1 = -\mu_1 \dot{x}^1,$$

implies the following:

$$x^1(\theta) = \ln \frac{c_1(1 - \varepsilon \text{sech}\theta)}{(1 - \varepsilon^2)\text{sech}\theta}, \quad c_1 \in R.$$

Since we are looking for the optimal trajectories starting from the initial point $(0, 1, 0)^t$, we have $x^1(0) = 0$ and:

$$\ln \frac{c_1}{1 + \varepsilon} = 0 \Rightarrow c_1 = 1 + \varepsilon,$$

which leads to:

$$x^1(\theta) = \ln \frac{1 - \varepsilon \text{sech}\theta}{(1 - \varepsilon)\text{sech}\theta} = \ln \frac{\cosh \theta - \varepsilon}{1 - \varepsilon}.$$

We also obtain the following:

$$\dot{\mu}_2 = \mu_1 \left(\dot{x}^2 - x^2 \frac{\partial \mathcal{H}}{\partial \mu_2} \right) = \mu_1 \dot{x}^2 + x^2 \dot{\mu}_1,$$

and, consequently, $\mu_2 = \mu_1 x^2 + c_2$. Furthermore:

$$x^2(\theta) = \frac{\sinh \theta}{\sqrt{1 - \varepsilon^2}} \pm \frac{c_2(1 - \varepsilon \text{sech}\theta)}{(1 - \varepsilon^2)\,\text{sech}\theta}.$$

From $x^2(0) = 1$, we obtain that $c_2 = 1 + \varepsilon$, and this yields:

$$x^2(\theta) = \frac{\sinh \theta}{\sqrt{1 - \varepsilon^2}} + \frac{\cosh \theta - \varepsilon}{1 - \varepsilon}.$$

In the same way, we obtain:

$$x^3(\theta) = \frac{\sinh \theta}{\sqrt{1 - \varepsilon^2}} \pm \frac{c_3(1 - \varepsilon \text{sech}\theta)}{(1 - \varepsilon^2)\,\text{sech}\theta}.$$

From $x^3(0) = 0$, we obtain that $c_3 = 0$ and it results in the following:

$$x^3(\theta) = \frac{\sinh \theta}{\sqrt{1 - \varepsilon^2}}.$$

The solution is optimal because the Hamiltonian is a convex function. Using (8), we have $u_2 = \dot{x}^1$, $u_1 = \dot{x}^3 - u_2 x^3 = \dot{x}^2 - u_2 x^2$, and by direct computation, we obtain the control variables:

$$u_2(\theta) = \frac{\sinh \theta}{\cosh \theta - \varepsilon}, \quad u_1(\theta) = \frac{1}{\sqrt{1 - \varepsilon^2}} \frac{1 - \varepsilon \cosh \theta}{\cosh \theta - \varepsilon}.$$

We have to remark that if $\varepsilon = 0$, then we obtain the case of driftless control affine systems with quadratic cost with the solution:

$$x^1(t) = \ln \cosh t, \quad x^2(t) = \sinh t + \cosh t, \quad x^3(t) = \sinh t,$$

and control variables

$$u_2(t) = \tanh t, \quad u_1(t) = \text{sech}t.$$

4. Economic Application

Let us consider that in a fixed period of time T, three types of products, P_1, P_2, P_3, must be manufactured in certain fixed quantities. It is assumed that the production rate of the product P_3 depends on production rates of P_1 and P_2 by a given law. We assume that the unit production costs for the first two products increase linearly with the level of production and the production operations costs for the third product are considered negligible (for example, being obtained by combining the first two products together with other external products). We have storage costs of holding inventory given by constants $(\beta_1, \beta_2, \beta_3)$ for each type of product, and there are no restrictions on production capacity. The goal is to find an optimal production plan such as to ensure the required quantities of each type of product on a fixed date but with minimal costs of production and storage.

We will consider the following notations: P_i are the products, $i = 1,2,3$; T is the fixed period of time to ensure the quantities of products; $x^i(t)$ are the accumulated quantities by time t; s_i are the final quantities required; $p^i(t)$ are the rates of production at time t; and c_i are the unit production costs.

We assume that the initial quantity of products is zero. If, however, there is a certain quantity of products, then it is deducted from the final quantities required, that is, $x^i(0) = 0$ and $x^i(T) = s_i$. The production costs increase linearly with the production level and are given by $c_i = \alpha_i p^i$, $\alpha_1, \alpha_2 > 0, i = 1,2$. The production rate for the last product is assumed to be given by the law $\dot{x}^3 = u_1 x^2 + u_2 x^1$, where $u_1, u_2 \geq 0$ are control variables with $\dot{x}^1 = u_1, \dot{x}^2 = u_2$. It is known that the inventory level is the accumulated past production $p^i = p^i(t)$. Considering $x^i(0) = 0$, we obtain:

$$x^i(t) = \int_0^t p^i(s)ds$$

and the rate of change of inventory level $\dot{x}^i$ is the production rate and we have $\dot{x}^i(t) = p^i(t)$. The unit production costs c_i increase linearly with the production level, $c_i = \alpha_i p^i$, where $\alpha_1, \alpha_2 > 0$ are positive constants, and it results that the total cost of production is given by $c_1 p^1 + c_2 p^2 = \alpha_1 (p^1)^2 + \alpha_2 (p^2)^2 = \alpha_1 (\dot{x}^1)^2 + \alpha_2 (\dot{x}^2)^2 = \alpha_1 u_1^2 + \alpha_2 u_2^2$. It results that the total cost of production and storage is $\alpha_1 u_1^2 + \alpha_2 u_2^2 + \beta_1 x^1 + \beta_2 x^2 + \beta_3 x^3$. Finally, we obtain the following optimal control problem:

$$\begin{aligned}
&\dot{x}^1 = u_1 \\
&\dot{x}^2 = u_2 \\
&\dot{x}^3 = u_1 x^2 + u_2 x^1 \\
&x^i(0) = 0, \quad x^i(T) = s_i \\
&u_1, u_2 \geq 0, \quad \alpha_1, \alpha_2 > 0, \quad \beta_1, \beta_2, \beta_3 \geq 0.
\end{aligned} \tag{17}$$

We are looking for a plan of production with the minimum total cost:

$$\min_u \int_0^T (\alpha_1 (u_1(t))^2 + \alpha_2 (u_2(t))^2 + \beta_1 x^1 + \beta_2 x^2 + \beta_3 x^3) dt.$$

From the conditions $u_1, u_2 \geq 0$, it results that $\dot{x}^1 \geq 0$ and $\dot{x}^2 \geq 0$ and $x^1(t)$, $x^2(t)$ are increasing functions, which together with the initial conditions $x^i(0) = 0$ ensures the economic conditions of positivity $x^i(t) \geq 0$, $i = 1,2$. In addition, $\dot{x}^3 = u_1 x^2 + u_2 x^1 \geq 0$, and using $x^3(0) = 0$, we obtain $x^3(t) \geq 0$. Some different mathematical models can be found in the papers [10,12–15]. It results that the system (17) is a driftless control affine system on $\mathbb{R}^3_+$, written in the following form:

$$\dot{x} = u_1 X_1 + u_2 X_2, \quad x = (x^1, x^2, x^3)^t \in \mathbb{R}^3_+$$
$$\min_u \int_0^T F(u(t), x(t))dt,$$

where we have denoted the vector fields and the total cost:

$$X_1 = \begin{pmatrix} 1 \\ 0 \\ x^2 \end{pmatrix}, \quad X_2 = \begin{pmatrix} 0 \\ 1 \\ x^1 \end{pmatrix}$$
$$F(u(t), x(t)) = \alpha_1(u_1(t))^2 + \alpha_2(u_2(t))^2 + \beta_1 x^1 + \beta_2 x^2 + \beta_3 x^3.$$

In the next, we are looking for the optimal trajectories of the dynamical system starting from the initial point $(0,0,0)$ and endpoint (s_1, s_2, s_3). In addition, the distribution $\Delta = span\{X_1, X_2\}$ generated by the vector fields X_1, X_2 has constant dimension, $\dim \Delta(x) = 2$, for all $x \in \mathbb{R}^3$. Moreover, in a natural basis $\left\{ \frac{\partial}{\partial x^1}, \frac{\partial}{\partial x^2}, \frac{\partial}{\partial x^3} \right\}$ of $\mathbb{R}^3$, the vector fields have the following expressions:

$$X_1 = \frac{\partial}{\partial x^1} + x^2 \frac{\partial}{\partial x^3}, \quad X_2 = \frac{\partial}{\partial x^2} + x^1 \frac{\partial}{\partial x^3}.$$

Using the Lie bracket formula $[fX, gY] = fg[X,Y] + fX(g)Y - gY(f)X$, it results in the following:

$$[X_1, X_2] = \left[\frac{\partial}{\partial x^1} + x^2 \frac{\partial}{\partial x^3}, \frac{\partial}{\partial x^2} + x^1 \frac{\partial}{\partial x^3} \right] = \frac{\partial}{\partial x^3} - \frac{\partial}{\partial x^3} = 0,$$

and we obtain that the distribution Δ is involutive. From the Frobenius theorem, it results that the distribution is integrable (holonomic), and it determines a foliation on $\mathbb{R}^3_+$. Two points can be joined by an optimal trajectory if and only if they are situated on the same leaf. In fact, the economical system is not controllable in the sense that we cannot reach any final stock quantity. Indeed, from the system (17), we obtain:

$$\dot{x}^3 = \dot{x}^1 x^2 + \dot{x}^2 x^1,$$

and it results, through integration, that $x^3 = x^1 x^2 + c, \quad c \in \mathbb{R}$, which are the surfaces in $\mathbb{R}^3_+$, which determine a foliation. Moreover, using $x^i(0) = 0$, we obtain the relation $x^3 = x^1 x^2$. From $x^i(T) = s_i$, it results that the problem has a solution (the system is controllable) if and only if the final product amounts satisfy the condition $s_3 = s_1 s_2$. In order to use the Pontryagin Maximum Principle in the framework of Lie algebroids, we will consider $E = \Delta$ (holonomic distribution with constant rank), where the anchor $\sigma : E \to TM$ is the inclusion and $[\cdot, \cdot]_E$ is the induced Lie bracket. In the case of the previous control system (17), the anchor σ has the local components given by:

$$\sigma^i_\alpha = \begin{pmatrix} 1 & 0 \\ 0 & 1 \\ x^2 & x^1 \end{pmatrix},$$

and we consider the Lagrangian function given by:

$$\mathcal{L} = F(u(t), x(t)) = \alpha_1(u_1(t))^2 + \alpha_2(u_2(t))^2 + \beta_1 x^1 + \beta_2 x^2 + \beta_3 x^3.$$

Using the Legendre transformation associated to the regular Lagrangian $\mathcal{L}$, we can find the Hamiltonian $\mathcal{H}$ on E^* in the form:

$$\mathcal{H}(\mu) = \frac{(\mu_1)^2}{4\alpha_1} + \frac{(\mu_2)^2}{4\alpha_2} - \beta_1 x^1 - \beta_2 x^2 - \beta_3 x^3. \tag{18}$$

Using the relations (6) and (7), we can calculate the Hamiltonian H on T^*M given by $H(x,p) = \mathcal{H}(\mu)$, $\mu = \sigma^\star(p)$, where

$$\begin{pmatrix} \mu_1 \\ \mu_2 \end{pmatrix} = \begin{pmatrix} 1 & 0 & x^2 \\ 0 & 1 & x^1 \end{pmatrix} \begin{pmatrix} p_1 \\ p_2 \\ p_3 \end{pmatrix}.$$

We find that $\mu_1 = p_1 + p_3 x^2$, $\mu_2 = p_2 + p_3 x^1$, and it results in the Hamiltonian function in the cotangent bundle:

$$H(x,p) = \frac{(p_1 + p_3 x^2)^2}{4\alpha_1} + \frac{(p_2 + p_3 x^1)^2}{4\alpha_2} - \beta_1 x^1 - \beta_2 x^2 - \beta_3 x^3. \tag{19}$$

However, the Equations (1) on T^*M with $H(x,p)$ from (19) lead to a complicated system of implicit differential equations. We will use the framework of a Lie algebroid. We will consider two cases:

4.1. The Case $\beta_1 = \beta_2 = \beta_3 = 0$

From an economic point of view, this means that we have no storage costs, for example, the products are delivered immediately after manufacture.

Theorem 4. *The optimal solution of the control system (17) in the case of zero storage costs has the following form for $0 \le t \le T$:*

$$x^1(t) = \frac{s_1 t}{T}, \quad x^2(t) = \frac{s_2 t}{T}, \quad x^3(t) = \frac{s_1 s_2 t^2}{T^2}, \tag{20}$$

where the production rates (control variables) are positive constants:

$$u_1(t) = \frac{s_1}{T}, \quad u_2(t) = \frac{s_2}{T}. \tag{21}$$

Proof. The Hamiltonian function (18) has, in this case, the form $\mathcal{H}(\mu) = \frac{(\mu_1)^2}{4\alpha_1} + \frac{(\mu_2)^2}{4\alpha_2}$. From the relation $[X_\alpha, X_\beta] = 0$ and Equation (5), we deduce the following:

$$\dot{x}^1 = \frac{\mu_1}{2\alpha_1}, \quad \dot{x}^2 = \frac{\mu_2}{2\alpha_2}, \quad \dot{x}^3 = \frac{\mu_1 x^2}{2\alpha_1} + \frac{\mu_2 x^1}{2\alpha_2}, \quad \dot{\mu}_1 = 0, \quad \dot{\mu}_2 = 0,$$

which leads to $\mu_1 = a_1$, $\mu_2 = a_2$, where $a_1, a_2 \in \mathbb{R}$. It results in $\dot{x}^1 = \frac{a_1}{2\alpha_1}$, $\dot{x}^2 = \frac{a_2}{2\alpha_2}$. Moreover, $x^1 = \frac{a_1 t}{2\alpha_1} + b_1$, $x^2 = \frac{a_2 t}{2\alpha_2} + b_2$. By using $x^i(0) = 0$, we obtain $x^1(t) = \frac{a_1 t}{2\alpha_1}$, $x^2(t) = \frac{a_2 t}{2\alpha_2}$ and $x^3(t) = \frac{a_1 a_2 t^2}{4\alpha_1\alpha_2}$. From $x^i(T) = s_i$, we obtain $s_1 = \frac{a_1 T}{2\alpha_1}$, $s_2 = \frac{a_2 T}{2\alpha_2}$, which leads to $a_1 = \frac{2\alpha_1 s_1}{T}$, $a_2 = \frac{2\alpha_2 s_2}{T}$, and we finally obtain, for $0 \le t \le T$, the results (20) and (21). The solution is optimal because the Hamiltonian is a convex function, which ends the proof. □

4.2. The Case $\beta_1, \beta_2, \beta_3 > 0$

From an economic point of view, this means that the products are stored and delivered at the end of the fixed period.

Theorem 5. *The optimal solution of the control system (17) is given by:*

$$x^1(t) = c_1 e^{\alpha t} + c_2 e^{-\alpha t} + c_3 \cos \alpha t + c_4 \sin \alpha t - \frac{\beta_2}{\beta_3}, \tag{22}$$

$$x^2(t) = \sqrt{\frac{\alpha_1}{\alpha_2}} (c_1 e^{\alpha t} + c_2 e^{-\alpha t} - c_3 \cos \alpha t - c_4 \sin \alpha t) - \frac{\beta_1}{\beta_3}, \tag{23}$$

$$x^3(t) = x^1(t) x^2(t),$$

with control variables:

$$u_1(t) = \frac{\sqrt{\beta_3}}{\sqrt{2\sqrt{\alpha_1 \alpha_2}}} (c_1 e^{\alpha t} - c_2 e^{-\alpha t} - c_3 \sin \alpha t + c_4 \cos \alpha t), \tag{24}$$

$$u_2(t) = \frac{\sqrt{\beta_3 \sqrt{\alpha_1}}}{\sqrt{2\alpha_2}} (c_1 e^{\alpha t} - c_2 e^{-\alpha t} + c_3 \sin \alpha t - c_4 \cos \alpha t), \tag{25}$$

where $\alpha = \frac{\sqrt{\beta_3}}{\sqrt{2\sqrt{\alpha_1 \alpha_2}}}$, $\beta = \alpha T$ *and constant coefficients*

$$c_1 = \frac{1}{e^{\beta} - e^{-\beta}} \left(\frac{s_1}{2} + \frac{s_2 \sqrt{\alpha_2}}{2\sqrt{\alpha_1}} + \left(\frac{\beta_2}{2\beta_3} + \frac{\beta_1 \sqrt{\alpha_2}}{2\beta_3 \sqrt{\alpha_1}} \right) \left(1 - e^{\beta} \right) \right), \tag{26}$$

$$c_2 = \frac{1}{e^{-\beta} - e^{\beta}} \left(\frac{s_1}{2} + \frac{s_2 \sqrt{\alpha_2}}{2\sqrt{\alpha_1}} + \left(\frac{\beta_2}{2\beta_3} + \frac{\beta_1 \sqrt{\alpha_2}}{2\beta_3 \sqrt{\alpha_1}} \right) \left(1 - e^{-\beta} \right) \right), \tag{27}$$

$$c_3 = \frac{\beta_2}{2\beta_3} - \frac{\beta_1 \sqrt{\alpha_2}}{2\beta_3 \sqrt{\alpha_1}}, \tag{28}$$

$$c_4 = \frac{1}{2 \sin \beta} \left(s_1 + \frac{\beta_2}{\beta_3} - \frac{\sqrt{\alpha_2}}{\sqrt{\alpha_1}} \left(s_2 + \frac{\beta_1}{\beta_3} \right) - \left(\frac{\beta_2}{\beta_3} - \frac{\beta_1}{\beta_3} \frac{\sqrt{\alpha_2}}{\sqrt{\alpha_1}} \right) \cos \beta \right). \tag{29}$$

Proof. The necessary conditions for optimality (5) and the Hamiltonian function (18) lead to the following differential equations:

$$\dot{x}^1 = \frac{\mu_1}{2\alpha_1}, \ \dot{x}^2 = \frac{\mu_2}{2\alpha_2}, \ \dot{x}^3 = x^2 \frac{\mu_1}{2\alpha_1} + x^1 \frac{\mu_2}{2\alpha_2}, \tag{30}$$

$$\dot{\mu}_1 = \beta_1 + x^2 \beta_3, \ \dot{\mu}_2 = \beta_2 + x^1 \beta_3. \tag{31}$$

From (30) and (31) by derivation, it results in the following:

$$\ddot{x}^1 = \frac{\dot{\mu}_1}{2\alpha_1} = \frac{\beta_1}{2\alpha_1} + x^2 \frac{\beta_3}{2\alpha_1}, \ \ddot{x}^2 = \frac{\dot{\mu}_2}{2\alpha_2} = \frac{\beta_2}{2\alpha_2} + x^1 \frac{\beta_3}{2\alpha_2}, \tag{32}$$

which leads, by twice derivation, to:

$$\ddddot{x}^1 = \ddot{x}^2 \frac{\beta_3}{2\alpha_1}, \ \ddddot{x}^2 = \ddot{x}^1 \frac{\beta_3}{2\alpha_2}, \tag{33}$$

and together with (32) yields the nonhomogeneous differential equations of order 4, given by:

$$\ddddot{x}^1 = x^1 \frac{\beta_3^2}{4\alpha_1 \alpha_2} + \frac{\beta_2 \beta_3}{4\alpha_1 \alpha_2}, \tag{34}$$

$$\ddddot{x}^2 = x^2 \frac{\beta_3^2}{4\alpha_1 \alpha_2} + \frac{\beta_1 \beta_3}{4\alpha_1 \alpha_2}. \tag{35}$$

Considering the homogeneous differential equation from (34)

$$\overset{....}{x}^1 = x^1 \frac{\beta_3^2}{4\alpha_1\alpha_2}, \tag{36}$$

with characteristic equation

$$\lambda^4 - \frac{\beta_3^2}{4\alpha_1\alpha_2} = 0,$$

we obtain the following solutions:

$$\lambda_{1,2} = \pm\frac{\sqrt{\beta_3}}{\sqrt{2\sqrt{\alpha_1\alpha_2}}},\ \lambda_{3,4} = \pm\frac{i\sqrt{\beta_3}}{\sqrt{2\sqrt{\alpha_1\alpha_2}}}.$$

Then, we obtain the general solution of the homogeneous differential Equation (36) in the following form:

$$x^1(t) = c_1 e^{\alpha t} + c_2 e^{-\alpha t} + c_3 \cos \alpha t + c_4 \sin \alpha t + d_1,$$

and the solution of the nonhomogeneous differential Equation (34) is given by (22).

In the same way, considering Equations (32) and (35), obtain (23). Using $u_1 = \dot{x}^1$, $u_2 = \dot{x}^2$, we obtain (24) and (25). The initial and final conditions $x^i(0) = 0$, $x^i(T) = s_i$, $i = 1,2$ lead to the following linear system:

$$\begin{cases} c_1 + c_2 + c_3 = \frac{\beta_2}{\beta_3} \\ c_1 + c_2 - c_3 = \frac{\beta_1\sqrt{\alpha_2}}{\beta_3\sqrt{\alpha_1}} \\ c_1 e^{\beta} + c_2 e^{-\beta} + c_3 \cos\beta + c_4 \sin\beta = s_1 + \frac{\beta_2}{\beta_3} \\ c_1 e^{\beta} + c_2 e^{-\beta} - c_3 \cos\beta - c_4 \sin\beta = \frac{\sqrt{\alpha_2}}{\sqrt{\alpha_1}}\left(s_2 + \frac{\beta_1}{\beta_3}\right) \end{cases}$$

By direct computation, we obtain the solution (26)–(29). The solution is optimal because the Hamiltonian is convex. □

5. Conclusions

In this paper, some topics of dynamical systems (distributional systems) using Lie geometric methods are studied. In the case of two driftless control affine systems with holonomic distributions, we proved that the framework of Lie algebroids is more sustainable than cotangent bundles in order to apply the Pontryagin Maximum Principle and find the optimal solutions. This approach significantly simplifies the study and shows once again the intrinsic link between geometry and the optimal control, in particular between Lie algebroids and distributional systems with holonomic distributions. In addition, an economical application is given. As further developments, we will try to use the framework of Lie algebroids in the case of nonholonomic distribution (in particular, strong bracket-generating distributions) and characterize the abnormal solutions using the geometry of Lie algebroids.

Author Contributions: Conceptualization, L.P.; methodology, L.P.; investigation, L.P., D.M., G.T.; writing, L.P. All authors have read and agreed to the published version of the manuscript.

Funding: The work of PhD student Gabriel Tică was supported by the grant POCU380/6/13/123990, co-financed by the European Social Fund within the Sectorial Operational Human Capital 2014–2020.

Institutional Review Board Statement: Not applicable.

Informed Consent Statement: Not applicable.

Data Availability Statement: Not applicable.

Conflicts of Interest: The authors declare no conflict of interest.

References

1. Anita, S.; Arnăutu, V.; Capasso, V. *An Introduction to Optimal Control Problems in Life Sciences and Economics*; Birkhäuser: New York, NY, USA, 2011.
2. Arrow, K.J. Applications of control theory of economic growth. In *Lectures in Applied Mathematics, Mathematics of the Decision Sciences Part 2*; American Mathematical Society: Providence, RI, USA, 1968; Volume 12.
3. Caputo, M. *Foundations of Dynamic Economic Analysis: Optimal Control Theory and Applications*; Cambridge University Press: Cambridge, UK, 2005.
4. Seierstad, A.; Sydsater, K. *Optimal Control Theory with Economic Applications*; North-Holland: Amsterdam, Switzerland, 1987.
5. Sethi, S.P.; Thompson, G.L. *Optimal Control Theory: Applications to Management Science and Economics*; Springer: New York, NY, USA, 2000.
6. Weber, T.A. *Optimal Control Theory with Applications in Economics*; MIT Press: Cambridge, MA, USA, 2011.
7. Benjaafar, S.; Gayon, J.P.; Tepe, S. Optimal control of a production-inventory system with customer impatience. *Oper. Res. Lett.* **2010**, *38*, 267–272. [CrossRef]
8. Li, B.; Arreola-Risa, A. Optimizing a production-inventory system under a cost target. *Comput. Oper. Res.* **2020**, *123*, 105015. [CrossRef]
9. Pando, V.; Sicilia, J. A new approach to maximize the profit/cost ratio in a stock-dependent demand inventory model. *Comput. Oper. Res.* **2020**, *120*, 104940. [CrossRef]
10. Kamien, M.I.; Schwartz, N.L. Dynamic Optimization, Second Edition: The Calculus of Variations and Optimal Control in Economics and Management. In *Dover Books on Mathematics*; North Holland: Amsterdam, The Netherlands, 2012.
11. AL-Khazraji, H.; Cole, C.; Guo, W. Optimization and Simulation of Dynamic Performance of Production–Inventory Systems with Multivariable Controls. *Mathematics* **2021**, *9*, 568. [CrossRef]
12. Popescu, L. Applications of driftless control affine systems to a problem of inventory and production. *Stud. Inform. Control* **2019**, *28*, 25–34. [CrossRef]
13. Popescu, L. Applications of optimal control to production planning. *Inf. Technol. Control.* **2020**, *49*, 89–99. [CrossRef]
14. Popescu, L.; Militaru, D.; Mituca, O. Optimal control applications in the study of production management. *Int. J. Comput. Commun. Control* **2020**, *15*, 3859. [CrossRef]
15. Popescu, L.; Dimitrov, R. Applications of maximum principle to optimization of production and storage costs. *Arch. Control Sci.* **2021**, *31*, 865–881.
16. Weinstein, A. Lagrangian mechanics and groupoids. *Fields Inst. Comm.* **1996**, *7*, 206–231.
17. Cortez, J.; de Leon, M.; Marrero, J.; de Diego, M.; Martinez, E. A survey of Lagrangian mechanics and control on Lie algebroids and groupoids. *Int. J. Geom. Methods Mod. Phys.* **2006**, *3*, 509. [CrossRef]
18. Martinez, E. Lagrangian mechanics on Lie algebroids. *Acta Appl. Math.* **2001**, *67*, 295–320. [CrossRef]
19. Martínez, E. Lie algebroids in classical mechanics and optimal control. *Symmetry Integr. Geom. Methods Appl.* **2007**, *3*, 050. [CrossRef]
20. Cortes, J.; Martinez, E. Mechanical control systems on Lie algebroids. *IMA Math. Control Inform.* **2004**, *21*, 457–492. [CrossRef]
21. Martinez, E. Reduction in optimal control theory. *Rep. Math. Phys.* **2004**, *53*, 79–90. [CrossRef]
22. Grabowski, J.; Jóźwikowski, M. Pontryagin maximum principle on almost Lie algebroids. *SIAM J. Control Optim.* **2011**, *49*, 1306–1357. [CrossRef]
23. Barbero-Liñán, M.; de Diego, D.M.; Muñoz-Lecanda, M.C. Lie algebroids and optimal control: Abnormality. *Aip Conf. Proc.* **2009**, *1130*, 113–119.
24. Abrunheiro, L.; Camarinha, M. Optimal control of affine connection control systems from the point of view of Lie algebroids. *Int. J. Geom. Methods Mod. Phys.* **2014**, *11*, 1450038. [CrossRef]
25. Mestdag, T.; Langerock, B. A Lie algebroid framework for nonholonomic systems. *J. Phys. A* **2005**, *38*, 1097–1111. [CrossRef]
26. Popescu, L. Mechanical Structures on Lie Algebroids. *Mediterr. J. Math.* **2018**, *15*, 154. [CrossRef]
27. Brocket, R. Lie algebra and Lie groups in control theory. In *Geometrical Methods in Systeml Theory*; Springer: Dordrecht, The Netherlands, 1973; pp. 43–82.
28. Jurdjevic, D. *Geometric Control Theory*; Cambridge University Press: Cambridge, UK, 1997.
29. Agrachev, A.; Sachkov, Y.L. Control theory from the geometric viewpoint. In *Encyclopedia of Mathematical Sciences*; Control Theory and Optimization, II; Springer: Berlin, Germany, 2004; Volume 87.
30. Hrimiuc, D.; Popescu, L. Geodesics of sub-Finslerian Geometry. *Differ. Geom. Its Appl. Proc. Conf. Prague* **2005**, *3*, 59–67.
31. Peyghan, E.; Popescu, L. Study on geometric structures on Lie algebroids with optimal control applications. *J. Nonlinear Math.* **2020**, *27*, 550–580. [CrossRef]
32. Peyghan, E.; Popescu, L. Geometric structures on Finsler Lie algebroids and applications to optimal control. *Filomat* **2022**, *36*, 39–71.
33. Popescu, L. The geometry of Lie algebroids and applications to optimal control. *Ann. Univ. Al. I. Cuza Iasi Ser. I Math.* **2005**, *51*, 155–170.
34. Popescu, L. Hamiltonian formalism on Lie algebroids and its applications. In *Differential Geometry and Its Applications*; World Scientific Publishing: Singapore, 2008; pp. 665–673.
35. Popescu, L. Lie algebroids framework for distributional systems. *Ann. Univ. Al. I Math.* **2009**, *55*, 257–274.

36. Popescu, L. Symmetries of second order differential equations on Lie algebroids. *J. Geom. Phys.* **2017**, *117*, 84–98. [CrossRef]
37. Boucetta, M. Riemannian geometry of Lie algebroids. *J. Egypt. Math. Soc.* **2011**, *19*, 57–70. [CrossRef]
38. Bellaiche, A.; Risler, J.J. (Eds.) *Sub-Riemannian Geometry*; Birkhäuser: New York, NY, USA, 1996; Volume 144.
39. Strichartz, R.S. Subrimannian geometry. *J. Diff. Geom.* **1984**, *24*, 221–263.
40. LaValle, S.M. *Planning Algorithms*; Cambridge University Press: Cambridge, UK, 2006.
41. Mackenzie, K. *Lie Groupoids and Lie Algebroids in Differential Geometry*; London Mathematical Society Lecture Note Series; Cambridge University Press: Cambridge, UK, 1987; Volume 124.
42. Higgins, P.J.; Mackenzie, K. Algebraic constructions in the category of Lie algebroids. *J. Algebra* **1990**, *129*, 194–230. [CrossRef]
43. Hrimiuc, D.; Popescu, L. Nonlinear connections on dual Lie algebroids. *Balk. J. Geom. Its Appl.* **2006**, *11*, 73.
44. de Leon, M.; Marrero, J.C.; Martinez, E. Lagrangian submanifolds and dynamics on Lie algebroids. *J. Phys. A* **2005**, *38*, 241–308. [CrossRef]
45. Hrimiuc, D.; Shimada, H. On the L-duality between Lagrange and Hamilton Manifold. *Nonlinear World* **1996**, *3*, 613–641.

mathematics

Article

Stability Analysis of Equilibria for a Model of Maintenance Therapy in Acute Lymphoblastic Leukemia

Irina Badralexi [1,*,†], Andrei-Dan Halanay [2,†] and Ragheb Mghames [3,†]

1 Department of Mathematical Methods and Models, Polytehnic University of Bucharest, 060042 Bucharest, Romania

2 Faculty of Mathematics and Informatics, University of Bucharest, 010014 Bucharest, Romania; halanay@fmi.unibuc.ro

3 Department of Mathematics and Physics, Lebanese International University, Beqaa Valley 1803, Lebanon; ragheb.mghames@liu.edu.lb

* Correspondence: irina.badralexi@gmail.com

† These authors contributed equally to this work.

Abstract: In this paper, we study two mathematical models, involving delay differential equations, which describe the processes of erythropoiesis and leukopoiesis in the case of maintenance therapy for acute lymphoblastic leukemia. All types of possible equilibrium points were determined, and their stability was analyzed. For some of the equilibrium points, conditions for parameters that imply stability were obtained. When this was not feasible, due to the complexity of the characteristic equation, we discuss the stability through numerical simulations. An important part of the stability study for each model is the examination of the critical case of a zero root of the characteristic equation. The mathematical results are accompanied by biological interpretations.

Keywords: delay differential equations; critical case for stability; acute lymphoblastic leukemia; maintenance therapy

Citation: Badralexi, I.; Halanay, A.-D.; Mghames, R. Stability Analysis of Equilibria for a Model of Maintenance Therapy in Acute Lymphoblastic Leukemia. *Mathematics* **2022**, *10*, 313. https://doi.org/10.3390/math10030313

Academic Editors: Ioannis G. Stratis and Dumitru Baleanu

Received: 4 November 2021
Accepted: 12 January 2022
Published: 20 January 2022

Publisher's Note: MDPI stays neutral with regard to jurisdictional claims in published maps and institutional affiliations.

Copyright: © 2022 by the authors. Licensee MDPI, Basel, Switzerland. This article is an open access article distributed under the terms and conditions of the Creative Commons Attribution (CC BY) license (https://creativecommons.org/licenses/by/4.0/).

1. Introduction

1.1. Mathematical Background

The primary goal of this paper is to present the stability analysis of two different mathematical models for the processes of erythropoiesis and leukopoiesis in the case of maintenance therapy in acute lymphoblastic leukemia (for more details, see [1–6]).

The models represent the original work of some of the authors and were first introduced in [7]. They consist of systems of delay differential equations (often used to capture cell dynamics). In [7], the authors presented the models and demonstrated the positivity of solutions and the existence of equilibria. The novelty of this paper is the thorough stability study of all of the equilibrium points.

A challenging situation was the presence of critical cases for stability analysis. For these, we used an original theorem proven in [8] by some of the authors of this paper.

1.2. Biological Background

The models presented in this paper depict the erythropoiesis and leukopoiesis processes. These are part of a bigger process called hematopoiesis: the process through which all blood cells are created. Hematopoietic stem cells become red blood cells (which transport oxygen), white blood cells (which fight infections), or platelets (which stop bleeding). There is a complex network of cytokines and growth factors that regulate the production of blood cells.

Hematopoietic stem cells generate two major progenitors cell lineages: myeloid and lymphoid. The myeloid line contains cells such as granulocytes, monocytes, erythrocytes, or platelets, while the lymphoid line contains cells associated with the immune system (i.e., natural killer cells, T-cells, and B-cells).

Erythropoiesisis the process through which red blood cells (also called erythrocytes) are produced. Erythropoietic stem cells mature to red blood cells when a decrease in oxygen levels is detected by the kidneys. The kidneys secrete a hormone called erythropoietin, which stimulates the production of erythrocytes. It is the action of this hormone (also considered in [2,9]), together with the effects of therapy, that complicates the stability study of equilibria. Since mortality rates are no longer constant, a new variable is introduced. This variable is the source of a zero root of the characteristic equation.

White blood cells are generated through the process of leukopoiesis. Depending on which progenitor line they come from, there are two groups of white blood cells: myelocytes and lymphocytes.

For more information about hematopoiesis and its underlying processes, please see [2,3,10,11], pp. 1–18.

Acute lymphoblastic leukemia (ALL) is a type of cancer that affects white blood cells [4]. In the case of ALL, blood cell production is disrupted, due to the presence of a large number of immature lymphocytes (called lymphoblasts) in the circulatory system. These immature cells do not function properly and overcrowd the healthy cells. Lymphoblasts eventually invade the liver, spleen, and lymph nodes. For more information, statistics, diagnoses, and prognoses, see [12], pp. 1556–1576, 1616–1636, and [10], pp. 173–192.

Maintenance therapy is conducted after an initial treatment is administered [6]. In patients with ALL, we consider that this therapy consists of oral administration of mercaptopurine (6-MP). The biologically inactive substance 6-MP metabolizes into the active 6-thioguanine nucleotide (6-TGN).

Studies have shown that maintenance therapy contributes greatly to the overall evolution and life expectancy of the patient ([12], p. 1566, [13]). The need to mathematically study the effects of this treatment is very important, as it may help to fill the existing gaps in knowledge [13].

The models studied in this paper capture the effects of this medication on the creation of red and white blood cells.

It is worth mentioning that, although we concentrated on patients with ALL, the same maintenance therapy is administrated for some autoimmune diseases. Thus, the models, computations, and discussions could also be suitable for those cases.

2. The Modeling of Erythropoiesis

2.1. The Mathematical Model

The mathematical model for erythropoiesis consists of seven delay differential equations (DDEs) with two delays. We considered the dynamics of the stem-like short-term erythroid cells (u_1), the mature erythrocytes (u_2), the concentration of erythropoietin (u_3), the amount of 6-MP in the gut (u_5), the amount of 6-MP in plasma (u_5), and the concentration of thioguanine nucleotide (6-TGN) in red blood cells (u_7) (see [7,9,14,15]). The variable u_4 represents the loss suffered during the cell cycle.

Each stem-like cell is considered to go through either asymmetric division (a fraction η_{1e} of the stem-like population), which means that one of the daughter cells re-enters the stem-like population, while the other goes on to differentiate, symmetric differentiation (a fraction η_{2e} of the stem-like population)—both daughter cells proceed to differentiate—or self-renewal (a fraction $1 - \eta_{1e} - \eta_{2e}$ of the stem-like population)—both daughter cells re-enter the stem-like pool. We used two feedback functions for the rates of self-renewal and differentiation:

$$\beta_e(u_1, u_3) = \beta_{0e}\frac{1}{1+u_1^m} \cdot \frac{u_3}{1+u_3}$$

$$k_e(u_3) = k_{0e}\frac{u_3}{1+u_3}$$

The time necessary for a stem-like cell to go through self-renewal or differentiation and re-enter the stem-like population is considered to be the same and is represented by the delay τ_1.

These aspects are included in the first equation of the model, which describes the evolution of stem-like short-term erythroid cells. As can be seen in this equation, both the erythropoietin and the medication have an impact on cell death. The drug concentration in the blood stream has a toxicity that increases the apoptosis rate of the stem-like cells (the second term). The erythropoietin regulates the process of erythropoiesis, either repressing (the first term) or stimulating (by influencing the rate of self-renewal and differentiation) the production of stem-like cells.

The constant $\tilde{R}_m$ represents the maximum drug effect on erythrocytes, and $\tilde{R}_{50}$ represents the drug saturation.

The healthy erythrocytes, illustrated by the second equation, have a constant mortality rate (the first term) and a supply controlled by the erythropoietin (the second term). We introduce the notation $\tilde{A}_e = A_e(2\eta_{1e} + \eta_{2e})$, where A_e is an amplification factor. The time necessary for the cells to mature is given by the delay τ_2.

The third equation captures the changes in the concentration of erythropoietin. This protein has a constant rate of clearance, and its production is influenced by the number of existing erythrocytes in the blood stream.

The fourth equation represents the cell loss sustained during the cell cycle. For more information regarding this equation, please consult [7].

The last three equations describe the dynamics of the drug after administration.

The model, which depicts the process of erythropoiesis under maintenance therapy, is:

$$\dot{u} = f_i(u, u_{\tau_j}), i = \overline{1,7}, j = \overline{1,2} \tag{1}$$

$$\begin{aligned}
\dot{u}_1 = & -\frac{\gamma_0}{1+u_3^{\alpha}}u_1 - \frac{\tilde{R}_m u_7}{\tilde{R}_{50}+u_7}u_1 - (\eta_{1e}+\eta_{2e})k_e(u_3)u_1 - (1-\eta_{1e}-\eta_{2e})\beta_e(u_1,u_3)u_1 \\
& +2u_4(1-\eta_{1e}-\eta_{2e})\beta_e(u_{1\tau_1},u_{3\tau_1})u_{1\tau_1} + \eta_{1e}u_4 k_e(u_{3\tau_2})u_{1\tau_1} \\
\dot{u}_2 = & -\gamma_2 u_2 + \tilde{A}_e k_e(u_{3\tau_2})u_{1\tau_2} \\
\dot{u}_3 = & -ku_3 + \frac{a_1}{1+u_2^n} \\
\dot{u}_4 = & \; u_4\left(-\frac{\gamma_0}{1+u_3^{\alpha}} - \frac{\tilde{R}_m u_7}{\tilde{R}_{50}+u_7} + \frac{\gamma_0}{1+u_{3\tau_1}^{\alpha}} + \frac{\tilde{R}_m u_{7\tau_1}}{\tilde{R}_{50}+u_{7\tau_1}}\right) \\
\dot{u}_5 = & -b_1u_5 + a_2 \\
\dot{u}_6 = & \; b_1u_5 - e_1u_6 - \frac{c_1(1-e_2)}{c_2+u_6}u_6 - \frac{m_2e_2}{m_1+u_6}u_6 \\
\dot{u}_7 = & \; \frac{b_2c_1(1-e_2)}{c_2+u_6}u_6 - e_3u_7.
\end{aligned}$$

2.2. The Equilibrium Points and Linearization

Using some elementary, but tedious calculations, the following types of equilibrium points corresponding to the model of erythropoiesis in ALL under treatment were obtained (see also [7]): $E_1 = (0, 0, \hat{u}_3, \hat{u}_4, \hat{u}_5, \hat{u}_6, \hat{u}_7)$, which corresponds to the "death of the patient", and $E_2 = (\hat{u}_1, \hat{u}_2, \hat{u}_3, \hat{u}_4, \hat{u}_5, \hat{u}_6, \hat{u}_7)$, which corresponds to a "chronic phase of the disease".

In order to study the stability of these equilibrium points, we performed a linearization of the nonlinear system (1). We denote the matrix of partial derivatives for the undelayed variables by $A = \frac{\partial f}{\partial u} = [a_{ij}]$ and the matrices of the partial derivatives with respect to the delayed variables by: $B = \frac{\partial f}{\partial u_{\tau_1}} = [b_{ij}]$ and $C = \frac{\partial f}{\partial u_{\tau_2}} = [c_{ij}]$. For a complete list of the matrix elements, please consult [7].

2.3. *Stability Analysis of the Equilibrium Point E_1*

Using the matrices defined above, the characteristic equation corresponding to E_1 is:

$$\lambda(\lambda - a_{22})(\lambda - a_{33})(\lambda - a_{55})(\lambda - a_{66})(\lambda - a_{77})(\lambda - a_{11} - b_{11}e^{-\lambda\tau_1}) = 0.$$

We notice that $\lambda = 0$ is a root, and so, we are in a critical case for the stability of the nonlinear system (1).

2.3.1. The Real Solutions of the Characteristic Equation

The real solutions of the characteristic equation are given by:

$$\lambda_1 = 0$$

$$\lambda_2 = a_{22} = -\gamma_2 < 0$$
$$\lambda_3 = a_{33} = -k < 0$$
$$\lambda_4 = a_{55} = -b_1 < 0$$
$$\lambda_5 = a_{66} = -e_1 - \frac{c_1(1 - e_2)c_2}{(c_2 + \hat{u}_7)^2} - \frac{m_2 e_2 m_1}{(m_1 + \hat{u}_7)^2} < 0$$
$$\lambda_7 = a_{77} = -e_3 < 0.$$

The existence of a zero eigenvalue implies a critical case for stability by the first approximation.

2.3.2. Analysis of the Critical Case

The following theorem, developed and proven by some of the authors in [8], gives stability criteria in the critical case of a zero eigenvalue.

Theorem 1 ([8], Theorem 2.1). *Consider the following nonlinear system with time delays:*

$$\begin{aligned} \dot{x}(t) &= A_0 x(t) + \sum_{j=1}^{m} A_j x(t - \tau_j) + F[x(t), x(t - \tau_1), ..., x(t - \tau_m), y(t)] \\ \dot{y}(t) &= G[x(t), x(t - \tau_1), ..., x(t - \tau_m), y(t)], \end{aligned} \tag{2}$$

where $A_j \in M_n(\mathbb{R})$, $\tau_j > 0$ for all $1 \leq j \leq m$, $G(0,0,...,0,y) = F(0,0,...,0,y) = 0$, $\forall y \in \mathbb{R}$, F takes values in $\mathbb{R}^n$, and G is scalar. F and G contain only powers of the variables with the sum greater than or equal to two. Then, for every $\delta > 0$, there exist $M_1(\delta)$ and $M_2(\delta)$ with $\lim_{\delta \to 0} M_1(\delta) = \lim_{\delta \to 0} M_2(\delta) = 0$ so that, whenever $||x(t)|| \leq \delta$, $||x(t - \tau_j)|| \leq \delta$, $1 \leq j \leq m$, $|y| \leq \delta$,

$$\begin{aligned} &\|F(x(t), x(t - \tau_1), \ldots, x(t - \tau_m), y(t))\| \leq \\ &\leq M_1(\delta)(\|x(t)\| + \|x(t - \tau_1)\| + \cdots + ||x(t - \tau_m)||) \\ &|G((x(t), x(t - \tau_1), \ldots, x(t - \tau_m), y(t))| \leq \\ &\leq M_2(\delta)(||x(t)|| + ||x(t - \tau_1)|| + \cdots + ||x(t - \tau_m)||). \end{aligned} \tag{3}$$

Now, consider the following system with time delays. Suppose that the linear system:

$$\dot{x}(t) = A_0 x(t) + \sum_{j=1}^{m} A_j x(t - \tau_j) \tag{4}$$

is asymptotically stable, that is, if λ is a root of the characteristic equation, then $Re(\lambda) < 0$. Then, the zero solution of (2) is simple stable, and if φ is the initial data of (2) in $C\left([-\tau,0];\mathbb{R}^{n+1}\right)$ with $\tau = \max_{1\le j\le m} \tau_j$, there exist $\delta > 0$ so that, if $\sup\{||\varphi(t)||_2/t \in [-\tau,0]\} < \delta$, then:

$$\lim_{t\longrightarrow\infty} x_i(t) = 0, i = 1,...,n \text{ and } \exists \lim_{t\longrightarrow\infty} y(t) = \tilde{y}.$$

In what follows, we transform our system so that Theorem 1 can be applied. We first perform a translation to zero: $y_i = u_i - \hat{u}_i$, for $i = \overline{3,7}$.

The new system becomes:

$$\dot{y} = f_i(y, y_{\tau_j}), i = \overline{1,7}, j = \overline{1,2} \tag{5}$$

$$\begin{aligned}
\dot{y}_1 = \;& -\frac{\gamma_0}{1+(y_3+\hat{u}_3)^\alpha}y_1 - \frac{\tilde{R}_m u_7}{\tilde{R}_{50}+y_7+\hat{u}_7}y_1 - (\eta_{1e}+\eta_{2e})k_e(y_3+\hat{u}_3)y_1 \\
& -(1-\eta_{1e}-\eta_{2e})\beta_e(y_1, y_3+\hat{u}_3)y_1 \\
& +2(y_4+\hat{u}_4)(1-\eta_{1e}-\eta_{2e})\beta_e(y_{1\tau_l}, y_{3\tau_l}+\hat{u}_{3\tau_l})y_{1\tau_l} \\
& +\eta_{1e}(y_4+\hat{u}_4)k_e(y_{3\tau_1}+\hat{u}_3)y_{1\tau_1} \\
\dot{y}_2 = \;& -\gamma_2 y_2 + \tilde{A}_e k_e(y_{3\tau_2}+\hat{u}_{3\tau_2})y_{1\tau_2} \\
\dot{y}_3 = \;& -k(y_3+\hat{u}_3) + \frac{\tilde{a}_1}{1+y_2^r} \\
\dot{y}_4 = \;& (y_4+\hat{u}_4)\left(-\frac{\gamma_0}{1+(y_3+\hat{u}_3)^\alpha} - \frac{\tilde{R}_m(y_7+\hat{u}_7)}{\tilde{R}_{50}+y_7+\hat{u}_7} + \frac{\gamma_0}{1+(y_{3\tau 1}+\hat{u}_3)^\alpha}\right. \\
& \left.+\frac{\tilde{R}_m(y_{7\tau_1}+\hat{u}_7)}{\tilde{R}_{50}+y_{7\tau_1}+\hat{u}_7}\right) \\
\dot{y}_5 = \;& -b_1(y_5+\hat{u}_5) + a_2 \\
\dot{y}_6 = \;& b_1(y_5+\hat{u}_5) - e_1(y_6+\hat{u}_6) - \frac{c_1(1-e_2)}{c_2+y_6+\hat{u}_6}(y_6+\hat{u}_6) - \\
& \frac{m_2 e_2}{m_1+y_6+\hat{u}_6}(y_6+\hat{u}_6) \\
\dot{y}_7 = \;& \frac{b_2 c_1(1-e_2)}{c_2+y_6+\hat{u}_6}(y_6+\hat{u}_6) - e_3(y_7+\hat{u}_7),
\end{aligned}$$

where we consider $\dot{y}_4 = g(y_3, y_4, y_7, y_{3\tau_1}, y_{7\tau_1})$, with $g(0) = 0$.

The matrices of partial derivatives into zero are, as before, $\tilde{A} = \frac{\partial f}{\partial y} = [\tilde{a}_{ij}]$, $\tilde{B} = \frac{\partial f}{\partial y_{\tau_1}} = [\tilde{b}_{ij}]$, $\tilde{C} = \frac{\partial f}{\partial y_{\tau_2}} = [\tilde{c}_{ij}]$.

The characteristic equation for the zero solution of the new system (5) has exactly the same form as the one for E_1. The terms that are different are:

$$\begin{aligned}
\frac{\partial g}{\partial y_3}(0) &= \frac{\hat{u}_4\gamma_0\alpha\hat{u}_3^{\alpha-1}}{(1+\hat{u}_3^\alpha)^2} \\
\frac{\partial g}{\partial y_7}(0) &= -\frac{\tilde{R}_m\tilde{R}_{50}}{(\tilde{R}_{50}+\hat{u}_7)^2} \\
\frac{\partial g}{\partial y_{3\tau_1}}(0) &= -\frac{\hat{u}_4\gamma_0\alpha\hat{u}_3^{\alpha-1}}{(1+\hat{u}_3^\alpha)^2} \\
\frac{\partial g}{\partial y_{7\tau_1}}(0) &= \frac{\tilde{R}_m\tilde{R}_{50}}{(\tilde{R}_{50}+\hat{u}_7)^2}.
\end{aligned}$$

Notice that Theorem 1 is not directly applicable since the linear part is not equal to zero. In order to apply Theorem 1, we need to rewrite the system (5).

Take $\eta = \alpha_1 y_1 + \cdots + \alpha_7 y_7$ where $\dot{y} = \tilde{A}y$. This means that:

$$\begin{bmatrix} \dot{y}_1 \\ \dot{y}_2 \\ \dot{y}_3 \\ \dot{y}_4 \\ \dot{y}_5 \\ \dot{y}_6 \\ \dot{y}_7 \end{bmatrix} = \begin{bmatrix} \tilde{a}_{11} & 0 & 0 & 0 & 0 & 0 & 0 & 0 \\ 0 & \tilde{a}_{22} & 0 & 0 & 0 & 0 & 0 & 0 \\ 0 & 0 & \tilde{a}_{33} & 0 & 0 & 0 & 0 & 0 \\ 0 & 0 & \tilde{a}_{43} & 0 & 0 & 0 & 0 & \tilde{a}_{47} \\ 0 & 0 & 0 & 0 & 0 & \tilde{a}_{55} & 0 & 0 \\ 0 & 0 & 0 & 0 & 0 & \tilde{a}_{65} & \tilde{a}_{66} & 0 \\ 0 & 0 & 0 & 0 & 0 & 0 & \tilde{a}_{76} & \tilde{a}_{77} \end{bmatrix} \begin{bmatrix} y_1 \\ y_2 \\ y_3 \\ y_4 \\ y_5 \\ y_6 \\ y_7 \end{bmatrix}$$

so,

$$\begin{bmatrix} \dot{y}_1 \\ \dot{y}_2 \\ \dot{y}_3 \\ \dot{y}_4 \\ \dot{y}_5 \\ \dot{y}_6 \\ \dot{y}_7 \end{bmatrix} = \begin{bmatrix} \tilde{a}_{11} y_1 \\ \tilde{a}_{22} y_2 \\ \tilde{a}_{33} y_3 \\ \tilde{a}_{43} y_3 + \tilde{a}_{47} y_7 \\ \tilde{a}_{55} y_5 \\ \tilde{a}_{65} y_5 + \tilde{a}_{66} y_6 \\ \tilde{a}_{76} y_6 + \tilde{a}_{77} y_7 \end{bmatrix}$$

We have $\dot{\eta} = \alpha_1 \dot{y}_1 + \cdots + \alpha_7 \dot{y}_7$.

Then,

$$\begin{aligned} \dot{\eta} \quad = \quad & \alpha_1 \tilde{a}_{11} y_1 + \alpha_2 \tilde{a}_{22} y_2 + \alpha_3 \tilde{a}_{33} y_3 + \alpha_4 (\tilde{a}_{43} y_3 + \tilde{a}_{47} y_7) + \\ & \alpha_5 \tilde{a}_{55} y_5 + \alpha_6 (\tilde{a}_{65} y_5 + \tilde{a}_{66} y_6) + \alpha_7 (\tilde{a}_{76} y_6 + \tilde{a}_{77} y_7). \end{aligned}$$

Next, by forcing $\dot{\eta} = 0$, it follows that:

$$\begin{aligned} 0 \quad = \quad & \alpha_1 \tilde{a}_{11} y_1 + \alpha_2 \tilde{a}_{22} y_2 + (\alpha_3 \tilde{a}_{33} + \alpha_4 \tilde{a}_{43}) y_3 + (\alpha_5 \tilde{a}_{55} + \alpha_6 \tilde{a}_{65}) y_5 \\ & + (\alpha_6 \tilde{a}_{66} + \alpha_7 \tilde{a}_{76}) y_6 + (\alpha_4 \tilde{a}_{47} + \alpha_7 \tilde{a}_{77}) y_7, \end{aligned}$$

from which we obtain:

$$\begin{aligned} & \alpha_1 = 0, \ \alpha_2 = 0 \\ & \alpha_3 \tilde{a}_{33} + \alpha_4 \tilde{a}_{43} = 0 \\ & \alpha_5 \tilde{a}_{55} + \alpha_6 \tilde{a}_{65} = 0 \\ & \alpha_6 \tilde{a}_{66} + \alpha_7 \tilde{a}_{76} = 0 \\ & \alpha_4 \tilde{a}_{47} + \alpha_7 \tilde{a}_{77} = 0. \end{aligned}$$

Thus, we have $\alpha_4 = 1$, $\alpha_3 = -\frac{\tilde{a}_{43}}{\tilde{a}_{33}}$, $\alpha_7 = -\frac{\tilde{a}_{47}}{\tilde{a}_{77}}$, $\alpha_6 = \frac{\tilde{a}_{76}\tilde{a}_{47}}{\tilde{a}_{66}\tilde{a}_{77}}$ and $\alpha_5 = -\frac{\tilde{a}_{65}\tilde{a}_{76}\tilde{a}_{47}}{\tilde{a}_{55}\tilde{a}_{66}\tilde{a}_{77}}$.

We remark that:

$$\begin{aligned} \dot{y}_{3\tau_1} &= \tilde{a}_{33} y_{3\tau_1} + R_{3\tau_1} \\ \dot{y}_{7\tau_1} &= \tilde{a}_{77} y_{7\tau_1} + R_{7\tau_1}, \end{aligned}$$

with $R_{3\tau_1}$ and $R_{7\tau_1}$ containing terms of order higher than or equal to two.

We introduce:

$$\eta_1 = \quad \alpha_3 y_3 + y_4 + \alpha_5 y_5 + \alpha_6 y_6 + \alpha_7 y_7 - \frac{\tilde{b}_{43}}{\tilde{a}_{33}} y_{3\tau_1} - \frac{\tilde{b}_{47}}{\tilde{a}_{77}} y_{7\tau_1}.$$

Then, we have:

$$\begin{aligned} \dot{\eta}_1 \quad &= \tilde{b}_{43} y_{3\tau_1} + \tilde{b}_{47} y_{7\tau_1} - \frac{\tilde{b}_{43}}{\tilde{a}_{33}} \dot{y}_{3\tau_1} - \frac{\tilde{b}_{47}}{\tilde{a}_{77}} \dot{y}_{7\tau_1} + R_4^1 \\ &= \tilde{b}_{43} y_{3\tau_1} + \tilde{b}_{47} y_{7\tau_1} - \frac{\tilde{b}_{43}}{\tilde{a}_{33}} (\tilde{a}_{33} y_{3\tau_1} + R_{3\tau_1}) - \frac{\tilde{b}_{47}}{\tilde{a}_{77}} (\tilde{a}_{77} y_{7\tau_1} + R_{7\tau_1}) \\ &= R_4^{(2)}(y, y_{\tau_1}). \end{aligned}$$

We replace the fourth equation in (5) with the equation of $\dot{\eta}_1$ from above, so that the equation has a zero linear part.

Next, we consider:

$$y_4 = \eta_1 - \alpha_3 y_3 - \alpha_5 y_5 - \alpha_6 y_6 - \alpha_7 y_7 + \frac{\tilde{b}_{43}}{\tilde{a}_{33}} y_{3\tau_1} + \frac{\tilde{b}_{47}}{\tilde{a}_{77}} y_{7\tau_1}$$

and we substitute y_4 in the equations of (5). The new system becomes:

$$\begin{aligned}
\dot{y}_1 &= -\frac{\gamma_0}{1+(y_3+\hat{u}_3)^{\alpha}} y_1 - \frac{\tilde{R}_m(y_7+\hat{u}_7)}{\tilde{R}_{50}+y_7+\hat{u}_7} y_1 - (\eta_{1l}+\eta_{2l})k_e(y_3+\hat{u}_3)y_1 \\
&\quad -(1-\eta_{1e}-\eta_{2e})\beta_e(y_1, y_3+\hat{u}_3)y_1 + \\
&\quad +2(\eta_1 - \alpha_3 y_3 - \alpha_5 y_5 - \alpha_6 y_6 - \alpha_7 y_7 + \frac{\tilde{b}_{43}}{\tilde{a}_{33}} y_{3\tau_1} + \frac{\tilde{b}_{47}}{\tilde{a}_{77}} y_{7\tau_1}) \\
&\quad \cdot (1-\eta_{1e}-\eta_{2e})\beta_e(y_{1\tau_1}, y_{3\tau_1}+\hat{u}_3)y_{1\tau_1} \\
&\quad +\eta_{1e}(\eta_1 - \alpha_3 y_3 - \alpha_5 y_5 - \alpha_6 y_6 - \alpha_7 y_7 + \frac{\tilde{b}_{43}}{\tilde{a}_{33}} y_{3\tau_1} + \frac{\tilde{b}_{47}}{\tilde{a}_{77}} y_{7\tau_1}) \\
&\quad \cdot k_e(y_{3\tau_1} + \hat{u}_3 y_{1\tau_1}) \\
\dot{y}_2 &= -\gamma_2 y_2 + \tilde{A}_e k_e(y_{3\tau_2}+\hat{u}_3)y_{1\tau_2} \\
\dot{y}_3 &= -k(y_3+\hat{u}_3) + \frac{\tilde{a}_1}{1+y_2^r} \\
\dot{\eta}_1 &= \tilde{b}_{43}y_{3\tau_1} + \tilde{b}_{47}y_{7\tau_1} - \frac{\tilde{b}_{43}}{\tilde{a}_{33}}\dot{y}_{3\tau_1} - \frac{\tilde{b}_{47}}{\tilde{a}_{77}}\dot{y}_{7\tau_1} + R_4^1 \\
&= \tilde{b}_{43}y_{3\tau_1} + \tilde{b}_{47}y_{7\tau_1} - \frac{\tilde{b}_{43}}{\tilde{a}_{33}}(\tilde{a}_{33}y_{3\tau_1} + R_{3\tau_1}) - \frac{\tilde{b}_{47}}{\tilde{a}_{77}}(\tilde{a}_{77}y_{7\tau_1} + R_{7\tau_1}) \\
&= R_4^{(2)}(y, y_{\tau_1}) \\
\dot{y}_5 &= -\tilde{b}_1(y_5+\hat{u}_5) + \tilde{a}_2 \\
\dot{y}_6 &= \tilde{b}_1(y_5+\hat{u}_5) - e_1(y_6+\hat{u}_6) - \frac{c_1(1-e_2)}{c_2+y_6+\hat{u}_6}(y_6+\hat{u}_6) - \\
&\quad -\frac{m_2 e_2}{m_1+y_6+\hat{u}_6}(y_6+\hat{u}_6) \\
\dot{y}_7 &= \frac{\tilde{b}_2 c_1(1-e_2)}{c_2+y_6+\hat{u}_6}(y_6+\hat{u}_6) - e_3(y_7+\hat{u}_7).
\end{aligned}$$

Consider:

$$\begin{aligned}
&f_1(y_1, y_3, y_4, \eta_1, y_6, y_7, y_{1\tau_1}, y_{3\tau_1}, y_{4\tau_1}, y_{7\tau_1}) = \\
&-\frac{\gamma_0}{1+(y_3+\hat{u}_3)^{\alpha}} y_1 - \frac{\tilde{R}_m(y_7+\hat{u}_7)}{\tilde{R}_{50}+y_7+\hat{u}_7} y_1 - \\
&-(\eta_{1e}+\eta_{2e})k_e(y_3+\hat{u}_3)y_1 - (1-\eta_{1e}-\eta_{2e})\beta_e(y_1, y_3+\hat{u}_3)y_1 + \\
&+[2(1-\eta_{1e}-\eta_{2e})\beta_e(y_{1\tau_1}, y_{3\tau_1}+\hat{u}_3)y_{1\tau_1} + \eta_{1e}k_e(y_{3\tau_1}+\hat{u}_3)y_{1\tau_1}]\eta_1 + B_1(y_{\tau 1})
\end{aligned}$$

and remark that the linear part of f_1 does not contain η_1.

Since the other equations, except that of η_1, do not contain η_1 at all, we conclude that Theorem 1 can be applied. Therefore, the stability of E_1 depends on the study of the transcendental terms in its characteristic equation.

2.3.3. The Transcendental Part of the Characteristic Equation

Consider the equation:

$$\lambda - a_{11} - b_{11}e^{-\lambda\tau_1} = 0. \tag{6}$$

The stability analysis of Equation (6) is classical (see, for example, [16]).

Proposition 1. *Let* $p_1 = \dfrac{\gamma_0}{1+\hat{u}_3^\alpha} + \dfrac{\tilde{R}_m \hat{u}_7}{\tilde{R}_{50}+\hat{u}_7}$. *Assume that the following condition is true:*

$$(\eta_{1e}\hat{u}_4 - \eta_{1e} - \eta_{2e})k_e(\hat{u}_3) < p_1 + (1-2\hat{u}_4)(1-\eta_{1e}-\eta_{2e})\beta_e(0,\hat{u}_3). \tag{7}$$

Then, Equation (6) is stable for $\tau_1 = 0$ and remains stable for $\tau_1 > 0$.

Proof. For the equilibrium point E_1, we have:

$$a_{11} = -p_1 - (\eta_{1e}+\eta_{2e})k_e(\hat{u}_3) - (1-\eta_{1e}-\eta_{2e})\beta_e(0,\hat{u}_3)$$

$$b_{11} = 2\hat{u}_4(1-\eta_{1e}-\eta_{2e})\beta_e(0,\hat{u}_3) + \eta_{1e}\hat{u}_3 k_e(z_4).$$

If $\tau_1 = 0$, Equation (6) becomes:

$$\lambda + p_1 + (1-2\hat{u}_4)(1-\eta_{1e}-\eta_{2e})\beta_e(0,\hat{u}_3) - (\eta_{1e}\hat{u}_4 - \eta_{1e} - \eta_{2e})k_e(\hat{u}_3) = 0$$

Equation (6) is stable for $\tau_1 = 0$ if:

$$(\eta_{1e}\hat{u}_4 - \eta_{1e} - \eta_{2e})k_e(\hat{u}_3) < p_1 + (1-2\hat{u}_4)(1-\eta_{1e}-\eta_{2e})\beta_e(0,\hat{u}_3).$$

Since $b_{11} > 0$, the following conditions from [16,17] must hold for stability when $\tau_1 > 0$:

1. $a_{11} < \dfrac{1}{\tau_1}$;
2. $a_{11} + b_{11} < 0$.

Since $a_{11} = -p_1 - (\eta_{1e}+\eta_{2e})k_e(\hat{u}_3) - (1-\eta_{1e}-\eta_{2e})\beta_e(0,\hat{u}_3) < 0 < \dfrac{1}{\tau_1}$, the first condition holds true.

For the second condition we must have:

$$-p_1 + (2\hat{u}_4 - 1)(1-\eta_{1e}-\eta_{2e})\beta_e(0,\hat{u}_3) + (\eta_{1e}\hat{u}_4 - \eta_{1e} - \eta_{2e})k_e(\hat{u}_3) < 0 \tag{8}$$

We remark that, if Condition (7) holds, Condition (8) will also hold. □

Remark 1. *The equilibrium point E_1 is stable if Condition (7) holds.*

2.4. Stability Analysis of the Equilibrium Point E_2

The characteristic equation corresponding to E_2 becomes:

$$\det(\lambda I - A - Be^{-\lambda\tau_1} - Ce^{-\lambda\tau_2}) = d_1(\lambda)d_2(\lambda),$$

where:

$$d_1(\lambda) = (\lambda - a_{77})(\lambda - a_{66})(\lambda - a_{55})$$

and:

$$\begin{aligned} d_2(\lambda) = \; & \lambda^4 - \lambda^3(a_{11}+a_{22}+a_{33}+b_{11}e^{-\lambda\tau_1}) - \\ & -\lambda^2[-a_{11}a_{22} - a_{11}a_{33} - a_{22}a_{33} - (a_{22}b_{11}+b_{11}a_{33})e^{-\lambda\tau_1} + a_{32}c_{23}e^{-\lambda\tau_2}] - \\ & -\lambda[a_{11}a_{22}a_{33} + a_{22}b_{11}a_{33}e^{-\lambda\tau_1} - (a_{11}a_{32}c_{23} - a_{13}a_{32}c_{21})e^{-\lambda\tau_2} - \\ & -(b_{11}a_{32}c_{23} - a_{32}b_{13}c_{21})e^{-\lambda\tau_1}e^{-\lambda\tau_2}] \\ & -a_{14}a_{32}a_{43}c_{21}e^{-\lambda\tau_2} - a_{14}a_{32}c_{21}b_{43}e^{-\lambda\tau_1}e^{-\lambda\tau_2}. \end{aligned}$$

Remark 2. *Since the characteristic equation corresponding to E_2 is complicated, the stability of E_2 is investigated using numerical procedures.*

2.5. Numerical Simulations

For the numerical simulations, we considered the time scale as days and consulted the available literature in order to set the values of the parameters. Table 1 shows the numerical values of the parameters presented in the erythropoiesis model (1).

Table 1. Parameter values for the erythropoiesis model.

Maximal value of the function β_e [9,14]	β_{0e}	1.5
Maximal value of the function k_e [9]	k_{0e}	0.1
Parameter for the death rate [14]	α	0.8
Loss of stem cells due to mortality [9]	γ_0	0.1
Rate of asymmetric/symmetric division [18]	η_{1e}, η_{2e}	0.3
Parameter in the Hill function [18]	m	2
Standard half-saturation (estimated)	a_1	3
Instant mortality of mature leukocytes [9]	γ_2	0.025
Amplification factor [9]	$\tilde{A}$	2400
Maximum effect of drug on erythrocytes [15]	$\tilde{R}_m$	0.0022
Saturation constant for drug on erythrocytes [15]	$\tilde{R}_{50}$	82.2
The supply rate of the 6-MP in the gut[15]	a_2	3.9×10^8
6-MP absorption rate from the gut[15]	b_1	4.8
6-MP elimination rate from plasma [15]	e_1	5
6-MP to 6-TCN conversion rate [15]	c_1	29.8
Activity of TPMT enzyme [15]	e_2	0.5
MM constant for 6-TGN [15]	c_2	4.04×10^5
MeMP elimination rate from erythrocytes [15]	m_2	0.06
MM constant for MeMP [15]	m_1	3.28×10^5
Stoichiometric coefficient for 6-TGN conversion [15]	v_{pt}	1
6-TGN elimination rate from erythrocytes [15]	e_3	0.0714
Self-renewal duration of erythrocytes [14]	τ_1	2.8
Differentiation duration of erythrocytes [14]	τ_2	6

Figure 1 depicts the equilibrium point E_1 in blue and a small perturbation from this equilibrium point in red. We obtained parameter conditions for the stability of this equilibrium point, but it would be preferable that it be unstable; remember that E_1 corresponds to the death of the patient.

We proved that the equilibrium point E_1 is stable if Condition (7) holds. For the parameter values in the table above, it is easy to verify that (7) does not hold. This can also be seen in Figure 1, where the evolution seems to be a favorable one for the patient.

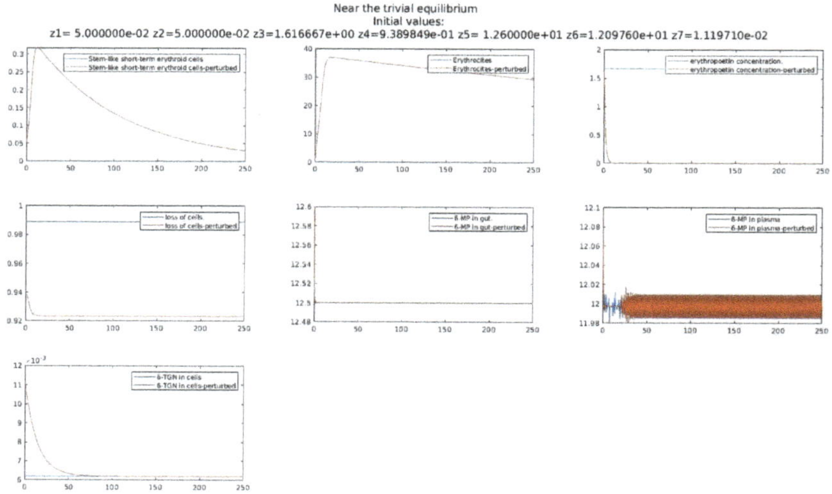

Figure 1. Stability of equilibrium point E_1.

Figure 2 depicts the equilibrium point E_2 in blue and a small perturbation from this equilibrium point in red. Although we could not obtain any parameter conditions for stability, it is easy to notice that the equilibrium point is unstable for the considered set of parameters. As E_2 corresponds to the chronic phase of the disease, the evolution obtained in Figure 2 is a desired one. The healthy cells grow in number, and the leukemic cells die out; this basically means that the patient recovers.

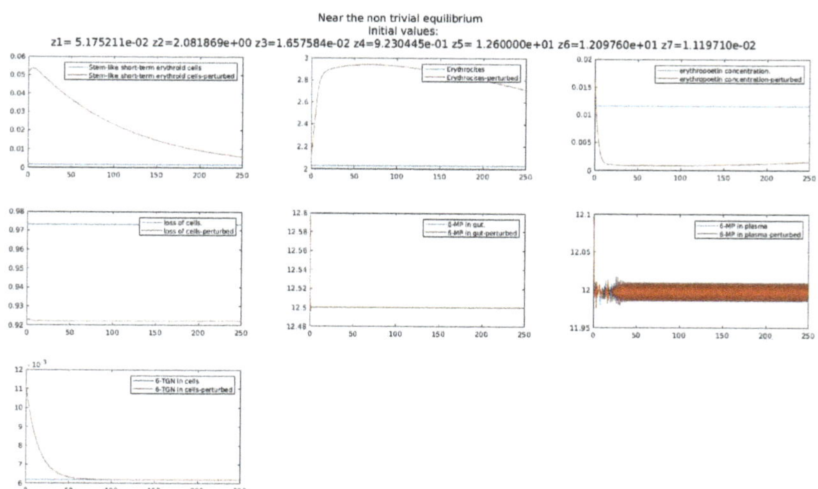

Figure 2. Stability of equilibrium point E_2.

3. The Leukopoiesis Model

3.1. The Mathematical Model

In this model, s_1 represents the concentration of short-term stem-like white blood cells' precursors and s_2 the adult leukocytes. The treatment is presented through the function l_1 (see [15]):

$$l_1(x_6) = \frac{x_6}{L_{1S0}+x_6}$$

The equations that describe these variables mirror those in the erythropoiesis model, as the evolution of the leukocytes has many points in common. The difference between the two is the influence erythropoietin has on the production of the cells.

A non-constant rate of the elimination of stem cells due to treatment is encountered, and this leads to the consideration of a new, auxiliary variable s_3.

Again, the last three equations depict the progress of the drug after administration.

The following model describes a compartment of leukopoiesis coupled with the dynamics of 6-MP used in the maintenance therapy:

$$\dot{s} = \tilde{f}_i(s, s_{\tau_j}), i = \overline{1,6}, j = \overline{3,4} \tag{9}$$

$$\begin{aligned}
\dot{s}_1 = & -\gamma_{1l}s_1 - T_1 l_1(s_6)s_1 - \eta_{1l}k_l(s_2)s_1 - \eta_{2l}k_l(s_2)s_1 - (1-\eta_{1l}-\eta_{2l})\beta_l(s_1)s_1 \\
& +2e^{-\gamma_{1l}\tau_1}s_3(1-\eta_{1l}-\eta_{2l})\beta_l(s_{1\tau_1})s_{1\tau_1} + \eta_{1l}e^{-\gamma_{1l}\tau_1}s_3k_l(s_{2\tau_1})s_{1\tau_1} \\
\dot{s}_2 = & -\gamma_{2l}s_2 + \tilde{A}_l k_l(s_{2\tau_2})s_{1\tau_2} \\
\dot{s}_3 = & s_3T_1[l_1(s_{6\tau_1}) - l_1(s_6)] \\
\dot{s}_4 = & -b_1s_4 + a_2 \\
\dot{s}_5 = & b_1s_4 - e_1s_5 - \frac{c_1(1-e_2)}{c_2+s_5}s_5 - \frac{m_2e_2}{m_1+s_5}s_5 \\
\dot{s}_6 = & \frac{b_2c_1(1-e_2)}{c_2+s_5}s_5 - c_3s_6.
\end{aligned}$$

3.2. The Equilibrium Points and Linearization

Following [7], we concluded that there are two types of equilibrium points corresponding to the model of leukopoiesis:

$$\tilde{E}_1 = (0, 0, \hat{s}_3, \hat{s}_4, \hat{s}_5, \hat{s}_6)$$

$$\tilde{E}_2 = (\hat{s}_1, \hat{s}_2, \hat{s}_3, \hat{s}_4, \hat{s}_5, \hat{s}_6).$$

When linearizing System (9), the matrix of partial derivatives with respect to undelayed variables is denoted as $\tilde{A} = \frac{\partial f}{\partial s} = [\tilde{a}_{ij}]$ and the matrices of the partial derivatives with respect to the delayed variables are denoted as $\tilde{B} = \frac{\partial f}{\partial s_{\tau_3}} = [\tilde{b}_{ij}]$ and $\tilde{C} = \frac{\partial f}{\partial s_{\tau_4}} = [\tilde{c}_{ij}]$. For a complete list of the coefficients, please consult [7].

3.3. Stability Analysis for Equilibrium Point $\tilde{E}_1$

The characteristic equation corresponding to $\tilde{E}_1 = (0, 0, \hat{s}_3, \hat{s}_4, \hat{s}_5, \hat{s}_6)$ is:

$$\lambda(\lambda - a_{22})(\lambda - a_{44})(\lambda - a_{55})(\lambda - a_{66})\left(\lambda - a_{11} - b_{11}e^{-\lambda\tau_1}\right) = 0.$$

The real solutions of the characteristic equation corresponding to $\tilde{E}_1$ are given by:

$$\lambda_1 = 0$$

$$\lambda_2 = a_{22} = -\gamma_{2l} < 0$$

$$\lambda_3 = a_{44} = -b_1 < 0$$

$$\lambda_4 = a_{55} = -e_1 - \frac{c_1(1-e_2)c_2}{(c_2+\hat{s}_5)^2} - \frac{m_2e_2m_1}{(m_1+\hat{s}_5)^2} < 0$$

$$\lambda_5 = a_{66} = -c_3 < 0.$$

The analysis of the critical case follows the same steps as in the previous case. As before, we can rewrite the system in a way that Theorem 1 can be applied. In order to avoid redundancy, we skip the computations. We concluded that the stability of $\tilde{E}_1$ relies on the zeros of the transcendental equation:

$$\lambda - a_{11} - b_{11}e^{-\lambda\tau_1} = 0. \tag{10}$$

Proposition 2. *Assume that the following condition is true:*

$$(2\hat{s}_3 - 1)(1 - \eta_{1l} - \eta_{2l})\beta_l(0) + (\hat{s}_3\eta_{1l} - \eta_{1l} - \eta_{2l})k_l(0) < \gamma_{1l} + T_1 l_1(\hat{s}_6). \tag{11}$$

Then, Equation (10) is stable for $\tau_1 = 0$, and it remains stable for all $\tau_1 > 0$.

The proof of Proposition 2 is similar to that of Proposition 1.

Remark 3. *The equilibrium point $\tilde{E}_1$ is stable if Condition (11) holds.*

3.4. Stability Analysis of the Equilibrium Point $\tilde{E}_2$

The characteristic equation corresponding to $\tilde{E}_2$ is:

$$\lambda(\lambda - a_{44})(\lambda - a_{66})(\lambda - a_{55}) \cdot \left[(\lambda - a_{11} - b_{11}e^{-\lambda\tau_1})(\lambda - a_{22} - c_{22}e^{-\lambda\tau_2}) - c_{21}e^{-\lambda\tau_2}(a_{12} + b_{12}e^{-\lambda\tau_1})\right] = 0.$$

3.4.1. The Real Solutions of the Characteristic Equation

The nonzero real solutions of the characteristic equation corresponding to $\tilde{E}_2$ are given by:

$$\lambda = a_{44} < 0$$

$$\lambda = a_{66} < 0$$

and

$$\lambda = a_{55} < 0$$

3.4.2. The Transcendental Part of the Characteristic Equation

The stability of the characteristic equation corresponding to $\tilde{E}_2$ depends on the stability of the following equation:

$$(\lambda - a_{11} - b_{11}e^{-\lambda\tau_1})(\lambda - a_{22} - c_{22}e^{-\lambda\tau_2}) - c_{21}e^{-\lambda\tau_2}(a_{12} + b_{12}e^{-\lambda\tau_1}) = 0 \tag{12}$$

The stability analysis of Equation (12) follows the approach in [19], Theorem (1).

Proposition 3. *Assume the following conditions hold:*

$$\begin{aligned} a_{11} + b_{11} + a_{22} + c_{22} < 0 \\ a_{11}a_{22} + a_{22}b_{11} + a_{11}c_{22} + b_{11}c_{22} - a_{12}c_{21} - b_{12}c_{21} > 0 \end{aligned} \tag{13}$$

then Equation (12) is stable for $\tau_1 = \tau_2 = 0$.

Proof. For $\tau_1 = \tau_2 = 0$, Equation (12) becomes:

$$\begin{aligned} &\lambda^2 + \lambda(-a_{11} - a_{22} - b_{11} - c_{22}) + a_{11}a_{22} + a_{22}b_{11} \\ &+a_{11}c_{22} + b_{11}c_{22} - a_{12}c_{21} - b_{12}c_{21} = 0 \end{aligned} \tag{14}$$

Suppose that λ_1 and λ_2 are two roots for Equation (14), then in order to have $Re(\lambda_1)$,$Re(\lambda_2) < 0$, the following conditions must hold:

$$\begin{aligned} a_{11} + b_{11} + a_{22} + c_{22} < 0 \\ a_{11}a_{22} + a_{22}b_{11} + a_{11}c_{22} + b_{11}c_{22} - a_{12}c_{21} - b_{12}c_{21} > 0 \end{aligned}$$

□

Now, let:

$$\begin{aligned} u_1 &= a_{11} + b_{11} + a_{22} \\ v_1 &= a_{11}a_{22} + a_{22}b_{11} = a_{22}(a_{11} + b_{11}) \\ u_2 &= -c_{22} \\ v_2 &= a_{11}c_{22} + b_{11}c_{22} - a_{12}c_{21} - b_{12}c_{21} = c_{22}(a_{11} + b_{11}) - c_{21}(a_{12} + b_{12}) \end{aligned}$$

Proposition 4. *Consider the following conditions:*

$$\left(u_1^2 - 2v_1 - v_2^2\right)^2 - 4\left(v_1^2 - u_2^2\right) > 0 \tag{15}$$

$$u_1^2 - 2v_1 - v_2^2 < 0. \tag{16}$$

If either Condition (15) or Condition (16) does not hold and if Equation (12) is stable for $\tau_1 = \tau_2 = 0$, *it will remain stable for* $\tau_1 = 0$ *and* $\tau_2 > 0$.

Proof. Suppose $\tau_1 = 0$ and $\tau_2 > 0$. Equation (12) becomes:

$$\lambda^2 - u_1\lambda + v_1 + e^{-\lambda\tau_2}(u_2\lambda + v_2) = 0 \tag{17}$$

The study of this equation follows the approach of Theorem 1 in [19]. Define:

$$\begin{aligned} P(\lambda) &= \lambda^2 - u_1\lambda + v_1 \\ Q(\lambda) &= u_2\lambda + v_2 \end{aligned}$$

Note that Conditions (i), (iv), and (v) of Theorem 1 in [19] are most likely to hold and Conditions (ii) and (iii) of Theorem 1 in [19] hold.

Then, the stability of Equation (17) depends on the zeros of the equation:

$$|P(iy)|^2 - |Q(iy)|^2 = 0 \tag{18}$$

If Equation (18) has no positive roots ($y > 0$) and if (17) is stable with $\tau_2 = 0$, it will be stable for all $\tau_2 > 0$. If Equation (18) has a least one $y > 0$ as a root and all the roots are simple, as τ_2 increases, there might be stability switches. Therefore, if Equation (17) is stable at $\tau_2 = 0$, it might become unstable when $\tau_2 = \tau_2^*$.

Now, consider $P(iy) = P_R(y) + iP_I(y)$ and $Q(iy) = Q_R(y) + iQ_I(y)$ where P_R, P_I, Q_R, and Q_I are real-valued functions.

Equation (18) can be written as:

$$|P(iy)|^2 - |Q(iy)|^2 = 0$$
$$|P(iy)|^2 = |Q(iy)|^2$$
$$P_R^2(y) + P_I^2(y) = Q_R^2(y) + Q_I^2(y).$$

We obtain the following polynomial:

$$y^4 + y^2(u_1^2 - 2v_1 - v_2^2) + v_1^2 - u_2^2 = 0, \tag{19}$$

By setting $\alpha = y^2$, we have:

$$\alpha^2 + \alpha(u_1^2 - 2v_1 - v_2^2) + v_1^2 - u_2^2 = 0. \tag{20}$$

In order that Equation (20) has at least one simple root $y > 0$, the following conditions must hold:

$$\left(u_1^2 - 2v_1 - v_2^2\right)^2 - 4\left(v_1^2 - u_2^2\right) > 0 \tag{21}$$

$$u_1^2 - 2v_1 - v_2^2 < 0 \tag{22}$$

Therefore, Equation (12) is stable if at least one of the conditions (21) or (22) is not met. □

Now, we consider $\tau_1 = \tau_1^*$ (fixed) and $\tau_2 > 0$.
Equation (12) becomes:

$$(\lambda - a_{11} - b_{11}e^{-\lambda\tau_1^*})(\lambda - a_{22} - c_{22}e^{-\lambda\tau_2}) - c_{21}e^{-\lambda\tau_2}(a_{12} + b_{12}e^{-\lambda\tau_1^*}) = 0 \tag{23}$$

Equation (23) can be written as:

$$P(\lambda) + Q(\lambda)e^{-\lambda\tau_2} = 0,$$

with

$$\begin{aligned} P(\lambda) &= \lambda^2 - (a_{11} + a_{22})\lambda + a_{11}a_{22} - (b_{11}\lambda + a_{22}b_{11})e^{-\lambda\tau_1^*} \\ Q(\lambda) &= -c_{22}\lambda + a_{11}c_{22} - a_{12}c_{21} + (b_{11}c_{22} - b_{12}c_{21})e^{-\lambda\tau_1^*}. \end{aligned}$$

Remark 4. *Suppose that the equation:*

$$|P(iy)|^2 - |Q(iy)|^2 = 0$$

has no positive real roots. Then, if Equation (23) is stable for $\tau_1 = \tau_2 = 0$, it will remain stable for $\tau_1 = \tau_1^*$ and all $\tau_2 > 0$.

Since $P(\lambda)$ and $Q(\lambda)$ are analytic functions, we can apply the result of Theorem 1 in [19]. Set $\lambda = iy$. We are interested in the roots of the equation:

$$F(y) = |P(iy)|^2 - |Q(iy)|^2.$$

We have:

$$\begin{aligned} F(y) = \; & y^4 + 2b_{11}y^3\sin(y\tau_1^*) + r_1y^2 + 2a_{11}b_{11}y^2\cos(y\tau_1^*) \\ & + r_2y\sin(y\tau_1^*) + r_3\cos(y\tau_1^*) + k_4 \end{aligned} \tag{24}$$

$$\begin{aligned}
r_1 = & \ a_{11}^2 + a_{22}^2 + b_{11}^2 - c_{22}^2 \\
r_2 = & \ 2a_{22}^2 b_{11} - 2b_{11}c_{22}^2 + 2b_{12}c_{21}c_{22} \\
r_3 = & \ 2a_{11}b_{12}c_{21}c_{22} + 2a_{11}a_{22}^2 b_{11} - 2a_{11}b_{11}c_{22}^2 - 2a_{12}b_{12}c_{21}^2 + 2a_{12}b_{11}c_{21}c_{22} \\
r_4 = & \ a_{11}^2 a_{22}^2 + a_{22}^2 b_{11}^2 - a_{11}^2 c_{22}^2 - a_{12}^2 c_{21}^2 - b_{11}^2 c_{22}^2 - b_{12}^2 c_{21}^2 + 2a_{11}a_{12}c_{21}c_{22} + \\
& +2b_{11}b_{12}c_{21}c_{22}
\end{aligned}$$

If the function (24) has no $y > 0$ as a root and if Equation (12) is stable with $\tau_1 = \tau_2 = 0$, it will remain stable for all $\tau_2 > 0$ and $\tau_1 = \tau_1^*$.

Remark 5. *The equilibrium point $\tilde{E}_2$ is stable if Equations (17) and (23) are stable.*

3.5. Numerical Simulations

Again, we considered the time scale as days and consulted the available literature in order to set the values of the parameters. Table 2 shows the numerical values of the parameters presented in the leukopoiesis model (9).

Table 2. Parameter values for the leukopoiesis model.

Maximal value of the function β_l [9,14]	β_{0l}	1.5
Maximal value of the function k_l [9]	k_{0l}	0.1
Loss of stem cells due to mortality [9]	γ_{1l}	0.1
Rate of asymmetric/ symmetric division [18]	η_{1l}, η_{2l}	0.3
Parameter in the Hill function [18]	m_l	2
Standard half-saturation (estimated)	a_1	3
Instant mortality of mature leukocytes [9]	γ_2	0.025
Amplification factor [9]	$\tilde{A}$	2400
Maximum effect of drug on leukocytes [15]	T_1	0.0782
The supply rate of the 6-MP in the gut[15]	a_2	3.9×10^8
6-MP absorption rate from the gut[15]	b_1	4.8
6-MP elimination rate from plasma [15]	e_1	5
6-MP to 6-TCN conversion rate [15]	c_1	29.8
Activity of TPMT enzyme [15]	e_2	0.5
MM constant for 6-TGN [15]	c_2	4.04×10^5
MeMP elimination rate from leukocytes [15]	m_2	0.06
MM constant for MeMP [15]	m_1	3.28×10^5
Stoichiometric coefficient for 6-TGN Conversion [15]	v_{pt}	1
6-TGN elimination rate from leukocytes [15]	e_3	0.1207
Self-renewal duration of leukocytes [14]	τ_1	1.4
Differentiation duration of leukocytes [14]	τ_2	3.5

Due to the fact that the drug dynamics after administration is similar to the one obtained in the erythropoiesis model, we only focus on the representation of the white blood cells' precursors, s_1, and the adult leukocytes, s_2.

We would prefer that the equilibrium point $\tilde{E_1}$ be unstable and perhaps attracted to a healthy state. Figure 3 depicts the equilibrium point $\tilde{E}_1$ in blue and a small perturbation from this equilibrium point in red. The evolution of both the white blood cells' precursors and adult leukocytes is a favorable one for the patient.

In Figure 3, the equilibrium point $\tilde{E_1}$ is clearly unstable. By checking Condition (11), we obtain the same result: the condition does not hold, and the equilibrium point is unstable.

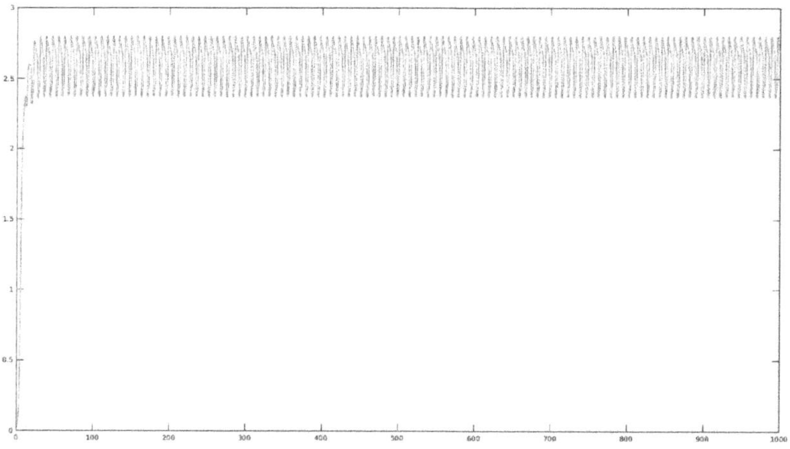

(a)

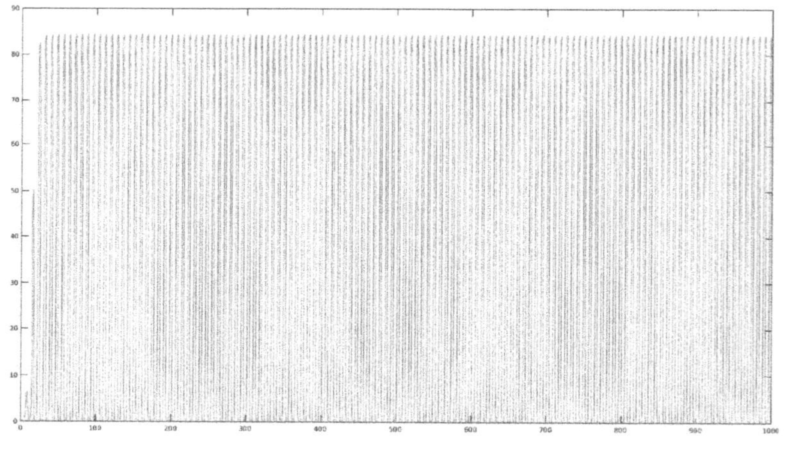

(b)

Figure 3. Stability of equilibrium point $\tilde{E_1}$: (**a**) Evolution of the white blood cells' precursors; (**b**) Evolution of the adult leukocytes

The equilibrium point $\tilde{E_2}$ corresponds to a healthy state of the patient. Figure 4 shows the equilibrium point $\tilde{E_2}$ in blue and a small perturbation from this equilibrium point in red. The equilibrium point is clearly unstable, but the evolution of the patient is still promising. The white blood cells' precursors and adult leukocytes both show an increase in number.

Oscillatory trajectories are common in blood cell evolution, as there is a natural inner oscillatory dynamics.

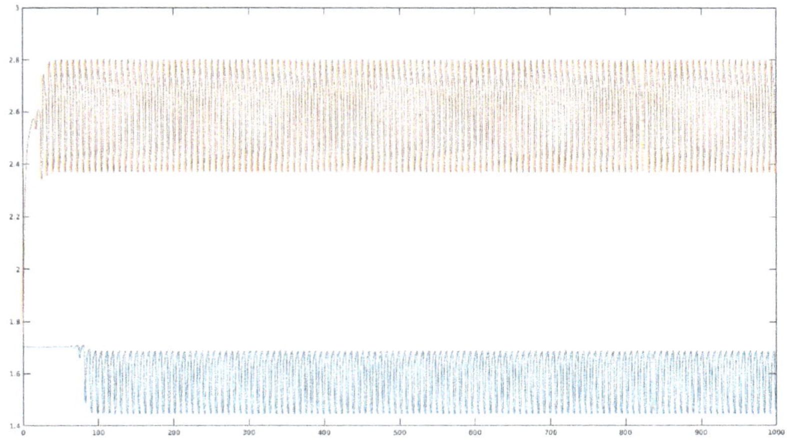

(**a**)

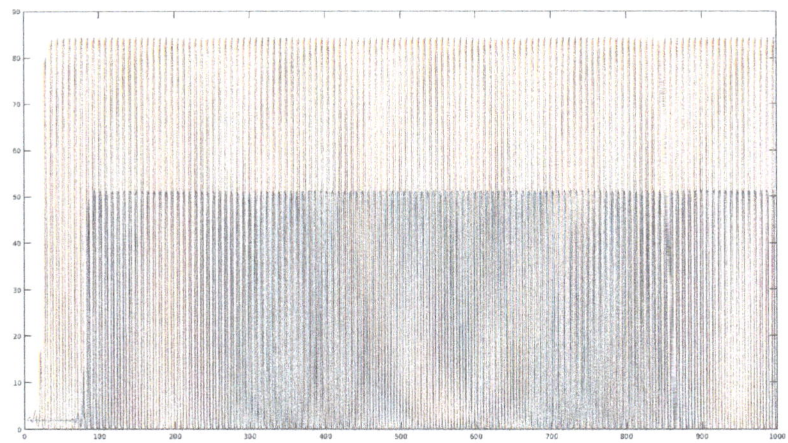

(**b**)

Figure 4. Stability of equilibrium point $\tilde{E_2}$: (**a**) Evolution of the white blood cells' precursors; (**b**) Evolution of the adult leukocytes

4. Conclusions

The objective of this article was to present the stability study of two mathematical models that describe the processes of erythropoiesis and leukopoiesis (which are responsible for the production of red and white blood cells) in the case of maintenance therapy for acute lymphoblastic leukemia. The models were developed by some of the authors and introduced in [7].

The stability of all the equilibrium points was thoroughly investigated. When possible, parameter conditions for stability were determined. Some provocative and interesting conclusions resulted from the study of critical cases. For this purpose, a theorem designed by some of the authors and introduced in [8] was used. Numerical simulations were also used to validate and extend the study in some cases where the characteristic equation was too complicated.

From a biological and medical point of view, these models were validated first by the fact that the equilibrium points make sense biologically, as they correspond to existing states in which a patient may be found. Secondly, the numerical simulations, which were obtained using parameter values from the relevant literature (thus as close to reality as we could manage), depicted genuine developments of a potential patient's health.

In the case of the erythropoiesis model, the steady-states correspond to either the death of the patient or a chronic phase of the disease. As neither is the desired state of a patient, it would be preferable that these equilibrium points be unstable. The numerical analysis yielded this favorable result.

In the leukopoiesis model, the equilibrium points depicted either the death of the patient or a healthy state of the patient. For the considered configuration of the parameters, both these steady-states were unstable, but presented a positive evolution of the patient's condition.

Mathematical models that capture complicated situations in some diseases can aid in designing an adequate treatment. After a comprehensive mathematical study, these models can help to determine the correct dose of drug that needs to be administered for each individual patient.

Author Contributions: Conceptualization, I.B.; formal analysis, R.M.; methodology, I.B. and R.M.; resources, R.M.; software, A.-D.H.; validation, A.-D.H.; visualization, A.-D.H.; writing—original draft, I.B. and R.M.; writing—review and editing, I.B. and A.-D.H. All authors have read and agreed to the published version of the manuscript.

Funding: This research received no external funding.

Institutional Review Board Statement: Not applicable.

Informed Consent Statement: Not applicable.

Data Availability Statement: Not applicable.

Conflicts of Interest: The authors declare no conflict of interest.

References

1. Abass, A.K.; Lichtman, A.H.; Pillai, S. *Cellular and Molecular Immunolgy*, 7th ed.; Elsevier: Amsterdam, The Netherlands, 2012.
2. Adimy, M.; Crauste, F.; Ruan, S. A mathematical study of the hematopoiesis process with application to chronic myelogenous leukemia. *SIAM J. Appl. Math.* **2005**, *65*, 1328–1352. [CrossRef]
3. Akushevich, I.; Veremeyeva, G.; Dimov, G.; Ukraintseva, S.; Arbeev, K.; Akleyev, A.; Yashin, A. Modeling hematopoietic system response caused by chronic exposure to ionizing radiation. *Radiat. Environ. Biophys.* **2011**, *50*, 299–311 . [CrossRef] [PubMed]
4. Childhood, A.; Collaborative Group. Duration and intensity of maintenance chemotherapy in acute lymphoblastic leukemia: Overview of 42 trials involving 12,000 randomized children. *Lancet* **1996**, *347* , 1783–1788.
5. Pui, C.H.; Evans, W.E. Treatment of acute lymphoblastic leukemia. *N. Engl. J. Med.* **2006**, *354* , 166–178. [CrossRef] [PubMed]
6. Lin, T.L.; Vala, M.S.; Barber, J.P.; Karp, J.E.; Smith, B.D.; Matsui, W.; Jones, R.J. Induction of acute lymphocytic leukemia differentiation by maintenance therapy. *Leukemia* **2007**, *21* , 1915–1920. [CrossRef] [PubMed]
7. Badralexi, I.; Halanay, A.; Mghames, R. A Delay Differential Equations model for maintenance therapy in acute lymphoblastic leukemia. *UPB Sci. Bull. Ser. A* **2020**, *82*, 13–24.
8. Amin, K.; Badralexi, I.; Halanay, A.; Mghames, R. A stability theorem for equilibria of delay differential equations in a critical case and some models of cell evolution. *Math. Model. Nat. Phenom.* **2021**, *16*, 36. [CrossRef]
9. Adimy, M.; Bourfia, Y.; Hbid, M.L.; Marquet, C. Age-structured model of hematopoiesis dynamics with growth factor-dependent coefficients. *Electr. J. Diff. Equ.* **2016**, *2016* , 1–20.
10. Munker, R.; Hiller, E.; Glass, J.; Paquette, R. *Modern Hematology. Biology and Clinical Management*, 2nd ed.; Humana Press: Berlin/Heidelberg, Germany, 2007.
11. Parajdi, L.G.; Precup, R.; Bonci, E.; Tomuleasa, C. A Mathematical Model of the Transition from Normal Hematopoiesis to the Chronic and Accelerated-Acute Stages in Myeloid Leukemia. *Mathematics* **2020**, *8*, 376. [CrossRef]
12. Greer, J.; Arber, D.; Glader, B.; List, A.; Means, R., Jr.; Paraskevas, F.; Rodgers, G. *Wintrobe's Clinical Hematology*, 13th ed.; LWW: Philadelphia, PA, USA , 2013.
13. Schmiegelow, K.; Nielsen, S.; Frandsen, T.; Nersting, J. Mercaptopurine/Methotrexate Maintenance Therapy of Childhood Acute Lymphoblastic Leukemia. Clinical Facts and Fiction. *J. Pediatr.* **2014**, *36*, 503–517.
14. Colijn, C.; Mackey, M.C. A Mathematical Model for Hematopoiesis: I. Periodic Chronic Myelogenous Leukemia. *J. Theor. Biol.* **2005**, *237* , 117–132. [CrossRef] [PubMed]

15. Jayachandran, D.; Rundell, A.E.; Hannemann, R.; Vik, T.A.; Ramkrishna, D. Optimal Chemotherapy for Leukemia: A model-Based Strategy for Individualized Treatment. *PLoS ONE* **2014**, *9*, e109623. [CrossRef]
16. Bellman, R.; Cooke, K.L. *Differential-Difference Equations*; Academic Press: New York, NY, USA, 1963.
17. Cooke, K.; Grossman, Z. Discrete Delay, Distribution Delay and Stability Switches. *J. Math. Anal. Appl.* **1982**, *86* , 592–627. [CrossRef]
18. Balea, S.; Halanay, A.; Neamtu, M. A feedback model for leukemia including, cell competition and the action of the immune system. In Proceedings of the ICNPAA World Congress, Narvik, Norway, 15–18 July 2014; Volume 1637, pp. 1316–1324.
19. Cooke, K.; van den Driessche, P. On Zeroes of Some Transcendental Equations. *Funkcialaj Ekvacioj* **1986**, *29*, 7790.

mathematics

MDPI

Article

Numerical Studies of Statistical Management Decisions in Conditions of Stochastic Chaos

Alexander Musaev [1] and Dmitry Grigoriev [2,*]

[1] St. Petersburg Institute for Informatics and Automation of the Russian Academy of Sciences, Saint-Petersburg State Institute of Technology (Technical University), 190013 St. Petersburg, Russia; amusaev@technolog.edu.ru
[2] Center of Econometrics and Business Analytics (CEBA), St. Petersburg State University, 199034 St. Petersburg, Russia
* Correspondence: d.a.grigoriev@spbu.ru

Citation: Musaev, A.; Grigoriev, D.; Numerical Studies of Statistical Management Decisions in Conditions of Stochastic Chaos. *Mathematics* **2022**, *10*, 226. https://doi.org/10.3390/math10020226

Academic Editors: Mihaela Neamţu, Eva Kaslik and Anca Rădulescu

Received: 2 December 2021
Accepted: 6 January 2022
Published: 12 January 2022

Publisher's Note: MDPI stays neutral with regard to jurisdictional claims in published maps and institutional affiliations.

Copyright: © 2022 by the authors. Licensee MDPI, Basel, Switzerland. This article is an open access article distributed under the terms and conditions of the Creative Commons Attribution (CC BY) license (https://creativecommons.org/licenses/by/4.0/).

Abstract: The research presented in this article is dedicated to analyzing the acceptability of traditional techniques of statistical management decision-making in conditions of stochastic chaos. A corresponding example would be asset management at electronic capital markets. This formulation of the problem is typical for a large number of applications in which the managed object interacts with an unstable immersion environment. In particular, this issue arises in problems of managing gas-dynamic and hydrodynamic turbulent flows. We highlight the features of observation series of the managed object's state immersed in an unstable interaction environment. The fundamental difference between observation series of chaotic processes and probabilistic descriptions of traditional models is demonstrated. We also present an additive observation model with a chaotic system component and non-stationary noise which provides the most adequate characterization of the original observation series. Furthermore, we suggest a method for numerically analyzing the efficiency of conventional statistical solutions in the conditions of stochastic chaos. Based on numerical experiments, we establish that techniques of optimal statistical synthesis do not allow for making effective management decisions in the conditions of stochastic chaos. Finally, we propose several versions of compositional algorithms focused on the adaptation of statistical techniques to the non-deterministic conditions caused by the specifics of chaotic processes.

Keywords: currency market; Forex risk control models; chaotic processes; trends prediction

1. Introduction

A distinctive feature of dynamic open systems that interact with unstable immersion environments is the presence of random perturbations associated with unpredictable physical, economical, political, natural, etc. factors. At each moment of time any given (often latent) factor may start to dominate, which makes many explicitly pronounced local trends appear. The presence of such trends and large areas of observation series that are oscillatory non-periodic processes exactly satisfies the description of chaotic processes [1–6]. The chaotic nature of observation series violates the established constraints that are required for statistical estimates to be consistent, efficient, and non-biased. In particular, this refers to the stationarity of the noise, independence of increments, normality (gaussianity) of value variations, and others. The full list of conditions that provide effectiveness and consistency of statistical solutions can be found in classical and modern textbooks on mathematical statistics [7–12]. At the same time, as dedicated statistical studies [5,6] show, the most adequate model for describing observation series of the state of unstable objects is an additive combination of a system component consisting of an oscillatory non-periodic process with many local trends, and a random component that is a non-stationary random process. It is obvious that in these conditions, aptly named stochastic chaos, the most important assumption about the repeatability of events which underlies the probabilistic-statistical paradigm has been violated. At the same time, statistical analysis methods remain the primary tool of management decision-making in the conditions of non-determinism. Hence, there is a problem of evaluating the effectiveness of statistical methods of data analysis in conditions in which they are obviously not optimal. Moreover, the estimates made in the conditions of stochastic chaos generally are not unbiased and effective. Therefore, the question about the effectiveness of management decisions based on them remains open. Currently, there are no analytical methods for assessing the quality of statistical estimates generated for chaotic data. The only available assessment means are numerical studies. Therefore, this is the methodology we use in the present article. Note that the specifics of chaotic dynamics do not allow us to obtain a stable result that determines the effectiveness of the estimation procedures themselves. However, considering the management task for which such estimates are made, it would be more sensible to assess the quality of the used algorithms via terminal performance indicators of the constructed strategies. We propose and examine the corresponding technique in this study.

2. Materials and Methods

2.1. Generalized Statistical Decision-Making

Decision-making via statistical synthesis requires constructing a statistical model of the trading situation state. A traditional example of such a model is the statement that an observation series of financial instrument quotations $y_1, y_2, \ldots, y_n$ are independent, identically distributed random variables with a total density $f(y,\ \theta), \quad y \in R^1$, where $\theta \in \Theta \subset R^m$ is the set of market parameters, R^m is an m-dimensional real Euclidean space, and D is the decision space. It is assumed that information about X is needed to select a trading decision $d \in D$ from a predetermined set of management decisions D. In general, d is the result of a certain predictive analysis procedure concerning the trading situation, which includes formalized data processing and interpretation of the obtained results. In practice, a trading decision d is made based on the predicted values of the trading situation state vector $\widehat{X}$.

A trading strategy (*a decision function*) is a map $S : Y \to D$, i.e., a rule according to which, within the framework of a given statistical model, a set of observations of a trading situation $y_1, y_2, \ldots y_n \in Y$ corresponds to a trading decision $d \in D$. Thus, a trading strategy is essentially a procedure for analyzing and processing observation series, and a rule for interpreting the result.

The fundamental task of statistical synthesis of management decisions is to choose a strategy that is optimal with respect to some specific quality measure. Such a choice requires the introduction of "more—less", "better—worse" order relations to the class of all strategies. Strategy S_0 is called *feasible* if it provides management decisions that satisfy a given class of constraints. Strategy S^* is optimal in a given class of feasible strategies $S \in \{S_0\}$ if, taking into account the existing constraints, it outperforms any other strategy from this class by a specified effectiveness measure $\mu(S)$: $S = S^* : \mu(S^*) > \mu(S), \forall S \in \{S_0\}$.

The most common method of ordering a set of strategies requires *a loss function* $w(X,d)$ that maps $\widehat{X}_\tau \times D \to \mathbb{R}^+$, where $\mathbb{R}^+$ is the positive semi-axis of the real axis $\mathbb{R}$, which represents the loss of a trading decision $d \in D$ based on forecast $\widehat{X}_\tau$ in conditions when the true state of the trading environment is described by a vector parameter X_τ, and τ is the forecasting interval. In estimation problems,the full square of the forecast error is often used as a loss function: $w(X_\tau,\ \widehat{X}_\tau) = (\widehat{X}_\tau - X_\tau)^2$, i.e., the square of the Euclidean distance between $\widehat{X}_\tau$ and X_τ. Sometimes the loss function is made more complex by adding a weight, thus becoming a weighted sum: $w(X_\tau,\ \widehat{X}_\tau) = \lambda(X_\tau)(\widehat{X}_\tau - X_\tau)^2$, where the weight $\lambda(X_\tau)$ is selected from the condition of a specific task.

2.2. Specifics of Asset Management in the Conditions of Dynamic Chaos

During the construction of Bayesian management strategies, each acceptable trading strategy $S \in \{S_0\}$ is mapped to an average loss or *risk* $R_S(X) = E\{w(X,\ d)\}$, where E is the symbol of mathematical expectation. In the tasks of proactive (i.e., forecast-based) management, which is typical for trading, the main cause of risk is errors in predicting the market situation state vector $\widehat{X}_\tau$ for the forecast interval τ chosen by the trader. In the case of a point state vector of a particular financial instrument based on an independent sample of observations $y_1, y_2, \ldots, y_N$ with a distribution density $f(Y)$, the average loss from the selected trading decision will have the form

$$R_S(x_\tau) = \int \ldots \int_{R^n} w(x_\tau, \widehat{x}_\tau(y_1, \ldots, y_n)) \prod_{i=1}^{n} f(y_i \mid x_\tau) dy. \quad (1)$$

In the conditions of chaotic dynamics, the observed process does not meet the conditions of stationarity and ergodicity [5,6], the distribution function continuously changes in time and it is not possible to restore it from the previous sample of observations. In this case, the quality of the forecast itself is often estimated based *on average square error* $R_S = E\{(\widehat{X}_\tau - X_\tau)^2\} = D\{\widehat{X}_\tau\} + (E\{\widehat{X}_\tau\} - X_\tau)^2$, where $D\{\widehat{X}_\tau\}$ is the variance of the forecast $\widehat{X}_\tau$. The trading risk from adopting a management strategy S is mainly related to the probabilistic scatter of $\widehat{X}_\tau$ near the true value X_τ for time $t + \tau$. For a normal distribution law, variance $\lim_{n \to \infty} E\{\widehat{X}_\tau - X\} = \lim_{n \to \infty} D\{\widehat{X}_\tau\} = \sigma_\tau^2$ is a natural scatter measure. However, for chaotic processes, there is no natural general measure of scatter, and the choice has to be made in accordance with the conditions of a specific task.

The risk $R_S(X)$ associated with the application of strategy S introduces a partial order in the set of strategies so that $S_1 \geq S_2 \Leftrightarrow R_{S_1}(X) \leq R_{S_2}(X) \ \forall X \in \{X\}$.

Generally, in chaotic conditions curves $R_{S_1}(X)$ and $R_{S_2}(X)$ intersect and, therefore, the corresponding strategies are generally incomparable. Only in very rare cases there may exist a curve in this class that lies uniformly below all the others at $\forall X \in \{X\}$, and thus corresponds to the optimal strategy according to the criterion of minimum trading risk.

2.3. Specifics of Applying Statistical Synthesis of Management Strategies in Asset Management

Making control decisions, according to the general theory of statistical decisions, is a procedure for choosing a decision $d \in D_0$ from a set of acceptable decision D_0 based on the results of observations of the state of the trading situation of interest to us, described by the process X_t, $t = 1, \ldots, n$. In a real situation, this vector process can not always be observed directly and a judgment about it is made on the basis of functionally related, technically accessible observations Y_t, $t = 1, \ldots, n$. In other words, an investor or trader usually only has access to direct or indirect observations that are functionally related to the considered trading situation. It would be interesting to consider a generalized representation of an asset's state phase trajectory in a visual coordinate system "profit P — risk R — state X" using posterior observation data. The trading situation is described by the respective estimates of the specified parameters $(P, R, X)_t$at the time $t = 1, \ldots, n$. In some cases, in particular forecasting changes in the trading situation, the estimated parameters include estimates of the speed and acceleration of changes in the expected profit and risk measures. Then, the vector of the estimated parameters of the phase trajectory at the time $t = 1, \ldots, n$ will have the form $(P, P', P'', R, R', R'', X, X', X'')_t$. Observations Y_t, $t = 1, \ldots, n$ are the values of quotation vector of financial instruments X_t, $t = 1, \ldots, n$ and, possibly, some other econometric, political, social and other values, the ensemble of which determines the dynamics of expected changes in profit and risk. The perturbations that a submersion environment exerts on a trading situation ensures that observations contain a random component in all cases. The association between the the parameter vector and observation series Y_t, $t = 1, \ldots, n$ can be expressed with a (generally non-linear) one-dimensional operator: $\Phi: \ y = \Phi(x, \ v), y \in Y_0, x \in X_0$.

A management decision $d \in D$ is chosen based on observations Y_t, $t = 1, \ldots, n$ and all prior information available to the trader. Thus, the procedure for making a management decision, which is the main subject of statistical synthesis, can be written in the form of an operator $S: \ d = S(Y, x)$. In the ideal case, this operator inverts Φ. The quality of a decision is substantially determined by the availability and reliability of prior information. The lack of reliable prior and current information relating to the trading situation defines the level of uncertainty at which a trading decision is made.

The most complete prior information consists of a probabilistic description of the observation and state spaces that uses a prior distribution of state $g(x)$ and a family of distributions of observations $f(y|x)$ for all $x \in X_0$. In this case, the management decision is chosen in accordance with the Bayesian ordering of strategies (1) described above, and the preference rule is based on Bayesian risk

$$R_S(f,\ g) = \int_\Theta R_S(x, f)g(x)dx = \int_X g(x) \int_{R^n} w(x_\tau, \hat{x}_\tau(y_1, \ldots, y_n)) \prod_{i=1}^{n} f(y_i, x)dy_i, \quad (2)$$

where $h(x)$ is the prior distribution density of the estimated parameter. This approach provides the most effective management decisions for conventional statistical synthesis problems. However, in the conditions of non-stationary dynamics of observation series, prior information turns out to be unsuitable for dynamic reconstruction of distribution functions. The transition to conventional non-Bayesian techniques eliminates the need to restore the prior distribution of the forecast parameter. However, statistical technologies remain inapplicable due to the non-stationarity of the observed process, i.e., the dependence of observation distribution density $f(Y \mid x)$ on time. In particular, maximum likelihood estimation (MLE) [1–4], which is based on minimization of the joint distribution of observations $L(X) = \prod_{i=1}^{n} f(y_i,\ x)$ and is widely used in statistical synthesis problems [1–4], turns out to be unsuitable for the same reason — the non-stationarity of observation series.

At the same time, it is known [1–6] that the purely random component of observations consisting of the difference between the values of observations, and the smoothed (or system) component of quotation dynamics, will tend to a Gaussian distribution by the central limit theorem. This convergence is approximate, and its degree remains undetermined, due to the impossibility to strictly separate the system component of observation series (deterministic chaos) and the random component.

Nevertheless, even for an approximate Gaussian scheme, using MLE (for a Gaussian model, this leads to the well-known least squares method (LSM)) it is possible to build constructive computational schemes intended to generate predictive management decisions. However, analytical estimates of the quality of such decisions are impossible due to the aforementioned uncertainty regarding the structure of the source data. The only possible way to analyze the problem is thus via numerical studies based on computational experiments. This article is dedicated to the investigation of these issues.

2.4. Observation Model in Asset Management

Figure 1 shows several examples of changes in various currency instrument quotations on a 250-day observation interval.

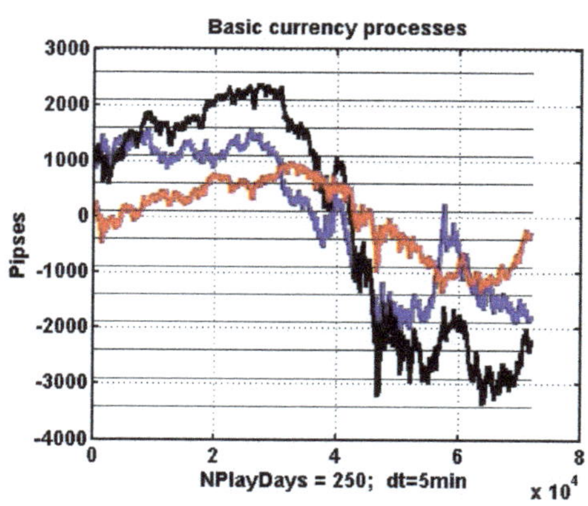

Figure 1. Change dynamics of EURUSD, EURJPY, and USDJPY during 250 observation days.

A visual examination of asset price observation series reveals that their nature is chaotic and can be described by an additive model of form:

$$y_k = x_k + v_k,\ k = 1, \ldots, n, \tag{3}$$

in which the systemic component of the observed process x_k, $k = 1, \ldots, n$ used to make a decision is an implementation of dynamic chaos [1–10,13], and usually is an oscillatory non-periodic process with a large quantity of false local trends of undetermined length. The noise v_k, $k = 1, \ldots, n$, as shown by applying statistical hypothesis tests [5,6], is a non-stationary heteroscedastic process with a non-degenerate time-dependent autocorrelation function.

Further on, we evaluate the feasibility of using decisions made based on conventional statistical methods in asset management described by the observation model (3). In order to do so, we have constructed a method based on the general theory of evaluating the effectiveness of *IT systems* presented in [14].

2.5. Analysing the Effectiveness of Statistical Decision-Making in Asset Management

According to the aforementioned theory, the effectiveness of IT systems (in this case, data analysis algorithms) should be assessed according to the performance of the meta-system, in interests of which it is utilized. In the trading case, the management decision is proactive, i.e., it is based on the forecast of the development of the observed process. Conventional statistical synthesis provides an ability to construct an optimal forecast based on well-known methods such as Bayesian estimation, maximum likelihood estimation, least squares method, etc. [1–4,7], as shown above. It is impossible to analytically estimate whether such methods are effective if the observed processes present chaotic and non-stationary dynamics. Considering that, we present numerical studies based on the terminal estimate of the efficiency of the forecast algorithms and proactive management decisions for real trading asset quotation dynamics processes.

In particular, we consider extrapolation-based computational schemes of optimal statistic forecasting, which estimate the parameters of a polynomial forecast model via the least squares method.

The movement model undergoes statistical identification on the segment of observations that corresponds to the time interval of a financial instrument's price changing by a specific value dL. This approach corresponds to the problem in which the entire range of price changes of a financial instrument is divided into uniform segments of width dL. A positive or correct decision based on statistical trend identification accounts for the ability of the considered process to preserve the previously identified trend from the level it was detected at to the crossing of the next level in its direction. In the cases when the process turns around and reaches the opposite level, the respective experiment fragment will be considered to be a false decision. This approach is quite compatible with the tasks of electronic trading, when the gain (TP, Take Profit) is achieved when the process reaches a given level before it reverses and reaches the level of limiting losses (SL, Stop Loss).

Thus, the criterion of decision effectiveness is defined as the estimated probability (frequency) of the process intersecting the trend presence and identification level earlier than the Stop Loss level.

It should be noted that visual posterior observation of chaotic processes reflecting the currency instrument quotation dynamics confirms the assumption that there are local areas with pronounced trends which can be described by low-order polynomials. If this can be proven mathematically, then we can say that the degree of chaos of the observed process decreases locally. This, in turn, opens up prospects for constructing effective trading strategies. Areas that preserve trend direction can be described via conventional statistical methods. If this assumption turns out to be incorrect, it will not be possible to construct effective management strategies based on statistical methods of identifying trends.

As a typical process with non-stationary dynamics, we will consider changes in the quotations of currency pairs at the electronic FOREX market. To study this question, a number of computational experiments for the most commonly used currency instruments are carried out in this paper.

3. Computational Experiments

3.1. Description

The area on which the studied process Y_t changes is divided into equal sectors that are dL points in percentage (pipses) long. Suppose that the process has moved from level dL to the level L_{k+1} above it. We can loosely consider this to be an appearance of a positive trend. Such a transition will be denoted as $L_k \to L_{k+1}$. Conversely, we can understand $L_k \to L_{k-1}$ as negative dynamics [15].

The main problem that we focus on consists in proving the persistence of process Y_t defined as it arriving at the next level of the detected trend.

We assess positive outcome probability, i.e., Y_t going from $L_k \to L_{k+1}$ after it has gone from $L_{k-1} \to L_k$. We define a negative outcome as a opposite transition to the level below $L_k \to L_{k+1}$ directly after an upwards transition $L_{k-1} \to L_k$. Due to symmetry, similar measurements work for downwards transitions as well. Therefore, the set of exhaustive events consists of two positive outcomes $(L_k \to L_{k+1} \mid L_{k-1} \to L_k)$,$(L_k \to L_{k-1} \mid L_{k+1} \to L_k)$ and two negative outcomes $(L_k \to L_{k-1} \mid L_{k-1} \to L_k)$, $(L_k \to L_{k+1} \mid L_{k+1} \to L_k)$.

As in illustration, Figure 2 presents an example of EURUSD quotation dynamics on a 10-day observation interval with segmentation boundaries and marks denoting boundary intersections.

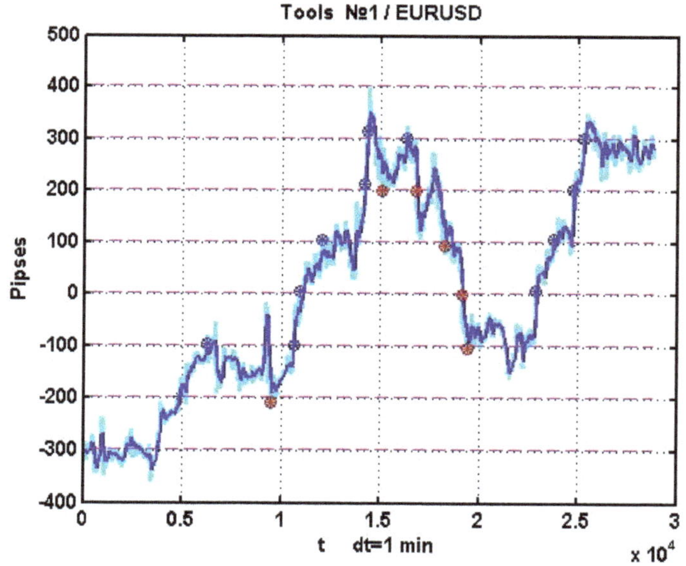

Figure 2. An example of EURUSD quotation dynamics on a 10-day observation interval.

The figure presents both the Y_t, $t = 1, \ldots, n$ process and its smoothed version $\tilde{Y}_t$, $t = 1, \ldots, n$. It was smoothed via a simple exponential filter $\tilde{Y}_t = \alpha Y_t + \beta \tilde{Y}_{t-1}$, $t = 1, \ldots, n$, in which $\alpha \in (0, 1)$, $\beta = 1 - \alpha$.

Examples of positive outcomes are shown in Figure 3, and negative ones in Figure 4.

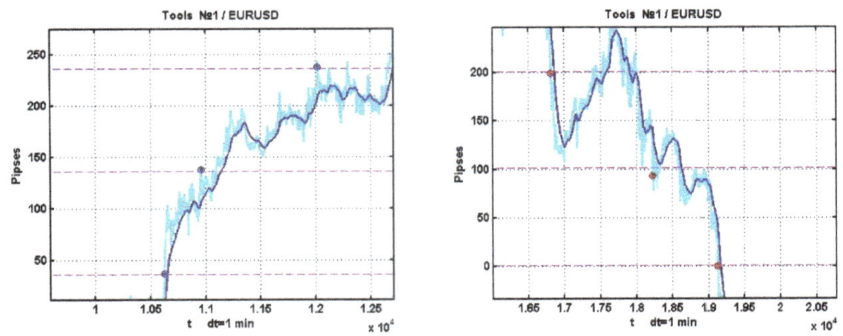

Figure 3. Examples of positive outcomes of management decisions.

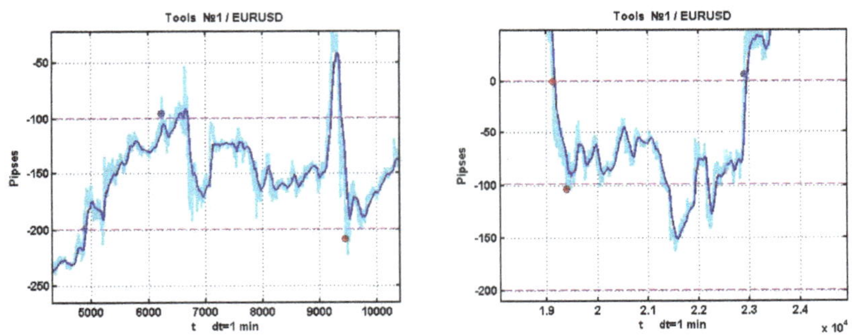

Figure 4. Examples of negative outcomes of management decisions.

Assume that N experiments were run, each of which has identified a transition from level to level, interpreted as a trend. If this direction persists until the process intersects the next level, this outcome will confirm the trend. Conversely, if the process reverses and reaches the previous level, it will point to trend absence. Suppose that as the result of N consecutive experiments m outcomes prove trend presence, and $N - m$ outcomes disprove it.

Thus, this can be considered as an alternative $H_1 : p \neq 0.5$ to the null hypothesis $H_0 : p = 0.5$ of trend absence, in which $p_n = \frac{m}{n}$ is the frequency of experiments that confirm trend presence. If an experiment is repeated a sufficient number of times, the frequency of observed event is considered to be an estimate of the probability of the corresponding assumption. To test the H_0 hypothesis, it is possible to use a well-known rule $u < u^*$, where $u = (f_n - 0.5)(0.25/n)^{1/2}$ [8–10]. The critical value u^* for the right-tailed criterion is found using the Laplace function Φ table of values, considering that $\Phi(u^*) = \frac{(1-2\gamma)}{2}$. Here, γ is the significance level of the null hypothesis.

3.2. Results

In order to cover as many variations of chaotic dynamics in electronic trading as possible, we have considered five 100-day segments for three most widespread financial instruments: *EURUSD*, *EURJPY*, and *USDJPY*. We have used the $dL = 100$ (pipses) as the size of the inter-level interval.

Probability is estimated using the frequency of positive outcomes, that is, the ratio of positive outcomes to the total number of experiments. The corresponding results of the computational experiment are presented in Table 1.

Table 1. Positive outcome frequency at $dL = 100$.

Time Interval, Days	EURUSD	EURJPY	USDJPY
1–100	0.552	0.484	0.444
101–200	0.507	0.536	0.465
201–300	0.533	0.552	0.560
301–400	0.494	0.452	0.465
401–500	0.446	0.545	0.444

The presented data clearly indicates complete absence of any inertia in quotation dynamics. This statement can be verified with statistical hypothesis testing. As mentioned above, the null hypothesis $H_0 : p = 0.5$ is tested against the alternative $H_1 : p \neq 0.5$.

As an example, consider an experiment for the 100-day interval of EURUSD observation. It resulted in $N = 76$ closes, out of which $m = 42$ corresponded to the inertia condition at $dL = 100$. The relative frequency $\frac{m}{N} = 0.552$ corresponds to the value of the decision statistic

$$u = \frac{(\frac{m}{N} - p_0)\sqrt{N}}{\sqrt{p_0 q_0}} = \frac{0.052\sqrt{76}}{\sqrt{0.25}} = 0.91.$$

Here, $q_0 = 1 - p_0, \quad u \in N(0,\ 1)$, i.e., it is subject to the standard Gaussian distribution with the $(0,1)$ parameters. The assumption about the normal distribution of the criterion follows from Laplace's theorem (for a sufficiently large n, the relative frequency can be approximately considered normally distributed with the mathematical expectation p and standard deviation $\sqrt{\frac{pq}{N}}$). In the general case, this assumption needs additional verification.

The critical area for the symmetrical competing hypothesis $H_1 : p \neq 0.5$ is determined based on the selected significance level $\alpha = 0.99$. For a two-sided critical area, u_{cr} is determined via the Laplace function value table according to $\Phi(u_{cr}) = \frac{1-\alpha}{2} = 0.005$. Using the distribution tables of the Laplace function, we determine $u_{cr} = 2.85$. Therefore, the calculated value of statistic $u = 0.91$ belongs to the area of hypothesis acceptance $H_0 : p = 0.5$, which means that statistical solutions are unsuitable for the considered processes.

The conclusion that inertia is absent in the previous experiment may be caused by an excessively large confirmation interval $dL = 100$ p. Let us check whether we can confirm the presence of inertia on smaller segmentation levels. Note that the considered process contains a significant completely random component. Considering the random value spread relatively to the smoothed process from $\tilde{Y}_t,\ t = 1, \ldots, n\ \alpha = 0.02$ then its SD on 100-day observation segments oscillates within 11–14 pipses for various currency instruments. Decreasing α to 0.01, the corresponding SD changes within 15–20 pipses, which is due to a lower degree of smoothness and therefore a smaller difference between the initial and smoothed processes.

The width of the spread is responsible for random decisions that do not correspond to systemic processes of quotation dynamics and therefore skews the conclusions on inertia presence. Thus, in order to obtain a correct conclusion on inertia, the size of the segmentation step (system dynamics) must be significantly larger than the random component.

As an example that illustrates the minimum feasible segmentation step for the presented SD values, consider the same task with $dL = 50$. The frequencies of positive outcomes that confirm process inertia can be found in Table 2.

Table 2. Positive outcome frequency at $dL = 50$.

Time Interval, Days	EURUSD	EURJPY	USDJPY
1–100	0.539	0.568	0.522
101–200	0.524	0.528	0.497
201–300	0.529	0.503	0.537
301–400	0.503	0.550	0.534
401–500	0.493	0.548	0.552

The results, similarly to the previous case, confirm the stability of the hypothesis $H_0 : p = 0.5$, which refutes the use of statistical management techniques in a chaotic environment. The positive asymmetry is too small to accept the null hypothesis on the significance of the difference between the frequency of positive outcomes and 50%. The difference between the following series of experiments is that the beginning of each stage of the management process is fixed when the segmentation level is crossed not by the process itself, but by its smoothed version. Positions are closed (i.e., establishing the fact of recognizing or not recognizing inertia in each experiment) by the process Y_t, $t = 1, \ldots, n$ itself.

Obviously, the higher the degree of smoothness, the less the result will depend on the fluctuating component of the process randomly crossing the levels. On the other hand, a higher degree of smoothness inevitably leads to a lag in the smoothed process relative to the original one, which skews the resulting estimates. As a compromise, we will use values $\alpha = 0.005 - 0.02$. The segmentation step, as in the first experiment, is equal to 100 p. The results of estimating the probability of a positive outcome confirming the alternative H_1 for five 100-day observation intervals and different values of the exponential filter transfer coefficient α are presented in Table 3.

Table 3. Positive outcome frequency: opening by smoothed curve.

Time Interval, Days	$\alpha = 0.005$	$\alpha = 0.01$	$\alpha = 0.01$
1–100	0.667	0.681	0.618
101–200	0.771	0.791	0.667
201–300	0.606	0.706	0.612
301–400	0.612	0.653	0.618
401–500	0.648	0.581	0.574

It can be seen from the above data that the smoothed version of the process has more inertia, which in general is suitable for making useful trading recommendations. However, one should keep in mind that negative decisions are more drastic in terms of loss, since in this case the dynamics of the quote reverse, and the departure of the process Y_t, $t = 1, \ldots, n$ during the time when the smoothed curve $\tilde{Y}_t$, $t = 1, \ldots, n$ crosses the opening level can be very large.

The final computational experiment is similar to the previous one, but both were carried out by a smoothed process $\tilde{Y}_t$, $t = 1, \ldots, n$ at the intersection of the corresponding level. The results of the experiment are shown in Table 4.

Table 4. Positive outcome frequency: opening and closing by smoothed curve.

Time Interval, Days	$\alpha = 0.005$	$\alpha = 0.01$	$\alpha = 0.01$
1–100	0.652	0.652	0.593
101–200	0.698	0.706	0.696
201–300	0.686	0.707	0.688
301–400	0.612	0.612	0.582
401–500	0.567	0.534	0.574

It is easy to see that the presented results are quite close to the corresponding estimates given in Table 3. In other words, using a smoothed curve did not change the final result. This is due to the fact that the probability of process Y_t, $t = 1, \ldots, n$ reaching the decision level will be higher both with a positive and negative outcome.

The disadvantage of the analysis method proposed above is that it does not take into account the quality of the transitions on which the local trend is detected. For example, transition $Y(k) \to Y(k) + dY$ can go on for a long time, with fluctuations and with a large negative "sagging" (it is only necessary that it does not turn around and does not reach level $Y(k) - dY$). Such a process is quite difficult to perceive as a trend. However, in accordance with the above formalization, such a transition will also be interpreted as a positive trend. In this regard, it makes sense to move on to a more complex trend detection criterion, based, for example, on the average rate of change in the state of the process on a sliding time window of size l:

$$\Delta Y(k,l) = Y(k-l+1,k) = [Y(k-l+1), Y(k-l+2), \ldots, Y(k)] \quad (4)$$

Trend detection in this case is exceeding the value of the linear approximation coefficient $a_1(k)$ calculated at the observation site $\Delta Y(k,l)$ of a certain critical value $a_1(k) > a^*$. This approach can be generalized to more complex trend detection rules. In particular, we can consider a version of trend detection based on linear approximation coefficients calculated on two observation windows of different lengths $\Delta Y(k,l_1)$ and $\Delta Y(k,l_2)$, $l_1 > l_2$, or a version that uses a sliding approximation by a second-order polynomial. The second half of the proposed effectiveness analysis approach, namely, confirming the existence of a trend, remains unchanged. The H_0 hypothesis of trend absence in the prolongation of the detected trend means that the process, after its detection at t_0, reaches threshold values $Y(t_0) + dL$ and $Y(t_0) - dL$ with the same probability, i.e., $H_0 : p = p_0 = 0.5$, p being the value of the observed process at the time of trend detection. An alternative hypothesis indicating the possibility of using such strategies in conditions of market chaos will have the form $H_1 : p \neq p_0 = 0.5$. As in previous experiments, in addition to the main chaotic process $Y(t)$, we will use its smoothed version $Y_s(t)$ (4) with a transmission coefficient. The process $Y_s(t)$, which simulates the system component of chaotic dynamics makes it possible to isolate the purely random component of the initial noise chaotic process $v(t) = Y(t) - Y_s(t)$, which is a centered random process with a distribution that is close to Gaussian. The variance of the residual process $D(v(t))$, in turn, allows us to estimate the lower bound of the parameter dL that determines the level of confirmation or denial of the trend presence hypothesis. The method of conducting computational experiments is close to its prototype described above. An observation series of a trading asset's state at various non-intersecting 100-day observation intervals is considered as the polygon of chaotic data. Next, we form a sliding observation window $\Delta Y(k,l)$ of size l, on which we calculate approximating polynomials $P(q,l)$, q being their degree. A decision about the presence of a trend is made based on the comparison of estimated coefficients of a with critical values a^* . The number of outcomes corresponding to the process reaching a predetermined level dL is calculated to statistically verify the effectiveness of management decisions made following trend-based strategies. Due to the symmetry of the task, the negative result consists in a trend reversal and reaching the $-dL$ level. If the ratio of reaching dL to the total number of position openings m/n (the frequency of the event) is close to 0.5, then this confirms the hypothesis H_0 that it is impossible to successfully implement trend-based management strategies. The parameters of the computational experiment are observation window l, degree of smoothing polynomial q, threshold values of trends a^*, and trend confirmation level dL . The simplest linear approximation scheme $\tilde{Y}(t) = a_0 + a_1 t$ is used on a sliding observation window $\Delta Y(k,l)$. A trend is confirmed when the linear approximation coefficient a_1 exceeds a pre-set value $a^* : a_1 \geq a^*$. Trend presence is either denied or confirmed when the condition $Y(t) = Y_0 \pm dL$ is met, in which $Y_0 = Y(t_0)$ is the value of the process at the time of trend detection t_0, and $dL = 30, 50, 100$ are the trend confirmation levels. The size of the observation window l varies in the (0.1–0.5) day range. A trend is detected if

$|a_1| \geq a^*$ is satisfied, and confirmed when $Y(t) = Y_0 \pm dL$ levels are reached. Estimates of the probability (frequency) of reaching the trend confirmation level for its various values $dL = 25, 50, 75, 100$, for the $l = 0.1, 0.25, 0.5$-day observation window and for threshold values $a^* = 0.05, 0.075, 0.1$ on a 100-day observation interval are shown in Table 5.

Table 5. Trend confirmation frequency for different parameter values.

a*	dL, n l, Days	0.025	0.05	0.075	0.1
0.025	25	0.50	0.48	0.49	0.49
0.025	50	0.51	0.50	0.51	0.50
0.025	75	0.50	0.50	0.50	0.51
0.025	100	0.50	0.51	0.51	0.51
0.05	25	0.50	0.48	0.48	0.50
0.05	50	0.50	0.50	0.51	0.50
0.05	75	0.50	0.50	0.51	0.50
0.05	100	0.50	0.51	0.51	0.51
0.075	25	0.50	0.49	0.49	0.49

The presented data clearly confirms that it is impossible to successfully prolongate $Y_s(t)$ in a wide range of changes in intensity values, fixation levels and trend confirmation levels. The disadvantage of this experiment is the fixed length of the sliding observation window l. A large window causes a significant delay in trend detection, which leads to a delayed decision and, as a result, to an incorrect assessment of the probability of trend confirmation. An example is shown in Figure 5.

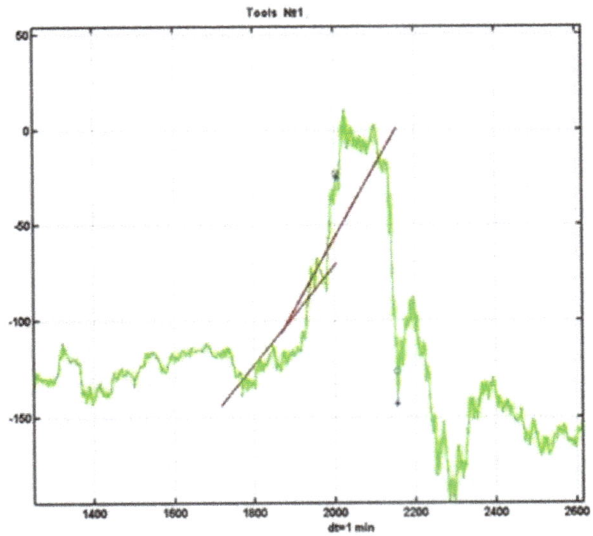

Figure 5. Examples of incorrect conclusions about the presence of a trend associated with a delay in decision-making.

A small window leads to an increased sensitivity of the trend detection procedure to the random component, which, in turn, leads to statistical errors of Type II (false alarms), that is, to the detection of a non-existent trend. In this regard, it makes sense to consider the problem of trend detection based on a complex criterion that uses two sliding observation windows of different sizes.

Unlike the previous experiment, in this one we consider two trends. In this case, linear approximations $q = 1$ are used for two sliding observation windows of size l_1 and

$l_2, l_1 > l_2$. It is obvious that the first trend has stronger smoothing characteristics, and the second is more sensitive to both systemic process changes and "false alarms". Let $l_1 = 300$, $l_2 = 90$ minute counts, critical values of the linear regression coefficient $a_1^* = 0.05, a_2^* = 0.1$, the level of trend confirmation $dL = 75$. The decision on the presence of a trend is made if both linear regression coefficients exceed their critical values by the absolute value. An example of the implementation of such a scheme is shown in Figure 6.

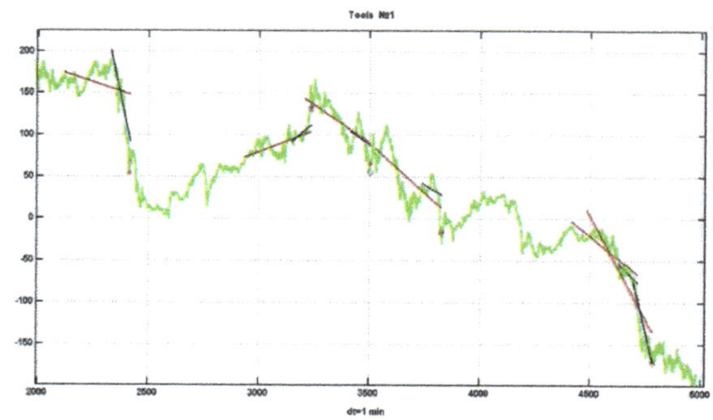

Figure 6. Example of a decision-making scheme with two trends

Longer trends correspond to larger observation windows. Let us consider the result of using this method for four 100-day intervals with different levels of trend confirmation $dL = 25 : 25 : 100$. The corresponding data is presented in Table 6. It is easy to see that the modification did not have a positive effect.

Table 6. Trend confirmation frequency for various time intervals.

dL, n / ΔT	1–100 Days	101–200 Days	201–300 Days	301–400 Days
25	0.48	0.49	0.48	0.47
50	0.53	0.46	0.48	0.48
75	0.53	0.50	0.49	0.45
100	0.55	0.50	0.52	0.47

Obviously, the issues of the previous version of this approach have not been resolved. Additionally, the method, as a rule, detects a trend at the time of confirmation (or denial) of the previous trend. At the same time, there is no new trend detection during the confirmation time: this would require a method that simultaneously analyzes several trends. The provided data clearly illustrates the extremely insignificant fluctuations in the frequency of trend confirmation relative to the 0.5 value. This conclusion is easily confirmed by testing the statistical hypothesis $H_0 : p = 0.5$ about the absence of a trend using U-statistics and confidence level $\alpha = 0.99$.

4. Discussion

The applicability of statistical techniques for making management decisions for chaotic processes with a non-stationary random component is a key question in applied problems of asset management. An example of such a problem are effective management strategies based on statistical trend analysis. It is impossible to make analytical estimates of unstable inertia-less system management strategies because the observed processes are unstable parameter-wise. Thus, such studies can be conducted as numerical studies exclusively, considering large observation intervals that cover the variety of stochastic chaos. The technique proposed in the article uses segmentation of the area of change of the considered random process. Thus, it allows us to build a visual system for analyzing the effectiveness of statistical management decisions. The use of the process itself to fix the levels of confirmation or denial of the management effectiveness does not lead to a stable solution due to the high degree of its stochastic spread. The process reaches decision-making levels more often due to high variance than as a result of an inertial trend. This leads to the idea of using a smoothed process for analysing inertia. In particular, good results are obtained with smoothing via an exponential filter with transfer coefficient $\alpha = 0.005 - 0.02$.

A better management result can be expected from more efficient computational schemes for the extraction of the system component in observation series. This issue is currently under study.

Trend analysis that we have conducted does not confirm the effectiveness of trend-based statistical decisions. The trend management strategy for areas with a weak trend of chaotic variations of the immersion environment is essentially non-profitable. Detecting a trend on such segments of price dynamics indicates an upcoming change in the near future and, therefore, it being appropriate to open a position in the opposite direction. In other words, the problem lies not in the statistical methods of estimation themselves, but in the infinite variety of dynamic forms generated by market chaos, which makes it impossible to utilize simple stationary observation models.

It should be noted that currency market inertia is still not confirmed with statistical significance. Some facts that are observed at financial markets confirm this point. If a trend is detected, it is highly likely that in the next moment it will reverse. This means strategies that were effective on backtest could accidentally be effective in the future. The inability of the trend to continue its development can be explained by the fact that most of the time the composition of trading participants is in an equilibrium balance. The influence of trend-following traders (momentum traders) is compensated by market makers and mean reversion traders [16]. At the same time, according to the inaction inertia theory [17], individual investors who missed the opportunity to enter a position at a good price prefer to hold back from a trade at a less favorable price, thereby preventing the trend from developing. However, these factors become insignificant when the fundamental valuation of the asset is changed. When this view is shared by the majority of traders, a continuous trend may develop [18].

This leads to the conclusion that the market situation needs multilateral analysis. In particular, the use of the considered types of statistical solutions combined with an automatic text analyzer, which would assess the general "mood" in relation to the financial instruments used, would significantly increase the effectiveness of the generated decisions.

Therefore, the most promising direction in the development of the task of automatic management of market assets is the development of trading robots based on multi-expert systems. Our current and planned publications will be dedicated to the development of the conceptual and practical aspects of such systems.

Author Contributions: Conceptualization, A.M. and D.G.; methodology, A.M.; software, A.M.; validation, A.M. and D.G.; formal analysis, A.M.; investigation, A.M.; resources, A.M. and D.G; data curation, A.M.; writing-original draft preparation, A.M.; writing—review and editing, D.G.; visualization, A.M.; supervision, A.M.; project administration, D.G.; funding acquisition, D.G. All authors have read and agreed to the published version of the manuscript.

Funding: The research of Alexander Musaev described in this paper is partially supported is partially supported by the Russian Foundation for Basic Research (grant 20-08-01046), state research FFZF-2022-0004. Dmitry Grigoriev research for this paper was funded by a Support from The Endowment Fund of St Petersburg University.

Institutional Review Board Statement: Not applicable.

Informed Consent Statement: Not applicable.

Data Availability Statement: The source of data—Finam.ru (https://www.finam.ru/, accessed on accessed on 1 December 2021).

Acknowledgments: The authors are grateful to participants at the Center for Econometrics and Business Analytics (ceba-lab.org, CEBA) seminar series for helpful comments and suggestions.

Conflicts of Interest: The authors declare no conflict of interest.

References

1. Peters, E.E. *Chaos and Order in the Capital Markets: A New View of Cycles, Prices, and Market Volatility*, 2nd ed.; John Wiley & Sons: New York, NY, USA, 1996.
2. Gregory-Williams, J.; Williams, B.M. *Trading Chaos: Maximize Profits with Proven Technical Techniques*, 2nd ed.; John Wiley & Sons: New York, NY, USA, 2004.
3. Bauwens, L.; Giot, P. *Econometric Modeling of Stock Market Intraday Activity*, 1st ed.; Springer Science & Business Media: New York, NY, USA, 2001.
4. Smith, L. *Chaos: A Very Short Introduction*; Oxford University Press: Oxford, UK, 2007.
5. Musaev, A.A. Quod est veritas. Views transformation at a system component of observable process. *Inform. Autom. SPIIRAS Proc.* **2010**, *15*, 53–74. [CrossRef]
6. Musaev, A.; Grigoriev, D. Analyzing, Modeling, and Utilizing Observation Series Correlation in Capital Markets. *Computation* **2021**, *9*, 88. [CrossRef]
7. Stuart, A.; Kendall, M.G. *The Advanced Theory of Statistics*; Griffin: London, UK, 1958; Volume 1.
8. Stuart, A.; Kendall, M.G. *The Advanced Theory of Statistics*; Griffin: London, UK, 1961; Volume 2.
9. Rao, C.R. *Linear Statistical Inference and its Applications*, 2nd ed.; Wiley: New York, NY, USA, 1973.
10. Bolch, B.W.; Huang, C.J. *Multivariate Statistical Methods for Business and Economics*; Prentice Hall: Hoboken, NJ, USA, 1973.
11. Singh, B.; Jalil, N.A.; Sharma, D.K.; Steffi, R.; Kumar, K. Computational systems overview and Random Process with Theoretical analysis. In Proceedings of the 7th International Conference on Advanced Computing and Communication Systems (ICACCS), Coimbatore, India, 19–20 March 2021; Volume 1.
12. Rawlings, J.O.; Pantula, S.G.; Dickey, D.A. *Applied Regression Analysis: A Research Tool*, 2nd ed.; Springer: New York, NY, USA, 2001.
13. Okolelova, E.Y.; Shulgina, L.V.; Trukhina, N.I.; Shibaeva, M.A.; Shulgin, A.V. The mechanism of evaluation under the conditions of uncertainty of innovational project as a random process. In *Perspectives on the Use of New Information and Communication Technology (ICT) in the Modern Economy*; Springer: Cham, Switzerland, 2017.
14. Yusupov, R.M.; Musaev, A.A. Efficiency of information systems and technologies: Features of estimation. *Proc. SPIIRAS* **2017**, *51*, 5–34. [CrossRef]
15. Yusupov, R.M.; Musaev, A.A.; Grigoriev, D.A. Evaluation of Statistical Forecast Method Efficiency in the Conditions of Dynamic Chaos. In Proceedings of the 2021 IV International Conference on Control in Technical Systems (CTS), Saint Petersburg, Russian Federation, 21–23 September 2021; pp. 178–180.
16. McGroarty, F.; Booth, A.; Gerding, E.; Chinthalapati, V.R. High frequency trading strategies, market fragility and price spikes: An agent based model perspective. *Ann. Oper. Res.* **2019**, *282*, 217–244. [CrossRef]
17. Tykocinski, O.; Israel, R.; Pittman, T.S. Inaction inertia in the stock market. *J. Appl. Soc. Psychol.* **2004**, *34*, 1166–1175. [CrossRef]
18. Schmitt, N.; Westerhoff, F. Heterogeneity, spontaneous coordination and extreme events within large-scale and small-scale agent-based financial market models. *J. Evol. Econ.* **2017**, *27*, 1041–1070. [CrossRef]

mathematics

MDPI

Article

An SIRS Epidemic Model Supervised by a Control System for Vaccination and Treatment Actions Which Involve First-Order Dynamics and Vaccination of Newborns

Santiago Alonso-Quesada *, Manuel De la Sen and Raúl Nistal

Department of Electricity and Electronics, Campus of Leioa, University of the Basque Country, UPV/EHU, Barrio Sarriena s/n, 48940 Leioa, Spain; manuel.delasen@ehu.eus (M.D.l.S.); raul.nistal@gmail.com (R.N.)
* Correspondence: santiago.alonso@ehu.eus

Abstract: This paper analyses an SIRS epidemic model with the vaccination of susceptible individuals and treatment of infectious ones. Both actions are governed by a designed control system whose inputs are the subpopulations of the epidemic model. In addition, the vaccination of a proportion of newborns is considered. The control reproduction number R_c of the controlled epidemic model is calculated, and its influence in the existence and stability of equilibrium points is studied. If such a number is smaller than a threshold value $\overline{R}_c$, then the model has a unique equilibrium point: the so-called disease-free equilibrium point at which there are not infectious individuals. Furthermore, such an equilibrium point is locally and globally asymptotically stable. On the contrary, if $R_c > \overline{R}_c$, then the model has two equilibrium points: the referred disease-free one, which is unstable, and an endemic one at which there are infectious individuals. The proposed control strategy provides several free-design parameters that influence both values R_c and $\overline{R}_c$. Then, such parameters can be appropriately adjusted for guaranteeing the non-existence of the endemic equilibrium point and, in this way, eradicating the persistence of the infectious disease.

Keywords: epidemic models; vaccination and treatment actions; feedback control; equilibrium points; stability

Citation: Alonso-Quesada, S.; De la Sen, M.; Nistal, R. An SIRS Epidemic Model Supervised by a Control System for Vaccination and Treatment Actions Which Involve First-Order Dynamics and Vaccination of Newborns. *Mathematics* **2022**, *10*, 36. https://doi.org/10.3390/math10010036

Academic Editors: Mihaela Neamţu, Eva Kaslik and Anca Rădulescu

Received: 4 November 2021
Accepted: 21 December 2021
Published: 23 December 2021

Publisher's Note: MDPI stays neutral with regard to jurisdictional claims in published maps and institutional affiliations.

Copyright: © 2021 by the authors. Licensee MDPI, Basel, Switzerland. This article is an open access article distributed under the terms and conditions of the Creative Commons Attribution (CC BY) license (https://creativecommons.org/licenses/by/4.0/).

1. Introduction

The propagation of epidemic diseases within a host population has been studied since several decades ago. Kermack and McKendrick developed one of the first works in the subject [1]. They proposed an SIR epidemic model where the host population is split in three categories depending on the status of the individuals with respect to the disease. In such a context, S, I, and R denote, respectively, the susceptible, infectious, and recovered subpopulations. Later, a lot of models have been proposed and analysed in the specialised literature. Such models consider some additional population categories and/or control actions for eradicating or, at least, diminishing the effects of the disease in the host population [2–5]. In this sense, the models can include the exposed subpopulation E composed by individuals without symptoms of the disease and without the capacity of transmitting the infection to a susceptible individual after a contagion. Such a model is referred to as the SEIR model [6]. The more usual control actions are the vaccination of susceptible individuals and the treatment of infectious individuals by antivirals, antibiotics, or other medicaments [7,8]. Additionally, other control actions such as quarantine of infectious individuals have also been proposed [9]. These control actions give place to include other subpopulations in the models as vaccinated, treated, and quarantined ones [10–13]. For instance, the authors in [13] propose a SVEIR model to analyse the impact of vaccination in the control of spread of poliomyelitis. Moreover, the control actions can be applied in continuous-time or impulsive ways during the vaccination campaigns and/or treatment procedures [14–16]. The social distancing is another alternative way to fight against the propagation of infectious diseases [17]. Such a measure may be crucial when there are asymptomatic infectious individuals and/or neither vaccines nor medicaments are available to maintain the disease propagation under control. The early phase of the COVID-19 pandemic is a clear example of such a fact [18]. In such situations, the susceptible subpopulation can be split in several categories according to its preventive care and responsible behaviour to avoid the disease contagion. For instance, the study in [19] considers two susceptible subpopulations according to its risk aversion. One of the categories contains individuals with self-protection awareness, and the other one is composed by people without such an awareness. Recently, new approaches from game theory have been proposed for analysing the spreading of infectious diseases [20,21]. The studies in [20] introduce the concept of a vaccination game so as to evaluate provisions other than vaccination, including protective measures as mask wearing. They also analyse the efficiency of quarantine compared with that of isolation policies or the efficiency of preventive versus late vaccination. The work in [21] proposes a double-layer game structure of vaccination and treatment. The vaccination game considers whether the vaccine is accepted or declined by the individual, while the treatment game depends on the antiviral resistance evolution and prescribing behaviour. In this way, the vaccination game deals with the presence of anti-vaccine behaviour, while the treatment game considers the antibiotic overuse.

In this paper, a controlled epidemic model with vaccination of newborns, vaccination of the susceptible, and treatment of the infectious individuals, as control actions, is proposed. The model is composed by a basic SIRS epidemic model and a first-order continuous-time control subsystem, which is based on the feedback of the state variables. Namely, the inputs of such a control subsystem are the susceptible, infectious, and recovered subpopulations so that the control subsystem acts under the knowledge of the state of the disease at each time. The control subsystem provides several free-design parameters that can be adjusted to eradicate the persistence of the disease within the host population. Namely, an appropriately adjustment of the control parameters guarantees the non-existence of the endemic equilibrium (EE) point of the controlled SIRS model. Such a fact implies the existence of a unique equilibrium point, namely, the disease-free equilibrium (DFE) point, which is a very advantageous tool to asymptotically eradicate the disease. Moreover, the local and global asymptotic stability of this DFE point is analytically proved under appropriate conditions relative to the adjustment of the control parameters. Under such conditions, the disease is eradicated from the host population after a transient period of propagation from the initial state until the DFE point is reached. Several modified SIRS-type models have been proposed and analysed in the epidemiological research. Some models study the dynamics of such models under the influence of control actions such as vaccinations of susceptible and/or the treatment of infectious individuals. Both control actions are applied either in a time continuous or impulsive way, and the intensity of each control action is usually proportional to the susceptible and/or infectious individuals. The SIRS model in our paper also proposes the combined application of vaccination to the susceptible and the treatment to the infectious subpopulation. The main novelties of the current paper related to the background literature are (i) the use of first-order dynamics in the vaccination and treatment controls and (ii) the availability of additional free-design control parameters to shape such vaccination and treatment actions. The first novelty provides some additional parameters derived from the fact that the vaccination and treatment actions are provided by a control subsystem instead of being proportional to the susceptible and/or infectious individuals. Concretely, two parameters arise from the dynamics of the control subsystem, and then, the control actions can be designed with two more free-design parameters than in the conventional research. Moreover, the control subsystem uses the information of the state of the propagation of the disease since the subpopulation variables are used as the inputs of such a subsystem. In this way, a feedback control strategy is used for obtaining the vaccination and treatment variables. In summary, the control designer can make use of the additional free-design parameters to shape the vaccination and treatment actions in a desired way, which is the second main novelty, to achieved the required objectives. One of the main objectives is to achieve the eradication of the disease or, at least, to minimise its influence within the host population. In this context, the control parameters governing the intensity of the vaccination of newborns and susceptible individuals influence on the control reproduction number R_c of the controlled SIRS model. Moreover, those parameters shaping the intensity of the treatment of the infectious individuals influence on a threshold value $\overline{R}_c$ of interest since if $R_c < \overline{R}_c$, then the EE point does not exist, so that the DFE point is the unique one. Furthermore, such a DFE point is locally and globally asymptotically stable under that condition. In this way, an appropriate choice of the control parameters associated with the vaccination actions allows the control reproduction number to reduce with respect to its value in absence of vaccination. Namely, R_c decreases by increasing the intensity of the vaccination. On the other hand, an appropriate choice of the control parameters associated with the treatment action allows the threshold value $\overline{R}_c$ to be strictly larger than the R_c value of the controlled SIRS model to guarantee the inexistence of the EE point and then the existence of a unique globally asymptotically stable DFE point. Namely, $\overline{R}_c$ increases by increasing the intensity of the treatment action. In this context, the infectious disease can be extinguished under an appropriate choice of the design parameters of the control laws. In summary, two options are available for such a purpose. The first one is increasing the intensity of vaccination in order to reduce the control reproduction number R_c, and the second one is increasing the intensity of treatment in order to augment the threshold value $\overline{R}_c$. Obviously, an appropriate solution to guarantee the extinction of the disease can be obtained by combining both options so that $R_c < \overline{R}_c$.

The rest of the paper is organized as follows. Section 2 describes the basic epidemic SIRS model as well as the subsystem providing the both proposed control actions, namely, the vaccination of the susceptible and treatment of the infectious subpopulations, respectively. The positivity of the controlled model, composed by combining the basic SIRS model with the control subsystem, is analysed. In addition, the study of the equilibrium points, as the existence as the stability properties, of the controlled model is dealt with in this section. Concretely, the conditions to be satisfied by the free-design control parameters in order to guarantee the non-existence of the EE point and then, the local and global asymptotic stability of the DFE point, which is the unique equilibrium point under such conditions, are established and mathematically proved. Finally, Section 3 illustrates the theoretical results by some simulation examples. An extended study of the influence of the control parameters in the dynamics of the controlled SIRS model is done. The results obtained with the proposed controlled SIRS model are compared with those obtained by an uncontrolled SIRS model. In addition, the results of the proposed model are compared with an SIRS model with only a control action, either the vaccination of the susceptible subpopulation or the treatment of the infectious one. These comparisons are interesting from the viewpoint of the available resources relative to the existence of vaccines and/or medicaments to fight against the propagation of the disease.

2. SIRS Epidemic Model under Vaccination and Treatment Controls

An SIRS epidemic model described by the following equations:

$$\dot{S}(t) = b(1-q) - \beta\frac{S(t)I(t)}{N(t)} + \rho R(t) - \mu S(t) - v(t) \tag{1}$$

$$\dot{I}(t) = \beta\frac{S(t)I(t)}{N(t)} - (\mu+\alpha+\gamma)I(t) - t_r(t) \tag{2}$$

$$\dot{R}(t) = bq + \gamma I(t) - (\mu+\rho)R(t) + v(t) + t_r(t) \tag{3}$$

with an initial condition given by $S(0) \geq 0$, $I(0) \geq 0$ and $R(0) \geq 0$ is proposed. The model considers the whole host population split into three categories depending on the state of the individuals with respect to the infectious disease, namely, susceptible (S), infectious (I), and recovered (R) subpopulations. Moreover, the model considers a constant recruitment rate for the population, which is represented by b [22,23]. The model includes three control actions: (i) a constant vaccination of newborn individuals defined by the control parameter $q \in [0, 1]$, (ii) a vaccination of the susceptible subpopulation into a vaccination rate $v(t)$, and (iii) a treatment of the infectious subpopulation into a treatment rate $t_r(t)$.

The variables $v(t)$ and $t_r(t)$ are the output variables of a first-order control system whose inputs are the SIRS model state variables. Then, both control actions are synthesised by the feedback of the epidemic model state. The equations of the control system providing both actions are as follows:

$$\dot{v}(t) = -c_1 v(t) + c_2\, S(t) + c_3\, I(t) + c_4 \frac{S(t)I(t)}{N(t)} \tag{4}$$

$$\dot{t_r}(t) = -c_5\, t_r(t) + c_6\, I(t). \tag{5}$$

Concretely, $v(t)$ and $t_r(t)$ are the number of vaccinated and treated individuals per time unit, respectively. Equations (4) and (5) regulate respectively the amount of vaccines and medicaments to be daily applied according to the time evolution of the disease propagation. In this context, if $\dot{v}(t) > 0$ ($\dot{t_r}(t) > 0$), then the amount of vaccines (medicaments) to be applied in the current day is larger than those applied in the previous one. Otherwise, if $\dot{v}(t) < 0$ ($\dot{t_r}(t) < 0$), then the amount of vaccines (medicaments) to be applied in the current day is smaller than those applied in the previous one. The time evolution of the amount of vaccines and medicaments daily applied to control the propagation of the disease depends on the current state of the epidemics, as it can be observed in (4) and (5) from the fact that the subpopulations $S(t)$ and $I(t)$ act as input in the control subsystem. Equations (1)–(5) compose the controlled epidemic model defined by two sets of parameters. The first set includes the parameters associated with the transmission of the disease and the features of the host population, namely:

- b is the natural birth rate of the host population;
- μ is the natural death rate of the host population;
- β is the transmission rate of the disease within the host population;
- ρ is the immunity loss rate within the recovered subpopulation, whose individuals become susceptible to the disease after losing the immunity;
- α is the death rate by causes related to the disease;
- γ is the recovery rate of the infectious subpopulation.

The second set of parameters is associated to the control actions, namely:

- $q \in [0, 1]$ is the proportion of newborn individuals who are vaccinated;
- c_i, for $i \in \{1, 2, 3, 4\}$, are the parameters for designing the vaccination of the susceptible subpopulation. Such control parameters allow us to weight the vaccination rate according to the state of the disease propagation considering the number of susceptible individuals S, infectious ones I, and/or the probability of contacts between them $\frac{S(t)I(t)}{N(t)}$ at each time t. Such parameters provide the availability of giving more importance to one term of (4) against the other ones in the design of the law for the vaccination $v(t)$. The unit of the parameter c_1 is (time)$^{-1}$, usually (day)$^{-1}$, and that of the parameters c_i, for $i \in \{2, 3, 4\}$ is (time)$^{-2}$ for coherence in (4).
- c_i, for $i \in \{5, 6\}$, are the parameters for designing the treatment of the infectious subpopulation. Such control parameters shape the law for the treatment $t_r(t)$ according to the number of infectious individuals. However, the values allowed for such parameters are constrained for the potency of the available medicaments. In this sense, a larger $\frac{c_6}{c_5}$ is less the recovery time interval for the treated infectious individuals. The unit of the parameter c_5 is (time)$^{-1}$, usually (day)$^{-1}$, and that of the parameters c_6 is (time)$^{-2}$ for coherence in (5).

All parameters are non-negative. The dynamics of the whole host population $N(t) = S(t) + I(t) + R(t)$ can be obtained by summing up Equations (1)–(3). In this way:

$$\dot{N}(t) = b - \mu N(t) - \alpha I(t). \tag{6}$$

The nature of the epidemic models requires the non-negativity of their solutions, so an analysis of the model positivity is developed in the following.

2.1. Positivity of the Controlled SIRS Epidemic Model

The result below establishes the non-negativity of all the subpopulations of the controlled model as well as the non-negativity of the vaccination and treatment control efforts under a set of sufficient conditions on the control parameters. The proof is written in Appendix A.

Theorem 1 ***(positivity of the model).*** *The solution of the model (1)–(6) is non-negative for all time and for any initial condition such that* $S(0) \geq 0$, $I(0) \geq 0$, $R(0) \geq 0$, $v(0) = 0$, *and* $t_r(0) = 0$ *provided that the control parameters* $c_i \geq 0$, *for* $i \in \{1, 2, \ldots, 6\}$, *and* $q \in [0, 1]$ *are chosen such that:*

$$(i)\ c_1 > \mu + \beta + 2\sqrt{c_2 + c_4} \quad ; \quad (ii)\ q \geq q_0 = 1 - \frac{\mu+\beta}{b} S(0)$$
$$(iii)\ c_3 \leq \frac{b(1-q)(\mu+\beta)}{I_{max}} \quad ; \quad (iv)\ c_5 > \mu + \alpha + \gamma + 2\sqrt{c_6}\,,$$

where $I_{max} = \max\limits_{0 \leq t < \infty} \{I(t)\}$.

Remark 1.

(i) *The conditions (i)–(iv) of Theorem 1 are only sufficient conditions, since the solutions of the model can be non-negative even if some of these conditions are not satisfied by the control parameters.*

(ii) *The birth rate of a host population b is close to its mortality rate* $\mu N(t)$ *for any* $t \geq 0$ *under normal conditions (in absence of a lethal disease). Then,* $b - \mu N(0)$ *is close to zero at the beginning of the propagation of an infectious disease. Then, typically* $b - (\mu + \beta)N(0) \approx -\beta N(0) < 0$. *Moreover, in the first stage of any epidemic disease propagation, the infectious and recovered subpopulations are much smaller than the susceptible one. Then,* $N(0) \approx S(0)$ *and, as a consequence, the condition (ii) is satisfied for any* $q \in [0, 1]$.

(iii) *The condition (iii) of Theorem 1 depends on the maximum value reached by the infectious subpopulation during the propagation of the disease. Such a value cannot be known 'a priori', and then, one cannot appropriately choose the values of the parameters* c_3 *and q to satisfy it. However, such a condition is fulfilled if* $c_3 = 0$ *for any* $q \in [0, 1]$. *Then, from continuity arguments, the condition is fulfilled for* $c_3 \in [0, \bar{c}_3]$ *with* $\bar{c}_3 = \bar{c}_3(q, I_{max}) = \frac{b(1-q)(\mu+\beta)}{I_{max}}$. *In summary, a value for* c_3 *small enough has to be chosen in order to satisfy the condition (iii) if a vaccination provided by (4) is proposed. Note that if* $q = 1$, *which means a vaccination of all the newborns, then* $c_3 = 0$ *has to be taken to guarantee the non-negativity of all the model variables for all the time.*

(iv) *In a real situation, the control actions to fight against an epidemic outbreak are taken after the disease is detected in the infectious individuals. Then, the constraint* $v(0) = t_r(0) = 0$ *in the initial condition for guaranteeing the non-negativity of the variables of the controlled epidemic model is coherent with such a fact.*

Corollary 1. *The feasible region* Γ *defined as:*

$$\Gamma = \left\{ [S(t)\ \ I(t)\ \ R(t)\ \ v(t)\ \ t_r(t)] \in \mathbb{R}^5_{0+} \middle| \begin{array}{c} 0 \leq min\left\{N(0), \frac{b}{\mu+\alpha}\right\} \leq N(t) \leq max\left\{N(0), \frac{b}{\mu}\right\} \\ v(t) \geq 0 \\ t_r(t) \geq 0 \end{array} \right\} \quad (7)$$

is positively invariant for models (1)–(5) under the conditions of Theorem 1, where $N(t) = S(t) + I(t) + R(t)$ *and* $\mathbb{R}^5_{0+}$ *denotes the non-negative five-dimension hyper-plane.*

Proof. From (6), it follows that:

$$b-(\mu+\alpha)N(t)\leq\dot{N}(t)\leq b-\mu N(t) \tag{8}$$

from the fact that $0\leq I(t)\leq N(t)\ \forall t\geq 0$ as a consequence of Theorem 1. Then, by direct calculations from (8), one obtains:

$$\frac{b}{\mu+\alpha}+\left(N(0)-\frac{b}{\mu+\alpha}\right)e^{-(\mu+\alpha)t}\leq N(t)\leq\frac{b}{\mu}+\left(N(0)-\frac{b}{\mu}\right)e^{-\mu t}. \tag{9}$$

From (4), it follows that $\dot{v}(t)\geq -c_1v(t)$ by taking into account that $S(t)\geq 0$, $I(t)\geq 0$ and $N(t)\geq 0\ \forall t\geq 0$ as a consequence of Theorem 1. Then, $v(t)\geq v(0)e^{-c_1t}\geq 0$. Finally, from (5), it follows that $\dot{t}_r(t)\geq -c_5t_r(t)$ by taking into account that $I(t)\geq 0\ \forall t\geq 0$ as a consequence of Theorem 1. Then, $t_r(t)\geq t_r(0)e^{-c_5t}\geq 0$. Equation (9) and the results $v(t)\geq 0$ and $t_r(t)\geq 0\ \forall t\geq 0$ lead to the conclusion that solutions for the model (1)–(5) with any initial condition within Γ persist in Γ for all time. □

From the positivity result of Theorem 1, the following assumption is established for the rest of the paper.

Assumption 1. *A choice of the control parameters satisfying the conditions of Theorem 1 is supposed in order to guarantee the positivity of the controlled SIRS epidemic model.*

Remark 2. *Assumption 1 mathematically guarantees that the model subpopulations as well as the control efforts do not take negative values for all time, as the nature of an epidemic model needs to be well defined. Such a fact justifies the adoption of such an assumption.*

2.2. Control Reproduction Number and Equilibrium Points of the Controlled SIRS Model

The controlled epidemic model given by (1)–(5) possesses two equilibrium points. One of them is a DFE point, and the other one is an EE point. They are obtained by setting $\dot{S}(t)=\dot{I}(t)=\dot{R}(t)=\dot{v}(t)=\dot{t}_r(t)=0$, since the equilibrium points are those at which the model variables do not change with time. In this way and by direct calculations, one obtains that the subpopulations and the values of the vaccination and treatment efforts at the DFE point are given by:

$$\begin{array}{c} S_{DFE}=\frac{bc_1[\mu(1-q)+\rho]}{\mu[c_1(\mu+\rho)+c_2]}\ ;\ I_{DFE}=0\ ;\ R_{DFE}=\frac{b(c_1q\mu+c_2)}{\mu[c_1(\mu+\rho)+c_2]} \\ v_{DFE}=\frac{bc_2[\mu(1-q)+\rho]}{\mu[c_1(\mu+\rho)+c_2]}\quad ;\quad t_{rDFE}=0. \end{array} \tag{10}$$

One can define the control reproduction number R_c for the model as the number of new infections that an infectious individual produces in a population at the DFE point. Such a number can be calculated by using the next-generation method [24]. For such a purpose, Equations (1)–(3) related to the dynamics of the epidemiological compartments of the controlled model (1)–(5) can be, equivalently, rewritten as:

$$\dot{x}=\mathcal{F}(x)-\mathcal{W}(x) \tag{11}$$

where $x=\begin{bmatrix} S & I & R\end{bmatrix}^T$ and $\mathcal{F}(x)$ and $\mathcal{W}(x)$ are given by:

$$\mathcal{F}=\begin{bmatrix} 0 & \beta\frac{SI}{N} & 0\end{bmatrix}^T\ ;\ \mathcal{W}=\begin{bmatrix} -b(1-q)+\beta\frac{SI}{N}-\rho R+\mu S+v \\ (\mu+\alpha+\gamma)I+t_r \\ -bq-v-\gamma I+(\mu+\rho)R-t_r \end{bmatrix}. \tag{12}$$

Each component of $\mathcal{F}(x)$ is the rate of appearance of new infections in the corresponding compartment. The new infections only appear in the infectious compartment; therefore, only the second component of $\mathcal{F}(x)$ is nonzero. On the other hand, the components of $\mathcal{W}(x)$

are the difference between the output and input transition rates corresponding, respectively, to the susceptible, infectious, and recovered compartments. The control reproduction number can be calculated by analysing the dynamics of the subsystem composed by the infected compartments. The proposed controlled model only has an infected compartment, namely, the infectious subpopulation I. The dynamics of such an infected subsystem around the DFE point is given by the both entries (2,2) of the matrices $F = \left.\frac{\partial \mathcal{F}}{\partial x}\right|_{x=x_{DFE}} \in \mathbb{R}^{3x3}$ and $W = \left.\frac{\partial \mathcal{W}}{\partial x}\right|_{x=x_{DFE}} \in \mathbb{R}^{3x3}$, where $x_{DFE} = \begin{bmatrix} S_{DFE} & I_{DFE} & R_{DFE} \end{bmatrix}^T$. By direct calculations, one obtains that such entries are, respectively, $f_{22} = \beta \frac{S_{DFE}}{N_{DFE}}$ and $w_{22} = \mu + \alpha + \gamma$. Then, the next generation matrix, which results a scalar for the current models, is $f_{2,2} \cdot (w_{22})^{-1}$. Finally, the control reproduction number under control efforts is obtained as the spectral radius of such a matrix, namely:

$$R_c = \sigma\left(f_{2,2} \cdot (w_{22})^{-1}\right) = \frac{f_{22}}{w_{22}} = \frac{\beta S_{DFE}}{(\mu+\alpha+\gamma)N_{DFE}} = \frac{\beta c_1[\mu(1-q)+\rho]}{(\mu+\alpha+\gamma)[c_1(\mu+\rho)+c_2]} \tag{13}$$

where $\sigma(M)$ denotes the spectral radius of the matrix M and the fact that $N_{DFE} = S_{DFE} + I_{DFE} + R_{DFE} = \frac{b}{\mu}$ and expressions in (10) have been used. Note that if the vaccination of newborn individuals is not applied, i.e., $q = 0$, while the vaccination of susceptible individuals is given by (4) with $c_2 = 0$, i.e., without the forced term depending on S, then the control reproduction number of the current controlled model is equal to the basic reproduction number $R_0 = \frac{\beta}{\mu+\alpha+\gamma}$ of a basic SIRS epidemic model. On the contrary, if $c_2 \neq 0$ and/or $q \neq 0$, then the control reproduction number is smaller than the basic reproduction number, i.e., $R_c < R_0$, since the control parameters are non-negative by definition. Such a fact is key to achieve the global stability of the DFE point of the controlled epidemic model (1)–(5), by means of an appropriate choice of the control parameters c_2 and q, in a situation where the DFE point of an SIRS model without vaccination is not globally stable. Note also that the control reproduction number depends on neither c_5 nor c_6, i.e., the application of a tréatment to the infectious subpopulation does not have influence on such a number.

On the other hand, one obtains that the subpopulations and the values of the vaccination and treatment efforts at the EE point are given by:

$$\begin{array}{c} S_{EE} = \frac{bk_1(k_2k_3R_c+k_1c_4)}{k_2c_5R_c(k_2k_4R_c+k_1k_5)} \quad ; \quad I_{EE} = \frac{bk_6[k_7(R_c-1)-c_6]}{k_2k_4R_c+k_1k_5} \\ R_{EE} = \frac{b\left(k_2^2k_8c_5R_c^2+k_1k_2k_9R_c-k_1^2c_4\right)}{k_2c_5R_c(k_2k_4R_c+k_1k_5)} \quad ; \quad t_{rEE} = \frac{bk_6c_6[k_7(R_c-1)-c_6]}{c_5(k_2k_4R_c+k_1k_5)} \\ v_{EE} = \frac{b\left(k_2^2c_3c_5^2[\mu(1-q)+\rho]R_c^2+k_1k_2k_{10}R_c-k_1^2c_4(\mu+\rho)\right)}{k_2c_5R_c(k_2k_4R_c+k_1k_5)} \end{array} \tag{14}$$

where:

$$\begin{array}{ll} k_1 = c_1[\mu(1-q)+\rho][c_5(\mu+\alpha+\gamma)+c_6] \ ; & k_2 = (\mu+\alpha+\gamma)[c_1(\mu+\rho)+c_2] \\ k_3 = c_1[c_5(\mu+\gamma+\rho+q\alpha)+c_6]+c_3c_5 \ ; & k_4 = c_1(c_5[\mu\gamma+(\mu+\alpha)(\mu+\rho)]+\mu c_6)+\mu c_3c_5 \\ k_5 = \mu c_4 - \alpha[c_1(\mu+\rho)+c_2] \ ; & k_6 = c_1[\mu(1-q)+\rho][c_1(\mu+\rho)+c_2] \\ k_7 = c_5(\mu+\alpha+\gamma) \ ; & k_8 = c_1(c_5[\gamma+q(\mu+\alpha)]+c_6)+c_3c_5 \\ k_9 = c_5[c_2-c_3+c_4-c_1(\alpha q+\gamma)]-c_1c_6 & \\ k_{10} = c_2[c_5(\mu+\gamma+\rho+q\alpha)+c_6]-c_3c_5(\mu+\rho)+c_4c_5[\mu(1-q)+\rho]. & \end{array} \tag{15}$$

The following very important result from the disease eradication viewpoint is proved.

Theorem 2 (*Non-existence of the EE point*). *If the control parameters are chosen such that:*

$$R_c < \overline{R}_c = 1 + \frac{c_6}{c_5(\mu+\alpha+\gamma)} \quad \Leftrightarrow \quad R_0 < \overline{R}_0 = \frac{[c_5(\mu+\alpha+\gamma)+c_6][c_1(\mu+\rho)+c_2]}{c_1c_5(\mu+\alpha+\gamma)[\mu(1-q)+\rho]} \tag{16}$$

then the EE point does not exist.

Proof. The condition (16) can be written as $R_c < 1 + \frac{c_6}{k_7}$ from the expression of k_7 in (15). Then, one obtains that $N_I = bk_6[k_7(R_c - 1) - c_6] < 0$, where N_I is the numerator of I_{EE}. Then, $I_{EE} < 0$ unless that $D_I = k_2k_4R_c + k_1k_5 < 0$, where D_I is the denominator of I_{EE}. Suppose that $D_I < 0$ so that $I_{EE} > 0$. Then, $D_S = k_2c_5R_c(k_2k_4R_c + k_1k_5) = k_2c_5R_cD_I < 0$ from the fact that c_5, k_2, and R_c are strictly positive, where D_S denotes the denominator of S_{EE}. Then, $S_{EE} < 0$, since $N_S = bk_1(k_2k_3R_c + k_1c_4) > 0$ since b, c_4, k_1, k_2, k_3 and R_c are strictly positive, where N_S denotes the numerator of S_{EE}. In summary, one of S_{EE} or I_{EE} is negative under the condition (16), which implies the non-existence of the EE point. □

Remark 3. *The proposed SIRS epidemic model defined by (1)–(3) has a globally asymptotically stable EE point whenever $R_0 > 1$, as is always the case in SIR-like epidemic models with demography [25]. The proposed control actions, namely, the vaccination of newborns, the vaccination of the susceptible subpopulation, given by (4), and the treatment of infectious individuals, given by (5), eliminate the existence of such an EE point under a choice of the control parameters, satisfying the condition (16). In this way, the SIRS epidemic model coupled with the control actions only has the DFE point, which is a crucial result in this paper.*

2.3. Local Stability of the Disease-Free Equilibrium Point

The study of the local stability of a nonlinear system around an equilibrium point can be done by analysing the eigenvalues of the Jacobian matrix corresponding to the linearisation of the system around such a point. For such a purpose, the controlled model (1)–(5) can be, equivalently, rewritten as $\dot{x}_c = f(x_c)$ where $x_c = \begin{bmatrix} S & I & R & v & t_r \end{bmatrix}^T$ is the extended state vector of the controlled SIRS model after including the control dynamics and:

$$f = \begin{bmatrix} b(1-q) - \beta\frac{SI}{N} + \rho R - \mu S - v \\ \beta\frac{SI}{N} - (\mu+\alpha+\gamma)I - t_r \\ bq + \gamma I - (\mu+\rho)R + v + t_r \\ -c_1 v + c_2 S + c_3 I + c_4\frac{SI}{N} \\ -c_5 t_r + c_6 I \end{bmatrix}. \quad (17)$$

The Jacobian matrix of the model around the DFE point is directly obtained as:

$$J_{DFE} = \left.\frac{\partial f}{\partial x_c}\right|_{x_c = x_{c,DFE}} = \begin{bmatrix} -\mu & j_{12} & \rho & -1 & 0 \\ 0 & j_{22} & 0 & 0 & -1 \\ 0 & \gamma & j_{33} & 1 & 1 \\ c_2 & j_{42} & 0 & -c_1 & 0 \\ 0 & c_6 & 0 & 0 & -c_5 \end{bmatrix} \quad (18)$$

where $x_{c,DFE} = \begin{bmatrix} S_{DFE} & I_{DFE} & R_{DFE} & v_{DFE} & t_{rDFE} \end{bmatrix}^T$ $j_{12} = -(\mu+\alpha+\gamma)R_c$, $j_{22} = (\mu+\alpha+\gamma)(R_c - 1)$, $j_{33} = -(\mu+\rho)$, and $j_{42} = c_3 + \frac{c_4}{\beta}(\mu+\alpha+\gamma)R_c$, and the expressions (10) and (13) for R_c have been used. The following result is proved.

Theorem 3 (*Local stability of the DFE point*). *If the control parameters are chosen such that:*

$$R_c < \overline{R}_c \quad \Leftrightarrow \quad R_0 < \overline{R}_0 \quad (19)$$

then the DFE point is locally asymptotically stable under Assumption 1, where $\overline{R}_c$ and $\overline{R}_0$ are given in (16).

Proof. The characteristic equation of the linearised model around the DFE point is given by $|\lambda\mathbb{I}_5 - J_{DFE}| = 0$, where $\mathbb{I}_5$ denotes the 5th-order unity matrix. By direct calculations, one obtains that:

$$|\lambda\mathbb{I}_5 - J_{DFE}| = (\lambda+\mu)\left(\lambda^2 + p_1\lambda + p_0\right)\left(\lambda^2 + q_1\lambda + q_0\right) \quad (20)$$

where:

$$
\begin{array}{ll}
p_1 = c_5 - (\mu + \alpha + \gamma)(R_c - 1) \;\; ; & p_0 = c_6 - c_5(\mu + \alpha + \gamma)(R_c - 1) \\
q_1 = \mu + \rho + c_1 \;\; ; & q_0 = (\mu + \rho)c_1 + c_2.
\end{array} \tag{21}
$$

Then, the eigenvalues of the matrix J_{DFE} are $-\mu$ and the roots of $P(\lambda) = \lambda^2 + p_1\lambda + p_0$ and $Q(\lambda) = \lambda^2 + q_1\lambda + q_0$. By applying the Routh–Hurwitz method to $Q(\lambda)$, one obtains that the real part of its roots is negative if $q_1 > 0$ and $q_2 > 0$. From (21), both conditions are satisfied, since the control parameters c_1 and c_2 as well as the model parameters μ and ρ are positive by definition. On the other hand, the roots of $P(\lambda)$ are given by:

$$
r_{1,2} = \frac{1}{2}\left((\mu + \alpha + \gamma)(R_c - 1) - c_5 \pm \sqrt{(c_5 + (\mu + \alpha + \gamma)(R_c - 1))^2 - 4c_6} \right) \tag{22}
$$

from (21), where the sign '+' corresponds to the root r_1 and the sign "$-$" corresponds to r_2. The control parameters c_5 and c_6 fulfil the condition (iv) of Theorem 1 provided Assumption 1. Then, it follows that $(c_5 + (\mu + \alpha + \gamma)(R_c - 1))^2 - 4c_6 > 0$, since $R_c > 0$ from (13). In this way, both roots r_1 and r_2 are real, and $r_2 < r_1$. By direct calculations from (22):

$$
\begin{array}{l}
r_1 < \frac{1}{2}\left((\mu + \alpha + \gamma)(\overline{R}_c - 1) - c_5 + \sqrt{(c_5 + (\mu + \alpha + \gamma)(\overline{R}_c - 1))^2 - 4c_6} \right) \\
= \frac{1}{2}\left(\frac{c_6}{c_5} - c_5 + \sqrt{\left(c_5 + \frac{c_6}{c_5}\right)^2 - 4c_6} \right) = \frac{1}{2c_5}\left(c_6 - c_5^2 + \sqrt{(c_6 + c_5^2)^2 - 4c_5^2 c_6} \right) = 0
\end{array} \tag{23}
$$

by using (19), the definition of $\overline{R}_c$ in (16) and the fact that $c_6 < c_5^2$ from the condition (iv) of Theorem 1. In summary, both roots of $P(\lambda)$ are strictly negative. Then, all the roots of the characteristic equation of the linearised model around the DFE point have a negative real part under Assumption 1, provided that the control parameters satisfy the condition (19). Then, the DFE point is locally asymptotically stable. □

Remark 4. *Note that the DFE point can be locally asymptotically stable although the control reproduction number R_c is larger than 1. Namely, such a property is proved if $R_c < \overline{R}_c$ independently of the particular value of R_c. Indeed, the local asymptotic stability of the DFE point is guaranteed if the transition rate from the infectious subpopulation to the recovered one is larger than the transition rate from the susceptible subpopulation to the infectious one. In this context, the number R_c is directly proportional to the transition rate from the susceptible subpopulation to the infectious one, while $\overline{R}_c$ is directly proportional to the transition rate from the infectious subpopulation to the recovered one. In this sense, such a transition rate from the infectious to the recovered subpopulation depends on several factors: potency of medicaments, amount of material and human resources in the health-care centres to treat the infectious individuals, and so on. Then, the availability of enough resources is crucial to avoid the persistence of a disease or, at least, diminish its effects on the population by increasing the value of $\overline{R}_c$.*

2.4. Global Stability of the Disease-Free Equilibrium Point

For the purpose of analysing the global stability of the DFE point, the controlled epidemic model is rewritten as [26,27]:

$$
\left\{ \begin{array}{l}
\dot{X}_n(t) = A(X_n(t) - X_{n,DFE}) + B(t)X_I(t) \\
\dot{X}_I(t) = C(t)X_I(t)
\end{array} \right. \tag{24}
$$

where $X_n(t) = [S(t)R(t)v(t)]^T$, $X_I = [I(t)t_r(t)]^T$, $X_{n,DFE} = [S_{DFE}R_{DFE}v_{DFE}]$ is the vector $X_n(t)$ at the DFE point, with its components given in (10), and:

$$A = \begin{bmatrix} -\mu & \rho & -1 \\ 0 & -(\mu+\rho) & 1 \\ c_2 & 0 & -c_1 \end{bmatrix}; \; B(t) = \begin{bmatrix} -\beta\frac{S(t)}{N(t)} & 0 \\ \gamma & 1 \\ c_3 + c_4\frac{S(t)}{N(t)} & 0 \end{bmatrix}$$
$$C(t) = \begin{bmatrix} \beta\frac{S(t)}{N(t)} - (\mu+\alpha+\gamma) & -1 \\ c_6 & -c_5 \end{bmatrix}. \tag{25}$$

The following result about the global stability of the DFE point is proved.

Theorem 4 ***(Global stability of the disease-free equilibrium point).*** *The DFE point is globally asymptotically stable under Assumption 1 provided that the control parameters are chosen such that $R_c < \overline{R}_c$ or, equivalently, $R_0 < \overline{R}_0$ and $q \in [0, q_c) \cap [0,1]$ so that $F(q) = \lambda_1 + \frac{(c_5+\lambda_1)\mu c_1 q\beta}{[c_1(\mu+\rho)+c_2](\lambda_1-\lambda_2)} < 0$, where q_c is such that $F(q_c) = 0$ while λ_1 and λ_2 are the eigenvalues of the matrix C_0 given by:*

$$C_0 = \begin{bmatrix} (\mu+\alpha+\gamma)(R_c-1) & -1 \\ c_6 & -c_5 \end{bmatrix}. \tag{26}$$

Proof. The subsystem $\dot{X}_I(t) = C(t)X_I(t)$ can be rewritten as:

$$\dot{X}_I(t) = C_0X_I(t) + C_1(t)X_I(t) \tag{27}$$

where C_0 is given by (26), and the matrix $C_1(t)$ is:

$$C_1(t) = \begin{bmatrix} \beta\left(\frac{S(t)}{N(t)} - \frac{S_{DFE}}{N_{DFE}}\right) & 0 \\ 0 & 0 \end{bmatrix} \tag{28}$$

and (13) has been used. From (27), it follows that:

$$\begin{bmatrix} I(t) \\ t_r(t) \end{bmatrix} = \phi(t)\begin{bmatrix} I(0) \\ t_r(0) \end{bmatrix} + \int_0^t \phi(t-\tau)C_1(\tau)X_I(\tau)d\tau \quad \forall t \geq 0, \tag{29}$$

with $\phi(t) = e^{C_0t} = L^{-1}\left\{(s\mathbb{I}_2 - C_0)^{-1}\right\}$. By direct calculations, one obtains that:

$$\phi_{11}(t) = \frac{(c_5+\lambda_1)e^{\lambda_1 t} - (c_5+\lambda_2)e^{\lambda_2 t}}{\lambda_1-\lambda_2}; \quad \phi_{12}(t) = -\frac{e^{\lambda_1 t} - e^{\lambda_2 t}}{\lambda_1-\lambda_2}; \quad \phi_{21}(t) = \frac{c_6\left(e^{\lambda_1 t} - e^{\lambda_2 t}\right)}{\lambda_1-\lambda_2}$$
$$\phi_{22}(t) = \frac{[\lambda_1-(\mu+\alpha+\gamma)(R_c-1)]e^{\lambda_1 t} - [\lambda_2-(\mu+\alpha+\gamma)(R_c-1)]e^{\lambda_2 t}}{\lambda_1-\lambda_2}, \tag{30}$$

where λ_1 and λ_2 are the eigenvalues of C_0. Note that these eigenvalues are, respectively, the roots r_1 and r_2, given in (22), of the polynomial $P(\lambda)$ defined in the proof of Theorem 3. Under Assumption 1, the condition (iv) of Theorem 1, and $R_c < \overline{R}_c$, it follows that the eigenvalues λ_1 and λ_2 are real and $\lambda_2 < \lambda_1 < 0$. From (29) and (30), it follows that:

$$I(t) = \phi_{11}(t)I(0) + \phi_{12}(t)t_r(0) + \int_0^t \phi_{11}(t-\tau)\beta\left(\frac{S(\tau)}{N(\tau)} - \frac{S_{DFE}}{N_{DFE}}\right)I(\tau)d\tau \quad \forall t \geq 0. \tag{31}$$

The fact that $\lambda_2 < \lambda_1 < 0$ implies that $e^{\lambda_1 t} \geq e^{\lambda_2 t} \geq 0 \; \forall t \geq 0$. Moreover, $c_5 + \lambda_2 \geq 0$, since $c_6 \geq 0$ from its definition. Then, $\phi_{11}(t) \leq \frac{(c_5+\lambda_1)e^{\lambda_1 t}}{\lambda_1-\lambda_2}$, $\phi_{12}(t) \leq 0 \; \forall t \geq 0$, and one obtains from (31) that:

$$I(t) \leq \frac{(c_5+\lambda_1)I(0)}{\lambda_1-\lambda_2}e^{\lambda_1 t} + \frac{(c_5+\lambda_1)\beta}{\lambda_1-\lambda_2}\int_0^t e^{\lambda_1(t-\tau)}\left(1-\frac{S_{DFE}}{N_{DFE}}\right)I(\tau)d\tau \quad \forall t \geq 0, \tag{32}$$

where the fact that $\frac{S(t)}{N(t)} \leq 1\ \forall t \geq 0$ from the positivity of the model has been used. Moreover, from the definition of S_{DFE} and N_{DFE} given in (10), it follows:

$$I(t) \leq \frac{c_5+\lambda_1}{\lambda_1-\lambda_2}e^{\lambda_1 t}\left(I(0) + \frac{\mu c_1 q\beta}{c_1(\mu+\rho)+c_2}\int_0^t e^{-\lambda_1\tau}I(\tau)d\tau\right) \quad \forall t \geq 0. \tag{33}$$

By applying the Bellman–Gronwall Lemma I [28] in (33), one obtains:

$$I(t) \leq \frac{c_5+\lambda_1}{\lambda_1-\lambda_2}e^{\lambda_1 t}\left(I(0) + \frac{\mu c_1 q\beta}{c_1(\mu+\rho)+c_2}\int_0^t \frac{c_5+\lambda_1}{\lambda_1-\lambda_2}I(0)e^{\int_\tau^t \frac{(c_5+\lambda_1)\mu c_1 q\beta}{[c_1(\mu+\rho)+c_2](\lambda_1-\lambda_2)}ds}d\tau\right) \tag{34}$$

that leads by direct calculations to:

$$I(t) \leq \frac{c_5+\lambda_1}{\lambda_1-\lambda_2}I(0)e^{(\lambda_1+\frac{(c_5+\lambda_1)\mu c_1 q\beta}{[c_1(\mu+\rho)+c_2](\lambda_1-\lambda_2)})t} = \frac{c_5+\lambda_1}{\lambda_1-\lambda_2}I(0)e^{F(q)t} \tag{35}$$

with $F(q) = \lambda_1 + \frac{(c_5+\lambda_1)\mu c_1 q\beta}{[c_1(\mu+\rho)+c_2](\lambda_1-\lambda_2)}$. Note that the eigenvalues λ_1 and λ_2 depend on R_c, which depends on q from its definition in (13). Furthermore, one obtains that $F(0) = \lambda_1(0) < 0$ since $\lambda_1(q) < 0\ \forall q \in [0,1]$ while:

$$\frac{dF}{dq} = \frac{2(\mu\beta c_1)^2 c_6 q}{[c_1(\mu+\rho)+c_2]^2\left([c_5+(\mu+\alpha+\gamma)(R_c-1)]^2-4c_6\right)^{\frac{3}{2}}}. \tag{36}$$

Then, $\frac{dF}{dq} = 0$ for $q = 0$ and $\frac{dF}{dq} > 0\ \forall q > 0$, since $c_6 < \frac{[c_5+(\mu+\alpha+\gamma)(R_c-1)]^2}{4}$ from the condition (iv) of Theorem 1. Such a fact implies that there exists a critical value $q_c > 0$ such that $F(q) < 0\ \forall q \in [0,\ q_c)$ by continuity of the function $F(q)$ with respect to q. As a consequence, one obtains that $\lim_{t\to\infty}\{I(t)\} = 0$ if $q \in [0,\ q_c) \cap [0,1]$. From (29) and (30), it follows that:

$$t_r(t) = \phi_{21}(t)I(0) + \phi_{22}(t)t_r(0) + \int_0^t \phi_{21}(t-\tau)\beta\left(\frac{S(\tau)}{N(\tau)} - \frac{S_{DFE}}{N_{DFE}}\right)I(\tau)d\tau \quad \forall t \geq 0. \tag{37}$$

The fact that $\lambda_2 < \lambda_1 < 0$ implies that $\phi_{21}(t) \leq \frac{c_6 e^{\lambda_1 t}}{\lambda_1-\lambda_2}\ \forall t \geq 0$. Moreover, from (22), it follows that $\lambda_2 - (\mu+\alpha+\gamma)(R_c-1) < \lambda_1 - (\mu+\alpha+\gamma)(R_c-1) < 0$ and then:

$$\phi_{22}(t) \leq -\frac{\lambda_2-(\mu+\alpha+\gamma)(R_c-1)}{\lambda_1-\lambda_2}e^{\lambda_2 t} \quad \forall t \geq 0. \tag{38}$$

One obtains from (37) that:

$$t_r(t) \leq \frac{c_6 I(0)}{\lambda_1-\lambda_2}e^{\lambda_1 t} + \frac{(\mu+\alpha+\gamma)(R_c-1)-\lambda_2}{\lambda_1-\lambda_2}t_r(0)e^{\lambda_2 t} + \frac{c_6\beta}{\lambda_1-\lambda_2}\left(1-\frac{S_{DFE}}{N_{DFE}}\right)\int_0^t e^{\lambda_1(t-\tau)}I(\tau)d\tau \tag{39}$$

where the fact that $\frac{S(t)}{N(t)} \leq 1\ \forall t \geq 0$ from the positivity of the model has been used. Moreover, from the definition of S_{DFE} and N_{DFE} given in (10), it follows:

$$t_r(t) \leq \frac{c_6}{\lambda_1-\lambda_2}e^{\lambda_1 t}\left(I(0) + \frac{\mu c_1 q\beta}{c_1(\mu+\rho)+c_2}\int_0^t e^{-\lambda_1\tau}I(\tau)d\tau\right) + \frac{(\mu+\alpha+\gamma)(R_c-1)-\lambda_2}{\lambda_1-\lambda_2}t_r(0)e^{\lambda_2 t}. \tag{40}$$

By introducing (35) in (40), one obtains:

$$t_r(t) \le \frac{c_6 I(0)}{\lambda_1 - \lambda_2} e^{\lambda_1 t} \left(1 + \frac{\mu c_1 q \beta (c_5 + \lambda_1)}{[c_1(\mu+\rho) + c_2](\lambda_1 - \lambda_2)} \int_0^t e^{(F(q)-\lambda_1)\tau} d\tau \right) + \frac{(\mu+\alpha+\gamma)(R_c - 1) - \lambda_2}{\lambda_1 - \lambda_2} t_r(0) e^{\lambda_2 t} \quad (41)$$

and finally:

$$t_r(t) \le \frac{c_6 I(0) e^{\lambda_1 t} + [(\mu+\alpha+\gamma)(R_c-1) - \lambda_2] t_r(0) e^{\lambda_2 t}}{\lambda_1 - \lambda_2} + \frac{\mu c_1 c_6 q \beta (c_5 + \lambda_1) I(0)}{[c_1(\mu+\rho) + c_2](\lambda_1 - \lambda_2)^2 (F(q) - \lambda_1)} \left(e^{F(q)t} - e^{\lambda_1 t} \right). \quad (42)$$

Then, one obtains that $\lim_{t\to\infty}\{t_r(t)\} = 0$ if $q \in [0,\ q_c) \cap [0,1]$ since $\lambda_1 < 0$ and $F(q) < 0$ under such a condition. On the other hand, if one applies the variable change $Y_n(t) = X_n(t) - X_{n,DFE}$ in the first equation of (24), then:

$$\dot{Y}_n(t) = AY_n(t) + B(t)X_I(t) \quad (43)$$

whose solution is:

$$Y_n(t) = \phi^A(t) Y_n(0) + \int_0^t \phi^A(t-\tau) B(\tau) X_I(\tau) d\tau \quad \forall t \ge 0 \quad (44)$$

where $\phi^A(t) = e^{At} = L^{-1}\left\{ (s\mathbb{I}_3 - A)^{-1} \right\}$. The eigenvalues of A are $\lambda_1^A = -\mu$ and the roots of the Hurwitz polynomial $Q(\lambda)$, which are defined in the proof of Theorem 3, namely:

$$\lambda_{2,3}^A = \frac{1}{2}\left(-(\mu+\rho+c_1)(R_c - 1) \pm \sqrt{(c_1 - (\mu+\rho))^2 - 4c_2} \right) \quad (45)$$

where the sign '+' corresponds to the root λ_2^A and the sign "−" corresponds to λ_3^A. Since the real part of such roots is negative, there exists some norm-dependent real constant $K_A \ge 1$ such that:

$$\|Y_n(t)\| \le K_A e^{\sigma_A t} (\|Y_n(0)\| + \int_0^t e^{\sigma_A \tau} \|B(\tau)\| \|X_I(\tau)\| d\tau) \quad \forall t \ge 0, \quad (46)$$

where $\sigma_A = max\{-\mu, \lambda_2^A\} < 0$ is the stability abscissa of A. From (35) and (42), one obtains that:

$$\|X_I(t)\| = \sqrt{I^2(t) + t_r{}^2(t)} \le I(t) + t_r(t) \le g_1 e^{\lambda_1 t} + g_2 e^{\lambda_2 t} + g_3 e^{F(q)t} \quad \forall t \ge 0, \quad (47)$$

where the positivity of the model has been used, and:

$$g_1 = \frac{c_6 I(0)}{\lambda_1 - \lambda_2}\left(1 - \frac{\mu c_1 q \beta (c_5 + \lambda_1)}{[c_1(\mu+\rho)+c_2](\lambda_1-\lambda_2)(F(q)-\lambda_1)}\right);\ g_2 = \frac{[(\mu+\alpha+\gamma)(R_c-1)-\lambda_2] t_r(0)}{\lambda_1 - \lambda_2}$$
$$g_3 = \frac{c_5+\lambda_1}{\lambda_1-\lambda_2} I(0)\left(1 + \frac{\mu c_1 c_6 q \beta}{[c_1(\mu+\rho)+c_2](\lambda_1-\lambda_2)(F(q)-\lambda_1)}\right). \quad (48)$$

From (25), it follows that:

$$\|B(t)\| = max\left\{1, c_3 + \gamma + (\beta + c_4)\frac{S(t)}{N(t)}\right\} \le max\{1, c_3 + \gamma + \beta + c_4\} = K_B \quad \forall t \ge 0 \quad (49)$$

where $\frac{S(t)}{N(t)} \le 1$ has been used. By introducing (47) and (49) in (46), one obtains:

$$\|Y_n(t)\| \le K_A e^{\sigma_A t} \left(\|Y_n(0)\| + K_B \int_0^t e^{-\sigma_A \tau} \left(g_1 e^{\lambda_1 \tau} + g_2 e^{\lambda_2 \tau} + g_3 e^{F(q)\tau} \right) d\tau \right). \quad (50)$$

By direct calculation from (50), it follows that:

$$\|Y_n(t)\| \leq K_A\|Y_n(0)\|e^{\sigma_A t}$$
$$+K_B\left[\frac{g_1}{\lambda_1-\sigma_A}\left(e^{\lambda_1 t}-e^{\sigma_A t}\right)+\frac{g_2}{\lambda_2-\sigma_A}\left(e^{\lambda_2 t}-e^{\sigma_A t}\right)+\frac{g_3}{F(q)-\sigma_A}\left(e^{F(q)t}-e^{\sigma_A t}\right)\right]. \tag{51}$$

Finally, from (51), one obtains that $\lim_{t\to\infty}\{\|Y_n(t)\|\}=0$ if $q\in[0,\,q_c)\cap[0,1]$, since $\lambda_1<0$, $\lambda_2<0$, $\sigma_A<0$, and $F(q)<0$ under such a condition. Then, $\lim_{t\to\infty}\{\|X_n(t)-X_{n,DFE}\|\}=0$ and, also, $\lim_{t\to\infty}\{X_n(t)\}=X_{n,DFE}$. In summary, the DFE point is globally asymptotically stable. □

Remark 5. *Note the following features:*

(i) *From Theorems 1–4, the model is positive and only has an equilibrium point, namely, the DFE point, which is locally and globally asymptotically stable provided that the control parameters fulfil the conditions of Theorem 1 and that $R_c<\overline{R}_c$ or, equivalently, $R_0<\overline{R}_0$.*

(ii) *The control parameters c_1 and c_2 influence the components of both the DFE and EE points, see (10) and (14), (15) respectively.*

(iii) *The control parameters c_3, c_4, c_5, and c_6 influence the components of the EE point.*

(iv) *The control parameters c_1, c_2, c_5, and c_6 influence the stability of the DFE point according to (19) and, also, they can imply the non-existence of the EE point under an appropriate choice of their values according to (16).*

(v) *The threshold value $\overline{R}_c$ given in (16) depends on the control parameters c_5 and c_6 associated with the treatment effort, while the control reproduction number R_c given in (13) depends on the control parameters q, c_1, and c_2 associated with the vaccination efforts so that the non-existence of the EE point can be guaranteed by a treatment strategy adapted to a designed vaccination campaign.*

(vi) *Neither R_c nor $\overline{R}_c$ depend on the control parameters c_3 and c_4 associated with the effort of the vaccination of susceptible individuals so that such parameters are not relevant for eradicating the disease. Such parameters affect the values of the subpopulations at the EE point if such a point is reached, which is intended to be avoided.*

(vii) *The expression for R_c can be equivalently written as $R_c=\frac{\beta[\mu(1-q)+\rho]}{(\mu+\alpha+\gamma)\left[\mu+\rho+\frac{c_2}{c_1}\right]}=\frac{\mu(1-q)+\rho}{\mu+\rho+\frac{c_2}{c_1}}R_0$. Then, one can see that R_c is inversely proportional to $\frac{c_2}{c_1}$ so that an increment in the value of $\frac{c_2}{c_1}$ results in a decrement of R_c, implying a small incidence of the infectious disease. In this context, a large value for $\frac{c_2}{c_1}$ is interesting for reducing the incidence of the disease on the host population. A large value for the relation $\frac{c_2}{c_1}$ can be obtained by considering small values for c_1. However, the condition (i) of Theorem 1 requires a lower bound for c_1, namely, $c_1>\mu+\beta+2\sqrt{c_2+c_4}$, for some prescribed values for μ, β, c_2, and c_4, in order to guarantee the non-negativity of the solutions of the controlled model under any non-negative initial condition. Then, the only way of increasing the relation $\frac{c_2}{c_1}$ is by means of an increment of c_2, which also implies an increment of c_1 to guarantee the condition $c_1>\mu+\beta+2\sqrt{c_2+c_4}$ for some prescribed values for μ, β, and c_4. In summary, the only practical way of increasing the relation $\frac{c_2}{c_1}$ is via increasing simultaneously the value of both parameters c_1 and c_2. However, a large value for c_2 can imply large values for the vaccination control effort, since it affects directly the forced term of Equation (4) for the dynamics of the vaccination law. In fact, the vaccination can be constrained to a number of available vaccines in a practical situation, which implies upper-bounds for the control parameters c_2, c_3, and c_4 of the forced terms of (4).*

(viii) *One can see that $\overline{R}_c$ is directly proportional to $\frac{c_6}{c_5}$ so that an increment in the value of $\frac{c_6}{c_5}$ implies an increase of $\overline{R}_c$. Moreover, note that the EE point of the controlled model does not exist if $R_c<\overline{R}_c$ so that an increment of $\overline{R}_c$ can be interesting in order to guarantee the non-existence of such an EE point for a prescribed value for R_c adjusted by values for q, c_1, and c_2 adapted to the number of available vaccines.*

(ix) *The influence of the parameter $q\in[0,1]$ on the value of the control reproduction number R_c is negligible if the value of the natural death rate μ is very small, as it happens in the case of* *humans. In such a case, the influence of $q \in [0,1]$ on the eigenvalues λ_1 and λ_2 of the matrix C_0 as well as on the function $F(q)$, both defined in the proof of Theorem 4, is also negligible. Such a fact implies that the DFE point is globally asymptotically stable $\forall q \in [0,1]$, since $q_c > 1$, provided that the control parameters c_i, for $i \in \{1, 2, \ldots, 6\}$, are chosen such that Assumption 1 and $R_c < \overline{R}_c$ are satisfied.*

3. Simulation Results

Some simulation examples based on a high infectivity disease illustrate the efficacy of the proposed control strategy. The examples have been developed by using the version R2020b of MATLAB. They have been simulated by using the solver "ode3 (Bogacki-Shampine)" with a fixed step of $\frac{1}{24}$ days, i.e., one hour. First, a SIRS model without vaccination and treatment is considered. Later, the same model with the proposed control actions is analysed to show their impact in the propagation of the disease within the host population. As a consequence, a drastic mortality reduction is exhibited due to the application of the proposed feedback control actions. In addition, a set of examples are used to analyse the influence of certain control parameters on the evolution of the disease spreading. Finally, the influence of the immunity loss rate is also studied.

3.1. Example 1: SIRS Model without Vaccination and Treatment

The model (1)–(3) with the values $b = 0.0384\, d^{-1}$, $\mu = 3.653 \times 10^{-5}\, d^{-1}$, $\beta = 0.65\, d^{-1}$, $\gamma = \frac{1}{22} = 0.0455\, d^{-1}$, $\rho = \frac{1}{120} = 0.0083\, d^{-1}$ and $\alpha = 5 \times 10^{-5}\, d^{-1}$, where d means the unit time "day", is analysed. In addition, the parameter $q = 0$ is considered, since vaccination on the newborn individuals is not applied. The main objective is to show the time evolution of the subpopulations and that of the whole population under the influence of the infectious disease. The basic reproduction number of this model is $R_0 = 14.2728$; that is, each primary infectious individual transmits the disease to more than 14 healthy individuals. Then, this example describes the propagation of a high infectivity disease. The DFE point of the model is unstable, while its EE point is globally asymptotically stable, as Remark 3 points out. Figure 1 shows the time evolution of the susceptible, infectious, and recovered subpopulations during the first 100 days, while Figure 2 displays that of the whole population in a very long time period. The considered initial condition is given by $S(0) = 985$, $I(0) = 10$ and $R(0) = 5$.

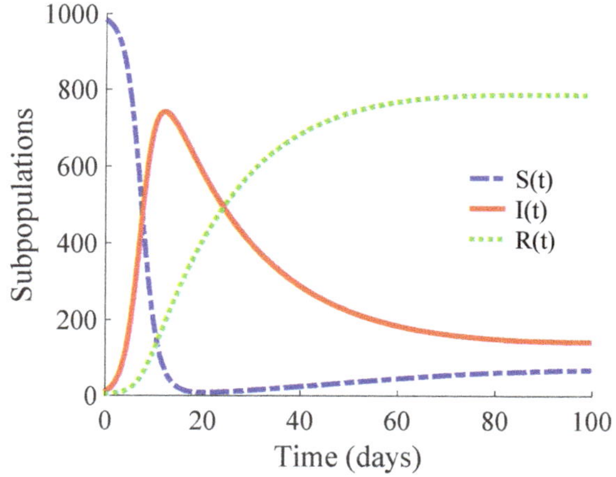

Figure 1. Time evolution of the subpopulations in the SIRS model without control actions.

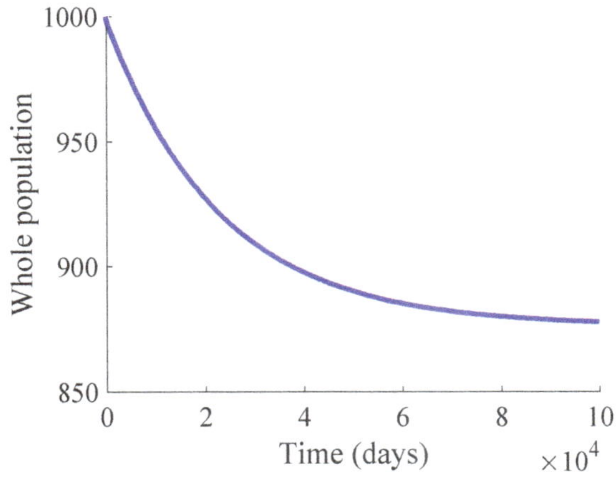

Figure 2. Time evolution of the whole population in the SIRS model without control actions.

In the longer simulation, one can see that the model tends to an EE point where the number of individuals is $N_{EE} = 877$ with $S_{EE} = 62$, $I_{EE} = 127$ and $R_{EE} = 688$ accordingly to (14), (15) with $c_i = 0$, for $i \in \{1, 2, \ldots, 6\}$. Although the mortality rate due to the disease $\alpha = 5 \times 10^{-5}\, d^{-1}$ is not very high, the influence of the disease is very important, since the whole population notably decreases from 1000 individuals to 877. Moreover, the percentage of infectious individuals with respect to the whole population in the EE point is 14.48%. In this situation, the application of control measures is crucial in order to eliminate the infection or, at least, to diminish its effect within the host population while reducing drastically the mortality.

3.2. Example 2: SIRS Model with Vaccination and Treatment

Models (1)–(3) with the same values for the parameters b, μ, β, γ, ρ and α as those given in Example 1 are considered. Moreover, a vaccination for newborn individuals together with the control actions (4) and (5) are applied in order to diminish the effects of the infectious disease. Concretely, the control actions are (i) a *vaccination* of a proportion $q = 0.1$ of the *newborn individuals*; i.e., 10% of newborns are vaccinated, (ii) a *vaccination of the susceptible* population given by (4) with the values $c_1 = 25$, $c_2 = 1$, $c_3 = 0.0001$ and $c_4 = 0.07$ for the associated parameters and (iii) a *treatment of the infectious* population given by (5) with the values $c_5 = 2$ and $c_6 = 0.27$ for the corresponding parameters. Such values satisfy the conditions of Theorem 1 so that the model subpopulations and the vaccination and treatment efforts are non-negative for all time. The initial condition is that of Example 1 together with $v(0) = t_r(0) = 0$ for the initial control efforts. Note that dynamics of the treatment effort, given by the control law (5), is of first order with a gain that reaches the value $\frac{c_6}{c_5} = 0.135$ in the stationary regime. Such a gain points out a transition rate from the infectious subpopulation to the recovered one via applied treatment to be added to the transition rate $\gamma = 0.0455$ between such subpopulations from the natural response of the immunity system of the individuals against the disease. This fact points out that an infectious individual under treatment needs an average time of 5.54 days to overcome the disease. In this way, the average time of recovering is reduced in 16.46 days, since such an interval time is $\gamma^{-1} = 22$ days if no treatment is applied. Figure 3 shows the time evolution of the susceptible, infectious, and recovered subpopulations during the first 100 days, while Figures 4 and 5 display, respectively, that of the whole population during the first 100 days and along 100,000 days.

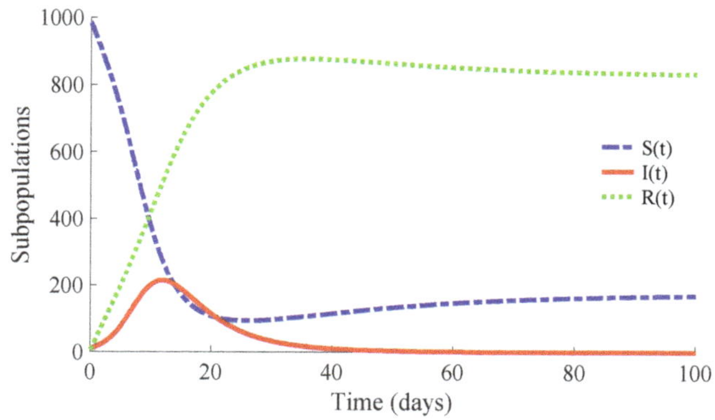

Figure 3. Time evolution of the subpopulations in the SIRS model with vaccination and treatment.

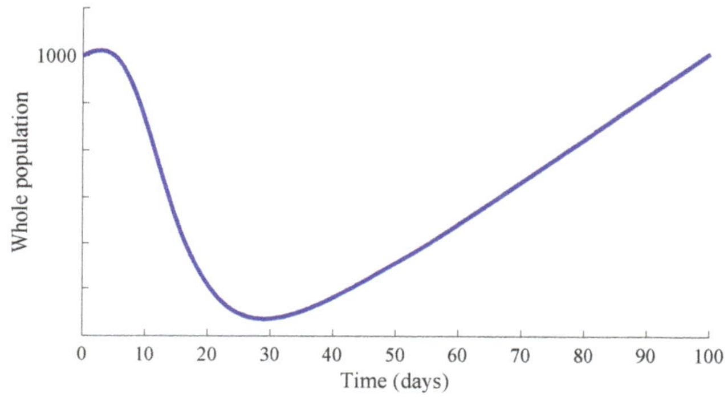

Figure 4. Time evolution of the whole population in the SIRS model with vaccination and treatment within the 100 first days.

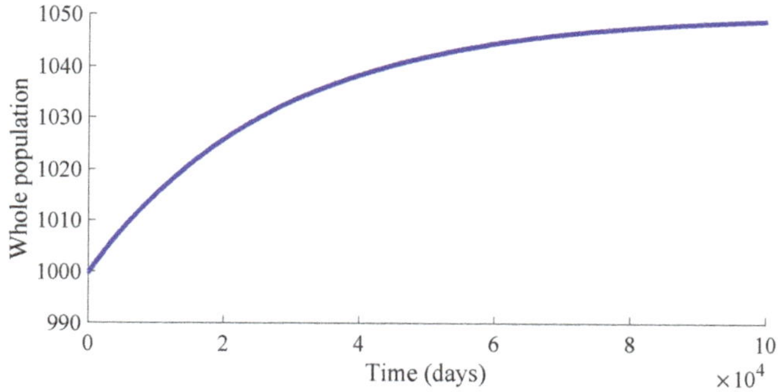

Figure 5. Time evolution of the whole population in the SIRS model with vaccination and treatment along 100,000 days.

In a longer time performed simulation, one can see that the model tends to a DFE point where the number of individuals is $N_{DFE} = 1050$ with $S_{DFE} = 182$, $I_{DFE} = 0$ and $R_{DFE} = 868$. This result is coherent with Theorems 3 and 4, since $R_c = 2.4687$, $\overline{R}_c = 3.9644$ and $F(q) = -0.0732 < 0$ for the used value $q = 0.1$ so that the DFE point is locally and globally asymptotically stable. Moreover, the choice of the control parameters satisfies the conditions of Theorem 2, which implies the non-existence of the EE point. Then, the DFE point is the unique equilibrium point for the controlled model. In summary, this example exhibits the high mortality reduction due to the application of the proposed feedback control actions in the propagation of an infectious disease of a high basic reproduction number R_0. Figure 6 displays the time evolution of the control efforts. One can see that the treatment effort reaches a peak value when the number of infectious individuals reaches its maximum value, and then, such an effort converges to zero as a consequence of the convergence to zero of the infectious subpopulation. On the other hand, the vaccination effort converges to a constant value $v_{DFE} = 7$.

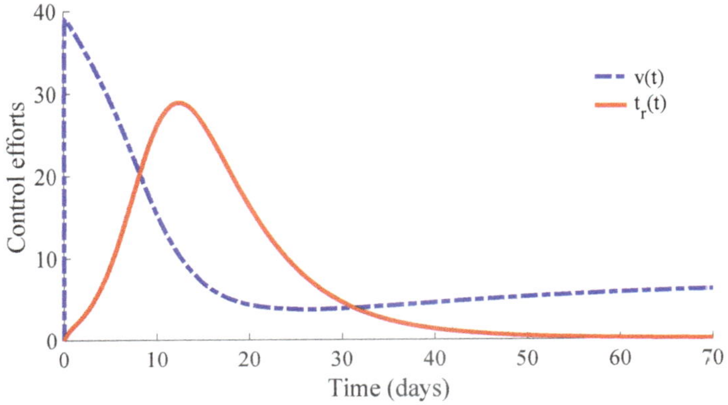

Figure 6. Time evolution of the control efforts: vaccination and treatment.

The vaccination and treatment efforts can be interpreted as the number of vaccines and antivirals, or some appropriated medicaments, applied to the susceptible and infectious individuals, respectively, during the transition from the initial state to the DFE point. In this context, one can consider that the DFE point is reached when the number of infectious individuals is less than 1. Moreover, although the vaccination effort has a constant value $v_{DFE} = 7$ given by (10) at the DFE point, one can stop the vaccination control effort once the number of infectious individuals is smaller than 1. In this way, the vaccination can be set $v(t) = 0$ once the DFE point is reached since the transition rate from the susceptible subpopulation to the infectious one is zero in the absence of infectious individuals due to such a rate depending on $\beta\frac{I(t)}{N(t)}$. In other words, there is no propagation of the disease in absence of infectious individuals, and then, the vaccination can be stopped. However, the vaccination with the value v_{DFE} could be maintained as a preventive measure against possible new outbreaks of the disease due to immigration or other causes. This preventive measure gives place to a transition from the susceptible subpopulation to the recovered one with a rate that compensates for the transition rate ρ from the recovered subpopulation to the susceptible one caused by the immunity loss. In case that the vaccination is stopped once the DFE point is reached, which is at the 64th day in this example, the number of vaccines and medicaments applied to control the propagation in the transient from the initial day until achieving the DFE point are, respectively, 578 and 483. Such numbers can be interpreted as the control cost to eradicate the persistence of the disease. Figure 7 displays the number of applied vaccines and medicaments each day during the transition to the DFE point. The numbers of vaccines and medicaments applied each day, respectively $n_v(t)$ and $n_{tr}(t)$, are obtained by evaluating $n_v(t) = \int_{t-1}^{t} v(\tau)d\tau$ and $n_{tr}(t) = \int_{t-1}^{t} t_r(\tau)d\tau$ for all integer $t \in \left[1, 2\ldots, t_f\right]$ with t_f denoting the day at which the DFE point is reached.

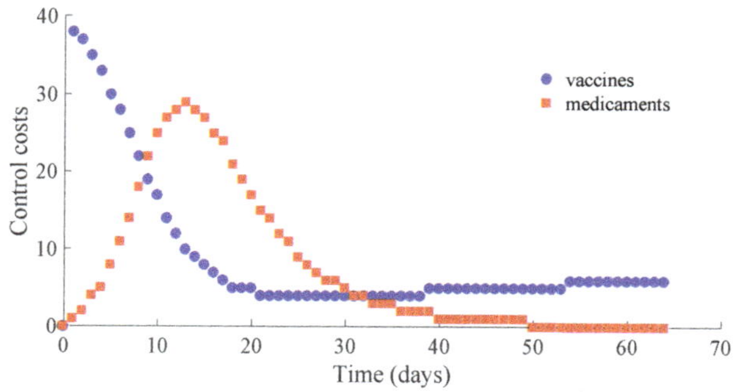

Figure 7. Number of vaccines and medicaments applied each day.

3.3. *Example 3: SIRS Model with a Defective Vaccination and Treatment*

The same values as those used in Example 2 for the parameters of the model are considered except that for c_2, which is replaced by $c_2 = 0.2$. In this way, $R_c = 7.2945$, while $\overline{R}_c = 3.9644$. Then, the conditions of Theorem 2 are not satisfied, which implies that the non-existence of the EE point cannot be guaranteed. In fact, the time evolution of the subpopulations under these conditions is displayed in Figure 8, and one can see that the EE point given by (14) and (15) is reached. Concretely, the whole number of individuals at the EE point is $N_{EE} = 1022$ with $S_{EE} = 284$, $I_{EE} = 21$, and $R_{EE} = 717$, and the vaccination and treatment control signals converge, respectively, to the values $v_{EE} = 2$ and $t_{rEE} = 3$.

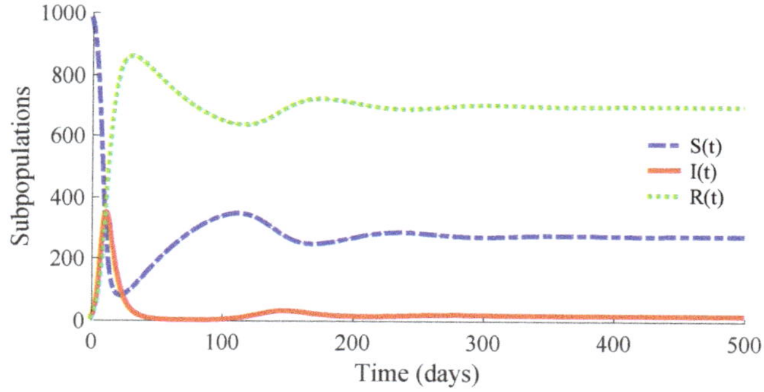

Figure 8. Time evolution of the subpopulations in the SIRS model under an insufficient vaccination and treatment.

The time evolution of the control efforts is displayed in Figure 9.

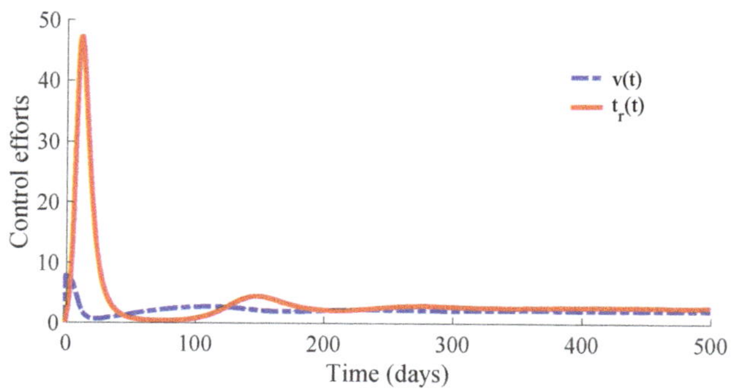

Figure 9. Time evolution of the control efforts under an insufficient vaccination and treatment.

3.4. Example 4: Study of the Influence of the Control Parameter c_2 on the Behaviour of the Controlled Model

Models (1)–(5) with the same values for the parameters as those given in Example 2, except that for c_2, are considered. In this way, $\overline{R}_c = 3.9644$ in all the analysed cases. Concretely, four cases are studied. Each one considers a different value for the parameter c_2 in order to change slightly the ratio $\frac{c_2}{c_1}$, which implies a different value for the control reproduction number R_c. The condition $R_c < \overline{R}_c$ as well as the conditions of the Theorems 1, 2, 3, and 4 are satisfied in the four cases so that the DFE point is the unique equilibrium point, and it is locally and globally asymptotically stable, while the solutions of the controlled model are non-negative for all time under any non-negative initial condition. The interest of this study is to analyse the influence of the parameter c_2 on the transition of the controlled model solutions from the initial state to the DFE point by evaluating the number of infectious individuals on each day as well as the cost of both vaccination and treatment control efforts. For such a purpose, the values for the control parameter c_2 and then that of R_c in the four studied cases are: (i) $c_2 = 1$, $R_c = 2.4687$, (ii) $c_2 = 0.9$, $R_c = 2.6912$, (iii) $c_2 = 0.8$, $R_c = 2.9579$, and (iv) $c_2 = 0.7$, $R_c = 3.2832$. Note that case (i) is that studied in Example 2, and it is used as the reference one for the current analysis. The initial condition is that used in Example 2 for all the cases. Figure 10 displays the time evolution of the infectious subpopulation in the four cases. Figures 11 and 12 show, respectively, the time evolution of the vaccination and treatment efforts in the four cases.

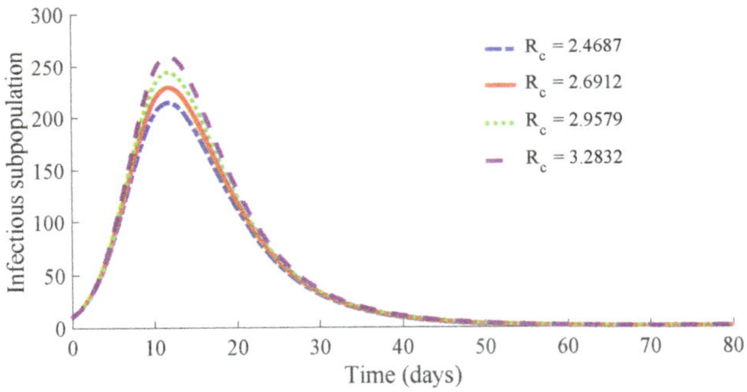

Figure 10. Time evolution of the infectious subpopulation for the four considered values of the control parameter c_2.

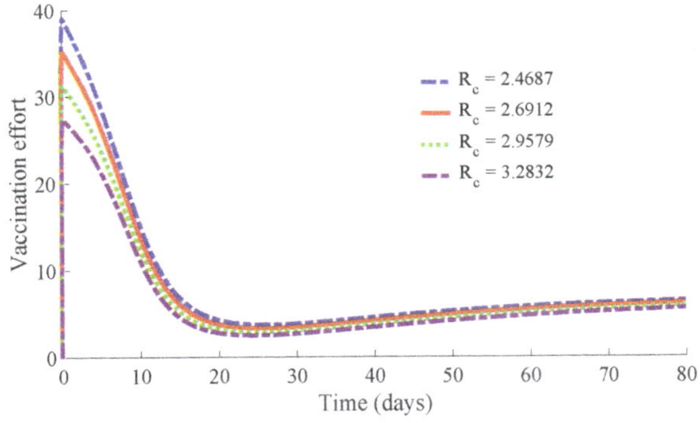

Figure 11. Time evolution of the vaccination effort for the four considered values of the control parameter c_2.

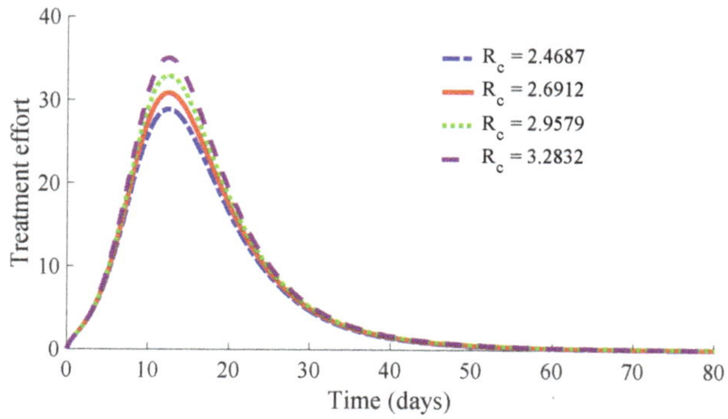

Figure 12. Time evolution of the treatment effort for the four considered values of the control parameter c_2.

One can see in these figures that the peaks in the infectious population, the vaccination and treatment efforts depend on the value of the control reproduction number R_c, which is established by the control parameters c_1 and c_2. Concretely, the peak of the infectious subpopulation as well as that of the treatment effort increases as R_c increases, while the peak of the vaccination effort decreases as R_c increases, while maintaining $R_c < \overline{R}_c$. Table 1 summarises the most relevant specifications, included the aforementioned ones, about the four cases studied in this subsection.

Table 1. Specifications of the model behaviour for the four considered values of the parameter c_2.

	$c_2 = 1$ $R_c = 2.4687$	$c_2 = 0.9$ $R_c = 2.6912$	$c_2 = 0.8$ $R_c = 2.9579$	$c_2 = 0.7$ $R_c = 3.2832$
Infectious peak	215	229	245	261
Transient duration (days)	64	66	69	72
Vaccination peak	39	35	31	27
Treatment peak	29	31	33	35
S_{DFE}	182	198	218	242
R_{DFE}	868	852	832	808
v_{DFE}	7	7	7	7
Vaccines	578	537	502	467
Medicaments	483	515	542	570

The results show that the duration of the transient from the initial state to the DFE point increases as R_c increases. However, the number of vaccines applied during the whole transient period decreases when the value of R_c increases. Such a result is due to the fact that the peak of the vaccination effort decreases and, also, the number of vaccines used per day, when R_c increases. In summary, the number of required vaccines decreases while that of the required medicaments increases as R_c increases. Then, a tradeoff between the cost of vaccination and the number of infectious individuals with its associated treatment cost has to be taken into account for adjusting the values for the control parameters c_1 and c_2. Such a tradeoff is going to depend on constraints about the number of available vaccines for the newborns and the susceptible individuals and/or medicaments for treatment of the infectious subpopulation. Finally, Figures 13 and 14 display the number of applied vaccines and medicaments each day during the transition to the DFE point for the four analysed cases.

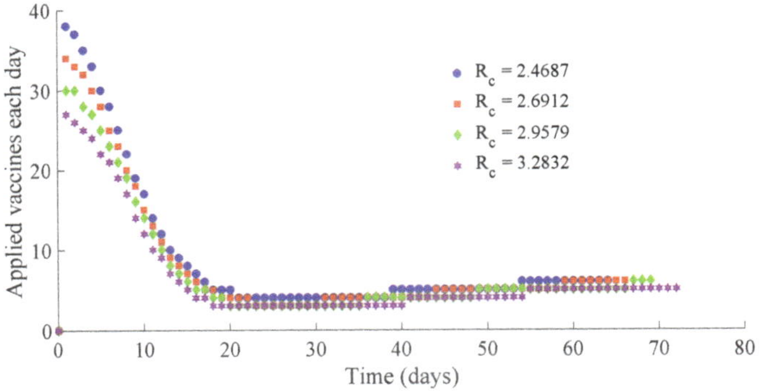

Figure 13. Number of applied vaccines each day for the four considered values of the control parameter c_2.

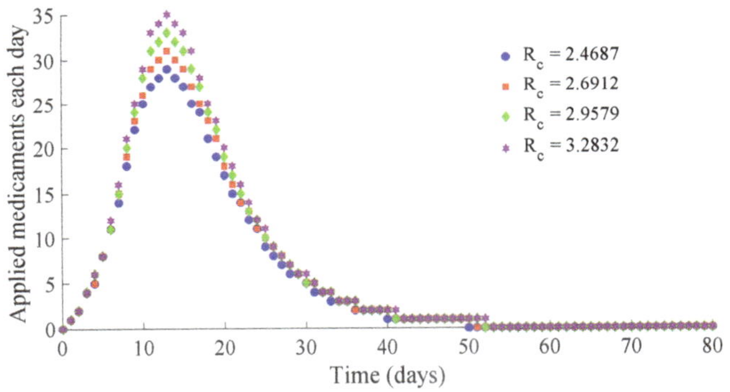

Figure 14. Number of applied medicaments each day for the four considered values of the control parameter c_2.

One can see by analysing the results in Table 1 and Figures 10–14 that the increase of the value of R_c, by decreasing the values of c_2, and maintaining the rest of the control parameters in the prescribed values, gives place to a decrease in the vaccination effort but both the infectious subpopulation and the applied treatment effort increase at the same time. The best of the studied cases from the viewpoint of the vaccination cost is case (iv), but it is not good enough from the viewpoint of both the treatment cost and the number of individuals who experience the infectious disease during the transition from the initial state until reaching the DFE point. In this context, it also seems interesting to examine the influence of the control parameter c_6, which take part in the dynamics of the treatment effort, on the specifications of the transient behaviour.

3.5. Example 5: Study of the Influence of the Control Parameter c_6 on the Behaviour of the Controlled Model

In the current example, the same values for the parameters b, μ, β, γ, ρ, α and q and for the initial condition as those given in Example 4 are considered. Furthermore, the values of $c_1 = 25$, $c_2 = 0.7$, and then $R_c = 3.2832$, $c_3 = 0.0001$ and $c_4 = 0.07$ corresponding with the case of Example 4 with the smallest vaccination cost, concretely case (iv), which requires the application of 467 vaccines as one can see in Table 1, are considered. Such a case is the worst scenario in Example 4 from the viewpoint of the time evolution of the infectious subpopulation and the cost in applied medicaments during the transient from the initial state to the DFE point; namely, 570 medicaments are required, as Table 1 points out. The objective is to analyse the influence of the control parameter c_6, which acts on the value of $\overline{R}_c$, on the specifications of such a transient time interval. Concretely, the value $c_5 = 2$ is going to be considered in four proposed cases while the value for the parameter c_6 will be modified in order to change slightly the ratio $\frac{c_6}{c_5}$. In this way, each case has a different transition rate from the infectious to recovered subpopulation due to the treatment action. Moreover, such a transition rate is added to the transition rate $\gamma = 0.0455\ d^{-1}$ from the infectious to the recovered subpopulation in the absence of treatment, i.e., that derived from the natural immunity system of the infectious individuals to fight against the disease. Such a fact implies a reduction in average for the recovering time of the infectious individuals. The analysed cases are (i) $c_6 = 0.24$, $\overline{R}_c = 3.635$, (ii) $c_6 = 0.27$, $\overline{R}_c = 3.9644$, (iii) $c_6 = 0.3$, $\overline{R}_c = 4.2937$, and (iv) $c_6 = 0.33$, $\overline{R}_c = 4.6231$. Note that $R_c < \overline{R}_c$ as well as the conditions of the Theorems 1, 2, 3 and 4 are satisfied in all the cases so that the DFE point is the unique equilibrium point while being locally and globally asymptotically stable. Note also that case (ii) of this current example is the same as case (iv) of Example 4, which is used as the reference one for the current analysis. In case (i), the recovering time is reduced on average from $\gamma^{-1} = 22$ days, in absence of treatment, to $\left(\gamma + \frac{c_6}{c_5}\right)^{-1} = 6.04$ days if a treatment is applied following the rule (5). For cases (ii), (iii), and (iv), the recovering time passes from 22 to 5.54, 5.12, and 4.75 days, respectively. The values of $S_{DFE} = 242$, $R_{DFE} = 808$, and $v_{DFE} = 7$ are reached at the DFE point in the four analysed cases, as it can be deduced from (10), since all cases have the same values for c_1 and c_2. Figure 15 displays the time evolution of the infectious subpopulation, while Table 2 summarises the specifications for these four studied cases.

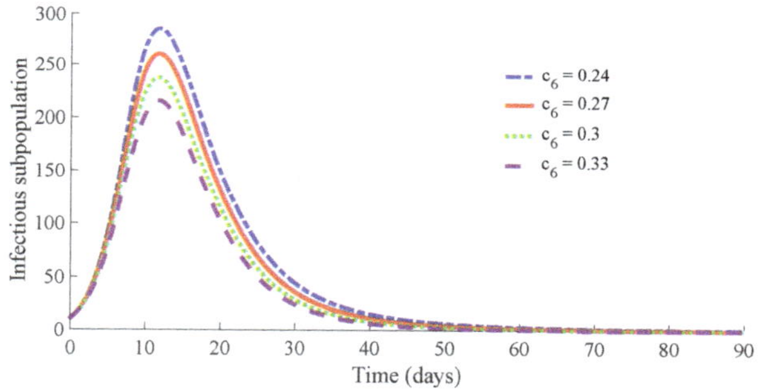

Figure 15. Time evolution of the infectious subpopulation for the four considered values of the control parameter c_6.

Table 2. Specifications of the model behaviour for the four considered values of the parameter c_6.

	c_6 = 0.24 $\overline{R}_c$ = 3.635	c_6 = 0.27 $\overline{R}_c$ = 3.9644	c_6 = 0.3 $\overline{R}_c$ = 4.2937	c_6 = 0.33 $\overline{R}_c$ = 4.6231
Infectious peak	286	261	237	215
Transient duration (days)	87	72	63	58
Vaccination peak	27	27	27	27
Treatment peak	34	35	35	35
Vaccines	525	467	439	428
Medicaments	573	570	565	554

One can see that the peak of the infectious subpopulation is decreasing as the value of c_6 and then also $\overline{R}_c$, increases. In addition, the number of required vaccines and medicaments during the transient decreases as the value for c_6, and then also $\overline{R}_c$, increases. Such a result is mainly due to the fact that the duration of the transient decreases as the value of c_6 increases, since there is a small duration on average for the transition of individuals from infectious to the recovered subpopulation. Note also that the peak in the vaccination effort is equal for the four studied cases, since such a peak depends on the values of the parameters c_1 and c_2, which are the same for the four cases. Case (iv) is the best of the analysed ones in the current study. It requires a strong treatment so that an infectious individual overcomes the infection after an average time of 4.75 days. In such a case, the transient from the initial state until arriving at the DFE point has the following characteristics:

- The evolution of the infectious population reaches a peak of 215 individuals, i.e., approximately 21% of the initial whole population.
- The vaccination cost during the transient supposes 428 vaccines, i.e., the percentage of the susceptible subpopulation to be vaccinated is around of the 43% of the initial population, assuming one vaccine per individual.
- The treatment cost during the transient is of 554 medicaments, i.e., the percentage of infectious subpopulation to be treated is around 55% of the initial population, assuming one medicament per individual.

If the applied treatment has a lower performance so that the average recovering time is 5.54 days, corresponding to case (ii), instead of the 4.75 days of case (iv), then the peak of the infectious population is of 261 individuals, i.e., around the 26% of the initial population. In such a situation, the costs of vaccination and treatment are, respectively, 467 vaccines and 570 medicaments. In the considered worst case, i.e., case (i), which requires an average time of 6.04 days for the transition from the infectious to the recovered subpopulation, the number of necessary vaccines and medicaments is 525 and 573, respectively. As a conclusion, the transient duration and the cost in vaccines and medicaments are related with the performance of the applied treatment. If the applied medicaments have a poor potency such that the reduction of the recovery time is of a few days, then the duration of the transient from the initial state until reaching the DFE point can be very long. In such a case, the transient can be very expensive in relation with the number of required vaccines and medicaments or, even, the disease evolution can converge to the EE point if the resources in the number of medicaments and/or vaccines is not enough.

3.6. Example 6: Study of the Behaviour of the Controlled Model under Vaccination

The objective of this subsection is to study the dynamics of the controlled model when there are not medicaments for treating the infectious subpopulation, and then the only measure to control the propagation of the disease is the planning of a vaccination campaign to the newborns and the susceptible subpopulation. For this purpose, models (1)–(5) with $c_6 = 0$ can be used with the initial condition $t_r(0) = 0$. In this way, $t_r(t) = 0$ for all time. Moreover, $\overline{R}_c = 1$ from (16) such that $R_c < 1$ is required to guarantee the non-existence of the EE point, and then, the DFE point is the unique equilibrium point while being locally and globally asymptotically stable. Since $R_c = \frac{\beta[\mu(1-q)+\rho]}{(\mu+\alpha+\gamma)\left[\mu+\rho+\frac{c_2}{c_1}\right]} = \frac{\mu(1-q)+\rho}{\mu+\rho+\frac{c_2}{c_1}}R_0$, it is necessary that $\frac{c_2}{c_1} > \left(\frac{c_2}{c_1}\right)_{min} = [\mu(1-q)+\rho]R_0 - (\mu+\rho)$ to guarantee that $R_c < 1$, and then, the eradication of the infectious disease can be achieved. Table 3 compares the results obtained for seven different cases of values for the pair c_1 and c_2, satisfying the conditions of the Theorems 1, 2, 3, and 4 so that the solutions of the controlled model are non-negative and the DFE point is the unique equilibrium point while being locally and globally asymptotically stable. The same values for the parameters b, μ, β, γ, ρ, α, q, c_3 and c_4 as those given in Example 2 are used, and the same initial condition is used as well.

Table 3. Specifications of the model behaviour for seven considered pairs of values of the control parameters c_1 and c_2 when there is not treatment to control the disease propagation.

	Infectious Peak	Transient Duration (Days)	Vaccines	DFE Point
Case 1 $c_1 = 25$; $c_2 = 10$ $R_c = 0.2924$	33	120	1869	$S_{DFE} = 22$ $R_{DFE} = 1028$ $v_{DFE} = 9$
Case 2 $c_1 = 22$; $c_2 = 10$ $R_c = 0.258$	29	110	1800	$S_{DFE} = 19$ $R_{DFE} = 1031$ $v_{DFE} = 9$
Case 3 $c_1 = 19$; $c_2 = 10$ $R_c = 0.2233$	25	101	1740	$S_{DFE} = 16$ $R_{DFE} = 1034$ $v_{DFE} = 9$
Case 4 $c_1 = 25$; $c_2 = 15$ $R_c = 0.1963$	22	94	1693	$S_{DFE} = 14$ $R_{DFE} = 1036$ $v_{DFE} = 9$
Case 5 $c_1 = 25$; $c_2 = 17$ $R_c = 0.1735$	20	88	1651	$S_{DFE} = 13$ $R_{DFE} = 1037$ $v_{DFE} = 9$
Case 6 $c_1 = 25$; $c_2 = 20$ $R_c = 0.1477$	18	82	1611	$S_{DFE} = 11$ $R_{DFE} = 1039$ $v_{DFE} = 9$
Case 7 $c_1 = 25$; $c_2 = 25$ $R_c = 0.1184$	16	76	1569	$S_{DFE} = 9$ $R_{DFE} = 1041$ $v_{DFE} = 9$

The results in Table 3 point out that the duration of the transient and also the peak in the infectious subpopulation decreases as the value for the parameter R_c decreases or, equivalently, as the ratio $\frac{c_2}{c_1}$ increases. Such a decrease in the transient duration implies a smaller cost in the number of vaccines as R_c decreases. However, even in the cheapest case shown in Table 3, i.e., case 7, the vaccination cost supposes the use of 1569 vaccines. Such a cost may be decreased even more by considering small values for R_c. In any case, the results show that the fight against the disease when there is not treatment and then, only the vaccination is available, is quite expensive. Moreover, the propagation of the disease can reach the EE point if there are not enough resources regarding vaccines. Note that the average rate for the transition from the susceptible subpopulation to the recovered one is approximately $\frac{c_2}{c_1}$ or, equivalently, the average time in the transition of vaccinated individuals from the susceptible subpopulation to the recovered one is $\frac{c_1}{c_2}$ days. In the cases studied in this subsection, such an average time is constrained between the values of 2.5 days, corresponding to case 1, and 1 day for case 7. Such a fact implies that the effect of the vaccination in the population is quite fast, less than 3 days, in any case. In this context, a more realistic scenario can be analysed by considering, for instance, $c_1 = 25$ and $c_2 = 5$. In such a case, $R_c = 0.5731$, and the average time in the transition from the susceptible to the recovered subpopulation is 5 days. The specifications in the transient for this case are as follows:

- Infectious peak: 92;
- Transient duration: 235 days;
- Number of vaccines: 2559;
- $S_{DFE} = 42$, $R_{DFE} = 1008$ and $v_{DFE} = 8$.

One can see that the transient is very large, as it is the number of vaccines to be applied. In this context, the slower the effect of the vaccination in the population is, the larger the value of the parameter R_c is as well as the duration of the transient and the number of vaccines to guarantee the eradication of the disease. Another alternative to deal with a more realistic scenario in the absence of treatment action could be the inclusion of a delay, corresponding to the reaction time of the vaccines, in the epidemic models (1)–(3).

3.7. Example 7: Study of the Behaviour of the Controlled Model under Treatment

The objective of this subsection is to study the dynamics of the controlled model when there are not vaccines for the newborns and the susceptible subpopulation, and then, the only measure to control the propagation of the disease is the application of some treatment to the infectious subpopulation. For this purpose, models (1)–(5) with $q = c_2 = c_3 = c_4 = 0$ can be used with the initial condition $v(0) = 0$. In this way, $v(t) = 0$ for all time. In this situation, $R_c = R_0 = \frac{\beta}{\mu+\alpha+\gamma}$ from (13), and such a value cannot be modified by control. However, a treatment procedure on the infectious population based on law (5) with appropriate values for the parameters c_5 and c_6 so that $R_c < \overline{R}_c$ can be designed. Since $\overline{R}_c = 1 + \frac{c_6}{c_5(\mu+\alpha+\gamma)}$, it is necessary that $\frac{c_6}{c_5} > \beta - (\mu + \alpha + \gamma)$ to guarantee that $R_c < \overline{R}_c$ and then the non-existence of the EE point and the local and global asymptotic stability of the DFE point, which is the unique one in that case. In this way, the eradication of the infectious disease can be achieved. The current subsection compares the results obtained for some different cases satisfying the conditions of Theorems 1, 2, 3, and 4 so that the solutions of the controlled models are non-negative and the DFE point is the unique equilibrium point while being locally and globally asymptotically stable. Note that $\frac{c_6}{c_5}$ is the rate of transition from the infectious subpopulation to the recovered one by means of treatment of infectious individuals and that such a transition rate is added to the average transition rate γ between such subpopulations in the absence of treatment. By considering the same values for the parameters μ, α, γ and β as those used in Example 1, then $R_c = R_0 = 14.2728$, and the transition rate $\frac{c_6}{c_5}$ from treatment actions has to be strictly larger than $\left(\frac{c_6}{c_5}\right)_{min} = 0.6045$ so that $R_c < \overline{R}_c$, and then, the DFE point is the unique equilibrium point while being locally and globally asymptotically stable. This supposes an average time strictly smaller than $\left(\gamma + \left(\frac{c_6}{c_5}\right)_{min}\right)^{-1} = 1.5387$ days for recovering from the infectious disease. Such a situation is not realistic, since the existence of medicaments with such an extraordinary performance is improbable. By assuming that the non-existence of the EE point is not secured only with the applied treatment to the infectious subpopulation, at least, such a treatment campaign can reduce the effects of the disease in the whole population. Such a result is illustrated by considering several cases, concretely, those studied in Example 5 but without applying the vaccination of either the susceptible subpopulation, i.e., $c_2 = c_3 = c_4 = 0$ with $v(0) = 0$, or the newborns, i.e., $q = 0$. The results are presented in Table 4 below.

Table 4. Specifications of the model behaviour for the four considered pairs of values of the control parameters c_5 and c_6 governing the treatment effort.

	c_5 = 2 c_6 = 0.24 $\overline{R}_c$ = 3.635	c_5 = 2 c_6 = 0.27 $\overline{R}_c$ = 3.9644	c_5 = 2 c_6 = 0.3 $\overline{R}_c$ = 4.2937	c_5 = 2 c_6 = 0.33 $\overline{R}_c$ = 4.6231
Infectious peak	424	395	369	342
Treatment peak	51	53	55	56
S_{EE}	255	280	304	329
I_{EE}	36	32	29	26
R_{EE}	710	694	677	659
N_{EE}	1001	1006	1010	1014
$\frac{I_{EE}}{N_{EE}} \times 100$	3.59%	3.2%	2.87%	2.59%
t_{rEE}	4	4	4	4

One can see the decrease of the value of $\overline{R}_c$ due to the decrease of the value of $\frac{c_6}{c_5}$, which implies an increase of the average time that an infectious individual stays in the infectious subpopulation before passing to the recovered one, leads to a decrease in the number N_{EE} of individuals in the whole population when the EE point is reached while increasing the percentage of infectious individuals at such an equilibrium point. In any case, the application of treatment reduces considerably the effects of the infectious disease if one compares the results of Table 4 with those of Example 1 where no treatment is applied, and then, the whole population at the EE point is of 877 individuals with 127 infectious, i.e., 14.48% of individuals are infectious. Thus, the application of treatment reduces notably the high mortality and maintains the percentage of infectious subpopulation in reasonably small numbers at the achieved EE point.

3.8. Example 8: Study of the Behaviour of the Controlled Model under Different Rates for the Loss of Immunity

The objective of this subsection is to study the dynamics of the controlled model under both control actions, vaccination and treatment, with different values of the rate ρ for the loss of immunity in the recovered subpopulation or, equivalently, different average periods of immunity for the recovered individuals. For this purpose, case (iv) of Example 4 can be taken as the reference one for this analysis, since it is one of the most convenient from the viewpoint of vaccination and treatment costs. The values of the model parameters are $b = 0.0384\ d^{-1}$, $\mu = 3.653 \times 10^{-5}\ d^{-1}$, $\beta = 0.65\ d^{-1}$, $\gamma = \frac{1}{22} = 0.0455\ d^{-1}$, $\rho = \frac{1}{120} = 0.0083\ d^{-1}$ and $\alpha = 5 \times 10^{-5}\ d^{-1}$, while those of the control actions are $q = 0.1$, $c_1 = 25$, $c_2 = 0.7$, $c_3 = 0.0001$, $c_4 = 0.07$, $c_5 = 2$ and $c_6 = 0.27$ in such a reference case. Other values for the parameter ρ are considered to analyse the influence of such a parameter on the transient specifications. Concretely, the following four cases are treated: (i) $\rho = \frac{1}{100}$, $R_c = 3.7647$ (ii) $\rho = \frac{1}{120}$, $R_c = 3.2832$, (iii) $\rho = \frac{1}{150}$, $R_c = 2.7554$, and (iv) $\rho = 0$, $R_c = 0.0167$. Note that case (ii) is used as a reference for this study and also note that the value of ρ has an influence on the value for R_c. In case (i), the average duration of the immunity is 100 days, and that of case (ii) is 120 days, while that of case (iii) is 150 days. Finally, there is no loss of immunity in case (iv). All the cases have the same values for $\overline{R}_c$, namely $\overline{R}_c = 3.9644$, so that $R_c < \overline{R}_c$ and then, the DFE point is locally and globally asymptotically stable for all the analysed cases. The results are presented in Table 5 below.

Table 5. Specifications of the model behaviour for the four considered values of the parameter ρ.

·	$\rho = \frac{1}{100}$	$\rho = \frac{1}{120}$	$\rho = \frac{1}{150}$	$\rho = 0$
Infectious peak	262	261	259	253
Transient duration (days)	91	72	64	49
Vaccination peak	27	27	27	27
Treatment peak	35	35	35	34
S_{DFE}	277	242	203	1
R_{DFE}	773	808	847	1049
v_{DFE}	8	7	6	0
Vaccines	615	467	400	290
Medicaments	589	570	556	512

One can see that the number of necessary vaccines and medicaments in the transient from the initial state to the DFE point decreases when the immunity period increases. Such a fact is mainly due to the transient period decreasing as the immunity period increases. Moreover, the number of recovery individuals at the DFE point increases when the immunity period increases.

4. Conclusions

An SIRS epidemic model with the vaccination of newborns and susceptible individuals and treatment of infectious ones has been investigated. The vaccination and the treatment are governed by a control subsystem containing several free-design parameters. This control subsystem provides two more free-design parameters comparing with those available in the more usual SIRS models with vaccination and treatment. In the proposed model, the intensity of both control actions is not directly proportional to the susceptible and/or infectious subpopulation, as it usually happens in the SIRS models. Here, both actions are provided by the control subsystem, and the parameters defining the dynamics of the controller are also available to shape the vaccination and treatment actions. This control strategy allows the achievement of several relevant results. First, an appropriate adjustment of the control parameters guarantees the positivity of the controller model, as it has been mathematically proved in Theorem 1. Note that such a property is required for coherence in epidemic models where all the subpopulations and the control actions, such as vaccination and treatment, have to be non-negative. Then, the equilibrium points of the proposed controlled epidemic model are calculated as a function of the system parameters. There are two equilibrium points: (i) the DFE point where the infectious subpopulation is zero and then the whole population is composed by susceptible and recovery subpopulations; and (ii) the EE point where the three subpopulations of the model are presented. The DFE one always exists while the existence of the EE point depends on the values of the control parameters. Moreover, the control reproduction number R_c of the controlled epidemic model and a threshold value $\overline{R}_c$ are mathematically obtained as functions of the control parameters. This result allows us to analyse the influence of the control parameters on both values R_c and $\overline{R}_c$. In summary, the values R_c and $\overline{R}_c$ and the existence of the EE point depends on the control parameters. As a consequence, the existence of the EE point can be related with R_c and $\overline{R}_c$. In this sense, values of the control parameters doing $R_c < \overline{R}_c$ guarantee the non-existence of the EE point, as it has been mathematically proved in Theorem 2. In such a situation, the proposed controlled SIRS model only possesses a unique equilibrium point, namely, the DFE point. On the other hand, the local and global asymptotic stability of the DFE point are mathematically proved in Theorems 3 and 4, respectively. Then, an appropriate adjustment of the control parameters so that $R_c < \overline{R}_c$ allows guaranteeing the positivity of the controlled epidemic model while ensuring the non-existence of the EE point. In such a situation, the DFE is the unique equilibrium point of the model, and then, the eradication of the infectious disease can be guaranteed.

Finally, the influence of the control parameters on the time evolution of the disease propagation has been studied by several simulation examples. The first example shows the time evolution of the propagation of the disease without applying control actions. One can see that the model converges to the EE point, and then, the disease is not eradicated. The second example shows the time evolution of the propagation of the disease if the proposed control actions are applied in an effective way, i.e., if the control parameters are chosen such that $R_c < \overline{R}_c$. In this case, one can see that the model converges to the DFE point, and the disease is eradicated. The third example shows the time evolution of the propagation of the disease if the proposed control actions are applied in a defective way, i.e., if the control parameters are chosen such that $R_c > \overline{R}_c$. One can see that the model converges to the EE point, and the disease is not eradicated. The fourth example analyses the influence of the controller parameter c_2 in the evolution of the propagation of the disease. Such a parameter acts in the vaccination control law, and it has an influence on the value of R_c. Concretely, the value of R_c is indirectly proportional to c_2. This example shows that the model converges to the DFE point whenever $R_c < \overline{R}_c$. The number of vaccines and medicaments associated to the control efforts obtained from the control subsystem to achieve the DFE point depends on c_2. One can see that by decreasing the value c_2, or equivalently increasing the value of R_c but guaranteeing that $R_c < \overline{R}_c$, the number of vaccines applied to the susceptible individuals decreases while that of medicaments applied to the infectious individuals increases. The fifth example analyses the influence of the controller parameter c_6 in the evolution of the propagation of the disease. Such a parameter acts in the treatment control law, and it has an influence on the value of $\overline{R}_c$. Concretely, the value of $\overline{R}_c$ is directly proportional to c_6. This example shows that the model converges to the DFE point whenever $R_c < \overline{R}_c$. The number of vaccines and medicaments associated to the control efforts obtained from the control subsystem to achieve the DFE point depends on c_6. One can see that by increasing the value c_6, or equivalently increasing the value of $\overline{R}_c$ but guaranteeing that $R_c < \overline{R}_c$, the number of both vaccines and medicaments applied to the respective subpopulations decreases. Example 6 analyses the behaviour of the model when only the vaccination action is available. Example 7 analyses the behaviour of the model when only the treatment action is available. Example 6 shows that the DFE point is not achieved with an appropriate number of vaccines. Example 7 requires a treatment with a great potency so that the recovery of the infectious individuals after being treated is faster than in a real situation. Finally, Example 8 studies the evolution of the propagation for different rates of loss of immunity. The number of vaccines and medicaments in the convergence of the model to the DFE point decreases as the rate of losing immunity decreases.

Author Contributions: Formal analysis, S.A.-Q. and M.D.l.S.; Funding acquisition, M.D.l.S.; Investigation, S.A.-Q.; Methodology, S.A.-Q.; Resources, R.N.; Software, S.A.-Q. and R.N.; Supervision, M.D.l.S.; Validation, M.D.l.S.; Writing—original draft, S.A.-Q.; Writing—review & editing, M.D.l.S. All authors have read and agreed to the published version of the manuscript.

Funding: This research received support from the Spanish Government through grant RTI2018-094336-B-I00 (MCIU/AEI/FEDER, UE).

Institutional Review Board Statement: Not applicable.

Informed Consent Statement: Not applicable.

Data Availability Statement: All data used are in the references and properly cited within the manuscript.

Acknowledgments: The authors are grateful to the Spanish Government for its support through grant RTI2018-094336-B-I00 (MCIU/AEI/FEDER, UE) and to the Basque Government for its support through grant IT1207-19. They also thank the Spanish Institute of Health Carlos III for its support through Grant COV 20/01213.

Conflicts of Interest: The authors declare no conflict of interest.

Appendix A

First, the non-negativity of $S(t)$ is proved by contradiction. Suppose that the result is false because there exists some $t_s > 0$ such that $S(t) \geq 0$, $I(t) \geq 0$, $R(t) \geq 0$, $v(t) \geq 0$, $t_r(t) \geq 0$ for all $0 \leq t < t_s$ and $S(t_s) < 0$. Equations (1) and (4) can be written as:

$$\begin{bmatrix} \dot{S}(t) \\ \dot{v}(t) \end{bmatrix} = \begin{bmatrix} -\mu - \beta\frac{I(t)}{N(t)} & -1 \\ c_2 + c_4\frac{I(t)}{N(t)} & -c_1 \end{bmatrix} \begin{bmatrix} S(t) \\ v(t) \end{bmatrix} + \begin{bmatrix} b(1-q) + \rho R(t) \\ c_3 I(t) \end{bmatrix}$$
$$= \begin{bmatrix} -(\mu+\beta) & -1 \\ c_2 + c_4 & -c_1 \end{bmatrix} \begin{bmatrix} S(t) \\ v(t) \end{bmatrix} + \begin{bmatrix} b(1-q) + \rho R(t) + \beta\frac{S(t)[S(t)+R(t)]}{N(t)} \\ c_3 I(t) - c_4\frac{S(t)[S(t)+R(t)]}{N(t)} \end{bmatrix} \quad \text{(A1)}$$

where the fact that $I(t) = N(t) - S(t) - R(t)$ has been used. Then, the subsystem (A1) may be compactly written as:

$$\begin{bmatrix} \dot{S}(t) \\ \dot{v}(t) \end{bmatrix} = A^{SV} \begin{bmatrix} S(t) \\ v(t) \end{bmatrix} + \begin{bmatrix} u_1(t) \\ u_2(t) \end{bmatrix} \quad \text{(A2)}$$

with:

$$A^{SV} = \begin{bmatrix} -(\mu+\beta) & -1 \\ c_2 + c_4 & -c_1 \end{bmatrix} \quad \text{(A3)}$$

and:

$$u_1(t) = b(1-q) + \rho R(t) + \beta\frac{S(t)[S(t)+R(t)]}{N(t)} \quad ; \quad u_2(t) = c_3 I(t) - c_4\frac{S(t)[S(t)+R(t)]}{N(t)}. \quad \text{(A4)}$$

From (A2), it follows that:

$$\begin{bmatrix} S(t) \\ v(t) \end{bmatrix} = \phi^{SV}(t) \begin{bmatrix} S(0) \\ v(0) \end{bmatrix} + \int_0^t \phi^{SV}(t-\tau) \begin{bmatrix} u_1(\tau) \\ u_2(\tau) \end{bmatrix} d\tau \quad \forall t \geq 0 \quad \text{(A5)}$$

with $\phi^{SV}(t) = e^{A^{SV}t} = L^{-1}\left\{\left(s\mathbb{I}_2 - A^{SV}\right)^{-1}\right\}$, where L^{-1} denotes the inverse of the Laplace transform, $\mathbb{I}_2$ is the second-order identity matrix and s is the Laplace variable in the complex domain. By direct calculations, one obtains that:

$$\begin{aligned} \phi_{11}^{SV}(t) &= \frac{(c_1+\lambda_1^{SV})e^{\lambda_1^{SV}t}-(c_1+\lambda_2^{SV})e^{\lambda_2^{SV}t}}{\lambda_1^{SV}-\lambda_2^{SV}} \quad ; \quad \phi_{12}^{SV}(t) = -\frac{e^{\lambda_1^{SV}t}-e^{\lambda_2^{SV}t}}{\lambda_1^{SV}-\lambda_2^{SV}} \\ \phi_{21}^{SV}(t) &= \frac{(c_2+c_4)\left(e^{\lambda_1^{SV}t}-e^{\lambda_2^{SV}t}\right)}{\lambda_1^{SV}-\lambda_2^{SV}} \quad ; \quad \phi_{22}^{SV}(t) = \frac{(\mu+\beta+\lambda_1^{SV})e^{\lambda_1^{SV}t}-(\mu+\beta+\lambda_2^{SV})e^{\lambda_2^{SV}t}}{\lambda_1^{SV}-\lambda_2^{SV}} \end{aligned} \tag{A6}$$

where:

$$\lambda_{1,2}^{SV} = \frac{1}{2}\left[-(\mu+\beta+c_1) \pm \sqrt{[c_1-(\mu+\beta)]^2 - 4(c_2+c_4)}\right] \tag{A7}$$

are the eigenvalues of A^{SV}, with λ_1^{SV} corresponding to the sign "+" and λ_2^{SV} corresponding to the sign "−". If the control parameters c_1, c_2, and c_4 fulfil condition (i), then λ_1^{SV} and λ_2^{SV} are real, and $\lambda_2^{SV} < \lambda_1^{SV} < 0$. From (A4) and (A5), it follows that:

$$\begin{aligned} S(t) = {} & \frac{(c_1+\lambda_1^{SV})e^{\lambda_1^{SV}t}-(c_1+\lambda_2^{SV})e^{\lambda_2^{SV}t}}{\lambda_1^{SV}-\lambda_2^{SV}}S(0) - \frac{e^{\lambda_1^{SV}t}-e^{\lambda_2^{SV}t}}{\lambda_1^{SV}-\lambda_2^{SV}}v(0) \\ & + \frac{1}{\lambda_1^{SV}-\lambda_2^{SV}}\int_0^t\left[(c_1+\lambda_1^{SV})e^{\lambda_1^{SV}(t-\tau)} - (c_1+\lambda_2^{SV})e^{\lambda_2^{SV}(t-\tau)}\right]u_1(\tau)d\tau \\ & - \frac{1}{\lambda_1^{SV}-\lambda_2^{SV}}\int_0^t\left[e^{\lambda_1^{SV}(t-\tau)} - e^{\lambda_2^{SV}(t-\tau)}\right]u_2(\tau)d\tau. \end{aligned} \tag{A8}$$

From (A4), (A7), (A8), and the condition $v(0) = 0$, one obtains:

$$\begin{aligned} S(t_s) \geq {} & \frac{(c_1+\lambda_1^{SV})e^{\lambda_1^{SV}t_s}-(c_1+\lambda_2^{SV})e^{\lambda_2^{SV}t_s}}{\lambda_1^{SV}-\lambda_2^{SV}}S(0) \\ & + \frac{b(1-q)}{\lambda_1^{SV}-\lambda_2^{SV}}\int_0^{t_s}\left[(c_1+\lambda_1^{SV})e^{\lambda_1^{SV}(t_s-\tau)} - (c_1+\lambda_2^{SV})e^{\lambda_2^{SV}(t_s-\tau)}\right]d\tau \\ & - \frac{c_3}{\lambda_1^{SV}-\lambda_2^{SV}}\int_0^{t_s}\left[e^{\lambda_1^{SV}(t_s-\tau)} - e^{\lambda_2^{SV}(t_s-\tau)}\right]I(\tau)d\tau, \end{aligned} \tag{A9}$$

where the facts that $S(t) \geq 0$, $I(t) \geq 0$, $R(t) \geq 0$, $N(t) \geq 0$, and $\frac{S(t)}{N(t)} \leq 1$ jointly to $e^{\lambda_1^{SV}t} > e^{\lambda_2^{SV}t}$, for all $0 \leq t < t_s$, provided that condition (i) is satisfied by the control parameters, have been used. Then, from (A9) and by direct calculations, one obtains:

$$\begin{aligned} S(t_s) \geq {} & \frac{(c_1+\lambda_1^{SV})e^{\lambda_1^{SV}t_s}-(c_1+\lambda_2^{SV})e^{\lambda_2^{SV}t_s}}{\lambda_1^{SV}-\lambda_2^{SV}}S(0) \\ & + \frac{b(1-q)}{\lambda_1^{SV}-\lambda_2^{SV}}\left[-\frac{c_1+\lambda_1^{SV}}{\lambda_1^{SV}}\left(1-e^{\lambda_1^{SV}t_s}\right) + \frac{c_1+\lambda_2^{SV}}{\lambda_2^{SV}}\left(1-e^{\lambda_2^{SV}t_s}\right)\right] \\ & - \frac{c_3 I_{max}}{\lambda_1^{SV}-\lambda_2^{SV}}\left[-\frac{1-e^{\lambda_1^{SV}t_s}}{\lambda_1^{SV}} + \frac{1-e^{\lambda_2^{SV}t_s}}{\lambda_2^{SV}}\right], \end{aligned} \tag{A10}$$

where $I_{max} = \max_{0 \leq t < \infty}\{I(t)\} \geq \max_{0 \leq t < t_s}\{I(t)\} \geq 0$. From (A10), it follows that:

$$\begin{aligned} S(t_s) \geq {} & \frac{(c_1+\lambda_1^{SV})\left[|\lambda_1^{SV}|S(0)-b(1-q)\right]+c_3 I_{max}}{(\lambda_1^{SV}-\lambda_2^{SV})|\lambda_1^{SV}|}e^{\lambda_1^{SV}t_s} \\ & - \frac{(c_1+\lambda_2^{SV})\left[|\lambda_2^{SV}|S(0)-b(1-q)\right]+c_3 I_{max}}{(\lambda_1^{SV}-\lambda_2^{SV})|\lambda_2^{SV}|}e^{\lambda_2^{SV}t_s} + \frac{b(1-q)c_1-c_3 I_{max}}{\lambda_1^{SV}\lambda_2^{SV}}. \end{aligned} \tag{A11}$$

Now, by introducing (A7) in (A11) and since $c_1 > \mu+\beta$ from condition (i), direct calculations lead to:

$$\begin{aligned} S(t_s) \geq {} & \frac{[(\mu+\beta)S(0)-b(1-q)]\sqrt{[c_1-(\mu+\beta)]^2-4(c_2+c_4)}}{2(\lambda_1^{SV}-\lambda_2^{SV})}\left(\frac{e^{\lambda_1^{SV}t_s}}{|\lambda_1^{SV}|} + \frac{e^{\lambda_2^{SV}t_s}}{|\lambda_2^{SV}|}\right) + \\ & \frac{[c_1-(\mu+\beta)][(\mu+\beta)S(0)-b(1-q)]+2[(c_2+c_4)S(0)+c_3 I_{max}]}{2(\lambda_1^{SV}-\lambda_2^{SV})}\left(\frac{e^{\lambda_1^{SV}t_s}}{|\lambda_1^{SV}|} - \frac{e^{\lambda_2^{SV}t_s}}{|\lambda_2^{SV}|}\right) + \frac{b(1-q)(\mu+\beta)-c_3 I_{max}}{\lambda_1^{SV}\lambda_2^{SV}}. \end{aligned} \tag{A12}$$

Then, $S(t_s) \geq 0$ from the conditions (i)–(iii) jointly with the facts that $\lambda_2^{SV} < \lambda_1^{SV} < 0$ and $e^{\lambda_1^{SV} t_s} > e^{\lambda_2^{SV} t_s}$. Such a result contradicts the existence of a time instant $t_s > 0$ such that $S(t_s) < 0$. Then, the non-negativity of $S(t)$ is proved. Now, the non-negativity of $I(t)$ is proved by contradiction. Suppose that the result is false because there exists some $t_I > 0$ such that $S(t) \geq 0$, $I(t) \geq 0$, $R(t) \geq 0$, $v(t) \geq 0$, $t_r(t) \geq 0$ for all $0 \leq t < t_I$ and $I(t_I) < 0$. Then, Equations (2) and (5) can be compactly written as:

$$\begin{bmatrix} \dot{I}(t) \\ \dot{t}_r(t) \end{bmatrix} = A^{IT} \begin{bmatrix} I(t) \\ t_r(t) \end{bmatrix} + \begin{bmatrix} 1 \\ 0 \end{bmatrix} u_3(t) \tag{A13}$$

where $u_3(t) = \beta \frac{S(t)I(t)}{N(t)}$ and:

$$A^{IT} = \begin{bmatrix} -(\mu+\alpha+\gamma) & -1 \\ c_6 & -c_5 \end{bmatrix}. \tag{A14}$$

From (A14), it follows that:

$$\begin{bmatrix} I(t) \\ t_r(t) \end{bmatrix} = \phi^{IT}(t) \begin{bmatrix} I(0) \\ t_r(0) \end{bmatrix} + \int_0^t \phi^{IT}(t-\tau) \begin{bmatrix} 1 \\ 0 \end{bmatrix} u_3(\tau) d\tau \quad \forall t \geq 0, \tag{A15}$$

with $\phi^{IT}(t) = e^{A^{IT} t} = L^{-1}\left\{ \left(s\mathbb{I}_2 - A^{IT} \right)^{-1} \right\}$. By direct calculations, one obtains that:

$$\begin{array}{ll} \phi_{11}^{IT}(t) = \frac{(c_5+\lambda_1^{IT}) e^{\lambda_1^{IT} t} - (c_5+\lambda_2^{IT}) e^{\lambda_2^{IT} t}}{\lambda_1^{IT} - \lambda_2^{IT}} \quad ; \quad \phi_{12}^{IT}(t) = -\frac{e^{\lambda_1^{IT} t} - e^{\lambda_2^{IT} t}}{\lambda_1^{IT} - \lambda_2^{IT}} \\ \phi_{21}^{IT}(t) = \frac{c_6 \left(e^{\lambda_1^{IT} t} - e^{\lambda_2^{IT} t} \right)}{\lambda_1^{IT} - \lambda_2^{IT}} \quad ; \quad \phi_{22}^{IT}(t) = \frac{(\mu+\alpha+\gamma+\lambda_1^{IT}) e^{\lambda_1^{IT} t} - (\mu+\alpha+\gamma+\lambda_2^{IT}) e^{\lambda_2^{IT} t}}{\lambda_1^{IT} - \lambda_2^{IT}} \end{array} \tag{A16}$$

where:

$$\lambda_{1,2}^{IT} = \frac{1}{2}\left[-(\mu+\alpha+\gamma+c_5) \pm \sqrt{[c_5 - (\mu+\alpha+\gamma)]^2 - 4c_6} \right] \tag{A17}$$

are the eigenvalues of A^{IT}, with λ_1^{IT} corresponding with the sign "+" and λ_2^{IT} corresponding with the sign "−". If the control parameters c_5 and c_6 fulfill condition (iv), then both eigenvalues are real, and $\lambda_2^{IT} < \lambda_1^{IT} < 0$. From (A15) and (A16), it follows that:

$$\begin{array}{l} I(t_I) = \frac{(c_5+\lambda_1^{IT}) e^{\lambda_1^{IT} t_I} - (c_5+\lambda_2^{IT}) e^{\lambda_2^{IT} t_I}}{\lambda_1^{IT} - \lambda_2^{IT}} I(0) - \frac{e^{\lambda_1^{IT} t_I} - e^{\lambda_2^{IT} t_I}}{\lambda_1^{IT} - \lambda_2^{IT}} t_r(0) \\ + \frac{\beta}{\lambda_1^{IT} - \lambda_2^{IT}} \int_0^{t_I} \left[(c_5+\lambda_1^{IT}) e^{\lambda_1^{IT}(t_I-\tau)} - (c_5+\lambda_2^{IT}) e^{\lambda_2^{IT}(t_I-\tau)} \right] \frac{S(\tau)I(\tau)}{N(\tau)} d\tau. \end{array} \tag{A18}$$

Condition (iv) about the control parameters c_5 and c_6 implies that $c_5 + \lambda_1^{IT} > 0$. Under such a condition jointly with $t_r(0) = 0$, one obtains that $I(t_I) \geq 0$ from (A18) since $e^{\lambda_1^{IT} t_I} > e^{\lambda_2^{IT} t_I} > 0$ and $\frac{S(t)I(t)}{N(t)} \geq 0$ for all $0 \leq t < t_I$. Then, such a result contradicts the existence of a time instant $t_I > 0$ such that $I(t_I) < 0$. Then, the non-negativity of $I(t)$ is proved. Now, the non-negativity of $R(t)$ is proved by contradiction. Suppose that the result is false because there exists some $t_R > 0$ such that $S(t) \geq 0$, $I(t) \geq 0$, $R(t) \geq 0$, $v(t) \geq 0$, and $t_r(t) \geq 0$ for all $0 \leq t < t_R$ and $R(t_R) < 0$. From (3), it follows that:

$$R(t) = e^{-(\mu+\rho)t} R(0) + \int_0^t e^{-(\mu+\rho)(t-\tau)} [bq + \gamma I(\tau) + v(\tau) + t_r(\tau)] d\tau \quad \forall t \geq 0. \tag{A19}$$

Then, it follows that $R(t_R) \geq 0$ for any $R(0) \geq 0$. Such a result contradicts the existence of a time instant $t_R > 0$ such that $R(t_R) < 0$. Then, the non-negativity of $R(t)$ is proved. Now, the non-negativity of $v(t)$ is proved by contradiction. Suppose that the result is false because there exists some $t_V > 0$ such that $S(t) \geq 0$, $I(t) \geq 0$, $R(t) \geq 0$, $v(t) \geq 0$, $t_r(t) \geq 0$ for all $0 \leq t < t_V$ and $v(t_V) < 0$. From (4), it follows that:

$$v(t) = e^{-c_1 t}v(0) + \int_0^t e^{-c_1(t-\tau)}\left[c_2 S(\tau) + c_3 I(\tau) + c_4 \frac{S(\tau)I(\tau)}{N(\tau)}\right]d\tau \quad \forall t \geq 0. \qquad \text{(A20)}$$

Then, it follows that $v(t_V) \geq 0$ for any $v(0) \geq 0$. Such a fact contradicts the existence of a time instant $t_V > 0$ such that $v(t_V) < 0$. Then, the non-negativity of $v(t)$ is proved. Finally, the non-negativity of $t_r(t)$ is proved by contradiction. Suppose that the result is false because there exists some $t_T > 0$ such that $S(t) \geq 0$, $I(t) \geq 0$, $R(t) \geq 0$, $v(t) \geq 0$, $t_r(t) \geq 0$ for all $0 \leq t < t_T$ and $t_r(t_T) < 0$. From (5), it follows that:

$$t_r(t) = e^{-c_5 t}t_r(0) + c_6 \int_0^t e^{-c_5(t-\tau)}I(\tau)d\tau \quad \forall t \geq 0. \qquad \text{(A21)}$$

Then, it follows that $t_r(t_T) \geq 0$ for any $t_r(0) \geq 0$, since $I(t) \geq 0\ \forall t \geq 0$. Such a result contradicts the existence of a time instant $t_T > 0$ such that $t_r(t_T) < 0$. Then, the non-negativity of $t_r(t)$ is proved. In summary, none of the controlled epidemic model variables take negative values.

References

1. Kermack, W.O.; McKendrick, A.G. Contributions to the mathematical theory of epidemics, part 1. *Proc. R. Soc. A* **1927**, *115*, 700–721.
2. Li, J.; Ma, Z. Global analysis of SIS epidemic models with variable total population size. *Math. Comput. Model.* **2004**, *39*, 1231–1242.
3. Pang, J.; Cui, J.; Zhou, X. Dynamical behavior of a hepatitis B virus transmission model with vaccination. *J. Theor. Biol.* **2010**, *265*, 572–578. [CrossRef]
4. Meng, X.; Chen, L.; Cheng, H. Two profitless delays for the SEIRS epidemic disease model with nonlinear incidence and pulse vaccination. *Appl. Math. Comput.* **2007**, *186*, 516–529. [CrossRef]
5. Cai, L.; Li, X.; Ghosh, M.; Guo, B. Stability analysis of an HIV/AIDS epidemic model with treatment. *J. Comput. Appl. Math.* **2009**, *229*, 313–323. [CrossRef]
6. De la Sen, M.; Alonso-Quesada, S.; Ibeas, A. On the stability of an SEIR epidemic model with distributed time-delay and a general class of feedback vaccination rules. *Appl. Math. Comput.* **2015**, *270*, 953–976. [CrossRef]
7. Trawicki, M.B. Deterministic SEIRS epidemic model for modeling vital dynamics, vaccinations, and temporary immunity. *Mathematics* **2017**, *5*, 7. [CrossRef]
8. Gbadamosi, B.; Ojo, M.M.; Oke, S.I.; Matadi, M.B. Qualitative analysis of a Dengue fever model. *Math. Comput. Appl.* **2018**, *23*, 33. [CrossRef]
9. Tulu, T.W.; Tian, B.; Wu, Z. Modeling the effect of quarantine and vaccination on Ebola disease. *Adv. Differ. Equ.* **2017**, *2017*, 178. [CrossRef]
10. Buonomo, B.; Della Marca, R. Oscillations and hysteresis in an epidemic model with information-dependent imperfect vaccination. *Math. Comput. Simul.* **2019**, *162*, 97–114. [CrossRef]
11. Ullah, S.; Khan, M.A.; Farooq, M.; Gul, T. Modeling and analysis of tuberculosis (TB) in Khyber Pakhtunkhwa, Pakistan. *Math. Comput. Simul.* **2019**, *165*, 181–199. [CrossRef]
12. Liu, Q.; Jiang, D.; Hayat, T.; Alsaedi, A. Dynamics of a stochastic multigroup SIQR epidemic model with standard incidence rates. *J. Frankl. Inst.* **2019**, *356*, 2960–2993. [CrossRef]
13. Nkamba, L.N.; Ntaganda, J.M.; Abboubakar, H.; Kamgang, J.C.; Castelli, L. Global stability of a SVEIR epidemic model: Application to poliomyelitis transmission dynamics. *Open J. Model. Simul.* **2017**, *5*, 98–112. [CrossRef]
14. Zaman, G.; Kang, Y.H.; Cho, G.; Jung, I.H. Optimal strategy of vaccination & treatment in an SIR epidemic model. *Math. Comput. Simul.* **2017**, *136*, 63–77.
15. Alonso-Quesada, S.; De la Sen, M.; Ibeas, A. On the discretization and control of an SEIR epidemic model with a periodic impulsive vaccination. *Commun. Nonlinear Sci. Numer. Simul.* **2017**, *42*, 247–274. [CrossRef]
16. Ling, L.; Jiang, G.; Long, T. The dynamics of an SIS epidemic model with fixed-time birth pulses and state feedback pulse treatments. *Appl. Math. Model.* **2015**, *39*, 5579–5591. [CrossRef]
17. Huang, H.; Chen, Y.; Yan, Z. Impacts of social distancing on the spread of infectious diseases with asymptomatic infection: A mathematical model. *Appl. Math. Comput.* **2021**, *398*, 1–13. [CrossRef]
18. Nadim, S.S.; Ghosh, I.; Chattopadhyay, J. Short-term predictions and prevention strategies for COVID-19: A model based study. *Appl. Math. Comput.* **2021**, *404*, 1–19. [CrossRef]
19. Han, D.; Shao, Q.; Li, D.; Sun, M. How the individuals' risk aversion affect the epidemic spreading. *Appl. Math. Comput.* **2020**, *369*, 1–10. [CrossRef]

20. Tanimoto, J. *Sociophysics Approach to Epidemics*; Springer: Singapore, 2021.
21. Kabir, K.M.A.; Tanimoto, J. Modelling and analysing the coexistence of dual dilemmas in the proactive vaccination game and retroactive treatment game in epidemic viral dynamics. *Proc. R. Soc. A* **2019**, *475*, 1–20. [CrossRef]
22. Moualeu, D.P.; Weiser, M.; Ehrig, R.; Deuflhard, P. Optimal control for a tuberculosis model with undetected cases in Cameroon. *Commun. Nonlinear Sci. Numer. Simul.* **2015**, *20*, 986–1003. [CrossRef]
23. Sharma, S.; Singh, F. Bifurcation and stability analysis of a cholera model with vaccination and saturated treatment. *Chaos Solitons Fractals* **2021**, *146*, 1–15. [CrossRef]
24. den Driessche, P.V.; Watmough, J. Reproduction numbers and sub-threshold endemic equilibria for compartmental models of disease transmission. *Math. Biosci.* **2002**, *180*, 29–48. [CrossRef]
25. Zhou, L.; Fan, M. Dynamics of an SIR epidemic model with limited medical resources revisited. *Nonlinear Anal. Real World Appl.* **2012**, *13*, 312–324. [CrossRef]
26. Dumont, Y.; Chiroleu, F.; Domerg, C. On a temporal model for the Chikungunea disease: Modeling, theory and numerics. *Math. Biosci.* **2008**, *213*, 80–91. [CrossRef] [PubMed]
27. Mwasunda, J.A.; Irunde, J.I.; Kajunguri, D.; Kuznetsov, D. Modeling and analysis of taeniasis and cysticercosis transmission dynamics in human, pigs and cattle. *Adv. Differ. Equ.* **2021**, *2021*, 176. [CrossRef]
28. Vidyasagar, M. *Nonlinear Systems Analysis*, 2nd ed.; Prentice Hall: Englewood Cliffs, NJ, USA, 1993.

mathematics

MDPI

Article

Global Stability Analysis of a Five-Dimensional Unemployment Model with Distributed Delay

Eva Kaslik [1,2,*,†], Mihaela Neamţu [3,*,†] and Loredana Flavia Vesa [1,2,*,†]

1 Department of Mathematics and Computer Science, West University of Timişoara, 300223 Timişoara, Romania
2 Institute for Advanced Environmental Research, West University of Timişoara, Bd. V. Pârvan nr. 4, 300223 Timişoara, Romania
3 Department of Economics and Business Administration, West University of Timişoara, 300223 Timişoara, Romania
* Correspondence: eva.kaslik@e-uvt.ro (E.K.); mihaela.neamtu@e-uvt.ro (M.N.); loredana.vesa@e-uvt.ro (L.F.V.)
† These authors contributed equally to this work.

Abstract: The present paper proposes a five-dimensional mathematical model for studying the labor market, focusing on unemployment, migration, fixed term contractors, full time employment and the number of available vacancies. The distributed time delay is considered in the rate of change of available vacancies that depends on the past regular employment levels. The non-dimensional mathematical model is introduced and the existence of the equilibrium points is analyzed. The positivity and boundedness of solutions are provided and global asymptotic stability findings are presented both for the employment free equilibrium and the positive equilibrium. The numerical simulations support the theoretical results.

Keywords: unemployment model; global stability; Lyapunov function; distributed delay

Citation: Kaslik, E.; Neamţu M.; Vesa, L.F. Global Stability Analysis of a Five-Dimensional Unemployment Model with Distributed Delay. *Mathematics* **2021**, *9*, 3037. https://doi.org/10.3390/math9233037

Academic Editor: Snezhana Hristova

Received: 11 November 2021
Accepted: 23 November 2021
Published: 26 November 2021

Publisher's Note: MDPI stays neutral with regard to jurisdictional claims in published maps and institutional affiliations.

Copyright: © 2021 by the authors. Licensee MDPI, Basel, Switzerland. This article is an open access article distributed under the terms and conditions of the Creative Commons Attribution (CC BY) license (https://creativecommons.org/licenses/by/4.0/).

1. Introduction

Over the time, one of the many challenges that a country faces is economical downturn fuelled by unemployment. The causes of appearance of this phenomena are countless and are based on country specificity. Problems ranging from the increasing population to slow economic growth are some of the leading factors linked to unemployment spanning out of control. One of the negative impact of unemployment pertains to the social side that puts the affected population at high risk, such as psychological and mental health problems, just to name a few [1]. Any responsible government cannot ignore the signals coming from the labour market and should take appropriate and immediate actions to improve the overall situation.

Therefore, the need to manage and handle the tendency of unwanted spread of unemployment brings about the necessity of studying the behaviour of complex mathematical models in conjunction with numerical simulations. In order to comprehend the dynamics of unemployment, using some concepts from [2], the following variables have been considered in a previous mathematical model investigated in [3,4]: number of unemployed persons, employed persons and new vacancies; time delay has been incorporated into the rate of change for creation of new vacancies. Moreover, by resorting to the skill development programs, Misra et al. [5] demonstrated the link between the betterment of workers' capabilities and the reduction of unemployment. The effect of training programs has also been studied in [6]. Furthermore, Harding and Neamţu [7] considered the migration as a contributing factor when defining policies pertaining to unemployment, including distributed time delay. The optimal control analysis has been completed in [8,9].

The motivation of the present paper is centered around the existing mathematical models, enabling the development of new ways for studying unemployment, based on the past history of the state variables. The potential impact of our findings is related to upcoming biological and economical mathematical models connected to population dynamics. The most important benefit could be related to the strategic policy makers of our society.

The novelty of this paper is the investigation of the interaction among the number of unemployed persons, immigrants, temporary employed persons, regularly employed persons and the number of available vacancies, in the set up of delay differential equations. The distributed time delay has been incorporated to reflect the dependence of the rate of change of available vacancies on the past regular employment levels. Basically, this type of delay is highlighted for acquiring the realistic sides of the economic process [10–16]. It is important to emphasize that there are many mathematical models including distributed time delays, originating from population biology, epidemiology and economics [17–33].

The paper is structured as follows: the mathematical model and its non-dimensional version are introduced in Section 2; the existence of the equilibrium points of the model is discussed in Section 3 and the positivity and boundedness of solutions are proved in Section 4; global asymptotic stability results are proved for the employment free equilibrium and the positive equilibrium in Sections 5 and 6, respectively; Section 7 includes numerical simulations to highlight the theoretical findings, followed by conclusions which are formulated in Section 8.

2. The Mathematical Model and Its Non-Dimensional Version

We consider the following five variables which describe the mathematical model for to control the unemployment: the number of unemployed persons $U(t)$, the number of immigrants $M(t)$, the number of temporary employed persons $T(t)$, the number of regularly employed persons $R(t)$, respectively and the number of available vacancies $V(t)$ at time t.

The system of differential equations is:

$$\begin{cases} \dot{U}(t) = a_1 - a_2U(t)V(t) + a_3R(t) - a_4U(t) + a_5T(t) - b_1U(t) \\ \dot{M}(t) = m_1 - m_2M(t)V(t) - b_2M(t) \\ \dot{T}(t) = a_4U(t) - a_5T(t) - c_1T(t)V(t) - b_3T(t) \\ \dot{R}(t) = a_2U(t)V(t) + m_2M(t)V(t) + c_1T(t)V(t) - a_3R(t) - b_4R(t) \\ \dot{V}(t) = b_4\int_0^\infty h(s)R(t-s)ds - b_5V(t) \end{cases} \tag{1}$$

where a_1, a_2, a_3, a_4 and a_5 are positive constants: $a_1$the constant growth rate of unemployed persons entering the labor market, a_2 the rate of hiring, a_3 the rate of firing, a_4 the rate of move to unemployment, a_5 the rate of move to temporary employment; b_1 the rate of migration of unemployed persons; b_2 is the rate of return or death, b_3 the rate of temporary employment, b_4 the rate of retirement, migration or death of employed persons; b_5 the rate of available vacancies; c_1 is the rate of good job finding; m_1 represent the exogenous increase in migration, m_2 is the migrant employment rate.

The bounded, piecewise continuous function $h : [0,\infty) \to [0,\infty)$ is the delay kernel with average time delay τ for available vacancies based on past employment levels. Hence, h is a probability density function satisfying the following properties:

$$\int_0^\infty h(s)ds = 1\,, \quad \tau = \int_0^\infty sh(s)ds < \infty. \tag{2}$$

With the changes of variables:

$$\begin{cases} x_1(t) = \frac{a_2 b_4}{a_5^2} U\left(\frac{t}{a_5}\right) \\ x_2(t) = \frac{a_2 b_4}{a_5^2} M\left(\frac{t}{a_5}\right) \\ x_3(t) = \frac{a_2 b_4}{a_5^2} T\left(\frac{t}{a_5}\right) \\ x_4(t) = \frac{a_2 b_4}{a_5^2} R\left(\frac{t}{a_5}\right) \\ x_5(t) = \frac{a_2}{a_5} V\left(\frac{t}{a_5}\right), \end{cases} \quad (3)$$

system (1) becomes a non-dimensional system, with fewer dimensionless parameters and dimensionless state variables:

$$\begin{cases} \dot{x}_1(t) = \gamma_1 - x_1(t)x_5(t) + \alpha_3 x_4(t) - \alpha_4 x_1(t) + x_3(t) - \beta_1 x_1(t), \\ \dot{x}_2(t) = \gamma_2 - \alpha_2 x_2(t)x_5(t) - \beta_2 x_2(t), \\ \dot{x}_3(t) = \alpha_4 x_1(t) - x_3(t) - \alpha_1 x_3(t)x_5(t) - \beta_3 x_3(t), \\ \dot{x}_4(t) = x_1(t)x_5(t) + \alpha_2 x_2(t)x_5(t) + \alpha_1 x_3(t)x_5(t) - \alpha_3 x_4(t) - \beta_4 x_4(t), \\ \dot{x}_5(t) = \int_0^\infty k(s)x_4(t-s)ds - \beta_5 x_5(t), \end{cases} \quad (4)$$

where the delay kernel is $k(s) = \frac{1}{a_5} h\left(\frac{s}{a_5}\right)$ and the coefficients are expressed as

$$\gamma_1 = \frac{a_1 a_2 b_4}{a_5^3}, \quad \gamma_2 = \frac{a_2 b_4 m_1}{a_5^3}$$

$$\alpha_1 = \frac{c_1}{a_2}, \quad \alpha_2 = \frac{m_2}{a_2}, \quad \alpha_3 = \frac{a_3}{a_5}, \quad \alpha_4 = \frac{a_4}{a_5}$$

$$\beta_1 = \frac{b_1}{a_5}, \quad \beta_2 = \frac{b_2}{a_5}, \quad \beta_3 = \frac{b_3}{a_5}, \quad \beta_4 = \frac{b_4}{a_5}, \quad \beta_5 = \frac{b_5}{a_5}$$

Initial conditions for system (4) are considered of the form

$$x_i(\theta) = \varphi_i(\theta) \quad , \forall\, \theta \in (-\infty, 0],\ \forall\, i = \overline{1,5},$$

where φ_i belong to the Banach space $C_{0,\mu}(\mathbb{R}_-, \mathbb{R})$ (where $\mu > 0$) of continuous real valued functions defined on $(-\infty, 0]$ such that $\lim_{t \to -\infty} e^{\mu t}\varphi(t) = 0$, considered with respect to the norm:

$$\|\varphi\|_{\infty,\mu} = \sup_{t \in (-\infty,0]} e^{\mu t}|\varphi(t)|.$$

The existence and uniqueness of solutions, as well as the continuous dependence of solutions on initial conditions, in the framework of the distributed delay system (4), are guaranteed by theoretical results presented in [34].

3. Equilibrium Points of the Model

The equilibrium points are constant solutions of system (4) and hence they satisfy the following algebraic equations:

$$\begin{cases} \gamma_1 - x_1 x_5 + \alpha_3 x_4 - \alpha_4 x_1 + x_3 - \beta_1 x_1 = 0 \\ \gamma_2 - \alpha_2 x_2 x_5 - \beta_2 x_2 = 0 \\ \alpha_4 x_1 - x_3 - \alpha_1 x_3 x_5 - \beta_3 x_3 = 0 \\ x_1 x_5 + \alpha_2 x_2 x_5 + \alpha_1 x_3 x_5 - \alpha_3 x_4 - \beta_4 x_4 = 0 \\ x_4 - \beta_5 x_5 = 0 \end{cases} \quad (5)$$

On one hand, we observe that system (5) has at least one solution, which corresponds to the case when $x_5 = 0$. Hence, we obtain the equilibrium point S^0 given by:

$$S^0 := (\delta_1(1+\beta_3), \delta_2, \delta_1\alpha_4, 0, 0)$$

where

$$\delta_1 = \frac{\gamma_1}{\beta_1 + \beta_1\beta_3 + \alpha_4\beta_3} \quad \text{and} \quad \delta_2 = \frac{\gamma_2}{\beta_2}.$$

This equilibrium point corresponds to the state of no regular employment and no available vacancies.

We now introduce the basic reproduction number R_0, which has the role of a threshold parameter that prognosticates whether the unemployment, immigration and temporary employment problems will increase or decrease. Using the next generation matrix method, we deduce:

$$R_0 = \frac{\delta_1(1+\beta_3+\alpha_1\alpha_4) + \delta_2\alpha_2}{(\alpha_3+\beta_4)\beta_5}$$

On the other hand, system (5) has at least one solution with the last component $x_5 > 0$, if and only if x_5 is the solution of the following cubic equation:

$$E_3x_5^3 + E_2x_5^2 + E_1x_5 + E_0 = 0, \tag{6}$$

where

$$\begin{aligned}
E_3 &= \alpha_1\beta_4\beta_5 \\
E_2 &= \alpha_1(\nu_2\beta_4\beta_5 - \gamma_1 - \gamma_2) + \beta_4\beta_5(1+\beta_3+\alpha_1\alpha_4) + \alpha_1\beta_1\mu \\
E_1 &= (\nu_2\beta_4\beta_5 - \gamma_1 - \gamma_2)(1+\beta_3+\alpha_1\alpha_4) - \nu_2\gamma_1\alpha_1 + \mu\beta_1\left(1+\beta_3+\alpha_1\alpha_4\frac{\nu_1}{\beta_1}\right) + (\mu\nu_2 - \gamma_2)\beta_1\alpha_1 \\
E_0 &= \frac{\mu\gamma_1\gamma_2}{\alpha_2\delta_1\delta_2}(1-R_0),
\end{aligned}$$

where R_0, δ_1 and δ_2 are given above and

$$\nu_1 = \frac{\beta_3}{\alpha_1}, \qquad \nu_2 = \frac{\beta_2}{\alpha_2}, \qquad \mu = (\alpha_3+\beta_4)\beta_5.$$

In this case, system (4) has at least one equilibrium point S^+ of the form

$$S^+ := \left(-\frac{Q(x_5)+\nu_1 d(x_5)}{(\beta_1-\nu_1)(x_5+\nu_2)}, \frac{\gamma_2}{\alpha_2(\nu_2+x_5)}, \frac{Q(x_5)+\beta_1 d(x_5)}{\alpha_1(\beta_1-\nu_1)(\nu_2+x_5)}, \beta_5x_5, x_5\right)$$

where

$$\begin{aligned}
Q(x_5) &= (\beta_4\beta_5x_5 - \gamma_1 - \gamma_2)(x_5+\nu_2) + \gamma_2\nu_2, \\
d(x_5) &= \mu(x_5+\nu_2) - \gamma_2.
\end{aligned}$$

In fact, we distinguish two cases:

Case 1: If $R_0 > 1$ then $E_0 < 0$ and system (4) has at least one positive equilibrum point. Moreover, if either $E_1 < 0$, or $E_1 > 0$ and $E_2 > 0$, Descartes' rule of signs guarantees the existence of a unique positive equilibrium point S^+.

Case 2: If $R_0 < 1$ then $E_0 > 0$, we may have either two or zero positive equilibrium points. For instance, if $E_1 > 0$ and $E_2 > 0$, system (4) does not have positive equilibrium points.

4. Positivity and Boundedness of Solutions

Theorem 1. *The open positive (nonnegative) orthant* $\mathbb{R}^5_+ = (0,\infty)^5$ *is invariant to the flow of system* (4)*. Furthermore, denoting* $\beta_m = \min(\beta_1, \beta_2, \beta_3, \beta_4)$*, the set*

$$\Omega = \left\{ (x_1, x_2, x_3, x_4, x_5) : 0 \le x_1 + x_2 + x_3 + x_4 \le \frac{\gamma_1 + \gamma_2}{\beta_m},\ 0 \le x_5 \le \frac{\gamma_1 + \gamma_2}{\beta_m \beta_5} \right\}$$

is a region of attraction for the system (1)*, attracting all the solutions with initial conditions in the open positive orthant* $\mathbb{R}^5_+$.

Proof. As a first step, we will show that the open positive (nonnegative) orthant of $\mathbb{R}^5$ is invariant to the flow of system (4). Considering initial conditions $\varphi_i : (-\infty, 0] \to (0,\infty)$, for $i = \overline{1,5}$, based on the continuity of the solutions, there exists $T > 0$ such that $x_i(t) > 0$, for any $t \in (0,T)$ and any $i = \overline{1,5}$.

Therefore, the fifth equation of system (4) provides

$$\dot{x}_5(t) \ge -\beta_5 x_5(t), \quad \forall\, t \in (0,T)$$

and an integration over the interval $(0,T)$ leads to:

$$x_5(T) \ge \varphi_5(0) e^{-b_5 T} > 0.$$

In a similar manner, it can be show that $x_i(T) > 0$, for any $i = \overline{1,4}$.

Moreover, from the first four equations of system (4) we observe that

$$\frac{d}{dt}[x_1(t) + x_2(t) + x_3(t) + x_4(t)] = \gamma_1 + \gamma_2 - \beta_1 x_1(t) - \beta_2 x_2(t) - \beta_3 x_3(t) - \beta_4 x_4(t),$$

and taking into account that $\beta_m = \min(\beta_1, \beta_2, \beta_3, \beta_4)$ we have

$$\frac{d}{dt}[x_1(t) + x_2(t) + x_3(t) + x_4(t)] \le \gamma_1 + \gamma_2 - \beta_m[x_1(t) + x_2(t) + x_3(t) + x_4(t)], \quad \forall\, t > 0.$$

Therefore, using basic differential inequality techniques, we obtain:

$$x_1(t) + x_2(t) + x_3(t) + x_4(t) \le e^{-\beta_m t}\left(x_1(0) + x_2(0) + x_3(0) + x_4(0) - \frac{\gamma_1 + \gamma_2}{\beta_m} \right) + \frac{\gamma_1 + \gamma_2}{\beta_m},$$

for any $t \ge 0$.

On one hand, if $x_1(0) + x_2(0) + x_3(0) + x_4(0) \le \frac{\gamma_1+\gamma_2}{\beta_m}$ it follows that $(x_1(0) + x_2(0) + x_3(0) + x_4(0) \le \frac{\gamma_1+\gamma_2}{\beta_m}$, for any $t > 0$. Otherwise, if $x_1(0) + x_2(0) + x_3(0) + x_4(0) > \frac{\gamma_1+\gamma_2}{\beta_m}$, we have

$$\limsup_{t\to\infty}[x_1(t) + x_2(t) + x_3(t) + x_4(t)] \le \frac{\gamma_1 + \gamma_2}{\beta_m}.$$

Solving the last equation of (4) for $x_5(t)$ leads to

$$x_5(t) = e^{-\beta_5 t} x_5(0) + e^{-\beta_5 t} \int_0^t e^{\beta_5 u} \left(\int_0^\infty k(s) x_4(u-s) ds \right) du.$$

Using the generalized l'Hospital rule (see Lemma 1.1 in [35]) in conjunction with Lemma 1 from [4], we deduce:

$$\begin{aligned}\limsup_{t\to\infty} x_5(t) &\le \frac{1}{\beta_5}\limsup_{t\to\infty}\int_0^\infty k(s)x_4(t-s)ds\\ &= \frac{1}{\beta_5}\limsup_{t\to\infty}\left(\int_0^t k(s)x_4(t-s)ds + \int_t^\infty k(s)\varphi_4(t-s)ds\right)\\ &\le \frac{1}{\beta_5}\left(\limsup_{t\to\infty} x_4(t) + \|\varphi_4\|_{\infty,\mu}\limsup_{t\to\infty}\int_t^\infty k(s)e^{-\mu(t-s)}ds\right)\\ &= \frac{1}{\beta_5}\left(\limsup_{t\to\infty} x_4(t) + \|\varphi_4\|_{\infty,\mu}\limsup_{t\to\infty} e^{-\mu t}\int_t^\infty k(s)e^{\mu s}ds\right)\\ &\le \frac{\gamma_1+\gamma_2}{\beta_5\beta_m},\end{aligned}$$

which concludes the proof. □

5. Global Stability Analysis for S^0

Theorem 2. *The equilibrium point S^0 is globally asymptotically stable in the open positive orthant $\mathbb{R}^5_+$, regardless of the delay kernel $k(s)$, if the following inequality holds:*

$$R_0 < \frac{\beta_4}{\alpha_3+\beta_4}. \tag{7}$$

Proof. Let us consider an arbitrary solution $x_i(t)$, $i = \overline{1,5}$, of system (4), with initial conditions $\varphi_i : (-\infty, 0] \to (0,\infty)$. From Theorem 1 it follows that $x_i(t) > 0$, for any $t > 0$ and $i = \overline{1,5}$. Hence, we consider the functions

$$L_k(t) = x_k(t) - x_k^0 - x_k^0 \ln\frac{x_k(t)}{x_k^0}\,,\ k = \overline{1,3}\,,$$

and we further denote:

$$E_k(t) = 1 - \frac{x_k(t)}{x_k^0} + \ln\frac{x_k(t)}{x_k^0} < 0 \quad , \quad \tilde{E_k}(t) = 1 - \frac{x_k^0}{x_k(t)} + \ln\frac{x_k^0}{x_k(t)} < 0\,,\ k = \overline{1,3}.$$

Therefore, we have:

$$\begin{aligned}L_1'(t) &= \dot{x}_1\left(1-\frac{x_1^0}{x_1}\right) = \left(1-\frac{x_1^0}{x_1}\right)[\gamma_1 - x_1x_5 + \alpha_3x_4 + x_3 - (\alpha_4+\beta_1)x_1]\\ &= \left(1-\frac{x_1^0}{x_1}\right)[x_1^0x_5^0 - \alpha_3x_4^0 - x_3^0 + (\alpha_4+\beta_1)x_1^0 - x_1x_5 + \alpha_3x_4 + x_3 - (\alpha_4+\beta_1)x_1]\\ &= \left(1-\frac{x_1^0}{x_1}\right)\left[(\alpha_4+\beta_1)x_1^0\left(1-\frac{x_1}{x_1^0}\right) - x_3^0\left(1-\frac{x_3}{x_3^0}\right) - x_1x_5 + \alpha_3x_4\right]\\ &= (\alpha_4+\beta_1)x_1^0\left(2-\frac{x_1^0}{x_1}-\frac{x_1}{x_1^0}\right) - x_3^0\left(1+\frac{x_1^0}{x_1}\cdot\frac{x_3}{x_3^0}-\frac{x_3}{x_3^0}-\frac{x_1^0}{x_1}\right)\\ &\quad - x_1x_5 + \alpha_3x_4 - \alpha_3x_4\frac{x_1^0}{x_1} + x_5x_1^0\\ &\le (\alpha_4+\beta_1)x_1^0\left(2-\frac{x_1^0}{x_1}-\frac{x_1}{x_1^0}\right) - x_3^0\left(2+\ln\frac{x_1^0}{x_1}+\ln\frac{x_3}{x_3^0}-\frac{x_3}{x_3^0}-\frac{x_1^0}{x_1}\right)\\ &\quad - x_1x_5 + \alpha_3x_4 + x_5x_1^0\\ &= (\alpha_4+\beta_1)x_1^0(E_1+\tilde{E_1}) - x_3^0(\tilde{E_1}+E_3) - x_1x_5 + x_1^0x_5 + \alpha_3x_4 - \alpha_3x_4\frac{x_1^0}{x_1}.\end{aligned}$$

Furthermore, we evaluate

$$\begin{aligned} L_2'(t) &= \dot{x}_2\Big(1-\frac{x_2^0}{x_2}\Big) = \Big(1-\frac{x_2^0}{x_2}\Big)(\gamma_2-\alpha_2x_2x_5-\beta_2x_2) \\ &= \Big(1-\frac{x_2^0}{x_2}\Big)(\alpha_2x_2^0x_5^0+\beta_2x_2^0-\alpha_2x_2x_5-\beta_2x_2) \\ &= \beta_2x_2^0\Big(2-\frac{x_2^0}{x_2}-\frac{x_2}{x_2^0}\Big)-\alpha_2x_2x_5+\alpha_2x_2^0x_5 \\ &= \beta_2x_2^0(E_2+\tilde{E_2})-\alpha_2x_2x_5+\alpha_2x_2^0x_5, \end{aligned}$$

and

$$\begin{aligned} L_3'(t) &= \dot{x}_3\Big(1-\frac{x_3^0}{x_3}\Big) = \Big(1-\frac{x_3^0}{x_3}\Big)(\alpha_4x_1-x_3-\alpha_1x_3x_5-\beta_3x_3) \\ &= \Big(1-\frac{x_3^0}{x_3}\Big)\Big[\alpha_4(x_1-x_1^0)-(1+\beta_3)(x_3-x_3^0)-\alpha_1(x_3x_5-x_3^0x_5^0)\Big] \\ &= \Big(1-\frac{x_3^0}{x_3}\Big)\Big[\alpha_4x_1^0(\frac{x_1}{x_1^0}-1)-x_3^0(1+\beta_3)(\frac{x_3}{x_3^0}-1)-\alpha_1x_3x_5\Big] \\ &= -\alpha_4x_1^0\left(1+\frac{x_3^0}{x_3}\cdot\frac{x_1}{x_1^0}-\frac{x_3^0}{x_3}-\frac{x_1}{x_1^0}\right)+x_3^0(1+\beta_3)(2-\frac{x_3}{x_3^0}-\frac{x_3^0}{x_3})-\alpha_1x_3x_5+\alpha_1x_3^0x_5 \\ &\le -\alpha_4x_1^0\left(2+\ln\frac{x_3^0}{x_3}+\ln\frac{x_1}{x_1^0}-\frac{x_3^0}{x_3}-\frac{x_1}{x_1^0}\right)+x_3^0(1+\beta_3)(2-\frac{x_3}{x_3^0}-\frac{x_3^0}{x_3})-\alpha_1x_3x_5+\alpha_1x_3^0x_5 \\ &= -\alpha_4x_1^0(E_1+\tilde{E_3})+(1+\beta_3)x_3^0(E_3+\tilde{E_3})-\alpha_1x_3x_5+\alpha_1x_3^0x_5. \end{aligned}$$

Moreover, considering

$$L_{4,5}(t) = x_4+cx_5+c\int_0^\infty\left(\hat{k}(s)\int_{t-s}^t x_4(r)dr\right)ds$$

we have:

$$\begin{aligned} L_{4,5}'(t) &= \dot{x}_4+c\dot{x}_5+c\int_0^\infty \hat{k}(s)[x_4(t)-x_4(t-s)]ds \\ &= x_1x_5+\alpha_2x_2x_5+\alpha_1x_3x_5-(\alpha_3+\beta_4-c)x_4-c\beta_5x_5. \end{aligned}$$

Further, considering:

$$L(t) = L_1(t)+L_2(t)+L_3(t)+L_{4,5}(t)$$

we compute

$$\begin{aligned} L'(t) &\le (\alpha_4+\beta_1)x_1^0(E_1+\tilde{E_1})-x_3^0(\tilde{E_1}+E_3)-x_1x_5+x_1^0x_5+\alpha_3x_4 \\ &\quad +\beta_2x_2^0(E_2+\tilde{E_2})-\alpha_2x_2x_5+\alpha_2x_2^0x_5 \\ &\quad -\alpha_4x_1^0(E_1+\tilde{E_3})+(1+\beta_3)x_3^0(E_3+\tilde{E_3})-\alpha_1x_3x_5+\alpha_1x_3^0x_5 \\ &\quad +x_1x_5+\alpha_2x_2x_5+\alpha_1x_3x_5-(\alpha_3+\beta_4-c)x_4-c\beta_5x_5 \\ &= \beta_1x_1^0E_1+\Big[(\alpha_4+\beta_1)x_1^0-x_3^0\Big]\tilde{E_1}+\beta_2x_2^0(E_2+\tilde{E_2})+\beta_3x_3^0E_3+\Big[(1+\beta_3)x_3^0-\alpha_4x_1^0\Big]\tilde{E_3} \\ &\quad -(\beta_4-c)x_4-(c\beta_5-x_1^0-\alpha_2x_2^0-\alpha_1x_3^0)x_5 \\ &= \beta_1x_1^0E_1+\beta_2x_2^0(E_2+\tilde{E_2})+\beta_3x_3^0E_3-(\beta_4-c)x_4-\beta_5[c-R_0(\alpha_3+\beta_4)]x_5. \end{aligned}$$

Therefore, as inequality $R_0 < \frac{\beta_4}{\alpha_3+\beta_4}$ holds, there exists $c > 0$ such that $L'(t) < 0$. By means of LaSalle's invariance principle [12,36], we deduce that the equilibrium equilibrium point S^0 of system (4) is globally asymptotically stable in $\mathbb{R}^5_+$. □

6. Global Stability Analysis for S^+

Theorem 3. *The positive equilibrium point S^+ is globally asymptotically stable in the open positive orthant $\mathbb{R}^5_+$, regardless of the delay kernel $k(s)$, if $a_3 = 0$.*

Proof. Let us consider an arbitrary solution $x_i(t)$, $i = \overline{1,5}$, of system (4), with initial conditions $\varphi_i : (-\infty, 0] \to (0, \infty)$. From Theorem 1 it follows that $x_i(t) > 0$, for any $t > 0$ and $i = \overline{1,5}$.

Taking into account that $x - 1 - \ln x \geq 0$, for any $x > 0$, we construct the Lyapunov function as follows:

$$L(t) = L_1(t) + L_2(t) + L_3(t) + L_4(t) + cL_5(t) + cL_{4,d}(t), \tag{8}$$

where

$$L_k(t) = x_k(t) - x_k^+ - x_k^+ \ln \frac{x_k(t)}{x_k^+}\,,\ k = \overline{1,5}\,, \tag{9}$$

and

$$L_{4,d}(t) = \int_0^\infty \left(k(s) \int_{t-s}^t \left(x_4(r) - x_4^+ - x_4^+ \ln \frac{x_4(r)}{x_4^+} \right) dr \right) ds$$

and we further denote:

$$E_k(t) = 1 - \frac{x_k(t)}{x_k^+} + \ln \frac{x_k(t)}{x_k^+} \leq 0 \quad , \quad \tilde{E_k}(t) = 1 - \frac{x_k^+}{x_k(t)} + \ln \frac{x_k^+}{x_k(t)} \leq 0 \quad , \quad k = \overline{1,5}.$$

With the notations defined above, taking into account that S^+ is an equilibrium point of system (4) we have:

$$\begin{aligned}
L_1'(t) &= \dot{x}_1\Big(1 - \frac{x_1^+}{x_1}\Big) = \Big(1 - \frac{x_1^+}{x_1}\Big)[\gamma_1 - x_1x_5 + \alpha_3x_4 + x_3 - (\alpha_4 + \beta_1)x_1] \\
&= \Big(1 - \frac{x_1^+}{x_1}\Big)[x_1^+x_5^+ - \alpha_3x_4^+ - x_3^+ + (\alpha_4 + \beta_1)x_1^+ - x_1x_5 + \alpha_3x_4 + x_3 - (\alpha_4 + \beta_1)x_1] \\
&= \Big(1 - \frac{x_1^+}{x_1}\Big)\Big[x_1^+x_5^+\Big(1 - \frac{x_1}{x_1^+}\frac{x_5}{x_5^+}\Big) - \alpha_3x_4^+\Big(1 - \frac{x_4}{x_4^+}\Big) - x_3^+\Big(1 - \frac{x_3}{x_3^+}\Big) \\
&\quad + (\alpha_4 + \beta_1)x_1^+\Big(1 - \frac{x_1}{x_1^+}\Big)\Big] \\
&= x_1^+x_5^+\Big(1 - \frac{x_1}{x_1^+}\frac{x_5}{x_5^+} - \frac{x_1^+}{x_1} + \frac{x_5}{x_5^+}\Big) - \alpha_3x_4^+\Big(1 - \frac{x_4}{x_4^+} - \frac{x_1^+}{x_1} + \frac{x_1^+}{x_1}\frac{x_4}{x_4^+}\Big) \\
&\quad - x_3^+\Big(1 - \frac{x_3}{x_3^+} - \frac{x_1^+}{x_1} + \frac{x_1^+}{x_1}\frac{x_3}{x_3^+}\Big) + (\alpha_4 + \beta_1)x_1^+\Big(2 - \frac{x_1}{x_1^+} - \frac{x_1^+}{x_1}\Big) \\
&\leq x_1^+x_5^+\Big(1 - \frac{x_1}{x_1^+}\frac{x_5}{x_5^+} - \frac{x_1^+}{x_1} + \frac{x_5}{x_5^+}\Big) - \alpha_3x_4^+\Big(2 + \ln\frac{x_1^+}{x_1} + \ln\frac{x_4}{x_4^+} - \frac{x_4}{x_4^+} - \frac{x_1^+}{x_1}\Big) \\
&\quad - x_3\Big(2 + \ln\frac{x_1^+}{x_1} + \ln\frac{x_3}{x_3^+} - \frac{x_1^+}{x_1} - \frac{x_3}{x_3^+}\Big) + (\alpha_4 + \beta_1)x_1^+\Big(E_1 + \tilde{E_1}\Big) \\
&= x_1^+x_5^+\Big(1 - \frac{x_1}{x_1^+}\frac{x_5}{x_5^+} - \frac{x_1^+}{x_1} + \frac{x_5}{x_5^+}\Big) - \alpha_3x_4^+\Big(\tilde{E_1} + E_4\Big) - x_3^+\Big(\tilde{E_1} + E_3\Big) \\
&\quad + (\alpha_4 + \beta_1)x_1^+\Big(E_1 + \tilde{E_1}\Big).
\end{aligned}$$

Furthermore, we get:

$$\begin{aligned}
L_2'(t) &= \dot{x}_2\Big(1-\frac{x_2^+}{x_2}\Big) = \Big(1-\frac{x_2^+}{x_2}\Big)(\gamma_2 - \alpha_2 x_2 x_5 - \beta_2 x_2) \\
&= \Big(1-\frac{x_2^+}{x_2}\Big)(\alpha_2 x_2^+ x_5^+ + \beta_2 x_2^+ - \alpha_2 x_2 x_5 - \beta_2 x_2) \\
&= \Big(1-\frac{x_2^+}{x_2}\Big)\Big[\alpha_2 x_2^+ x_5^+\Big(1-\frac{x_2}{x_2^+}\frac{x_5}{x_5^+}\Big) + \beta_2 x_2^+\Big(1-\frac{x_2}{x_2^+}\Big)\Big] \\
&= \alpha_2 x_2^+ x_5^+\Big(1-\frac{x_2}{x_2^+}\frac{x_5}{x_5^+} - \frac{x_2^+}{x_2} + \frac{x_5}{x_5^+}\Big) + \beta_2 x_2^+\Big(2-\frac{x_2^+}{x_2}-\frac{x_2}{x_2^+}\Big) \\
&= \alpha_2 x_2^+ x_5^+\Big(1-\frac{x_2}{x_2^+}\frac{x_5}{x_5^+} - \frac{x_2^+}{x_2} + \frac{x_5}{x_5^+}\Big) + \beta_2 x_2^+\Big(E_2 + \tilde{E}_2\Big).
\end{aligned}$$

Moreover, we obtain:

$$\begin{aligned}
L_3'(t) &= \dot{x}_3\Big(1-\frac{x_3^+}{x_3}\Big) = \Big(1-\frac{x_3^+}{x_3}\Big)(\alpha_4 x_1 - x_3 - \alpha_1 x_3 x_5 - \beta_3 x_3) \\
&= \Big(1-\frac{x_3^+}{x_3}\Big)\Big[\alpha_4(x_1 - x_1^+) - (1+\beta_3)(x_3 - x_3^+) - \alpha_1(x_3x_5 - x_3^+x_5^+)\Big] \\
&= \Big(1-\frac{x_3^+}{x_3}\Big)\Big[\alpha_4 x_1^+\left(\frac{x_1}{x_1^+}-1\right) - x_3^+(1+\beta_3)\left(\frac{x_3}{x_3^+}-1\right) - \alpha_1 x_3^+ x_5^+\Big(\frac{x_3}{x_3^+}\frac{x_5}{x_5^+}-1\Big)\Big] \\
&= \alpha_4 x_1^+\left(\frac{x_1}{x_1^+} - 1 - \frac{x_3^+}{x_3}\cdot\frac{x_1}{x_1^+} + \frac{x_3^+}{x_3}\right) + x_3^+(1+\beta_3)\left(2-\frac{x_3}{x_3^+}-\frac{x_3^+}{x_3}\right) \\
&\quad + \alpha_1 x_3^+ x_5^+\Big(1-\frac{x_3^+}{x_3} - \frac{x_3}{x_3^+}\frac{x_5}{x_5^+} + \frac{x_5}{x_5^+}\Big) \\
&\le \alpha_4 x_1^+\left(-2 - \ln\frac{x_3^+}{x_3} - \ln\frac{x_1}{x_1^+} + \frac{x_3^+}{x_3} + \frac{x_1}{x_1^+}\right) + x_3^+(1+\beta_3)\left(2-\frac{x_3}{x_3^+}-\frac{x_3^+}{x_3}\right) \\
&\quad + \alpha_1 x_3^+ x_5^+\Big(1-\frac{x_3^+}{x_3} - \frac{x_3}{x_3^+}\frac{x_5}{x_5^+} + \frac{x_5}{x_5^+}\Big) \\
&= \alpha_4 x_1^+(-E_1 - \tilde{E}_3) + (1+\beta_3)x_3^+(E_3 + \tilde{E}_3) + \alpha_1 x_3^+ x_5^+\Big(1-\frac{x_3^+}{x_3} - \frac{x_3}{x_3^+}\frac{x_5}{x_5^+} + \frac{x_5}{x_5^+}\Big),
\end{aligned}$$

and finally:

$$\begin{aligned}
L_4'(t) &= \dot{x}_4\left(1-\frac{x_4^+}{x_4}\right) = \left(1-\frac{x_4^+}{x_4}\right)(x_1x_5 + \alpha_2 x_2 x_5 + \alpha_1 x_3 x_5 - \alpha_3 x_4 - \beta_4 x_4) \\
&= \left(1-\frac{x_4^+}{x_4}\right)[x_1^+x_5^+\left(\frac{x_1x_5}{x_1^+x_5^+}-1\right) + \alpha_2 x_2^+ x_5^+\left(\frac{x_2x_5}{x_2^+x_5^+}-1\right) + \alpha_1 x_3^+ x_5^+\left(\frac{x_3x_5}{x_3^+x_5^+}-1\right) \\
&\quad - (\alpha_3+\beta_4)x_4^+\left(\frac{x_4}{x_4^+}-1\right)] \\
&= x_1^+x_5^+\left(\frac{x_1x_5}{x_1^+x_5^+} - 1 - \frac{x_4^+x_1x_5}{x_4x_1^+x_5^+} + \frac{x_4^+}{x_4}\right) + \alpha_2 x_2^+ x_5^+\left(\frac{x_2x_5}{x_2^+x_5^+} - 1 - \frac{x_4^+x_2x_5}{x_4x_2^+x_5^+} + \frac{x_4^+}{x_4}\right) \\
&\quad + \alpha_1 x_3^+ x_5^+\left(\frac{x_3x_5}{x_3^+x_5^+} - 1 - \frac{x_4^+x_3x_5}{x_4x_3^+x_5^+} + \frac{x_4^+}{x_4}\right) + (\alpha_3+\beta_4)x_4^+\left(2-\frac{x_4}{x_4^+}-\frac{x_4^+}{x_4}\right) \\
&\le x_1^+x_5^+\left(\frac{x_1x_5}{x_1^+x_5^+} - 2 - \ln\frac{x_4^+}{x_4} - \ln\frac{x_1}{x_1^+} - \ln\frac{x_5}{x_5^+} + \frac{x_4^+}{x_4}\right) \\
&\quad + \alpha_2 x_2^+ x_5^+\left(\frac{x_2x_5}{x_2^+x_5^+} - 2 - \ln\frac{x_4^+}{x_4} - \ln\frac{x_2}{x_2^+} - \ln\frac{x_5}{x_5^+} + \frac{x_4^+}{x_4}\right) \\
&\quad + \alpha_1 x_3^+ x_5^+\left(\frac{x_3x_5}{x_3^+x_5^+} - 2 - \ln\frac{x_4^+}{x_4} - \ln\frac{x_3}{x_3^+} - \ln\frac{x_5}{x_5^+} + \frac{x_4^+}{x_4}\right) + (\alpha_3+\beta_4)x_4^+(E_4 + \tilde{E}_4).
\end{aligned}$$

Furthermore, taking into account the algebraic relations satisfied by the coordinates of the equilibrium point S^+, we obtain:

$$
\begin{aligned}
&L_1'(t)+L_2'(t)+L_3'(t)+L_4'(t)\leq\\
&\leq x_1^+x_5^+\left(1-\frac{x_1^+}{x_1}+\frac{x_5}{x_5^+}-2+\ln\frac{x_1^+}{x_1}-\ln\frac{x_4^+}{x_4}-\ln\frac{x_5}{x_5^+}+\frac{x_4^+}{x_4}\right)\\
&+\alpha_2x_2^+x_5^+\left(1-\frac{x_2^+}{x_2}+\frac{x_5}{x_5^+}-2-\ln\frac{x_4^+}{x_4}+\ln\frac{x_2^+}{x_2}-\ln\frac{x_5}{x_5^+}+\frac{x_4^+}{x_4}\right)\\
&+\alpha_1x_3^+x_5^+\left(1-\frac{x_3^+}{x_3}+\frac{x_5}{x_5^+}-2-\ln\frac{x_4^+}{x_4}+\ln\frac{x_3^+}{x_3}-\ln\frac{x_5}{x_5^+}+\frac{x_4^+}{x_4}\right)\\
&-\alpha_3x_4^+(\tilde{E}_1+E_4)-x_3^+(\tilde{E}_1+E_3)+(\alpha_4+\beta_1)x_1^+(E_1+\tilde{E}_1)\\
&+\beta_2x_2^+(\tilde{E}_2+E_2)-\alpha_4x_1^+(E_1+\tilde{E}_3)+(1+\beta_3)x_3^+(E_3+\tilde{E}_3)\\
&+(\alpha_3+\beta_4)x_4^+(E_4+\tilde{E}_4)\\
&=x_1^+x_5^+(\tilde{E}_1-E_5-\tilde{E}_4)+\alpha_2x_2^+x_5^+(\tilde{E}_2-E_5-\tilde{E}_4)\\
&+\alpha_1x_3^+x_5^+(\tilde{E}_3-E_5-\tilde{E}_4)-\alpha_3x_4^+(\tilde{E}_1+E_4)-x_3^+(\tilde{E}_1+E_3)\\
&+(\alpha_4+\beta_1)x_1^+(E_1+\tilde{E}_1)+\beta_2x_2^+(\tilde{E}_2+E_2)-\alpha_4x_1^+(E_1+\tilde{E}_3)\\
&+(1+\beta_3)x_3^+(E_3+\tilde{E}_3)+(\alpha_3+\beta_4)x_4^+(E_4+\tilde{E}_4)\\
&=\tilde{E}_1\left[x_1^+x_5^+-\alpha_3x_4^+-x_3^++(\alpha_4+\beta_1)x_1^+\right]+\beta_1x_1^+E_1\\
&+(\alpha_2x_2^+x_5^++\beta_2x_2^+)\tilde{E}_2+\beta_2x_2^+E_2\\
&+\tilde{E}_3[\alpha_1x_3^+x_5^+-\alpha_4x_1^++(1+\beta_3)x_3^+]+\beta_3x_3^+E_3\\
&+\tilde{E}_4[-x_1^+x_5^+-\alpha_2x_2^+x_5^+-\alpha_1x_3^+x_5^++(\alpha_3+\beta_4)x_4^+]+\beta_4x_4^+E_4\\
&+E_5[-x_1^+x_5^+-\alpha_2x_2^+x_5^+-\alpha_1x_3^+x_5^+]\\
&=\gamma_1\tilde{E}_1+\gamma_2\tilde{E}_2+\beta_1x_1^+E_1+\beta_2x_2^+E_2+\beta_3x_3^+E_3+\beta_4x_4^+E_4\\
&-(\alpha_3+\beta_4)x_4^+E_5.
\end{aligned}
$$

Denoting $D[x_4]=\int\limits_0^\infty k(s)x_4(t-s)ds$ and employing the inequality $x\geq\ln x+1$, for any $x>0$, we have:

$$
\begin{aligned}
L_5'(t)&=\dot{x}_5\left(1-\frac{x_5^+}{x_5}\right)=\left(1-\frac{x_5^+}{x_5}\right)(D[x_4]-\beta_5x_5)\\
&=D[x_4]-\beta_5x_5-D[x_4]\frac{x_5^+}{x_5}+\beta_5x_5^+\\
&=(D[x_4]-x_4)+x_4^+\frac{x_4}{x_4^+}+\beta_5x_5^+\left(1-\frac{x_5}{x_5^+}\right)-x_4^+\frac{D[x_4]x_5^+}{x_4^+x_5}\\
&\leq(D[x_4]-x_4)+x_4^+\left(\frac{x_4}{x_4^+}+1-\frac{x_5}{x_5^+}\right)-x_4^+\left(\ln\frac{D[x_4]}{x_4}+\ln\frac{x_4}{x_4^+}+\ln\frac{x_5^+}{x_5}+1\right)\\
&=(D[x_4]-x_4)+x_4^+\left(\frac{x_4}{x_4^+}-\ln\frac{x_4}{x_4^+}-\frac{x_5}{x_5^+}+\ln\frac{x_5}{x_5^+}\right)-x_4^+\ln\frac{D[x_4]}{x_4}\\
&=(D[x_4]-x_4)+x_4^+(E_5-E_4)-x_4^+\ln\frac{D[x_4]}{x_4}.
\end{aligned}
$$

Moreover:

$$L'_{4,d}(t) = \int_0^\infty k(s)\left(x_4(t) - x_4^+ - x_4^+ \ln\frac{x_4(t)}{x_4^+} - x_4(t-s) + x_4^+ + x_4^+ \ln\frac{x_4(t-s)}{x_4^+}\right)ds$$
$$= (x_4 - D[x_4]) + x_4^+ \int_0^\infty k(s)\ln\frac{x_4(t-s)}{x_4(t)}ds$$

Therefore, combining the previous relations, we have:

$$\begin{aligned}L'(t) &= L'_1(t) + L'_2(t) + L'_3(t) + L'_4(t) + c\cdot L'_5(t) + c\cdot L'_{4,d}(t)\\
&\le \gamma_1\tilde{E}_1 + \gamma_2\tilde{E}_2 + \beta_1 x_1^+ E_1 + \beta_2 x_2^+ E_2 + \beta_3 x_3^+ E_3 + \beta_4 x_4^+ E_4 - (\alpha_3+\beta_4)x_4^+ E_5\\
&\quad + c x_4^+(E_5 - E_4) - c x_4^+\left(\ln\frac{D[x_4]}{x_4} - \int_0^\infty k(s)\ln\frac{x_4(t-s)}{x_4(t)}ds\right)\\
&= \gamma_1\tilde{E}_1 + \gamma_2\tilde{E}_2 + \beta_1 x_1^+ E_1 + \beta_2 x_2^+ E_2 + \beta_3 x_3^+ E_3 + (\beta_4 - c)x_4^+ E_4 + (c-\alpha_3-\beta_4)x_4^+ E_5\\
&\quad - c x_4^+\left(\ln\int_0^\infty k(s)x_4(t-s)ds - \int_0^\infty k(s)\ln x_4(t-s)ds\right).\end{aligned}$$

For $\alpha_3 = 0$, choosing $c = \beta_4$, it follows that:

$$\begin{aligned}L'(t) &\le \gamma_1\tilde{E}_1 + \gamma_2\tilde{E}_2 + \beta_1 x_1^+ E_1 + \beta_2 x_2^+ E_2 + \beta_3 x_3^+ E_3\\
&\quad - \beta_4 x_4^+\left(\ln\int_0^\infty k(s)x_4(t-s)ds - \int_0^\infty k(s)\ln x_4(t-s)ds\right).\end{aligned}$$

Jensen's inequality for probability density functions applied to the concave logarithmic function [37] provides that

$$\ln\int_0^\infty k(s)x_4(t-s)ds \ge \int_0^\infty k(s)\ln x_4(t-s)ds,$$

and hence, we obtain that $L'(t) \le 0$, for any $t \ge 0$. Hence, by means of LaSalle's invariance principle [12,36], we deduce that the equilibrium equilibrium point S^+ of system (4) is globally asymptotically stable in $\mathbb{R}^5_+$. □

7. Numerical Simulations

7.1. Scenario 1: Global Asymptotic Stability of S^0

In this case, we first consider the following parameter values: $a_1 = 687.5$, $a_2 = 0.0000152588$, $a_3 = 1$, $a_4 = 0.9375$, $a_5 = 0.125$, $b_1 = 0.5$, $b_2 = 0.51$, $b_3 = 0.03125$, $b_4 = 0.5$, $b_5 = 0.5$, $c_1 = 0.0000152588$, $m_1 = 25.5$, $m_2 = 0.0078125$. For these parameter values, system (1) has a unique equilibrium point:

$$S^0 = (U_0, M_0, T_0, R_0, V_0) = (1000, 50, 6000, 0, 0).$$

We remark that the basic reproduction number R_0 satisfies inequality (7), and hence, based on Theorem 2, the equilibrium point S^0 is globally asymptotically stable, for any delay kernel considered in system (1).

In fact, in this scenario, the very low hiring rates for unemployed persons, immigrants and temporarily employed persons and the low value of the basic reproduction number, justify the existence of only one globally asymptotically stable equilibrium point, S^0. For the numerical simulations shown in Figure 1, the following initial condition has been considered: $U(0) = 280, M(0) = 50, T(0) = 2000, R(0) = 6400, V(0) = 40$, which represents a plausible situation in a stable economic environment. However, due to the low hiring rates mentioned above, the trajectories of system (1) quickly converge to the "crisis" equilibrium S^0, corresponding to no regular employment and no available vacancies.

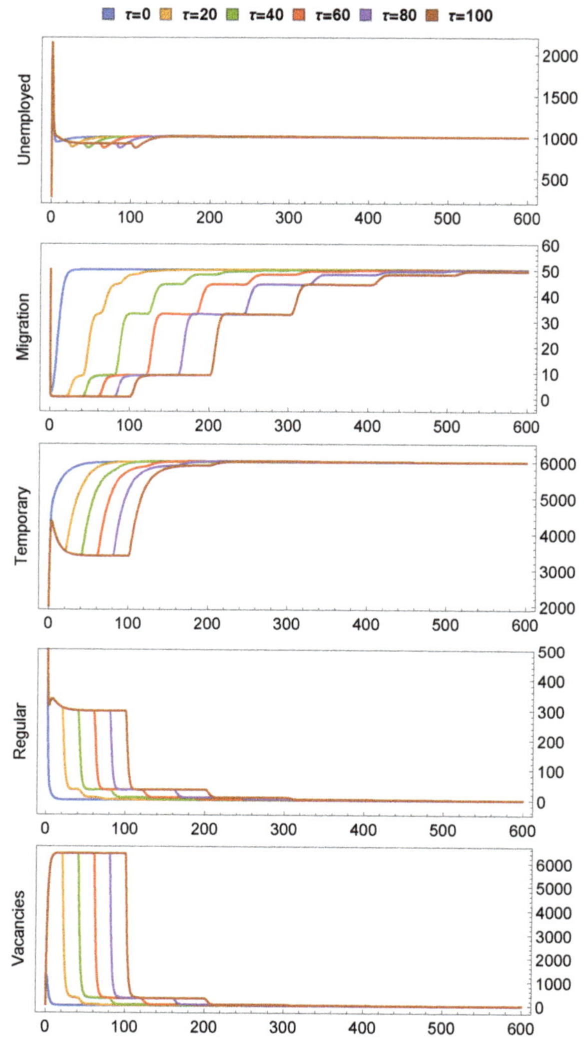

Figure 1. Evolution of the state variables $U(t)$, $M(t)$, $T(t)$, $R(t)$ and $V(t)$ in Scenario 1, with fixed initial conditions: $U(0) = 280, M(0) = 50, T(0) = 2000, R(0) = 6400, V(0) = 40$ and a discrete time delay $\tau \in [0, 100]$; the trajectories converge to the equilibrium point S^0=(1000, 50, 6000, 0, 0).

7.2. Scenario 2: Global Asymptotic Stability of S^+

For the numerical simulations we considered: $a_1 = 300$, $a_2 = 0.0303571$, $a_3 = 0.0582031$, $a_4 = 0.5$, $a_5 = 0.035$, $b_1 = 0.9375$, $b_2 = 0.5$, $b_3 = 0.02625$, $b_4 = 0.003125$, $b_5 = 0.5$, $c_1 = 0.00021875$, $m_1 = 60$, $m_2 = 0.0175$. For these values of the parameters, we obtain a unique positive equilibrium point

$$S^+ = (U_+, M_+, T_+, R_+, V_+) = (280, 50, 2000, 6400, 40).$$

In fact, in this scenario, we observe that the hiring rate a_3 is very small, and hence, global asymptotic stability can be expected for the equilibrium S^+, in line with the theoretical results presented in Theorem 3. Indeed, in the numerical simulations shown in Figure 2, convergence to the positive equilibrium S^+ is observed, starting from a "crisis" initial condition: $U(0) = 260, M(0) = 120, T(0) = 2100, R(0) = 1, V(0) = 1$, with almost no regular employment or available vacancies. We emphasize that in this scenario, the convergence is very slow. Moreover, a larger time delay is associated with an even slower convergence to the positive equilibrium.

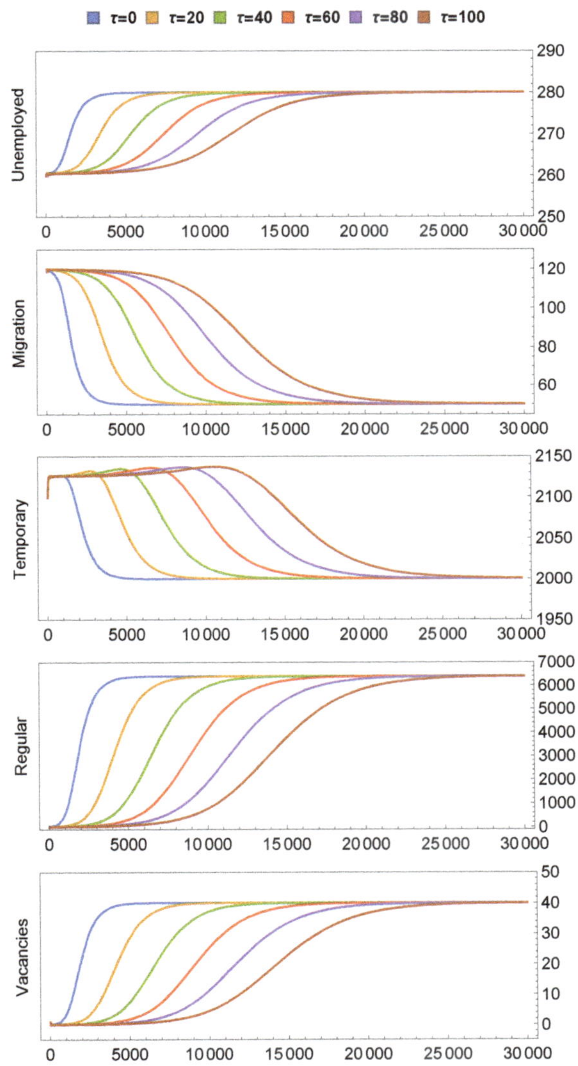

Figure 2. Evolution of the state variables $U(t)$, $M(t)$, $T(t)$, $R(t)$ and $V(t)$ in Scenario 2, with initial conditions: $U(0) = 260, M(0) = 120, T(0) = 2100, R(0) = 1, V(0) = 1$ and a discrete time delay $\tau \in [0, 100]$; the trajectories converge to $S^+ = (280, 50, 2000, 6400, 40)$.

8. Conclusions

The present paper introduced a five-dimensional mathematical model that facilitates the understanding of new ways for studying the labour market while observing the levels of unemployment, migration, fixed term contractors, full time employment and the number of available vacancies. The distributed time delay has been incorporated to reflect the dependence of the rate of change of available vacancies on the past regular employment levels.

The positivity and boundedness of solutions have been provided. Using the basic reproduction number, the existence of equilibrium points has been discussed. As a consequence, the employment free equilibrium and positive equilibrium came into focus, for which we have undertaken a global stability analysis, regardless of the delay kernel included in the mathematical model.

Through the numerical simulations, the theoretical findings have been emphasized as well, further revealing that in an economic crisis scenario (Scenario 1), a fast convergence to the "crisis" equilibrium (with no regular employment or vacancies) is observed. However, in Scenario 2, if the initial conditions correspond to a crisis situation and the parameters of the system are adjusted, the convergence to the positive equilibrium is very slow, which reflects the slow recovery of the job market.

The findings of the paper could be used as input for future developments linked to population dynamics, and the decision making entities of the society could also use these findings in various strategies.

The proposed mathematical model is open ended that allows the introduction of future variables depicting the effects of the unemployment, such as poverty and unsocial behaviours that can lead to crime, as in [38].

Author Contributions: Conceptualization, E.K., M.N. and L.F.V.; methodology, E.K., M.N. and L.F.V.; software, E.K., M.N. and L.F.V.; validation, E.K., M.N. and L.F.V.; formal analysis, E.K., M.N. and L.F.V.; investigation, E.K., M.N. and L.F.V.; resources, E.K., M.N. and L.F.V.; data curation, E.K., M.N. and L.F.V.; writing—original draft preparation, E.K., M.N. and L.F.V.; writing—review and editing, E.K., M.N. and L.F.V.; visualization, E.K., M.N. and L.F.V.; supervision, E.K., M.N. and L.F.V.; project administration, E.K., M.N. and L.F.V.; funding acquisition, E.K., M.N. and L.F.V. All authors have read and agreed to the published version of the manuscript.

Funding: This research received no external funding.

Institutional Review Board Statement: Not applicable.

Informed Consent Statement: Not applicable.

Data Availability Statement: Not applicable.

Conflicts of Interest: The authors declare no conflict of interest.

References

1. Singh, A.K.; Singh, P.K.; Misra, A.K. Combating unemployment through skill development. *Nonlinear Anal. Model. Control* **2020**, *25*, 919–937. [CrossRef]
2. Nikolopoulos, C.; Tzanetis, D. A model for housing allocation of a homeless population due to a natural disaster. *Nonlinear Anal. Real. World Appl.* **2003**, *4*, 561–579. [CrossRef]
3. Misra, A.; Singh, A.K. A delay mathematical model for the control of unemployment. *Differ. Equations Dyn. Syst.* **2013**, *21*, 291–307. [CrossRef]
4. Kaslik, E.; Neamţu, M.; Vesa, L.F. Global stability analysis of an unemployment model with distributed delay. *Math. Comput. Simul.* **2021**, *185*, 535–546. [CrossRef]
5. Misra, A.; Singh, A.K.; Singh, P.K. Modeling the Role of Skill Development to Control Unemployment. In *Differential Equations and Dynamical Systems*; Springer Nature: Cham, Switzerland, 2017; pp. 1–13.
6. Al-Maalwi, R.; Al-Sheikh, S.; Ashi, H.; Asiri, S. Mathematical modeling and parameter estimation of unemployment with the impact of training programs. *Math. Comput. Simul.* **2021**, *182*, 705–720. [CrossRef]
7. Harding, L.; Neamţu, M. A dynamic model of unemployment with migration and delayed policy intervention. *Comput. Econ.* **2018**, *51*, 427–462. [CrossRef]
8. Munoli, S.; Gani, S.; Gani, S. A Mathematical Approach to Employment Policies: An Optimal Control Analysis. *Int. J. Stat. Syst.* **2017**, *12*, 549–565.
9. Mallick, U.; Biswas, M. Optimal Analysis of Unemployment Model taking Policies to Control. *Adv. Model. Optim.* **2018**, *20*, 303–312.
10. Corduneanu, C.; Lakshmikantham, V. Equations with unbounded delay: A survey. *Nonlinear Anal. Theory Methods Appl.* **1980**, *4*, 831–877. [CrossRef]

11. Gripenberg, G.; Londen, S.O.; Staffans, O. *Volterra Integral and Functional Equations*; Cambridge University Press: Cambridge, UK, 1990; Volume 34.
12. Hale, J.K.; Lunel, S.M.V. *Introduction to Functional Differential Equations*; Applied Mathematical Sciences; Springer: New York, NY, USA, 1991; Volume 99.
13. Hino, Y.; Murakami, S.; Naito, T. *Functional Differential Equations with Infinite Delay*; Lecture Notes in Mathematics Sereies; Springer: New York, NY, USA, 1991; Volume 1473.
14. Diekmann, O.; Van Gils, S.A.; Lunel, S.M.; Walther, H.O. *Delay Equations: Functional-, Complex-, and Nonlinear Analysis*; Applied Mathematical Sciences; Springer: New York, NY, USA,1995; Volume 110.
15. Diekmann, O.; Gyllenberg, M. Equations with infinite delay: Blending the abstract and the concrete. *J. Differ. Equ.* **2012**, *252*, 819–851. [CrossRef]
16. Staffans, O.J. Hopf bifurcation for functional and functional differential equations with infinite delay. *J. Differ. Equ.* **1987**, *70*, 114–151. [CrossRef]
17. Kuang, Y. *Delay Differential Equations: With Applications in Population Dynamics*; Academic Press: New York, NY, USA, 1993; Volume 1913.
18. Liu, S.; Chen, L. Necessary-sufficient conditions for permanence and extinction in Lotka-Volterra system with distributed delays. *Appl. Math. Lett.* **2003**, *16*, 911–917. [CrossRef]
19. Adimy, M.; Crauste, F.; Halanay, A. Neamţu, M.; Opriş, D. Stability of limit cycles in a pluripotent stem cell dynamics model. *Chaos Solitons Fractals* **2006**, *27*, 1091–1107. [CrossRef]
20. Berezansky, L.; Braverman, E.; Idels, L. Nicholson's blowflies differential equations revisited: Main results and open problems. *Appl. Math. Model.* **2010**, *34*, 1405–1417. [CrossRef]
21. Cushing, J.M. *Integrodifferential Equations and Delay Models in Population Dynamics*; Springer Science & Business Media: Berlin/Heidelberg, Germany, 2013; Volume 20.
22. Huang, G.; Liu, A. A note on global stability for a heroin epidemic model with distributed delay. *Appl. Math. Lett.* **2013**, *26*, 687–691. [CrossRef]
23. Chen, S.S.; Cheng, C.Y.; Takeuchi, Y. Stability analysis in delayed within-host viral dynamics with both viral and cellular infections. *J. Math. Anal. Appl.* **2016**, *442*, 642–672. [CrossRef]
24. Liu, Q.; Jiang, D. Stationary distribution and extinction of a stochastic predator–prey model with distributed delay. *Appl. Math. Lett.* **2018**, *78*, 79–87. [CrossRef]
25. Bichara, D.M. Global analysis of multi-host and multi-vector epidemic models. *J. Math. Anal. Appl.* **2019**, *475*, 1532–1553. [CrossRef] [PubMed]
26. Hallegatte, S.; Ghil, M.; Dumas, P.; Hourcade, J.C. Business cycles, bifurcations and chaos in a neo-classical model with investment dynamics. *J. Econ. Behav. Organ.* **2008**, *67*, 57–77. [CrossRef]
27. Matsumoto, A.; Szidarovszky, F. Delay differential nonlinear economic models. In *Nonlinear Dynamics in Economics, Finance and Social Sciences*; Springer: Berlin/Heidelberg, Germany, 2010; pp. 195–214.
28. Matsumoto, A.; Szidarovszky, F. Delay differential neoclassical growth model. *J. Econ. Behav. Organ.* **2011**, *78*, 272–289. [CrossRef]
29. Yu, J.; Peng, M. Stability and bifurcation analysis for the Kaldor-Kalecki model with a discrete delay and a distributed delay. *Phys. A Stat. Mech. Appl.* **2016**, *460*, 66–75. [CrossRef]
30. Fanelli, V.; Maddalena, L. A nonlinear dynamic model for credit risk contagion. *Math. Comput. Simul.* **2020**, *174*, 45–58. [CrossRef]
31. Chen, W.; Wu, W.; Teng, Z. Complete dynamics in a nonlocal dispersal two-strain SIV epidemic model with vaccinations and latent delays. *Appl. Comput. Math.* **2020**, *19*, 360–391.
32. Khalsaraei, M.M.; Shokri, A.; Ramos, H.; Heydari, S. A positive and elementary stable nonstandard explicit scheme for a mathematical model of the influenza disease. *Math. Comput. Simul.* **2021**, *182*, 397–410. [CrossRef]
33. Shokri, A.; Khalsaraei, M.M.; Molayi, M. Dynamically Consistent NSFD Methods for Predator-prey System. *J. Appl. Comput. Mech.* **2021**, *25*, 1565–1574.
34. Kolmanovskii, V.; Myshkis, A. *Introduction to the Theory and Applications of Functional Differential Equations*: Kluwer Academic Publishers: Dordrecht, The Netherlands, 1999; Volume 463, .
35. Manojlovic, J.; Maric, V. An asymptotic analysis of positive solutions of Thomas-Fermi type sublinear differential equations. *Mem. Differ. Equ. Math. Phys* **2012**, *57*, 75–94.
36. Smith, H.L. *An Introduction to Delay Differential Equations with Applications to the Life Sciences*; Springer New York: New York, NY, USA, 2011; Volume 57.
37. Durrett, R. *Probability: Theory and Examples*; Cambridge University Press: Cambridge, UK, 2019; Volume 49.
38. Soemarsono, A.; Fitria, I.; Nugraheni, K.; Hanifa, N. *Analysis of Mathematical Model on Impact of Unemployment Growth to Crime Rates*; Journal of Physics: Conference Series; IOP Publishing: Bristol, UK, 2021; Volume 1726, p. 012003. [CrossRef]

mathematics

MDPI

Article

Global Dynamics for an Age-Structured Cholera Infection Model with General Infection Rates

Xin Jiang

College of Science, North China University of Technology, Beijing 100144, China; jiangx@buaa.edu.cn

Abstract: This paper studies the global dynamics of a cholera model incorporating age structures and general infection rates. First, we explore the existence and point dissipativeness of the orbit and analyze the asymptotical smoothness. Then, we perform rigorous mathematical analysis on the existence and local stability of equilibria. Based on the uniform persistence, we further investigate the global behavior of the cholera infection model. The results of theoretical analysis are well confirmed by numerical simulations. This research generalizes some known results and provides deeper insights into the dynamics of cholera propagation.

Keywords: age structure; cholera infection model; general infection rates; stability analysis

MSC: 92D30; 34K20

Citation: Jiang, X. Global Dynamics for an Age-Structured Cholera Infection Model with General Infection Rates. *Mathematics* **2021**, *9*, 2993. https://doi.org/10.3390/math9232993

Academic Editors: Mihaela Neamţu, Eva Kaslik and Anca Rădulescu

Received: 31 October 2021
Accepted: 16 November 2021
Published: 23 November 2021

Publisher's Note: MDPI stays neutral with regard to jurisdictional claims in published maps and institutional affiliations.

Copyright: © 2021 by the author. Licensee MDPI, Basel, Switzerland. This article is an open access article distributed under the terms and conditions of the Creative Commons Attribution (CC BY) license (https://creativecommons.org/licenses/by/4.0/).

1. Introduction

Cholera is a serious infectious disease that is caused by the bacterium *Vibrio cholera*. Due to the cholera toxin produced by the bacterium *Vibrio cholera*, it is characterized by severe symptoms, including acute diarrhea, vomiting, hypotension and a weak pulse. Without proper medical treatment, it can cause dehydration and death within hours. This disease peaks in summer and its propagation among humans depends on direct person-to-person contact, as well as indirect contact through contaminated food and water [1,2]. Due to the lack of clean food and water, cholera can spread quickly in regions with poor sanitation conditions and has long been a threat to the public health of human society. In 2018, it was estimated that there were 2.9 million burden cases worldwide, with a death toll of around 95,000, which corresponds to dozens of countries and regions [3].

A mathematical model of cholera propagation was first proposed to study cholera infection in 1973 around the Mediterranean region [4]. Henceforward, there have been numerous studies on the dynamics of cholera infection models. Tien and Earn [2] established a water-borne infectious disease model including multiple propagation paths. Control strategies, such as vaccination, were also considered in cholera models in Posny et al. [5] to inhibit the propagation of epidemics. Recently, by combining the cholera infection around aquatic regions, as well as the interaction between the bacteriophage and the cholera bacterium, researchers constructed a refined cholera infection model and provided reasonable cholera control strategies [6]. Considering environmental uncertainties and stochastic factors, researchers also studied a cholera system with respect to the *Itô* stochastic differential equation and confirmed the decisive effect of the stochastic basic reproduction number on the system [7].

Age structure, incorporating the age structure of the pathogen the infection age of individuals, is a significant characteristic in the cholera model [8–11]. A cholera model with bilinear incidence rates including two age structures was introduced and discussed in the work of Brauer et al. [12] and was further investigated in the work of Wang and Zhang [13]. The relative compactness of the orbits and the uniform persistence of the system were explored in [13]. The local stability of disease-free equilibrium and endemic equilibrium was analyzed in [13] and global stability was studied in [12]. Furthermore, a cholera transmission model incorporating vaccination age was analyzed in [14]. Actually, incidence rates are influenced by the complicated connections between susceptible individuals and the infected individuals/pathons. Various nonlinear incidence rates have been considered by researchers [15–19].

Inspired by the above works, we aim to discuss an age-structured cholera model. At time t, let $S(t)$ and $i(t,a)$ stand for the number of susceptible individuals and infected individuals with infection age a, with $p(t,b)$ representing the quantity of aquatic cholera pathogens at the age of b. Then, the infectivity of infected individuals and the total infectivity of the cholera pathogen at time t can be measured by $J(t) = \int_0^\infty k(a)i(t,a)da$ and $Q(t) = \int_0^\infty q(b)p(t,b)db$, in which kernel functions denote the infectivity of infected individuals and pathogens at corresponding ages. In this manuscript, we consider the following cholera model, taking general incidence rates into account, which is a generation of the model in Brauer et al. [12].

$$\begin{cases} \dfrac{dS(t)}{dt} = \Lambda - \mu S(t) - S(t)f(J(t)) - S(t)g(Q(t)), \\ \dfrac{\partial i(t,a)}{\partial t} + \dfrac{\partial i(t,a)}{\partial a} = -\delta(a)i(t,a), \\ \dfrac{\partial p(t,b)}{\partial t} + \dfrac{\partial p(t,b)}{\partial b} = -\gamma(b)p(t,b), \end{cases} \tag{1}$$

with boundary conditions

$$\begin{aligned} i(t,0) &= S(t)f(J(t)) + S(t)g(Q(t)), \quad t > 0, \\ p(t,0) &= \int_0^\infty \xi(a)i(t,a)da := P(t), \quad t > 0, \end{aligned} \tag{2}$$

and the initial condition

$$X_0 := (S(0), i(0,\cdot), p(0,\cdot)) = (S_0, i_0(\cdot), p_0(\cdot)) \in \chi_+, \tag{3}$$

where $\chi_+ = \mathbb{R}_+ \times \mathcal{L}^1_+(0,\infty) \times \mathcal{L}^1_+(0,\infty)$ is a functional space equipped with the norm

$$\|(\ell, \varphi, \phi)\|_{\chi_+} = |\ell| + \int_0^\infty |\varphi(a)|da + \int_0^\infty |\phi(b)|db.$$

In model (1), $\Lambda \in \mathbb{R}_+$ denotes the recruitment of the susceptible, and $\mu \in \mathbb{R}_+$ represents the natural death rate of individuals. $\gamma(b)$ describes the removal rate of pathogens at age b and $\xi(a)$ describes the pathogen shedding rate of an infected patient with infection age a. $\delta(a) = \mu + \delta_1(a) + \delta_2(a)$, where $\delta_1(a)$ is the disease-related death rate and $\delta_2(a)$ accounts for the recovery rate of infected individuals at infection age a. $S(t)f(J(t))$ and $S(t)g(Q(t))$ represent the direct and indirect transmission of cholera. For system (1), we make the following assumptions.

Assumption 1. *(I) The functions $\delta(a)$, $\gamma(b)$, $\xi(a)$, $k(a)$, $q(b) \in \mathcal{L}^\infty_+(0,+\infty)$ are bounded, integrable and Lipschitz-continuous. Denote $\bar{r} = \text{ess.}\sup_{a\in\mathbb{R}_+} r(a)$ and $\underline{r} = \text{ess.}\inf_{a\in\mathbb{R}_+} r(a)$ as essential upper and lower bound of $r(a)$ for $a \in \mathbb{R}_+$.*

(II) There exists one positive constant $\mathfrak{a}$ satisfying $i(t,a) = 0$ for $a \in [\mathfrak{a}, +\infty)$.

(III) $f(\ell)$ and $g(\ell)$ are Lipschitz-continuous on $\mathbb{R}_+$ with $f(0) = g(0) = 0$, $\frac{f(\ell)}{\ell} \geq f'(\ell) \geq 0$, $\frac{g(\ell)}{\ell} \geq g'(\ell) \geq 0$ and $f''(\ell) \leq 0, g''(\ell) \leq 0$, for $\ell \in \mathbb{R}_+$.

In this paper, for an age-infection model, we analyze the qualitative behavior by means of the Lyapunov functional method [20–22]. By considering the routes of the spread from the pathogen to the susceptible group and from the infected group to the susceptible group spread with generalized infection functions, we form a unified theoretical structure to present the propagation features of the epidemic. The basic reproduction number $\mathfrak{R}_0$ is defined as the threshold value, determining whether the epidemic dies out or not. Specifically, the cholera epidemic withers away if $\mathfrak{R}_0 < 1$, whereas if $\mathfrak{R}_0 > 1$, the disease persists at the endemic level.

The plan of this article is as follows. We give some preliminaries in the next section. In Section 3, we explore the existence and local stability of equilibria. In Section 4, we construct Lyapunov functionals to discuss the global stability of equilibria. In Section 5, we perform numerical simulations. Section 6 presents brief conclusions and a discussion.

2. Preliminaries

2.1. Existence and Uniqueness of Solutions

The standard theory for age-dependent models [8,11] can be applied to establish the existence and uniqueness of solutions for system (1) with boundary conditions (2) and initial condition (3). For this, we introduce the following Banach spaces

$$\begin{aligned}
\chi =&\mathbb{R} \times \mathbb{R} \times \mathcal{L}^1(\mathbb{R}_+,\mathbb{R}) \times \mathbb{R} \times \mathcal{L}^1(\mathbb{R}_+,\mathbb{R}),\\
\chi_0 =&\mathbb{R} \times \{0\} \times \mathcal{L}^1(\mathbb{R}_+,\mathbb{R}) \times \{0\} \times \mathcal{L}^1(\mathbb{R}_+,\mathbb{R}),\\
\chi_+ =&\mathbb{R}_+ \times \mathbb{R}_+ \times \mathcal{L}^1(\mathbb{R}_+,\mathbb{R}) \times \mathbb{R}_+ \times \mathcal{L}^1(\mathbb{R}_+,\mathbb{R}),\\
\chi_{0+} =&\chi_+ \cap \chi_0 = \mathbb{R}_+ \times \{0\} \times \mathcal{L}^1(\mathbb{R}_+,\mathbb{R}) \times \{0\} \times \mathcal{L}^1(\mathbb{R}_+,\mathbb{R}).
\end{aligned}$$

In order to formulate system (1) as an abstract Cauchy problem [23], we define the following linear operator, where $Dom(F_0) = \mathbb{R} \times \{0\} \times w^{1,1}(0,\infty) \times \{0\} \times w^{1,1}(0,\infty)$,

$$F_0: \ Dom(F_0) \subset \chi \to \chi,$$

$$F_0\begin{pmatrix} \phi_1 \\ \begin{pmatrix} 0 \\ \varphi_1 \end{pmatrix} \\ \begin{pmatrix} 0 \\ \varphi_2 \end{pmatrix} \end{pmatrix} = \begin{pmatrix} -\mu\phi_1 \\ \begin{pmatrix} -\varphi_1(0) \\ -\delta(\cdot)\varphi_1 - \varphi_1' \end{pmatrix} \\ \begin{pmatrix} -\varphi_2(0) \\ -\gamma(\cdot)\varphi_2 - \varphi_2' \end{pmatrix} \end{pmatrix},$$

and the nonlinear operator

$$F: \ \overline{Dom(F_0)} \subset \chi \to \chi,$$

$$F\begin{pmatrix} \phi_1 \\ \begin{pmatrix} 0 \\ \varphi_1 \end{pmatrix} \\ \begin{pmatrix} 0 \\ \varphi_2 \end{pmatrix} \end{pmatrix} = \begin{pmatrix} \Lambda - \phi_1 f(\int_0^\infty k(a)\varphi_1(a)da) - \phi_1 g(\int_0^\infty q(b)\varphi_2(b)db) \\ \begin{pmatrix} \phi_1 f(\int_0^\infty k(a)\varphi_1(a)da) + \phi_1 g(\int_0^\infty q(b)\varphi_2(b)db) \\ 0 \end{pmatrix} \\ \begin{pmatrix} \int_0^\infty \xi(a)\varphi_1(a)da \\ 0 \end{pmatrix} \end{pmatrix}.$$

Similarly to the proof process in [24,25], we can verify that operator F_0 is a Hille–Yosida operator [23].

Let $u(t) = (S(t),(0,i(t,\cdot))^T,(0,p(t,\cdot))^T)^T \in \chi_{0+}$. System (1) can be expressed by the following abstract cauchy problem:

$$\begin{cases} \dfrac{du(t)}{dt} = F_0 u(t) + F(u(t)), \quad \forall t \geq 0, \\ u(0) = u_0 \in \chi_0 \bigcap \chi_{0+}. \end{cases}$$

Let $(S(t),(0,i(t,\cdot))^T,(0,p(t,\cdot))^T)^T \in \chi_{0+}$. We have the following theorem by [23,26]:

Theorem 1. *There exists one unique determined semiflow $\{\mathcal{U}(t)\}_{t\geq 0}$ on χ_{0+} such that for any X_0, there exists one unique continuous map $\mathcal{U} \in C([0,\infty],\chi_{0+})$, acting as an integrated solution of the Cauchy problem, that is*

$$\begin{cases} \int_0^t \mathcal{U}(s)X_0 ds \in Dom(F_0), & \forall t \geq 0, \\ \mathcal{U}(t)X_0 = X_0 + F_0 \int_0^t \mathcal{U}(s)X_0 ds + \int_0^\infty F(\mathcal{U}(s)X_0)ds, & \forall t \geq 0. \end{cases}$$

2.2. Point Dissipativeness

Let

$$\Xi := \{(S(t), i(t,a), p(t,b)) \in \chi_{0+} | \\ S(t) + \int_0^\infty i(t,a)da \leq \frac{\Lambda}{\min\{\mu, \underline{\delta}\}}, \int_0^\infty p(t,b)db \leq \frac{\Lambda \bar{\xi}}{\min\{\mu, \underline{\delta}\}\underline{\gamma}}\}.$$

Then we have the following proposition:

Theorem 2. *Ξ is a positive invariant set under the semiflow $\{\mathcal{U}(t)\}_{t\geq 0}$. Moreover, the semiflow $\{\mathcal{U}(t)\}_{t\geq 0}$ is point-dissipative and attracts all the positive solutions of system (1) in χ_{0+}.*

Proof. From the first equation of (1), we have $\frac{dS(t)}{dt} \leq \Lambda - \mu S(t)$. Due to $S(0) \leq \frac{\Lambda}{\mu}$, we have $S(t) \leq \frac{\Lambda}{\mu}$. Note that

$$\begin{aligned} \frac{d}{dt}\int_0^\infty i(t,a)da &= \int_0^\infty (-\frac{\partial}{\partial a} i(t,a) - \delta(a)i(t,a))da \\ &= - i(t,a)|_0^\infty - \int_0^\infty \delta(a)i(t,a)da \\ &\leq i(t,0) - \int_0^\infty \delta(a)i(t,a)da \\ &= S(t)f(J(t)) + S(t)g(Q(t)) - \int_0^\infty \delta(a)i(t,a)da. \end{aligned}$$

Combining the first equation of (1), one yields

$$\begin{aligned} \frac{d}{dt}\left(S(t) + \int_0^\infty i(t,a)da\right) &\leq \Lambda - \mu S(t) - \int_0^\infty \delta(a)i(t,a)da \\ &\leq \Lambda - \mu S(t) - \underline{\delta}\int_0^\infty i(t,a)da \\ &\leq \Lambda - \min\{\mu, \underline{\delta}\}\left(S(t) + \int_0^\infty i(t,a)da\right). \end{aligned}$$

Since $S(0) + \int_0^\infty i(0,a)da \leq \frac{\Lambda}{\min\{\mu,\underline{\delta}\}}$, we have $S(t) + \int_0^\infty i(t,a)da \leq \frac{\Lambda}{\min\{\mu,\underline{\delta}\}}$. Thus, it follows that

$$\begin{aligned} \frac{d}{dt}\left(\int_0^\infty p(t,b)db\right) &= \int_0^\infty \left[-\frac{\partial}{\partial b}p(t,b) - \gamma(b)p(t,b)\right]db \\ &\leq p(t,0) - \int_0^\infty \gamma(b)p(t,b)db \\ &\leq \int_0^\infty \xi(a)i(t,a)da - \underline{\gamma}\int_0^\infty p(t,b)db \\ &\leq \frac{\Lambda\bar{\xi}}{\min\{\mu,\underline{\delta}\}} - \underline{\gamma}\int_0^\infty p(t,b)db. \end{aligned}$$

This implies that $\int_0^\infty p(t,b)db \leq \frac{\Lambda\bar{\xi}}{\min\{\mu,\underline{\delta}\}\underline{\gamma}}$.

Hence, $\mathcal{U}(t)\Xi \subset \Xi$ and this implies that Ξ is a positively invariant set and attracts all positive solutions of (1). □

From Theorem 2, we obtain the following result.

Proposition 1. *If $X_0 \in \chi_+$ and $\|X_0\|_\chi \le B$ with some sufficiently large constant B, then for $t \in \mathbb{R}_+$, we have the following propositions*

(i) $0 \le S(t), \int_0^\infty i(t,a)da, \int_0^\infty p(t,b)db \le B$;

(ii) $i(t,0) \le (\bar{k}f'(0) + \bar{q}g'(0))B^2$, $p(t,0) \le \bar{\xi}B$.

Proof. From the boundedness of system (1), we can find a constant B such that proposition (i) holds. Due to Assumption 1 (III), we further have

$$\begin{aligned} i(t,0) =& S(t)f(J(t)) + S(t)g(Q(t)) \\ \le& f'(0)S(t)J(t) + g'(0)S(t)Q(t) \\ \le& f'(0)S(t)\bar{k}\int_0^\infty i(t,a)da + g'(0)S(t)\bar{q}\int_0^\infty p(t,b)db \\ \le& (f'(0)\bar{k} + g'(0)\bar{q})B^2 \end{aligned}$$

and

$$p(t,0) = \int_0^\infty \xi(a)i(t,a)da \le \bar{\xi}\int_0^\infty i(t,a)da \le \bar{\xi}B.$$

This completes the proof. □

2.3. Asymptotical Smoothness and Global Attractor

From Equations (2) and (3), using the method presented in [8] to integrate the second and the third equations in (1) along the characteristic lines $t - a = const.$, we have

$$i(t,a) = \begin{cases} i(t-a,0)\varpi_1(a), & 0 \le a \le t, \\ i(0,a-t)\dfrac{\varpi_1(a)}{\varpi_1(a-t)}, & 0 \le t \le a, \end{cases} \tag{4}$$

and

$$p(t,b) = \begin{cases} p(t-b,0)\varpi_2(b), & 0 \le b \le t, \\ p(0,b-t)\dfrac{\varpi_2(b)}{\varpi_2(b-t)}, & 0 \le t \le b, \end{cases} \tag{5}$$

where

$$\varpi_1(a) = e^{-\int_0^a \delta(\tau)d\tau} \text{ and } \varpi_2(b) = e^{-\int_0^b \gamma(\tau)d\tau} \tag{6}$$

denote the fraction at which an infected cell and virus survive up to age a and b.

In order to explore the existence of an attractor, we first analyze the asymptotical smoothness of semiflow $\mathcal{U}(t)$. For this, we present the following proposition.

Proposition 2. *The functions $J(t)$, $Q(t)$ and $P(t)$ are Lipschitz-continuous.*

Proof. Here we give the proof of $J(t)$ being Lipschitz-continuous. From Assumption 1, there exists a positive constant M_k such that $|k(a+l) - k(a)| \le M_k l$. Then, combining Proposition 1, it holds that

$$\begin{aligned} &|J(t+l) - J(t)| \\ \le& \int_0^l k(a)i(t+l-a,0)\varpi_1(a)\mathrm{d}a + \left|\int_l^\infty k(a)i(t+l,a)\mathrm{d}a - \int_0^\infty k(a)i(t,a)\mathrm{d}a\right| \\ \le& \bar{k}(\bar{k}f'(0) + \bar{q}g'(0))B^2 l + \left|\int_l^\infty k(a)i(t+l,a)\mathrm{d}a - \int_0^\infty k(a)i(t,a)\mathrm{d}a\right|. \end{aligned}$$

Let $o = a - l$; we have

$$\begin{aligned}
&|J(t+l) - J(t)| \\
\leq& \bar{k}(\bar{k}f'(0) + \bar{q}g'(0))B^2 l + \left| \int_0^\infty k(o+l)i(t+l, o+l)\mathrm{d}o - \int_0^\infty k(a)i(t,a)\mathrm{d}a \right| \\
=& \bar{k}(\bar{k}f'(0) + \bar{q}g'(0))B^2 l + \left| \int_0^\infty \left(k(a+l)\frac{\varpi_1(a+l)}{\varpi_1(a)} - k(a) \right) i(t,a)\mathrm{d}a \right| \\
=& \bar{k}(\bar{k}f'(0) + \bar{q}g'(0))B^2 l + \left| \int_0^\infty k(a+l)\left(e^{-\int_a^{a+l}\delta(\tau)\mathrm{d}\tau} - 1 \right) i(t,a)\mathrm{d}a \right| \\
&+ \left| \int_0^\infty (k(a+l) - k(a)) i(t,a)\mathrm{d}a \right| \\
\leq& \bar{k}(\bar{k}f'(0) + \bar{q}g'(0))B^2 l + \bar{k}\bar{\delta}lB + M_k lB := M_J.
\end{aligned}$$

Hence, $J(t)$ is Lipschitz-continuous with the coefficient M_J. Through similar verification, the functions $Q(t)$ and $P(t)$ are Lipschitz-continuous with coefficients M_Q and M_P. □

For the asymptotical smoothness of the semiflow, the following lemma [27] is necessary.

Lemma 1. *The semiflow $\mathcal{U} : \mathbb{R}_+ \times \mathcal{X}_+ \to \mathcal{X}_+$ is asymptotically smooth if there are maps Ψ, $\Theta : \mathbb{R}_+ \times \mathcal{X}_+ \to \mathcal{X}_+$ such that $\mathcal{U}(t,x) = \Psi(t,x) + \Theta(t,x)$ and the following holds for any bounded closed set $\mathbb{B} \subset \mathcal{X}_+$, which is forward invariant under $\mathcal{U}$:* (i) $\lim_{t\to\infty} \mathrm{diam}\Theta(t,\mathbb{B}) = 0$; (ii) *There exists $t_{\mathbb{B}} \geq 0$ such that $\Psi(t,\mathbb{B})$ has compact closure for each $t \geq t_{\mathbb{B}}$.*

For condition (ii) of Lemma 1, we introduce the following lemma [27].

Lemma 2. *A set $\mathbb{B} \in \mathcal{L}^1_+(0,\infty)$ has a compact closure if the following conditions hold:* (i) $\sup_{f\in\mathbb{B}} \int_0^\infty f(\ell)\mathrm{d}\ell < \infty$; (ii) $\lim_{r\to\infty} \int_r^\infty f(\ell)\mathrm{d}\ell \to 0$ *uniformly in $f \in \mathbb{B}$;* (iii) $\lim_{h\to 0^+} \int_0^\infty |f(\ell+h) - f(\ell)|\mathrm{d}\ell \to 0$ *uniformly in $f \in \mathbb{B}$;* (iv) $\lim_{h\to 0^+} \int_0^h f(\ell)\mathrm{d}\ell \to 0$ *uniformly in $f \in \mathbb{B}$.*

Based on Lemmas 1 and 2, we investigate the asymptotical smoothness.

Theorem 3. *The semiflow $\mathcal{U}$ generated by (1) is asymptotically smooth.*

Proof. Define the maps Ψ and Θ such that $\mathcal{U} = \Psi + \Theta$, with

$$\begin{cases} \Psi(t,x_0) = (S(t), \check{i}(t,\cdot), \check{p}(t,\cdot)), \\ \Theta(t,x_0) = (0, \check{\varphi}_i(t,\cdot), \check{\varphi}_p(t,\cdot)), \end{cases}$$

where

$$\begin{aligned}
\check{i}(t,a) &= \begin{cases} i(t-a,0)\varpi_1(a), & 0 \leq a \leq t, \\ 0, & 0 \leq t \leq a, \end{cases} \\
\check{p}(t,b) &= \begin{cases} p(t-b,0)\varpi_2(b), & 0 \leq b \leq t, \\ 0, & 0 \leq t \leq b, \end{cases} \\
\check{\varphi}_i(t,a) &= \begin{cases} 0, & 0 \leq a \leq t, \\ i(0,a-t)\dfrac{\varpi_1(a)}{\varpi_1(a-t)}, & 0 \leq t \leq a, \end{cases} \\
\check{\varphi}_p(t,b) &= \begin{cases} 0, & 0 \leq b \leq t, \\ p(0,b-t)\dfrac{\varpi_2(b)}{\varpi_2(b-t)}, & 0 \leq t \leq b. \end{cases}
\end{aligned}$$

Firstly, we show that map Θ satisfies condition (i) of Lemma 1. For $X_0 \in \Gamma$ satisfying $\|X_0\|_\mathcal{X} \le r$, letting $a - t = \epsilon_1$ and $b - t = \epsilon_2$, we have

$$\begin{aligned}
&\|\Theta(t, X_0)\|_\mathcal{X} \\
=&\int_t^\infty \left| i(0, a-t)\frac{\varpi_1(a)}{\varpi_1(a-t)} \right| \mathrm{d}a + \int_t^\infty \left| p(0, b-t)\frac{\varpi_2(b)}{\varpi_2(b-t)} \right| \mathrm{d}b \\
=&\int_0^\infty \left| i(0, \epsilon_1)e^{-\int_{\epsilon_1}^{\epsilon_1+t}\delta(\tau)\mathrm{d}\tau} \right| \mathrm{d}\epsilon + \int_0^\infty \left| p(0, \epsilon_2)e^{-\int_{\epsilon_2}^{\epsilon_2+t}\gamma(\tau)\mathrm{d}\tau} \right| \mathrm{d}\epsilon \\
\le& e^{-\underline{\delta}t}\int_0^\infty |i(0,\epsilon_1)|\mathrm{d}\epsilon + e^{-\underline{\gamma}t}\int_0^\infty |p(0,\epsilon_2)|\mathrm{d}\epsilon \\
\le& e^{-\min\{\underline{\delta},\underline{\gamma}\}t}r, \qquad t \in \mathbb{R}_+.
\end{aligned}$$

This shows that $\|\Theta(t, X_0)\|_\mathcal{X} \to 0$ as $t \to \infty$, which indicates that $\|\Theta(t, X_0)\|_\mathcal{X}$ approaches 0 with uniform exponential speed. Thus, the proof of Lemma 1 (i) is completed.

Then, we verify that Lemma 2 holds. Using Proposition 1, we can verify that conditions (i), (ii) and (iv) of Lemma 2 hold since

$$0 \le \check{i}(t,a) \le S(t-a)[f'(0)J(t-a) + g'(0)Q(t-a)]\varpi_1(a) \le [f'(0)\bar{k} + g'(0)\bar{q}]B^2 e^{-\underline{\delta}a}.$$

Finally, we focus on condition (iii) of Lemma 2. For sufficiently small $h \in (0,t)$, we have

$$\int_0^\infty |\check{i}(a+h,t) - \check{i}(a,t)|\mathrm{d}a \le \zeta_1 + \zeta_2 + \zeta_3, \tag{7}$$

where

$$\begin{aligned}
\zeta_1 =& \int_0^{t-h} |S(t-a-h)f(J(t-a-h))\varpi_1(a+h) - S(t-a)f(J(t-a))\varpi_1(a)|\mathrm{d}a, \\
\zeta_2 =& \int_0^{t-h} |S(t-a-h)g(Q(t-a-h))\varpi_1(a+h) - S(t-a)g(Q(t-a))\varpi_1(a)|\mathrm{d}a, \\
\zeta_3 =& f'(0)\int_{t-h}^{t} |S(t-a)J(t-a)\varpi_1(a)|\mathrm{d}a + g'(0)\int_{t-h}^{t} |S(t-a)Q(t-a)\varpi_1(a)|\mathrm{d}a \\
&\le (f'(0)\bar{k} + g'(0)\bar{q})B^2 h := \zeta_{3M}.
\end{aligned}$$

Note that

$$\begin{aligned}
\zeta_1 =& \int_0^{t-h} |S(t-a-h)f(J(t-a-h))(\varpi_1(a+h) - \varpi_1(a))\mathrm{d}a \\
&+ \int_0^{t-h} S(t-a-h)|f(J(t-a-h)) - f(J(t-a))|\varpi_1(a)\mathrm{d}a \\
&+ \int_0^{t-h} |S(t-a-h) - S(t-a)|f(J(t-a))\varpi_1(a)\mathrm{d}a \\
\le& f'(0)\bar{k}B^2\int_0^{t-h} |\varpi_1(a+h) - \varpi_1(a)|\mathrm{d}a \\
&+ f'(0)\int_0^{t-h} S(t-a-u)|J(t-a-h) - J(t-a)|\varpi_1(a)\mathrm{d}a \\
&+ f'(0)\int_0^{t-h} J(t-a)|S(t-a-h) - S(t-a)|\varpi_1(a)\mathrm{d}a.
\end{aligned} \tag{8}$$

Let $M_S = \Lambda + \mu B + (\bar{k}f'(0) + \bar{q}g'(0))B^2$ be the Lipschitz coefficient of $S(t)$. Then, the following holds:

$$\zeta_1 \le f'(0)\bar{k}B^2 h + f'(0)BM_J h^2 + f'(0)\bar{k}BM_s h^2 := \zeta_{1M}. \tag{9}$$

Similarly, we have

$$\zeta_2 \le g'(0)\bar{q}B^2h + g'(0)BM_Qh^2 + g'(0)\bar{q}BM_sh^2 := \zeta_{2M}. \tag{10}$$

Combining equations (9) and (10) with (7), we obtain

$$\int_0^\infty |\check{i}(a+h,t) - \check{i}(a,t)|da \le \zeta_{1M} + \zeta_{2M} + \zeta_{3M}.$$

ζ_{iM}, $i = 1,2,3$, does not rely on the initial condition X_0. Thus, Lemma 2 holds. Hence, $\check{i}(t,a)$ remains in a pre-compact subset in $\mathcal{L}^1_+(0,\infty)$, and so does $\check{p}(t,b)$. We thus accomplish the proof. □

Based on the above preparations, the following results hold due to Theorem 3.4.6 of Hale [28].

Theorem 4. *The semi-flow $\mathcal{U}(t)$ has a global attractor $\mathcal{A}$ in χ_+, which attracts all bound subsets of χ_+.*

3. Existence and Local Stability of Equilibria

3.1. Equilibria and Basic Reproductive Number

System (1) possesses two equilibria at most in Θ. Besides the infection-free equilibrium $E^0 = (S_0, 0, 0)$ with $S_0 = \Lambda/\mu$, there possibly exists an infection equilibrium $E^* = (S^*, i^*(a), p^*(b))$ in Θ, satisfying the following equations

$$\begin{cases} \Lambda = \mu S^* + S^* f(J^*) + S^* g(Q^*), \\ \dfrac{\partial i^*(a)}{\partial a} = -\delta(a) i^*(a), \\ \dfrac{\partial p^*(b)}{\partial b} = -\gamma(b) p^*(b), \\ i^*(0) = S^* f(J^*) + S^* g(Q^*), \\ p^*(0) = \displaystyle\int_0^\infty \xi(a) i^*(a) da, \end{cases} \tag{11}$$

where $J^* = \int_0^\infty k(a) i^*(a) da$ and $Q^* = \int_0^\infty q(b) p^*(b) db$.

From the second and third equations of system (11), we have

$$i^*(a) = i^*(0)\varpi_1(a), \;\; p^*(b) = p^*(0)\varpi_2(b).$$

Let

$$\Pi_1 = \int_0^\infty k(a)\varpi_1(a)da, \; \Pi_2 = \int_0^\infty q(b)\varpi_2(b)db \text{ and } \Pi_3 = \int_0^\infty \xi(a)\varpi_1(a)da. \tag{12}$$

We can further obtain

$$\begin{aligned} J^* &= \int_0^\infty k(a) i^*(0)\varpi_1(a)da = \int_0^\infty k(a)[S^* f(J^*) + S^* g(Q^*)]\varpi_1(a)da \\ &= [S^* f(J^*) + S^* g(Q^*)]\Pi_1 \end{aligned} \tag{13}$$

and

$$\begin{aligned} Q^* &= \int_0^\infty q(b) p^*(0)\varpi_2(b)db = \int_0^\infty \xi(a) i^*(a)da \int_0^\infty q(b)\varpi_2(b)db \\ &= \Pi_2 \int_0^\infty \xi(a) i^*(0)\varpi_1(a)da = \Pi_2\Pi_3(S^* f(J^*) + S^* g(Q^*)) = \frac{\Pi_2\Pi_3}{\Pi_1} J^*. \end{aligned} \tag{14}$$

Thus, combining equations $S^* = \Lambda/(\mu + f(J^*) + g(Q^*))$, (13) and (14), we have

$$J^* = \frac{\Lambda\Pi_1[f(J^*) + g(Q^*)]}{\mu + f(J^*) + g(Q^*)} = \frac{\Lambda\Pi_1[f(J^*) + g(\frac{\Pi_2\Pi_3}{\Pi_1}J^*)]}{\mu + f(J^*) + g(\frac{\Pi_2\Pi_3}{\Pi_1}J^*)}.$$

Let $h(J) = \mu J + [J - \Lambda\Pi_1][f(J) + g(\frac{\Pi_2\Pi_3}{\Pi_1}J)]$. Then, we yield $h(0) = 0$, $h(\Lambda\Pi_1) = \mu\Lambda\Pi_1$ and

$$\begin{aligned} h'(0) &= \{\mu + [f(J) + g(\frac{\Pi_2\Pi_3}{\Pi_1}J)] + [J - \Lambda\Pi_1][f'(J) + \frac{\Pi_2\Pi_3}{\Pi_1}g'(\frac{\Pi_2\Pi_3}{\Pi_1}J)]\}|_{J=0} \\ &= \mu[1 - \frac{\Lambda\Pi_1}{\mu}(f'(0) + \frac{\Pi_2\Pi_3}{\Pi_1}g'(0))]. \end{aligned}$$

Define the basic reproduction number of system (1) as

$$\Re_0 = \frac{\Lambda\Pi_1}{\mu}(f'(0) + \frac{\Pi_2\Pi_3}{\Pi_1}g'(0)). \tag{15}$$

When $\Re_0 > 1$, $h'(0) < 0$ and there exists at least one E^*. Then, we obtain

$$h'(J^*) = \mu + [f(J^*) + g(\frac{\Pi_2\Pi_3}{\Pi_1}J^*)] + [J^* - \Lambda\Pi_1][f'(J^*) + \frac{\Pi_2\Pi_3}{\Pi_1}g'(\frac{\Pi_2\Pi_3}{\Pi_1}J^*)]$$

and

$$h''(J^*) = 2[f'(J^*) + \frac{\Pi_2\Pi_3}{\Pi_1}g'(\frac{\Pi_2\Pi_3}{\Pi_1}J^*)] + [J^* - \Lambda\Pi_1][f''(J^*) + \frac{\Pi_2\Pi_3}{\Pi_1}g''(\frac{\Pi_2\Pi_3}{\Pi_1}J^*)] > 0.$$

Thus, there exists one unique positive equilibrium E^*. This yields the following theorem.

Theorem 5. *System (1) always exists a disease-free steady state $E^0 = (S_0, 0, 0)$. Furthermore, another endemic steady state $E^* = (T^*, i^*(a), V^*)$ exists if $\Re_0 > 1$.*

3.2. Local Stability of Equilibria

The global asymptotical stability of equilibria is conducive to forecasting the trends of epidemics [29–35]. For this, we first focus on the local stability by exploring the corresponding characteristic equations.

Theorem 6. *The infection-free equilibrium is locally asymptotically stable when $\Re_0 < 1$. The infection equilibrium is locally asymptotically stable when $\Re_0 > 1$.*

Proof. The characteristic equation for the linearized part of system (1) with boundary conditions (2) on $(S_0, 0, 0)$ is

$$(\lambda + \mu)(-1 + S_0 f'(0)\pi_1(\lambda) + S_0 g'(0)\pi_2(\lambda)\pi_3(\lambda)) = 0, \tag{16}$$

where

$$\begin{aligned} \pi_1(\lambda) &= \int_0^\infty k(a)e^{-\lambda a}\varpi_1(a)da, \\ \pi_2(\lambda) &= \int_0^\infty q(b)e^{-\lambda b}\varpi_2(b)db, \\ \pi_3(\lambda) &= \int_0^\infty \xi(a)e^{-\lambda a}\varpi_1(a)da. \end{aligned}$$

Then, if $\mathfrak{R}_0 < 1$, all roots of the characteristic equation (16) have negative parts. If not, that is, if there exists a λ_0 such that $Re\lambda_0 \geq 0$, then

$$\begin{aligned} &-1 + S_0 f'(0)\pi_1(\lambda) + S_0 g'(0)\pi_2(\lambda)\pi_3(\lambda) \\ \leq &\frac{\Lambda\Pi_1}{\mu}[f'(0) + \frac{\Pi_2\Pi_3}{\Pi_1}g'(0)] - 1 = \mathfrak{R}_0 - 1 < 0. \end{aligned}$$

This is a contradiction with equation (16). Thus, the infection-free equilibrium is locally asymptotically stable when $\mathfrak{R}_0 < 1$.

Similarly, for $\mathfrak{R}_0 > 1$, combining the linearization of the system on $(S^*, i^*(a), p^*(b))$, the corresponding characteristic equation of the linearization for system (1) is

$$(\lambda + \mu + f(J^*) + g(Q^*))/(\lambda + \mu) = S^* f'(J^*)\pi_1(\lambda) + S^* g'(Q^*)\pi_2(\lambda)\pi_3(\lambda). \quad (17)$$

Now we assume that system (17) has one characteristic root with a positive real root. Since $J^* = i^*(0)\Pi_1$, $Q^* = p^*(0)\Pi_2$, $\Pi_3 = \int_0^\infty \xi(a)\frac{i^*(a)}{i^*(0)}da$ and $p^*(0) = \int_0^\infty \xi(a)i^*(a)da$, we have

$$\begin{aligned} |S^* f'(J^*)\pi_1(\lambda) + S^* g'(Q^*)\pi_2(\lambda)\pi_3(\lambda)| &\leq \left| S^*\frac{f(J^*)}{J^*}\Pi_1(\lambda) + S^*\frac{g(Q^*)}{Q^*}\Pi_2(\lambda)\Pi_3(\lambda) \right| \\ &\leq \left| \frac{S^* f(J^*)}{i^*(0)} + \frac{S^* g(Q^*)}{p^*(0)}\Pi_3 \right| \\ &= 1. \end{aligned}$$

This is a contradiction with Equation (17). Thus, E^* is locally asymptotically stable when $\mathfrak{R}_0 > 1$. □

4. Global Stability of Equilibria

For the proof of the global attractiveness of equilibria, we apply the Lyapunov functional method. For the invariance principle, we have investigated the relative compactness of the orbits. For the well-posedness of Lyapunov functionals, the uniform persistence of system should also be discussed.

4.1. Uniform Persistence

In this section, we aim to investigate the uniform persistence of system (1). Define

$$M = \{(S, (0, i), (0, p)) \in \Xi : S(t) + \int_0^\infty i(t, a)da + \int_0^\infty p(t, b)db > 0\}$$

and $\partial M = \Xi \backslash M$.

Lemma 3. *The subsets M and ∂M are both positively invariant under the semiflow $\{\mathcal{U}(t)\}_{t\geq 0}$ generated by system (1) on χ_{0+}, that is,*

$$\mathcal{U}(t)M \subset M, \mathcal{U}(t)\partial M \subset \partial M.$$

Moreover, for each $\xi \in \partial M, \mathcal{U}(t)\xi \to E^0$ as $t \to +\infty$.

Proof. Let $G(t) = \int_0^\infty i(t,a)da + \int_0^\infty p(t,b)db$. For any $\varsigma = (S(t),(0,i(t,a)),(0,p(t,b))) \in M$, we have

$$\begin{aligned}\frac{dG(t)}{dt} &= \int_0^\infty \left(-\delta(a)i(t,a) - \frac{\partial i(t,a)}{\partial a}\right)da + \int_0^\infty \left(-\gamma(b)p(t,b) - \frac{\partial p(t,b)}{\partial b}\right)db \\ &= -\int_0^\infty \delta(a)i(t,a)da + i(t,0) - \int_0^\infty \gamma(b)p(t,b)db + p(t,0) \\ &\geq -\overline{\delta}\int_0^\infty i(t,a)da - \overline{\gamma}\int_0^\infty p(t,b)db \\ &= -\max\{\overline{\delta},\overline{\gamma}\}G(t).\end{aligned}$$

For any $\varsigma = (S(t),(0,i(t,a)),(0,p(t,b))) \in M$, we have $G(0) > 0$. Thus, $G(T) \geq G(0)e^{-\max\{\overline{\delta},\overline{\gamma}\}} > 0$ and then we have $\mathcal{U}(t)M \subset M$. Thus, M is positively invariant.

In the following, we try to prove that $\mathcal{U}(t)\partial M \subset \partial M$. For any $\varsigma = (S_0(t),(0,i_0(t,a)),(0,p_0(t,b))) \in \partial M$, we have

$$0 \leq \int_0^\infty i(t,a)da = \int_0^t i(t-a,0)\varpi_1(a)da + \int_t^\infty i(0,a-t)\frac{\varpi_1(a)}{\varpi_1(a-t)}da \leq 0$$

and

$$0 \leq \int_0^\infty p(t,b)db = \int_0^t p(t-b,0)\varpi_2(b)db + \int_t^\infty p(0,b-t)\frac{\varpi_2(b)}{\varpi_2(b-t)}db \leq 0.$$

Thus, $\int_0^\infty i(t,a)da = 0$ and $\int_0^\infty p(t,b)db = 0$. □

Then we obtain the following theorem by means of [36].

Theorem 7. *If $\mathfrak{R}_0 > 1$, then the semiflow $\{\mathcal{U}(t)\}_{t\geq 0}$ is uniformly persistent with respect to the pair $(\partial M, M)$, that is, there exists $\epsilon > 0$, such that $\liminf_{t\to\infty} d(\mathcal{U}(t)\xi, \partial M) \geq \epsilon$, $\forall \xi \in M$.*

Proof. We need to verify that $W^s(E^0)\bigcap M = \emptyset$, where

$$W^s(E^0) = \{\xi \in \Omega : \lim_{t\to\infty} \mathcal{U}(t)\xi = E^0\}.$$

Suppose there exists $\xi_0 \in W^s(E^0)\bigcap M$. Then, there exists a t_1, such that

$$\int_0^\infty i(t_1,a)da + \int_0^\infty p(t_1,b)db > 0.$$

Since M is an invariant set, we have

$$\int_0^\infty i(t,a)da + \int_0^\infty p(t,b)db > 0, \ \forall t > t_1.$$

Since $\xi_0 \in W^s(E^0)$, we have $\lim_{t\to\infty} S(t) = S^0$. Thus, for $\epsilon_0 > 0$, there exists t_2, such that

$$S(t) > S^0 - \epsilon_0, \ \forall t \geq t_2.$$

Let $H(t) = \int_0^\infty \sigma(a)i(t,a)da + \int_0^\infty \varrho(b)p(t,b)db$, where

$$\begin{aligned}\sigma(a) &= \int_a^\infty [S_0 f'(0)k(u) + \varrho(0)\xi(u)]e^{-\int_a^u \delta(\tau)d\tau}du, \\ \varrho(b) &= \int_b^\infty S_0 g'(0)q(v)e^{-\int_b^v \gamma(\tau)d\tau}dv.\end{aligned}$$

Then we have

$$\begin{aligned}H'(t) =& \int_0^{\infty} \sigma(a)\frac{\partial i(t,a)}{\partial t}da + \int_0^{\infty} \varrho(b)\frac{\partial p(t,b)}{\partial b}db \\ =&\sigma(0)i(t,0) + \int_0^{\infty} i(t,a)[\sigma'(a) - \sigma(a)\delta(a)]da \\ &+ \varrho(0)p(t,0) + \int_0^{\infty} p(t,b)[\varrho'(b) - \varrho(b)\gamma(b)]db \\ =&\sigma(0)i(t,0) - \int_0^{\infty} i(t,a)[S_0 f'(0)k(a) - \varrho(0)\xi(a)]da \\ &+ \varrho(0)p(t,0) - \int_0^{\infty} p(t,b)S_0 g'(0)q(b)db.\end{aligned}$$

Since $p(t,0) = \int_0^{\infty} i(t,a)\xi(a)da$, $i(t,0) = Sf(J) + Sg(Q)$ and

$$\sigma(0) = S_0 f'(0)\Pi_1 + \varrho(0)\Pi_3 = \frac{\Lambda}{\mu}\Pi_1[f'(0) + \frac{\Pi_2\Pi_3}{\Pi_1}g'(0)] = \Re_0,$$

we further have for a sufficiently large t,

$$\begin{aligned}H'(t) =&\sigma(0)i(t,0) - S_0 f'(0)\int_0^{\infty} i(t,a)k(a)da - S_0 g'(0)\int_0^{\infty} p(t,b)q(b)db \\ =&\sigma(0)[Sf(J) + Sg(Q)] - S_0 f'(0)J - S_0 g'(0)Q \\ =&[\sigma(0)Sf(J) - S_0 f'(0)J] + [\sigma(0)Sg(Q) - S_0 g'(0)Q] \\ \geq&[\sigma(0)(S_0 - \epsilon_0)f(J) - S_0 f'(0)J] + [\sigma(0)(S_0 - \epsilon_0)g(Q) - S_0 g'(0)Q] \\ =&S_0[\sigma(0)(1 - \frac{\epsilon_0}{S_0})f(J) - f'(0)J] + S_0[\sigma(0)(1 - \frac{\epsilon_0}{S_0})g(Q) - g'(0)Q] \\ \geq&0.\end{aligned}$$

This indicates that $H(t)$ is a non-decreasing function for a sufficiently large t. Hence, for a sufficiently large t, $H(t) > 0$, which prevents the orbits from converging to E^0 as $t \to +\infty$. This contradicts $\xi_0 \in W^s(E^0)$. □

4.2. Global Stability of the Infection-Free Equilibrium

This subsection explores the global stability of the infection-free equilibrium E^0.

Theorem 8. *E^0 is globally asymptotically stable when $\Re_0 < 1$.*

Proof. Define the Liapunov function $L(t) = L_1(t) + L_2(t) + L_3(t)$, with

$$L_1(t) = S(t) - S_0 - S_0\ln(\frac{S(t)}{S_0}),\ L_2(t) = \int_0^{\infty} \sigma(a)i(t,a)da,\ L_3(t) = \int_0^{\infty} \varrho(b)p(t,b)db.$$

Then, calculating the derivatives of $L_i(t)$, $i = 1,2,3$, along the trajectories of system (1) gives

$$\begin{aligned}\frac{dL_1}{dt} =& -\frac{\mu}{S(t)}(S(t) - S_0)^2 - i(t,0) + S_0 f(J(t)) + S_0 g(Q(t)) \\ \leq& -\frac{\mu}{S(t)}(S(t) - S_0)^2 - i(t,0) \\ &+ S_0 f'(0)\int_0^{\infty} k(a)i(t,a)da + S_0 g'(0)\int_0^{\infty} q(b)p(t,b)db,\end{aligned} \tag{18}$$

and

$$\begin{aligned}\frac{dL_2}{dt} &= \int_0^\infty \sigma(a)(-\delta(a)i(t,a) - \frac{\partial i(t,a)}{\partial a})da \\ &= -\int_0^\infty \sigma(a)\delta(a)i(t,a)da - \int_0^\infty \sigma(a)di(t,a) \\ &= \sigma(0)i(t,0) + \int_0^\infty i(t,a)(\sigma'(a) - \sigma(a)\delta(a))da,\end{aligned} \tag{19}$$

and similarly

$$\frac{dL_3}{dt} = \varrho(0)p(t,0) + \int_0^\infty p(t,b)(\varrho'(b) - \varrho(b)\gamma(b))db. \tag{20}$$

Since $\sigma'(a) = -[S_0 f'(0)k(a) + \varrho(0)\xi(a)] + \delta(a)\sigma(a)$ and $\varrho'(b) = -S_0 g'(0)q(b) + \gamma(b)\varrho(b)$, we further have

$$\begin{aligned}\frac{dL}{dt} =& \frac{dL_1}{dt} + \frac{dL_2}{dt} + \frac{dL_3}{dt} \\ \leq& -\frac{\mu}{S(t)}(S(t) - S_0)^2 - i(t,0) + \sigma(0)i(t,0) \\ &+ \int_0^\infty i(t,a)[S_0 f'(0)k(a) + \sigma'(a) - \sigma(a)\delta(a) + \varrho(0)\xi(a)]da \\ &+ \int_0^\infty p(t,b)[S_0 g'(0)q(b) + \varrho'(b) - \varrho(b)\gamma(b)]db \\ =& -\frac{\mu}{S(t)}(S(t) - S_0)^2 + i(t,0)[\sigma(0) - 1].\end{aligned} \tag{21}$$

Thus, when $\Re_0 = \sigma(0) < 1$, $\frac{dL}{dt} \leq 0$. The largest invariant set of $\{\frac{dL}{dt} = 0\}$ is singleton $\{E^0\}$. Hence, due to the invariance principle [37], E^0 is globally asymptotically stable when $\Re_0 < 1$. □

4.3. Global Stability of the Infection Equilibrium

In this subsection, we focus on the global stability of the infection equilibrium E^*. To this end, we introduce a function h defined by

$$h(z) = z - 1 - \ln z,\ \ z \in R_+.$$

In order to ensure that $h\left(\frac{i(t,a)}{i^*(a)}\right)$ and $h\left(\frac{p(t,b)}{p^*(b)}\right)$ are well-defined, we have shown that $i(t,a)/i^*(a)$ and $p(t,b)/p^*(b)$ are bounded below and above through the above uniform persistence analysis. In the following, we prove the following result.

Theorem 9. *The infection equilibrium E^* is globally asymptotically stable when $\Re_0 > 1$.*

Proof. Define a Lyapunov function $V(t) = V_1(t) + V_2(t) + V_3(t)$, where

$$V_1(t) = S^* h(\frac{S}{S^*})i^*(0),\ V_2(t) = \int_0^\infty \Gamma(a)i^*(a)h(\frac{i}{i^*})di,\ V_3(t) = \frac{1}{\Pi_3}\int_0^\infty \Upsilon(b)p^*(b)h(\frac{p}{p^*})dp,$$

with

$$\begin{aligned}\Gamma(a) =& \frac{1}{\Pi_1}\int_a^\infty S^* f(J^*)k(u)e^{-\int_a^u \delta(\tau)d\tau}du + \frac{1}{\Pi_3}\int_a^\infty \Upsilon(0)\xi(u)e^{-\int_a^u \delta(\tau)d\tau}du, \\ \Upsilon(b) =& \frac{1}{\Pi_2}\int_b^\infty S^* g(Q^*)q(v)e^{-\int_b^v \gamma(\tau)d\tau}dv.\end{aligned}$$

Here, we make some preparations. Firstly, since $i_a^*(a) = -i^*(a)\delta(a)$, we have

$$\begin{aligned} i^*(a)\frac{d}{da}[\frac{i(t,a)}{i^*(a)} - 1 - \ln\frac{i(t,a)}{i^*(a)}] &= (1 - \frac{i^*(a)}{i(t,a)})\frac{i_a(t,a)i^*(a) - i(t,a)i_a^*(a)}{i^*(a)} \\ &= (1 - \frac{i^*(a)}{i(t,a)})[i_a(t,a) + i(t,a)\delta(a)]. \end{aligned}$$

Thus,

$$(1 - \frac{i^*(a)}{i(t,a)})i_a(t,a) = i^*(a)\frac{d}{da}[h(\frac{i(t,a)}{i^*(a)})] - \delta(a)[i(t,a) - i^*(a)]. \tag{22}$$

Similarly, we have

$$(1 - \frac{p^*(b)}{p(t,b)})p_b(t,b) = p^*(b)\frac{d}{db}[h(\frac{p(t,b)}{p^*(b)})] - \gamma(a)[p(t,b) - p^*(b)]. \tag{23}$$

Then, calculating the derivative of V_1 along system (1) gives

$$\begin{aligned} \frac{dV_1}{dt} =& [-\frac{\mu}{S}(S - S^*)^2 + S^*f(J^*) + S^*g(Q^*) - Sf(J) - Sg(Q) \\ & - \frac{S^*}{S}S^*f(J^*) - \frac{S^*}{S}S^*g(Q^*) + S^*f(J) + S^*g(Q)]i^*(0). \end{aligned} \tag{24}$$

Because of equation (22), we obtain

$$\begin{aligned} \frac{dV_2}{dt} &= -\int_0^\infty \Gamma(a)(1 - \frac{i^*(a)}{i(t,a)})[\frac{\partial i(t,a)}{\partial a} + \delta(a)i(t,a)]da \\ &= \int_0^\infty -\Gamma(a)i^*(a)\frac{d}{da}[h(\frac{i(t,a)}{i^*(a)})]da, \\ &= -\Gamma(a)i^*(a)h(\frac{i(t,a)}{i^*(a)})|_0^\infty + \int_0^\infty h(\frac{i(t,a)}{i^*(a)})[\Gamma'(a)i^*(a) + \Gamma(a)i_a^*(a)]da \\ &= -\Gamma(\infty)i^*(\infty)h(\frac{i(t,\infty)}{i^*(\infty)}) + \Gamma(0)i^*(0)h(\frac{i(t,0)}{i^*(0)}) \\ &\quad + \int_0^\infty h(\frac{i(t,a)}{i^*(a)})[\Gamma'(a)i^*(a) + \Gamma(a)i_a^*(a)]da. \end{aligned}$$

Due to the fact that $\Gamma'(a) = -\frac{1}{\Pi_1}S^*f(J^*)k(a) - \frac{1}{\Pi_3}\Upsilon(0)\xi(a) + \delta(a)\Gamma(a)$ and $i_a^*(a) = -i^*(a)\delta(a)$, we have

$$\frac{dV_2}{dt} \leq \Gamma(0)i^*(0)h(\frac{i(t,0)}{i^*(0)}) - \int_0^\infty i^*(a)h(\frac{i(t,a)}{i^*(a)})[\frac{1}{\Pi_1}S^*f(J^*)k(a) + \frac{1}{\Pi_3}S^*g(Q^*)\xi(a)]da.$$

Since

$$\Gamma(0)i^*(0) = \frac{1}{\Pi_1}\int_0^\infty S^*f(J^*)i^*(a)k(a)da + \frac{1}{\Pi_3}\int_0^\infty S^*g(Q^*)i^*(a)\xi(a)da,$$

we further have

$$\begin{aligned} \frac{dV_2}{dt} \leq& \frac{1}{\Pi_1}\int_0^\infty S^*f(J^*)i^*(a)k(a)[h(\frac{i(t,0)}{i^*(0)}) - h(\frac{i(t,a)}{i^*(a)})]da \\ & + \frac{1}{\Pi_3}\int_0^\infty S^*g(Q^*)i^*(a)\xi(a)[h(\frac{i(t,0)}{i^*(0)}) - h(\frac{i(t,a)}{i^*(a)})]da. \end{aligned} \tag{25}$$

Similarly, we have

$$\frac{dV_3}{dt} \leq \frac{1}{\Pi_2\Pi_3}\int_0^\infty S^*g(Q^*)p^*(b)q(b)[h(\frac{p(t,0)}{p^*(0)}) - h(\frac{p(t,b)}{p^*(b)})]db. \tag{26}$$

We introduce

$$\begin{aligned}\mathcal{A}_0 :=& \frac{1}{\Pi_1}\int_0^\infty S^*f(J^*)k(a)i^*(a)[1 - \frac{i^*(0)Sf(J)}{i(t,0)S^*f(J^*)}]da \\ &+ \frac{1}{\Pi_3}\int_0^\infty S^*g(Q^*)\xi(a)i^*(a)[1 - \frac{i^*(0)Sg(Q)}{i(t,0)S^*g(Q^*)}]da.\end{aligned}$$

Then we can verify that $\mathcal{A}_0 = 0$. Combining equations (24), (25) and (26), we can transfer $\frac{dV}{dt}$ as follows:

$$\frac{dV}{dt} = \frac{dV}{dt} + \mathcal{A}_0 \leq \mathcal{A}_1 + \mathcal{A}_2 + \mathcal{A}_3 + \mathcal{A}_4 + \mathcal{A}_5,$$

where

$$\begin{aligned}\mathcal{A}_1 :=& [-\frac{\mu}{S}(S-S^*)^2 - Sf(J) - Sg(Q)]i^*(0) \\ &+ \frac{1}{\Pi_1}\int_0^\infty S^*f(J^*)i^*(a)k(a)\frac{i(t,0)}{i^*(0)}da + \frac{1}{\Pi_3}\int_0^\infty S^*g(Q^*)i^*(a)\xi(a)\frac{i(t,0)}{i^*(0)}da, \\ \mathcal{A}_2 :=& -\frac{1}{\Pi_3}\int_0^\infty S^*g(Q^*)i^*(a)\xi(a)\frac{i(t,a)}{i^*(a)}da + \frac{1}{\Pi_2\Pi_3}\int_0^\infty S^*g(Q^*)p^*(b)q(b)\frac{p(t,0)}{p^*(0)}db, \\ \mathcal{A}_3 :=& [S^*f(J^*) - \frac{S^*}{S}S^*f(J^*) + S^*f(J)]i^*(0) \\ &+ \frac{1}{\Pi_1}\int_0^\infty S^*f(J^*)i^*(a)k(a)[-\frac{i(t,a)}{i^*(a)} - \ln\frac{i(t,0)}{i^*(0)} + \ln\frac{i(t,a)}{i^*(a)} + 1 - \frac{i^*(0)Sf(J)}{i(t,0)S^*f(J^*)}]da, \\ \mathcal{A}_4 :=& [S^*g(Q^*) - \frac{S^*}{S}S^*g(Q^*)]i^*(0) \\ &+ \frac{1}{\Pi_3}\int_0^\infty S^*g(Q^*)i^*(a)\xi(a)\ln\frac{i(t,a)}{i^*(a)}da - \frac{1}{\Pi_2\Pi_3}\int_0^\infty S^*g(Q^*)p^*(b)q(b)\ln\frac{p(t,0)}{p^*(0)}db, \\ \mathcal{A}_5 :=& S^*g(Q)i^*(0) + \frac{1}{\Pi_2\Pi_3}\int_0^\infty S^*g(Q^*)p^*(b)q(b)[-\frac{p(t,b)}{p^*(b)} + \ln\frac{p(t,b)}{p^*(b)}]db \\ &+ \frac{1}{\Pi_3}\int_0^\infty S^*g(Q^*)\xi(a)i^*(a)[1 - \frac{i^*(0)Sg(Q)}{i(t,0)S^*g(Q^*)} - \ln\frac{i(t,0)}{i^*(0)}]da.\end{aligned}$$

Since $Sf(J) + Sg(Q) = i(t,a)$ and $i^*(a) = \varpi_1(a)i^*(0)$, we have

$$\begin{aligned}\mathcal{A}_1 =& -\frac{\mu}{S}(S-S^*)^2 i^*(0) - i(t,0)i^*(0) \\ &+ \frac{1}{\Pi_1}S^*f(J^*)i(t,0)\int_0^\infty \varpi_1(a)k(a)da + \frac{1}{\Pi_3}S^*g(Q^*)i(t,0)\int_0^\infty \varpi_1(a)\xi(a)da \\ =& -\frac{\mu}{S}(S-S^*)^2 i^*(0) \leq 0.\end{aligned} \tag{27}$$

Due to $\int_0^\infty \xi(a)i(t,a)da = p(t,0)$ and $\int_0^\infty \varpi_2(b)q(b)db = \Pi_2$, we obtain

$$\mathcal{A}_2 = -\frac{1}{\Pi_3}S^*g(Q^*)\int_0^\infty \xi(a)i(t,a)da + \frac{1}{\Pi_2\Pi_3}S^*g(Q^*)p(t,0)\int_0^\infty \varpi_2(b)q(b)db = 0. \tag{28}$$

Since $i^*(0) = \frac{1}{\Pi_1}\int_0^\infty k(a)\varpi_1(a)i^*(0)da = \frac{1}{\Pi_1}\int_0^\infty k(a)i^*(a)da = \frac{J^*}{\Pi_1}$, we obtain

$$\mathcal{A}_3 = \frac{1}{\Pi_1}\int_0^\infty S^*f(J^*)k(a)i^*(a)[-h(\frac{S^*}{S}) - h(\frac{i^*(0)Sf(J)}{i(t,0)S^*f(J^*)}) - h(\frac{i(t,a)}{i^*(a)}) + h(\frac{f(J)}{f(J^*)})],$$

and

$$\frac{1}{\Pi_1}\int_0^\infty S^*f(J^*)k(a)i^*(a)h(\frac{i(t,a)}{i^*(a)})da = S^*f(J^*)\int_0^\infty \frac{k(a)i^*(a)}{\int_0^\infty k(a)\varpi_1(a)da}h(\frac{i(t,a)}{i^*(a)})da$$
$$=S^*f(J^*)i^*(0)\int_0^\infty \frac{k(a)i^*(a)}{J^*}h(\frac{i(t,a)}{i^*(a)})da \geq S^*f(J^*)i^*(0)h(\frac{J(t)}{J^*}).$$

Then, we have

$$\begin{aligned}\mathcal{A}_3 \leq&\frac{1}{\Pi_1}\int_0^\infty S^*f(J^*)k(a)i^*(a)[-h(\frac{S^*}{S})-h(\frac{i^*(0)Sf(J)}{i(t,0)S^*f(J^*)})]da\\&+S^*f(J^*)i^*(0)[h(\frac{f(J)}{f(J^*)})-h(\frac{J}{J^*})]\leq 0.\end{aligned} \quad (29)$$

Due to $p^*(0)=\int_0^\infty \xi(a)i^*(a)da$, $\Pi_2=\int_0^\infty \varpi_2(b)q(b)db$ and

$$\int_0^\infty i^*(a)\xi(a)[1-\frac{i(t,a)p^*(0)}{i^*(a)p(t,0)}]da = \int_0^\infty i^*(a)\xi(a)da - p^*(0)\frac{1}{p(t,0)}\int_0^\infty \xi(a)i(t,a)da = 0,$$

we have

$$\begin{aligned}\mathcal{A}_4 =&\frac{1}{\Pi_3}\int_0^\infty S^*g(Q^*)i^*(a)\xi(a)[-h(\ln\frac{S^*}{S})-h(\ln\frac{i(t,a)p^*(0)}{i^*(a)p(t,0)})+\ln\frac{S}{S^*}]da\\ \leq&\frac{1}{\Pi_3}\int_0^\infty S^*g(Q^*)i^*(a)\xi(a)\ln\frac{S}{S^*}da.\end{aligned}$$

Thus, combining $\mathcal{A}_4$ and $\mathcal{A}_5$ yields

$$\begin{aligned}\mathcal{A}_4+\mathcal{A}_5 \leq&\frac{1}{\Pi_3}\int_0^\infty S^*g(Q^*)i^*(a)\xi(a)\frac{g(Q)}{g(Q^*)}da\\&+\frac{1}{\Pi_3}\int_0^\infty S^*g(Q^*)i^*(a)\xi(a)[1-\frac{i^*(0)Sg(Q)}{i(t,0)S^*g(Q^*)}+\ln\frac{S}{S^*}-\ln\frac{i(t,0)}{i^*(0)}]da\\&+\frac{1}{\Pi_2\Pi_3}\int_0^\infty S^*g(Q^*)p^*(b)q(b)[\ln\frac{p(t,b)}{p^*(b)}-\frac{p(t,b)}{p^*(b)}]db.\end{aligned}$$

Then, due to $\frac{1}{\Pi_2}\int_0^\infty p^*(b)q(b)db = \frac{1}{\Pi_2}\int_0^\infty p^*(0)\varpi_2(b)q(b)db = \int_0^\infty i^*(a)\xi(a)da$ and

$$h(\frac{p(t,b)}{p^*(b)}) = \int_0^\infty \frac{q(b)p^*(b)}{\Pi_2 p^*(0)}h(\frac{p(t,b)}{p^*(b)})db \geq h\left(\frac{\int_0^\infty q(b)p(t,b)db}{\int_0^\infty q(b)p^*(b)db}\right) = h(\frac{Q}{Q^*}),$$

we further obtain

$$\begin{aligned}\mathcal{A}_4+\mathcal{A}_5 \leq&\frac{1}{\Pi_2\Pi_3}\int_0^\infty S^*g(Q^*)p^*(b)q(b)\times\\&[-h(\frac{i^*(0)Sg(Q)}{i(t,0)S^*g(Q^*)})+h(\frac{g(Q)}{g(Q^*)})-h(\frac{p(t,b)}{p^*(b)})]db\\=&\frac{1}{\Pi_2\Pi_3}\int_0^\infty S^*g(Q^*)p^*(b)q(b)[-h(\frac{i^*(0)Sg(Q)}{i(t,0)S^*g(Q^*)})]db\\&+\frac{1}{\Pi_3}S^*g(Q^*)p^*(0)[h(\frac{g(Q)}{g(Q^*)})-h(\frac{p(t,b)}{p^*(b)})]db\\ \leq&0.\end{aligned} \quad (30)$$

From Equations (27)–(30), we have $\frac{dV}{dt} \leq 0$ and the largest invariant subset of set $\left\{\frac{dV}{dt} = 0\right\}$ is $\{E^*\}$. Due to the invariance principle [37], we conclude that E^* is globally asymptotically stable if it exists. □

5. Numerical Simulations

In this section, as a special case for the age-infection model (1), we consider the following model:

$$\begin{cases} \dfrac{dS(t)}{dt} = \Lambda - \mu S(t) - \dfrac{S(t)\int_0^\infty k(a)i(t,a)da}{\int_0^\infty k(a)i(t,a)da + A} - S(t)\displaystyle\int_0^\infty q(b)p(t,b)db, \\ \dfrac{\partial i(t,a)}{\partial t} + \dfrac{\partial i(t,a)}{\partial a} = -\delta(a)i(t,a), \\ \dfrac{\partial p(t,b)}{\partial t} + \dfrac{\partial p(t,b)}{\partial b} = -\gamma(b)p(t,b), \end{cases} \tag{31}$$

with the initial condition (3) and the following boundary conditions

$$i(t,0) = \frac{S(t)\int_0^\infty k(a)i(t,a)da}{\int_0^\infty k(a)i(t,a)da + A} + S(t)\int_0^\infty q(b)p(t,b)db, \ t > 0,$$
$$p(t,0) = \int_0^\infty \xi(a)i(t,a)da, \ t > 0.$$

Following (15), the basic reproduction number of system (31) is

$$\Re_1 = \frac{\Lambda}{A\mu}\Pi_1 + \frac{\Lambda}{\mu}\Pi_2\Pi_3.$$

From Theorems 8 and 9, we obtain the following corollary:

Corollary 1. *When $\Re_1 < 1$, model (31) generates unique infection-free equilibrium E_1^0, which is globally asymptotically stable. When $\Re_1 > 1$, model (31) has E_1^0 and a globally asymptotically stable infection equilibrium E_1^*.*

To verify the result, we perform numerical simulations. Following [6,7] and references therein, with some assumptions, we adopt the following coefficients, for $0 \leq a, b \leq 10$,

$$\Lambda = 1000, \ \mu = 10^{-5}, \ A = 10^5, \ \xi(a) = 1 + \sin\frac{(a-5)\pi}{10},$$
$$\delta(a) = 0.2\left(1 + \sin\frac{(a-5)\pi}{10}\right), \ \gamma(b) = 0.3\left(1 + \sin\frac{(b-5)\pi}{10}\right),$$
$$k(a) = k\left(1 + \sin\frac{(a-5)\pi}{10}\right), \ q(b) = q\left(1 + \sin\frac{(b-5)\pi}{10}\right).$$

Let $k = 10^{-5}$and observe the dynamical behavior of the model when q varies. Let $q = 10^{-4}$ decrease to $q = 10^{-10}$. The globally asymptotically stable E_1^* changes to be unstable and the epidemic is inhibited effectively, which can be seen in Figures 1 and 2.

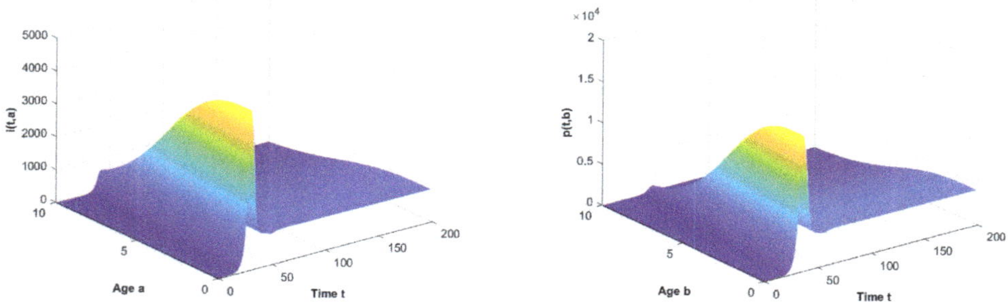

Figure 1. The long-term dynamical behavior of $i(t,a)$ and $p(t,b)$ as $q = 10^{-4}$.

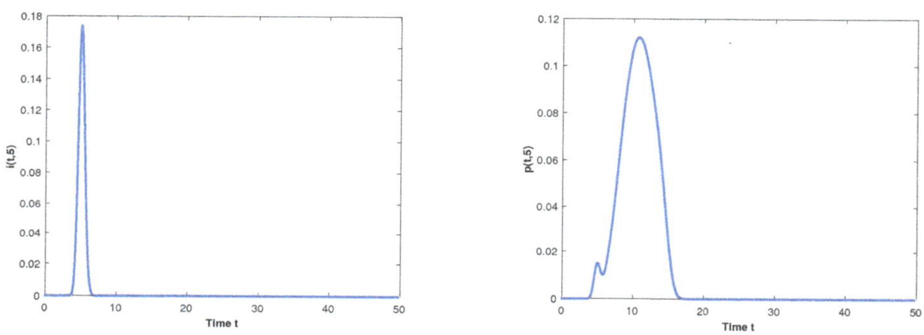

Figure 2. The long-term dynamical behavior of $i(t,a)$ and $p(t,b)$ for $a = b = 5$ as $q = 10^{-10}$.

6. Conclusions and Discussion

In this paper, an age-structured model of cholera infection was explored. By considering general infection functions, the discussion provided in this paper serves as a generalization and supplement to the work presented in F. Brauer et al. [12]. We applied the Lyapunov functional method to show that the global stability of equilibria are determined by the basic reproduction number $\Re_0$. The infection-free equilibrium is globally asymptotically stable if $\Re_0$ is less than one, whereas a globally asymptotically stable infection equilibrium emerges if $\Re_0$ is greater than one. This shows that both the direct contact with infected individuals and indirect pathogen infection have vital effects on cholera epidemics. It is significant to implement effective treatment for infected individuals and to clean pathogens from contaminated water in a timely fashion. More specifically, for the critical case when $\Re_0$ equals one, further bifurcation studies are needed.

In our model, vaccinated individuals and vaccination age have not been incorporated, which play vital effects on the spread of cholera. Furthermore, the immigration of infected individuals plays a significant role in the outbreak and infection of cholera. For the actual control and elimination of cholera, it is necessary to take into account the effects of vaccination and immigration [5,38]. Thus, our future work will consider these factors and focus on their effects on cholera transmission. In addition to qualitative analyses, tremendous amounts of works on numerical methods have been proposed and developed to deal with various epidemic models [39–41], which provide us with more aspects and methods to analyze in relation to this model.

Funding: This research was funded by Fundamental Research Funds of Beijing Municipal Education Commission (Grant Number: 110052972027/141) and North China University of Technology Research Fund Program for Young Scholars (Grant Number: 110051360002).

Institutional Review Board Statement: Not applicable.

Informed Consent Statement: Not applicable.

Data Availability Statement: Not applicable.

Conflicts of Interest: The author declares no conflict of interest.

References

1. Hartley, D.M.; Morris, J.G., Jr.; Smith, D.L. Hyperinfectivity: A critical element in the ability of V. cholerae to cause epidemics? *PLoS Med.* **2006**, *3*, 63–69. [CrossRef]
2. Tien, J.H.; Earn, D.J.D. Multiple transmission pathways and disease dynamics in a waterborne pathogen model. *Bull. Math. Biol.* **2010**, *72*, 1506–1533. [CrossRef] [PubMed]
3. World Health Organization. Cholera Prevention and Control, May 2018. Available online: www.who.int (accessed on 16 April 2021).
4. Capasso, V.; Paveri-Fontana, S.L. A mathematical model for the 1973 cholera epidemic in the european mediterranean region. *Revue d'épidémiologie et de Santé Publiqué* **1979**, *27*, 121.
5. Posny, D.; Wang, J.; Mukandavire, Z.; Modnak, C. Analyzing transmission dynamics of cholera with public health interventions. *Math. Biosci.* **2015**, *264*, 38–53. [CrossRef]
6. Misra, A.K.; Gupta, A.; Venturino, E. Cholera dynamics with bacteriophage infection: A mathematical study. *Chaos Solitons Fractals* **2016**, *91*, 610–621. [CrossRef]
7. Phan, T.A.; Tian, J.P.; Wang, B. Dynamics of cholera epidemic models in fluctuating environments. *Stoch. Dyn.* **2021**, *21*, 2150011. [CrossRef]
8. Brauer, F.; Driessche, P.v.; Wu, J. (Eds.) *Mathematical Epidemiology*; Lecture Notes in Mathematics; Springer: Berlin/Heidelberg, Germany, 2008; Volume 1945.
9. Iannelli, M. *Mathematical Theory of Age-Structured Population Dynamics*; Appl. Math. Monogr. CNR 7; Giadini Editori e Stampator: Pisa, Italy, 1994.
10. Kuniya, T. Global stability analysis with a discretization approach for an age-structured multigroup SIR epidemic model. *Nonlinear Anal. RWA* **2011**, *12*, 2640–2655. [CrossRef]
11. Webb, G.F. *Theory of Nonlinear Age-Dependent Population Dynamics*; Marcel Dekker: New York, NY, USA; Basel, Switzerland, 1985.
12. Brauer, F.; Shuai, Z.; van den Driessche, P. Dynamics of an age-of-infection cholera model. *Math. Biosci. Eng.* **2013**, *10*, 1335–1349.
13. Wang, J.; Zhang, R.; Kuniya, T. A note on dynamics of an age-of-infection chorela model. *Math. Biosci. Eng.* **2016**, *13*, 227–247. [CrossRef] [PubMed]
14. Cai, L.; Fan, G.; Yang, C.; Wang, J. Modeling and analyzing cholera transmission dynamics with vaccination age. *J. Frankl. Inst.* **2020**, *357*, 8008–8034. [CrossRef]
15. Huang, G.; Takeuchi, Y.; Ma, W.; Wei, D. Global stability for delay SIR and SEIR epidemic models with nonlinear incidence rate. *Bull. Math. Biol.* **2010**, *72*, 1192–1207. [CrossRef] [PubMed]
16. Huang, G.; Yokoi, H.; Takeuchi, Y.; Kajiwara, T.; Sasaki, T. Impact of intracellular delay, immune activation delay and nonlinear incidence on viral dynamics. *Jpn. J. Ind. Appl. Math.* **2011**, *28*, 383–411. [CrossRef]
17. Korobeinikov, A.; Maini, P.K. Nonlinear incidence and stability of infectious disease models. *Math. Med. Biol.* **2005**, *22*, 113–128. [CrossRef]
18. Xu, S. Global stability of the virus dynamics model with Crowley-Martin functional response. *Electron. J. Qual Theory Differ. Equ.* **2012**, *9*, 1–10. [CrossRef]
19. Zhou, X.; Cui, J. Global stability of the viral dynamics with Crowley-Martin functional response. *Bull. Korean Math. Soc.* **2011**, *48*, 555–574. [CrossRef]
20. Huang, G.; Liu, X.; Takeuchi, Y. Lyapunov functions and global stability for age-structured HIV infection model. *SIAM J. Appl. Math.* **2012**, *72*, 25–38. [CrossRef]
21. Magal, P.; McCluskey, C.C.; Webb, G.F. Lyapunov functional and global asymptotic stability for an infection-age model. *Appl. Anal.* **2010**, *89*, 1109–1140. [CrossRef]
22. Melnik, A.V.; Korobeinikov, A. Lyapunov functions and global stability for SIR and SEIR models with age-dependent susceptibility. *Math. Biosci. Eng.* **2013**, *10*, 369–378.
23. Magal, P.; Ruan, S. *Theory and Applications of Abstract Semilinear Cauchy Problems*; Springer: Berlin/Heidelberg, Germany, 2018.
24. Liu, S.; Zhang, R. On an Age-Structured Hepatitis B Virus Infection Model with HBV DNA-Containing Capsids. *Bull. Malays. Math. Sci. Soc.* **2021**, *44*, 1345–1370. [CrossRef]
25. Zhang, R.; Liu, S. Global dynamics of an age-structured within-host viral infection model with cell-to-cell transmission and general humoral immunity response. *Math. Biosci. Eng.* **2020**, *17*, 1450–1478. [CrossRef]

26. Magal, P.; Thieme, H. Eventual compactness for a semiflow generated by an age-structured models. *Commun. Pure Appl. Anal.* **2004**, *3*, 695–727. [CrossRef]
27. Smith, H.L.; Thieme, H.R. *Dynamical Systems and Population Persistence*; American Mathematical Society: Providence, RI, USA, 2011.
28. Hale, J.K. *Asymptotic Behavior of Dissipative Systems, Mathematical Surveys and Monographs*; American Mathematical Society: Providence, RI, USA, 1988; Volume 25.
29. Georgescu, P.; Hsieh, Y.H. Global stability for a virus dynamics model with nonlinear incidence of infection and removal. *SIAM J. Appl. Math.* **2006**, *67*, 337–353. [CrossRef]
30. Huang, G.; Ma, W.; Takeuchi, Y. Global properties for virus dynamics model with Beddington-DeAngelis functional response. *Appl. Math. Lett.* **2009**, *22*, 1690–1693. [CrossRef]
31. Korobeinikov, A. Global properties of basic virus dynamics models. *Bull. Math. Biol.* **2004**, *66*, 879–883. [CrossRef] [PubMed]
32. Leenheer, P.D.; Smith, H.L. Virus dynamics: A global analysis. *SIAM J. Appl. Math.* **2003**, *63*, 1313–1327.
33. McCluskey, C.C. Global stability for an SEI model of infectious disease with age structure and immigration of infecteds. *Math. Biosci. Eng.* **2016**, *13*, 381–400. [CrossRef]
34. Shuai, Z.; van den Driessche, P. Global dynamics of cholera models with differential infectivity. *Math. Biosci.* **2010**, *234*, 118–126. [CrossRef]
35. Vargas-De-León, C.; Esteva, L.; Korobeinikov, A. Age-dependency in host-vector models: The global analysis. *Appl. Math. Comput.* **2014**, *243*, 969–981. [CrossRef]
36. Hale, J.; Waltman, P. Persistence in infinite-dimensional systems. *SIAM J. Math. Anal.* **1989**, *20*, 388–395. [CrossRef]
37. Walker, J.A. *Dynamical Systems and Evolution Equations*; Plenum Press: New York, NY, USA; London, UK, 1980.
38. Brauer, F.; van den Driessche, P. Models for transmission of disease with immigration of infectives. *Math. Biosci.* **2001**, *171*, 143-154. [CrossRef]
39. Khalsaraei, M.M.; Shokri, A.; Ramos, H.; Heydari, S. A positive and elementary stable nonstandard explicit scheme for a mathematical model of the influenza disease. *Math. Comput. Simul.* **2021**, *182*, 397–410. [CrossRef]
40. Jódar, L.; Villanueva, R.J.; Arenas, A.J.; González, G.C. Nonstandard numerical methods for a mathematical model for influenza disease. *Math. Comput. Simul.* **2008**, *79*, 622–633. [CrossRef]
41. Mickens, R.E. Numerical integration of population models satisfying conservation laws: NSFD methods. *Biol. Dyn.* **2007**, *1*, 1751–1766. [CrossRef] [PubMed]

mathematics

MDPI

Article

Graph, Spectra, Control and Epidemics: An Example with a SEIR Model

Giacomo Aletti [†], Alessandro Benfenati [*,†] and Giovanni Naldi [†]

Environmental Science and Policy Department, Università degli Studi di Milano, 20133 Milan, Italy; giacomo.aletti@unimi.it (G.A.); giovanni.naldi@unimi.it (G.N.)
* Correspondence: alessandro.benfenati@unimi.it
† The three authors are members of the Italian Group GNCS of the Italian Institute "Istituto Nazionale di Alta Matematica", and of the ADAMSS Center of the Università degli Studi di Milano (Italy).

Abstract: Networks and graphs offer a suitable and powerful framework for studying the spread of infection in human and animal populations. In the case of a heterogeneous population, the social contact network has a pivotal role in the analysis of directly transmitted infectious diseases. The literature presents several works where network-based models encompass realistic features (such as contacts networks or host–pathogen biological data), but analytical results are nonetheless scarce. As a significant example, in this paper, we develop a multi-group version of the epidemiological SEIR population-based model. Each group can represent a social subpopulation with the same habits or a group of geographically localized people. We consider also heterogeneity in the weighting of contacts between two groups. As a simple application, we propose a simple control algorithm in which we optimize the connection weights in order to minimize the combination between an economic cost and a social cost. Some numerical simulations are also provided.

Keywords: epidemic spread; multi-group models; network based model; control of spread dynamics

Citation: Aletti, G.; Benfenati, A.; Naldi, G. Graph, Spectra, Control and Epidemics: An Example with a SEIR Model. *Mathematics* **2021**, *9*, 2987. https://doi.org/10.3390/math9222987

Academic Editors: Mihaela Neamţu, Eva Kaslik and Anca Rădulescu

Received: 31 October 2021
Accepted: 19 November 2021
Published: 22 November 2021

Publisher's Note: MDPI stays neutral with regard to jurisdictional claims in published maps and institutional affiliations.

Copyright: © 2021 by the authors. Licensee MDPI, Basel, Switzerland. This article is an open access article distributed under the terms and conditions of the Creative Commons Attribution (CC BY) license (https://creativecommons.org/licenses/by/4.0/).

1. Introduction

The epidemiological modeling of infectious disease transmission has a long history in mathematical biology, for humans [1–7], animals [3,8] and plants [9–11]. In recent years it has had an increasing influence on the theory and practice of disease management and control, e.g., [12–18]. Indeed the forecast of the spread of an infectious disease is critical to public health decision making.

The proper modeling and analysis of the dynamics of infectious diseases has been a long-standing area of research among many different fields, including economics, social sciences, mathematical biology, physics, computer science and engineering [5]. In the classical population approach, the underlying common factor is the partitioning of the population into "compartments"; we assume that the populations in the various compartments are homogenous in the sense that all individuals behave similarly. The two most common compartments that exist in almost all epidemic models are susceptible (S) and infected (I) [2,3]. The subpopulation S represents individuals who are healthy but susceptible to becoming infected, while I represents individuals who became infected but are able to recover. If the model contains only these two compartments, a given population is initially divided into them. From this basic compartmentalization, there are numerous ways for introducing different interactions within the population. Most of these models for the disease evolution make two basic assumptions. The first assumption states that the population is well-mixed. In such a population, each individual has the same probability of encountering other infected individuals, and thus the resulting force of infection is equal for all. The second assumption states that there are a priori constraints upon the biological process, whilst gradual but random mutation of disease traits (such as transmission rate and infectious period) could occur. More refined epidemic models are required; the entire population cannot just be divided into two or more compartments/groups which are defined by a single quantity. In this paper, we consider the effect of the heterogeneity in the weighting of contacts between two individuals. Moreover, we focus on a meta-population model where the population is previously subdivided into subpopulations that can consist in spatially distinct groups of individuals (neighborhoods, towns, cities, etc.) or groups of individuals with different features. The resulting model is described by a dynamic system defined on a network (graph). We have also added the possibility of varying the weight of the connections between groups in order to formalize the problem of controlling the spread of the epidemic on the network. This generality could also allow the changing of disease features.

The paper is organized as follows: Section 2 introduces the model and the analysis of the corresponding dynamical system. Moreover, as an application of this approach, we will also discuss a new definition of control problem about the spread of epidemics on the network. Section 3 is devoted to numerical experiments using a reduced network with different features. Following the findings of the case study and of previous analysis, the conclusions are presented in Section 4.

2. A Meta-Population Model on a Network

The transmission of infectious diseases raises many important questions. In some instances, the average behavior of a large population with respect to the time is sufficient to provide useful insight from the available data. However, the spatial component of many transmission systems has been recognized to be of pivotal importance in the recent years. Due to this, spatially heterogeneous interventions must be included in the model, and hence it is essential to properly represent the transmission pattern. A reasonable hypothesis may consider that the spatial aspects of transmission heavily influence the aggregation characteristic of epidemic influence. Hence, we need to investigate data by using models that include such spatial connections. For example, the understanding of human mobility and the developing of qualitative and quantitative theories is of key importance for the modeling and for the comprehension of human infectious disease dynamics on geographical scales of different size.

2.1. Spatial Heterogeneity in Epidemiological Models

Ideally, the model should be able to account for the states of all N individuals in the population in an independent manner and, at the same time, it should allow for arbitrary interactions among them. The analysis of these models is a difficult task, and the computational cost of numerical simulations is very onerous and the extraction of the collective behaviors very complex. Although studies on the temporal dynamics of diseases proved insightful, incorporating space explicitly into epidemiological models revealed various emergent properties [19]. The phenomenon of the spatial spread of infection involves several components and scale [20]. Indeed, small region/group models can incorporate spatial heterogeneity, and more general models allowing for larger households with continuous or discrete time can be developed. Other typical approaches encompassing spatial variation in epidemic models involve partial differential equations (PDE). There exist, nonetheless, cases and scenarios where the latter type of spatial approach may not reliably model the phenomenon. Consider, for example, a human specific disease which is spread only by person-to-person contact and consider a geographical context consisting in a large country with a small number of large cities and a very sparse (or even non-existent) rural population [21]. The travel of individuals between discrete geographical regions and/or cities plays a pivotal role in the disease spreading. The depicted situation is easily described by a directed graph, where the vertices represent the cities (or discrete geographical regions/patches) and the arcs represent the links between such cities [22].

The main approaches for spatial models concern a different scale: an individual-based simulation, a meta-population model or a network model (see, e.g., [14]). Individual-based models explicitly represent every individual i with a state $X_i(t)$ at time t, e.g., $X_i(t) = 1$

indicates that i is infected while $X_i(t) = 0$ indicates that i is healthy at time t. Infected nodes can transmit the disease to neighboring nodes and an individual becomes infected with a certain probability based on the status of neighboring individuals. The meta-population framework consists of dividing the whole population into distinct subpopulations, each having independent internal dynamics. In addition, at the same time, there is a limited interaction between the subpopulations. This approach has been used to great effect within ecological systems [23,24]. The individuals in such subpopulations belong to a particular state (e.g., susceptible or infectious) which can change during the time. For large networks, a general approach consists of merging some network information in a relatively small set of statistics and then studying the impact of such statistics on the infection spread [25].

In the following section, we describe our meta-population model, which considers communities as the aggregate unit that may represent either subpopulations in different areas or distinct groups with similar characteristics (e.g., students on a campus and citizens of a neighborhood or high school students or office workers). Then, each subpopulation is partitioned according to a particular state of individuals with respect to disease. Finally, connections and mobility between different communities are introduced. We point out that various proposed models encompass the geography of spread of the disease, but they do not present a mathematical analysis of their main properties, while presenting realistic simulations and an appropriate identification of the parameters involved, e.g., [26].

2.2. A Prototype: SEIR Model on a Direct Graph

We introduce a prototype model that can be generalized considering several states related to a given disease. Our analysis can, therefore, easily be extended to these more complex models. We partition a population of N individuals into subpopulations (groups, patches, communities, etc.) without taking into account any biological interpretation they have but considering spatially segregated large subpopulations. In this way, we can encompass a more realistic contact structure into epidemic models, since it usually preserves analytic tractability (in stochastic and in deterministic models), but at the same time it also captures the most important structural inhomogeneity in contact patterns in several applied contexts. The subpopulations and the interactions/connections between them are modeled through a weighted direct graph $\mathcal{G} = (\mathcal{V}, \mathcal{E})$ with n vertices (nodes, regions, patches, subpopulations) and m edges (connections). Each edge is described by an ordered pairs of nodes (u, v), where $u, v \in \mathcal{V}$. We label the nodes with an integer index; two vertices i and j of the directed graph are joined or adjacent if and only if there exists an edge from i to j or from j to i. If such an edge exists, then i and j are called its endpoints. If there is an edge from i to j then i and j are often called tail and head, respectively. The $(n \times n)$ *adjacency matrix* A^d associated to the graph is constructed as follows: if there exists an edge from node i to node j, then the entry at row i and column j is set to 1 in the matrix A^d: $a^d_{ij} = 1$.

In node i, the corresponding subpopulation possesses N_i individuals, and $\sum_{i=1}^{n} N_i = N$. We hypothesize that individuals can move to a different node, interact with people in that node and then return to the original one. If $a^d_{ij} = 1$, there is an interaction between node i and node j, but not all the subpopulation N_i from node i interacts with the population in node j: we denote by a_{ij} the total amount of the subpopulation i that "goes" to node j and interacts with the people in that node. We call A the *routing matrix* with entries a_{ij}, so that $\sum_{j=1}^{n} a_{ij} = N_i$, $i = 1, \dots, n$. Associated to A, let P^o the *probability outgoing matrix* with entries p^o_{ij}, where we denote by p^o_{ij} the percentage (probability) of the subpopulation i that "goes" to node j. In addition, we denote by P^i the *probability incoming matrix* with entries p^i_{ij}, where p^i_{ij} is now the percentage (probability) of the subpopulation in j that "arrived" from i. Finally, let $M_i = \sum_{j=1}^{n} a_{ji}$ be the total amount of people arrived in node $i = 1, \dots, n$, so that $\sum_{i=1}^{n} M_i = N$ again. Then, for any $i = 1, \dots, n$, $\sum_{j=1}^{n} p^o_{ij} = \sum_{j=1}^{n} p^i_{ji} = 1$. Moreover we have

$$A = \mathrm{Diag}(N_1, N_2, \dots, N_n) P^o = P^i\, \mathrm{Diag}(M_1, M_2, \dots, M_n)$$

where $\text{Diag}(x_1, x_2, \ldots, x_n)$ is the diagonal matrix with the vector $(x_1, x_2, \ldots, x_n)^T \in \mathbb{R}^n$ on the main diagonal.

Four different discrete classes are considered for statuses of individuals in each: susceptible, exposed, infectious and recovered (SEIR model) [2]. All individuals are born as susceptible: a susceptible individual in contact with an infectious one may become exposed; the probability depends on the particular strain of the disease. Exposed individuals are infected but not yet infectious: individuals experience a long incubation duration. With a suitable incubation rate, latent individuals become infectious. Finally, a reliable assumption is that the immune system of infectious individuals combats the infection and then they move directly into the recovered class, which refers to individuals that are no longer infectious and have gained full immunity from further infection. Let $S(t)$, $E(t)$, $I(t)$, $R(t)$ the number of individuals in a node at time t, $S(t) + E(t) + I(t) + R(t) = N$: we consider a time interval in which we can neglect demographics. Without any interaction with other nodes, within a deterministic approach of the compartmental models, with continuous time t, the epidemic dynamics can be described by the system of differential equations in (1):

$$\begin{aligned} \dot{S}(t) &= -\lambda\, S(t) \\ \dot{E}(t) &= \lambda\, S(t) - \mu E(t) \\ \dot{I}(t) &= \mu E(t) - \gamma I(t) \\ \dot{R}(t) &= \gamma I(t) \end{aligned} \tag{1}$$

where the parameter λ is the force of infection, γ is the recovery rate and $1/\mu$ is an average latent period.

With respect to the behavior of an epidemic, λ is the rate at which susceptible individuals become infected or exposed and it is a function depending on the number of infectious individuals; it contains information about the interactions between individuals that concur to the infection transmission. If we suppose that the population of N individuals mixes at random, meaning that all pairs of individuals have the same probability of interacting, the force of infection may be computed as:

$$\begin{aligned} \lambda &= \text{transmission rate} \\ &\quad \times \text{effective number of contacts per unit time} \\ &\quad \times \text{proportion of contacts infectious} \\ &\sim \tau \times n_c \times \tfrac{I}{N} = \beta \tfrac{I}{N} \end{aligned}$$

Then the dynamics state,

$$\begin{aligned} \dot{S}(t) &= -\beta \tfrac{I}{N} S(t) \\ \dot{E}(t) &= \beta \tfrac{I}{N} S(t) - \mu E(t) \end{aligned} \tag{2}$$

where β is the infectious rate. Rescaling the quantities S, E, I, R dividing by N we obtain,

$$\begin{aligned} \dot{s}(t) &= -\beta\, \imath(t) s(t) \\ \dot{e}(t) &= \beta\, \imath(t) s(t) - \mu e(t) \\ i(t) &= \mu e(t) - \gamma\, \imath(t) \\ \dot{r}(t) &= \gamma\, \imath(t) \end{aligned} \tag{3}$$

Pay attention to the fact that i stands for the derivative of the function $\imath$. Now, we take a node j that is connected to the other nodes as encoded in matrix A. Then, $S_j(t)$ can change due to the contribution of susceptible people from j that reached an adjacent node k and met infectious people in that node, wheresoever they came from. Then the contribution to $\dot{S}_j$ due to the interactions in node k is given by the $p^o_{jk} S_j = a_{jk} s_j$ susceptible people that met a population in node k with a proportion of infectious people given by

$$\frac{\#\{\text{infectious people in node } k\}}{\#\{\text{total people in node } k\}} = \frac{\sum_{l=1}^{n} p^{o}_{lk} I_l}{\sum_{l=1}^{n} a_{lk}} = \frac{\sum_{l=1}^{n} p^{o}_{lk} I_l}{M_k}$$
$$= \frac{\sum_{l=1}^{n} a_{lk} \imath_l}{\sum_{l=1}^{n} a_{lk}} = \sum_{l=1}^{n} p^{i}_{lk} \imath_l.$$

Let the vectors $\mathbf{X}(t) = (x_1(t), x_2(t), \ldots, x_n(t))^T \in \mathbb{R}^n$, with $X = S, E, I, R$, the $SEIR$ model on the graph $\mathcal{G}$ is the following

$$\begin{aligned} \dot{\boldsymbol{S}}(t) &= -\beta \operatorname{Diag}(\boldsymbol{S}(t))\hat{B}\boldsymbol{I}(t) & \dot{\boldsymbol{s}}(t) &= -\beta \operatorname{Diag}(\boldsymbol{s}(t))B\boldsymbol{\imath}(t) \\ \dot{\boldsymbol{E}}(t) &= \beta \operatorname{Diag}(\boldsymbol{S}(t))\hat{B}\boldsymbol{I}(t) - \mu\boldsymbol{E}(t) & \dot{\boldsymbol{e}}(t) &= \beta \operatorname{Diag}(\boldsymbol{s}(t))B\boldsymbol{\imath}(t) - \mu\boldsymbol{e}(t) \\ \dot{\boldsymbol{I}}(t) &= \mu\boldsymbol{E}(t) - \gamma\boldsymbol{I}(t) & \dot{\boldsymbol{\imath}}(t) &= \mu\boldsymbol{e}(t) - \gamma\boldsymbol{\imath}(t) \\ \dot{\boldsymbol{R}}(t) &= \gamma\boldsymbol{I}(t) & \dot{\boldsymbol{r}}(t) &= \gamma\boldsymbol{\imath}(t) \end{aligned} \tag{4}$$

where $\hat{B} = P^o \operatorname{Diag}(M_1, \ldots, M_n)^{-1} P^{o\top}$, $B = P^o P^{i\top}$, and the equations on the right side have been obtained by a premultiplication with $\operatorname{Diag}(N_1, N_2, \ldots, N_n)^{-1}$.

Remark 1. *We have assumed that the parameter β is the same in all nodes. It is possible to easily introduce a different parameter for each node considering more heterogeneity in the model.*

In the following, we adopt the notations $\mathbf{1} = [1, 1, \ldots, 1]^\top$, $\mathbf{0} = [0, 0, \ldots, 0]^\top$, for any vectors $x, y \in \mathbb{R}^n$, $x \ll y \Leftrightarrow x_i < y_i$, $i = 1, 2, \ldots, n$; $x \leq y \Leftrightarrow x_i \leq y_i$, $i = 1, 2, \ldots, n$ (and $x < y$ if $x \leq y$ but $x \neq y$).

We suppose that the directed graph $\mathcal{G}$ is strongly connected, i.e., there exists a path in each direction between each pair of vertices of the graph, then the matrices A and P are irreducible. This means that we cannot divide the nodes of the graph into two subsets such that there are no connections between the nodes of the two subsets but only within each subset. It also follows that the matrix B is a non-negative irreducible $n \times n$ matrix; by the Perron–Frobenius theorem [27] we deduce:

- B has a positive real eigenvalue equal to its spectral radius $\rho(B)$;
- There exists an eigenvector $\boldsymbol{v} \gg 0$ corresponding to $\rho(B)$;
- $\rho(B)$ increases when any entry of B increases;
- $\rho(B)$ is a simple eigenvalue of B;
- Collatz–Wielandt formula: $\rho(B) = \min_{\boldsymbol{x}>\mathbf{0}} \max_{i:x_i>0} \frac{[\boldsymbol{x}^\top B]_i}{x_i} = \max_{\boldsymbol{x}>\mathbf{0}} \min_{i:x_i>0} \frac{[\boldsymbol{x}^\top B]_i}{x_i}$ that are reached identically on every component of the eigenvector: $\rho(B) = \frac{[\boldsymbol{v}^\top B]_i}{v_i}$, for any $i = 1, \ldots, n$;
- There is no other, unless rescaled, non-negative eigenvector of B, different from v.

Let $B_{\boldsymbol{s}}(t) = \operatorname{Diag}(\boldsymbol{s}(t))\, B$, $\lambda_{\boldsymbol{s}}(t) = \rho(B_{\boldsymbol{s}}(t))$ the dominant eigenvalue of $B_{\boldsymbol{s}}(t)$ and $v_{\boldsymbol{s}}(t) \gg 0$ the corresponding positive left eigenvector.

It is easy to prove that this system of differential equations has a local solution by standard argument by the Cauchy–Lipschitz–Picard–Lindelöf theorem. Furthermore, if $\boldsymbol{\imath}(0) > 0$ and $\boldsymbol{s}(0) \gg 0$ then

- $\boldsymbol{\imath}(t) > 0$, and $\boldsymbol{s}(t) \gg 0$ for all $t > 0$;
- $\forall t_1 > t_2 > 0$, $\boldsymbol{s}(t_2) \gg \boldsymbol{s}(t_1)$;
- $\lambda_{\boldsymbol{s}}(t)$ is monotone decreasing.

Then, the solution is non-negative. Moreover, it is easy to check that if $s_j(0) + e_j(0) + i_j(0) + r_j(0) = 1$ then $s_j(t) + e_j(t) + i_j(t) + r_j(t) = 1$ and the solution is bounded, so there is a global solution for any time $t > 0$.

About the behavior of the epidemic dynamics, we will analyze the most important epidemiological properties:

- The *threshold phenomenon* that states that, under some condition, an epidemic propagates, in the sense that the introduction of a percentage of infected in the population triggers the contamination of many other individuals, otherwise the epidemic fades off.
- The asymptotic profiles of the steady states in order to understand if an *endemic* level can be reach.

Theorem 1 (Asymptotic behavior). *Denote by $x_s(t)$ the normalized version of $v_s(t)$: $x_s(t)^\top B_s(t) = \lambda_s(t)x_s(t)^\top$ and $x_s(t)^\top x_s(t) = 1$. For any initial condition, $\lambda_s(t)$ is a continuous function, and there exist the limits of the above quantities: $s_\infty = \lim_{t\to\infty} s(t)$, $\lim_{t\to\infty} B_s(t) = B_{s_\infty} = \mathrm{Diag}(s_\infty)B$, $\lim_{t\to\infty}\lambda_s(t) = \lambda_{s_\infty}$, where $\rho(B_{s_\infty}) = \lambda_{s_\infty}$.*

If $s_\infty > 0$, then $\rho(B_{s_\infty}) = \lambda_\infty > 0$ and any converging subsequence of $x_s(t)$ converges to a λ_∞-eigenvector.

If, in addition, $s_\infty \gg 0$, λ_∞ is simple, and then $\lim_{t\to\infty} x_s(t) = x_{s_\infty} \gg 0$, where $x_{s_\infty}^\top B_{s_\infty} = \lambda_{s_\infty} x_{s_\infty}^\top$.

Proof. The existence of the limit of $s(t)$ is obvious, since $s(t)$ is a continuous monotone function. Accordingly, $B_s(t) = \mathrm{Diag}(s_\infty)B$ converges to $\mathrm{Diag}(s_\infty)B$. From now on, to simplify the notations in the proof, we will use $B_t = B_s(t)$, $\lambda_t = \lambda_s(t)$ and $x_t = x_s(t)$ for any $t \in \mathbb{R}_+$. Now, let t fixed and $\delta > 0$ sufficiently small so that $(1-\epsilon)\,\mathrm{Diag}(\mathbf{1}) \le \mathrm{Diag}(s_v)\,\mathrm{Diag}(s_t)^{-1} \le (1+\epsilon)\,\mathrm{Diag}(\mathbf{1})$ for $v \in (t-\delta, t+\delta)$. Then, for $t \le v < t+\delta$,

$$
\begin{aligned}
0 \le \lambda_t - \lambda_v &\le \lambda_t - \min_i \frac{[x_t^\top B_v]_i}{[x_t]_i} \\
&= \lambda_t - \min_i \frac{[x_t^\top \mathrm{Diag}(s_v)B]_i}{[x_t]_i} \\
&= \lambda_t - \min_i \frac{[x_t^\top \mathrm{Diag}(s_v)\,\mathrm{Diag}(s_t)^{-1}\,\mathrm{Diag}(s_t)B]_i}{[x_t]_i} \\
&\le \lambda_{t_1} - \min_i (1-\epsilon)\frac{[x_t^\top (B_t)]_i}{[x_t]_i} \\
&\le \epsilon\lambda_t,
\end{aligned}
$$

while, for $t-\delta < v \le t$,

$$
\begin{aligned}
0 \le \lambda_v - \lambda_t &\le \max_i \frac{[x_t^\top B_v]_i}{[x_t]_i} - \lambda_t \\
&= \max_i \frac{[x_t^\top \mathrm{Diag}(s_v)\,\mathrm{Diag}(s_t)^{-1} B_t]_i}{[x_t]_i} - \lambda_t \\
&= \max_i (1+\epsilon)\frac{[x_t^\top (B_t)]_i}{[x_t]_i} - \lambda_t \\
&\le \epsilon\lambda_t,
\end{aligned}
$$

which implies that $\lambda_s(t)$ is a continuous monotone function that must have a limit; denote it by λ_∞. If $s_\infty = \mathbf{0}$, then $\mathbf{1}^\top B_t \to \mathbf{0}^\top$, which implies that $\lambda_t \to 0$. From now on, we then assume $\mathbf{1}^\top s_\infty = |s_\infty|_1 > 0$, so that $\lambda_\infty > 0$. Let x_{t_n} any converging subsequence, call $\tilde{x}$ its limit. Then, is a λ_∞ non-negative eigenvector of B_∞, since

$$
\begin{aligned}
\tilde{x}^\top B_\infty &= \bullet(\tilde{x}^\top - x_{t_n})B_\infty + x_{t_n}^\top(B_\infty - B_{t_n}) + x_{t_n}^\top B_{t_n} \\
&= \epsilon(n)\mathbf{1}^\top + \lambda_{t_n} x_{t_n}^\top B_{t_n} \\
&\to \lambda_\infty \tilde{x}^\top.
\end{aligned}
$$

Now, if we add the hypothesis that $s_\infty \gg 0$, then B_∞ is still a Perron matrix, and hence there exists a unique positive eigenvector of B_∞, whence $x_t \to x_\infty \gg 0$. □

Theorem 2 (Threshold). *Consider the SEIR model* (4) *on a strongly connected graph G, let $\lambda_{\mathbf{s}}(t)$ be the dominant eigenvalue of $B_{\mathbf{s}}(t)$ and let $\boldsymbol{v}_{\mathbf{s}}(t) \gg 0$ be the corresponding positive left eigenvector.*

(1) *If for a time $\tau \geq 0$, $\beta\lambda_{\mathbf{s}}(\tau) < \gamma$ then $q_{\tau}^{(\epsilon)}(t) = \boldsymbol{v}_{\mathbf{s}}(\tau)^{\top}(\boldsymbol{e}(t)(1+\varepsilon) + \boldsymbol{\imath}(t))$ decreases exponentially to zero for $t \geq \tau$ and any $\epsilon \in \left(0, \frac{\gamma}{\beta\lambda_{\mathbf{s}}(\tau)} - 1\right)$.*

(2) *If $\beta\lambda_{\mathbf{s}}(0) > \gamma$, $\boldsymbol{\imath}(0) > \mathbf{0}$ and $\mathbf{s}(0) \gg \mathbf{0}$, then $\exists t^* > 0$ such that $q_0(t) = \boldsymbol{v}_{\mathbf{s}}(0)^{\top}(\boldsymbol{e}(t) + \boldsymbol{\imath}(t))$ increases for $t \in (0, t^*)$. Moreover, $\boldsymbol{e}(t) \to \mathbf{0}$ and $\boldsymbol{\imath}(t) \to \mathbf{0}$.*

Proof. Take $\epsilon \in \left(0, \frac{\gamma - \beta\lambda_{\mathbf{s}}(\tau)}{\beta\lambda_{\mathbf{s}}(\tau)}\right)$, so that

$$\beta\lambda_{\mathbf{s}}(\tau)(1+\epsilon) - \gamma < 0,$$

and define

$$c_{\epsilon} = \min\left(\gamma - \beta\lambda_{\mathbf{s}}(\tau)(1+\epsilon), \frac{\epsilon\mu}{1+\epsilon}\right) > 0. \tag{5}$$

Multiplying the weighted sum of the second equation of exposed and the third equation of the infected by $\boldsymbol{v}_{\mathbf{s}}(\tau)$,

$$\begin{aligned}\boldsymbol{v}_{\mathbf{s}}(\tau)^{\top}(\dot{\boldsymbol{e}}(t)(1+\epsilon) + \dot{\boldsymbol{\imath}}(t)) &= \boldsymbol{v}_{\mathbf{s}}(\tau)^{\top}(\beta\,\mathrm{Diag}(\mathbf{s}(t))B\boldsymbol{\imath}(t)(1+\epsilon) - \epsilon u\boldsymbol{e}(t) - \gamma\boldsymbol{\imath}(t)) \\ &= \boldsymbol{v}_{\mathbf{s}}(\tau)^{\top}(\beta B_{\mathbf{s}}(t)\boldsymbol{\imath}(t)(1+\epsilon) - \epsilon\mu\boldsymbol{e}(t) - \gamma\boldsymbol{\imath}(t)),\end{aligned}$$

then, for $t \geq \tau$,

$$\frac{d\,\boldsymbol{v}_{\mathbf{s}}(\tau)^{\top}(\boldsymbol{e}(t)(1+\epsilon) + \boldsymbol{\imath}(t))}{dt} \leq \boldsymbol{v}_{\mathbf{s}}(\tau)^{\top}(\beta B_{\mathbf{s}}(t)(1+\epsilon)\boldsymbol{\imath}(t) - \epsilon\mu\boldsymbol{e}(t) - \gamma\boldsymbol{\imath}(t)).$$

Now, $\beta\boldsymbol{v}_{\mathbf{s}}(\tau)^{\top}B_{\mathbf{s}}(\tau) = \boldsymbol{v}_{\mathbf{s}}(\tau)^{\top}\beta\lambda_{\mathbf{s}}(\tau)$, then

$$\begin{aligned}\frac{d\,\boldsymbol{v}_{\mathbf{s}}(\tau)^{\top}(\boldsymbol{e}(t)(1+\epsilon) + \boldsymbol{\imath}(t))}{dt} &\leq \boldsymbol{v}_{\mathbf{s}}(\tau)^{\top}((\beta\lambda_{\mathbf{s}}(\tau)(1+\epsilon) - \gamma)\boldsymbol{\imath}(t) - \epsilon\mu\boldsymbol{e}(t)) \\ &\leq -\boldsymbol{v}_{\mathbf{s}}(\tau)^{\top}c_{\epsilon}(\boldsymbol{\imath}(t) + (1+\epsilon)\boldsymbol{e}(t)),\end{aligned}$$

where c_{ϵ} is defined in (5). Using the previous differential inequality, Gronwall lemma implies that

$$\boldsymbol{v}_{\mathbf{s}}(\tau)^{\top}(\boldsymbol{e}(t)(1+\epsilon) + \boldsymbol{\imath}(t)) \leq \boldsymbol{v}_{\mathbf{s}}(\tau)^{\top}(\boldsymbol{e}(\tau)(1+\epsilon) + \boldsymbol{\imath}(\tau))e^{-c_{\epsilon}(t-\tau)}.$$

To prove (2), we start by noticing that

$$\frac{d}{dt}\left(\boldsymbol{v}_{\mathbf{s}}(0)^{\top}(\boldsymbol{e}(t) + \boldsymbol{\imath}(t))\right)_{t=0} = (\beta\lambda_{\mathbf{s}}(0) - \gamma)\boldsymbol{v}_{\mathbf{s}}(0)^{\top}\boldsymbol{\imath}(0).$$

Since $(\beta\lambda_{\mathbf{s}}(0) - \gamma) > 0$, $\boldsymbol{\imath}(0) > \mathbf{0}$, and $\boldsymbol{v}_{\mathbf{s}}(0) \gg \mathbf{0}$, then

$$\frac{d}{dt}\left(\boldsymbol{v}_{\mathbf{s}}(0)^{\top}(\boldsymbol{e}(t) + \boldsymbol{\imath}(t))\right)_{t=0} > 0.$$

We have that the solution is a continuous differentiable function, then exists $t^* > 0$ such that for $s \in (0, t^*)$

$$\frac{d}{dt}\left(\boldsymbol{v}_{\mathbf{s}}(0)^{\top}(\boldsymbol{e}(t) + \boldsymbol{\imath}(t))\right)_{t=s} > 0,$$

which implies that $q_0(t) = \boldsymbol{v}_{\mathbf{s}}(0)^{\top}(\boldsymbol{e}(t) + \boldsymbol{\imath}(t))$ increases for $t \in (0, t^*)$.

Moreover, it is $\dot{\mathbf{s}} \ll \mathbf{0}$ and $\mathbf{s}(t) \gg \mathbf{0}$, then exist $\lim_{t\to+\infty}\mathbf{s}(t) = \mathbf{s}_{\infty}$. Define $\boldsymbol{b}_j^{\top}$ the j–th row of B, so that $\dot{s}_j = \beta s_j\boldsymbol{b}_j^{\top}\boldsymbol{\imath}$. Since

$$\ddot{s}_j = \beta^2 s_j(\boldsymbol{b}_j^{\top}\boldsymbol{\imath})^2 + \beta s_j\boldsymbol{b}_j^{\top}(\mu\boldsymbol{e} - \gamma\boldsymbol{\imath})$$

then $\|\ddot{s}\| \leq K$, and hence the boundedness and monotonicity of $s(t)$ implies $\lim_{t\to+\infty} \dot{s}(t) = \mathbf{0}$, which implies $\mathbf{1}^\top \dot{s}(t) \to 0$. Then, for any $\epsilon > 0$ we have for t sufficiently large that

$$\begin{aligned}\frac{d}{dt}\mathbf{1}^\top e(t) &\leq \frac{d}{dt}\Big(\mathbf{1}^\top(e(t)+s(t))\Big)+\epsilon \\ &\leq \epsilon - \mu\mathbf{1}^\top e(t).\end{aligned}$$

Then $e(t) \to \mathbf{0}$, that also implies $\mathbf{1}^\top \dot{e}(t) \to 0$. As a consequence, for t sufficiently large,

$$\begin{aligned}\frac{d}{dt}\mathbf{1}^\top \imath(t) &\leq \frac{d}{dt}\Big(\mathbf{1}^\top(e(t)+s(t)+\imath(t))\Big)+\epsilon \\ &\leq \epsilon - \gamma\mathbf{1}^\top \imath(t)\end{aligned}$$

so that $\imath(t) \to \mathbf{0}$. □

2.3. A Control Problem

We have adopted a network framework that explicitly accounts for the interactions structure among individuals and group of individuals, in order to provide insights regarding the spread of a disease. If the proposed model describes the epidemiological phenomenon sufficiently well, some problems relating to the behavior and the forecast of the epidemic itself can be addressed.

First, we would like to prevent an epidemic. This is achieved when condition (1) in Theorem 2 holds at $t = 0$. Before the epidemic starts, the fractions of infected/exposed individuals are negligible, for viral infections the recovery rate γ is usually out of control. Then, the only way to satisfy the no-epidemic requirement is either: control the transmission (which means to reduce β and/or interactions) or immunization (meaning to increase $r(0)$). Second, we aim to limit the economic and social impact as the epidemic occurs. The supply of healthcare services is inelastic in the short run. Thus, it is important to maintain the maximum infection rate below the capacity of the existing healthcare system. This may be achieved by lowering the transmission rate, by controlling the inflow and the outflow of individuals from and into a node.

We point out that only recent works, e.g., [28–30], started investing the trade-off between epidemic and economic costs with some analysis. The aim of our applications would like to be a new step in this direction inside a well based framework. We introduce the following diagonal matrix

$$\boldsymbol{U} = \mathrm{Diag}(u_{loc(i)})_{i=1}^n, \quad \boldsymbol{V} = \mathrm{Diag}(v_{loc(i)})_{i=1}^n, \tag{6}$$

where $u_{loc(i)} \in (0,1]$, $i = 1,\dots,n$ are the control variables for the incoming individuals into the node i, while $v_{loc(i)} \in [0,1]$, $i = 1,\dots,n$ are the control variables of the outgoing individuals from the node i to other nodes. Then, the routing matrix A, and its associated matrices (see their definitions in Section 2.2), changes as follows

$$\begin{aligned}\tilde{A}_{uv} = \boldsymbol{U}\,A\,\boldsymbol{V} = \mathrm{Diag}(\tilde{N}_1,\,\dots,\,\tilde{N}_n)\widetilde{P^o} = \widetilde{P^i}\,\mathrm{Diag}(\tilde{M}_1,\dots,\tilde{M}_n), \\ \widetilde{\mathcal{B}} = \widetilde{P^o}\,\mathrm{Diag}(\tilde{M}_1,\dots,\tilde{M}_n)^{-1}\widetilde{P^o}^\top, \qquad \tilde{B} = \widetilde{P^o}\widetilde{P^i}^\top,\end{aligned}$$

so that the $SEIR$ model on the "controlled" graph $\mathcal{G}$ becomes

$$\begin{aligned}&\dot{\boldsymbol{S}}(t) = -\beta\,\mathrm{Diag}(\boldsymbol{S}(t))\widetilde{\mathcal{B}}\boldsymbol{I}(t) && \dot{s}(t) = -\beta\tilde{I}\,\mathrm{Diag}(s(t))B\imath(t) \\ &\dot{\boldsymbol{E}}(t) = \beta\,\mathrm{Diag}(\boldsymbol{S}(t))\widetilde{\mathcal{B}}\boldsymbol{I}(t) - \mu\boldsymbol{E}(t) && \dot{e}(t) = \beta\tilde{I}\,\mathrm{Diag}(s(t))B\imath(t) - \mu e(t) \\ &\dot{\boldsymbol{I}}(t) = \mu\boldsymbol{E}(t) - \gamma\boldsymbol{I}(t) && \dot{\imath}(t) = \mu e(t) - \gamma\imath(t) \\ &\dot{\boldsymbol{R}}(t) = \gamma\boldsymbol{I}(t) && \dot{r}(t) = \gamma\imath(t)\end{aligned}$$

where $\tilde{I} = \mathrm{Diag}(\tilde{N}_1/N_1,\,\dots,\,\tilde{N}_n/N_n)$.

In order to study the "lockdown policies" applied to the various groups (nodes), we combine a measure of social cost (i.e., hospitalization cost) and, above all, loss of life and the economic loss. The first objective consists of minimizing both total (excess) deaths during the epidemic and the public health cost. We suppose that it is possible estimating the severity of the epidemic in a time interval $[0,T]$, by weighing the total number of infected

$$\text{social cost} = \boldsymbol{C}^T \int_0^T \boldsymbol{I}(t)dt$$

where $\boldsymbol{C}$ is a vector of positive weights, and the integral is to be understood component by component. The lockdown of individuals affects the economic activities; we model the economic loss through the evaluation of the reduction in flow of individuals between nodes with a linear cost function

$$\text{economic loss} = \boldsymbol{W}^T \int_0^T [A - \tilde{A}_{uv}] \cdot \mathbf{1}dt.$$

The goal is finding an optimal trade-off between the total economic loss and the total social cost. Then, the optimal strategy is obtained by minimizing the following cost function

$$\min_{\boldsymbol{U},\boldsymbol{V}} \left(\boldsymbol{C}^T \int_0^T \boldsymbol{I}(t)dt - \boldsymbol{W}^T \int_0^T [A - \tilde{A}_{uv}] \cdot \mathbf{1}dt \right), \tag{7}$$

where T is the time for which a certain strategy is applied.

Remark 2. *If the meta-population model represents a non-geographical subdivision, but instead it is dependent on certain individuals' characteristics (such as age, profession, habits, etc.), the weights* $\boldsymbol{C}$ *and* $\boldsymbol{W}$ *can include information linked to these characteristics (e.g., propensity for mobility, disease mortality). In this case, the lockdown strategy can change based on the vulnerability of each group.*

3. Numerical Tests

This section is devoted to numerical experiments. These experiments were carried out on a laptop equipped with Linux 19.04, with an Intel(R) Core(TM) i5–8250U CPU (1.60 GHz), 16 GiB RAM memory (Intel, Santa Clara, CA, USA) and under MATLAB R2020b environment (MathWorks, Natick, MA, USA).

In order to use our framework, two types of parameters are needed:

(BP) biological parameters related to the different epidemiological features of the disease (parameters γ, β, μ in (4));

(MP) mobility data for the probability outgoing matrix P^o and the probability incoming matrix P^i.

For the first set of parameters, we have referred to a recent work [31] in which the authors applied a SEIR epidemiological model to the recent SARS-CoV-2 outbreak in the world. Moreover, they focused on the application of a stochastic approach in fitting the biological model parameters analyzing the official data and the predicted evolution of the epidemic in the Italian regions, Spain and South Korea. We considered two different scenarios,

(A) The parameters of the disease are $\gamma = 0.14$, $\beta = 0.74$, $\mu = 0.5$.

(B) The parameters of the disease are $\gamma = 0.22$, $\beta = 1.0$, $\mu = 0.03$.

For the topology of the directed graph $\mathcal{G}(V,E)$, we did not refer to any particular geographic area but we reproduced a realistic situation. The network consists of three large agglomerates, each one representing a city. The nodes of the graph are the neighborhoods of the cities and the edges represent the connections between such neighborhoods: these edges encompass the social and working movements between the nodes; hence, they are not simply geographical connections (see Section 2.2). The number of the nodes is 20, 10 and 5, respectively, meaning that the largest city has 20 neighborhoods, the second one 10 and the last one just 5 neighborhoods. This toy model considers a social cost $\boldsymbol{C}$ which is ten times higher than the economic cost $\boldsymbol{W}$: normalizing such costs leads to set $\boldsymbol{C} = \mathbf{1}, \boldsymbol{W} = 10 \cdot \mathbf{1}$. The matrix A is set starting from the adjacency matrix E and from the population of the nodes: the number of individuals, the subpopulations and the matrices P^o and P^i were randomly selected using suitable probability distributions.

All scenarios started with the initial distribution for susceptible, infected and exposed individuals: the epidemic starts from $1/5$ of randomly selected neighborhoods of the largest city. Once a lockdown strategy is decided by optimizing Equation (7), it is applied for 14 days: after such period the new distributions for S, I and E are checked and a new optimization is carried on. The last time interval has a longer duration for observing the effects of the overall strategy on the long period. This scheme is applied three times in the numerical simulations.

Figure 1a presents the optimized strategy for Scenario A, while Figure 1b shows the strategy for Scenario B. We have not represented the whole network but a part of it considering nodes that represent agglomerations with a different number of individuals. Furthermore, the strategy that optimizes our objective function is reproduced by showing the values of the vector $\boldsymbol{V}$, see (6), for some cities/agglomerations present in the network. Values close to 1 mean that there are no particular restrictions on mobility, values close to 0 mean strong restrictions on movement. We point out that the value 0 is not allowed because it is not realistic to consider a total block of each movement in this context.

In the case of Scenario A, the parameter of the disease induced a light lockdown (85%) on the large city (blue line), whilst the other two are almost completely open. On the second case, the disease is more infectious: the large city is forced to adopt a severe lockdown, while the strategy on the other two suggests a mild lockdown. As soon as the epidemic spreads, due to the characteristics of the disease, even the smallest cities are forced to adopt a more severe strategy until the number of infected individuals decreases. We can observe in Scenario B that, after a period of severe lockdown, when the last chosen strategy is applied for a longer period, then a further severe approach must be adopted in order to contain the epidemic.

In both scenarios, we can observe that there is a converging behavior of the strategies to be applied on the three different cities. Check, for example, the period 50–100 days in Scenario A: the lockdown strategy for the largest city (blue line) is increasing, while the strategies for the other cities (orange and yellow lines) are decreasing. Eventually, due to the disease parameters, they converge to 1, meaning that the epidemic threat is no more. This behavior is more evident in Scenario B.

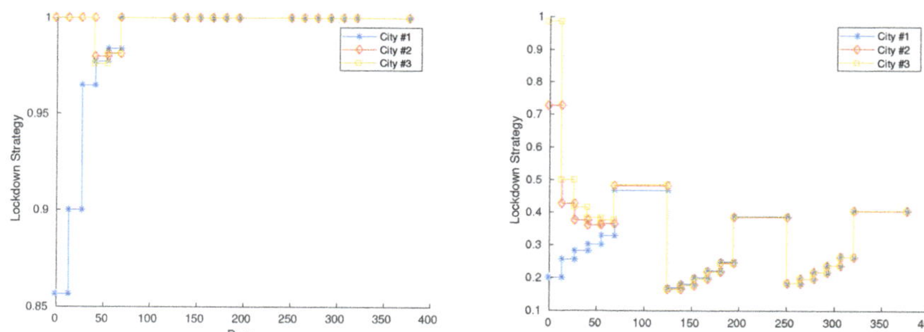

(a) (b)

Figure 1. (**a**,**b**): Lockdown strategies for Scenario A and B, respectively. We show the optimal strategy $\boldsymbol{V}$ that minimizes (7) the closer is to 0, the more severe the restrictions are. On the other hand, values close to 1 denote very mild restrictions on mobility.

Figure 2 presents a visual representation of the evolution of the strategies for Scenario B at $t = 0, 20, 40$ and 179 days. The color of the nodes represents the number of infected people in that node, while the color of the edges represents the percentage of people blocked. At the beginning, the optimal strategy suggests to mainly block the outgoing from the large city (the agglomerate on the top left corner), while the connection between and inside the other two are open. Letting the epidemic spread in a controlled way, in order to maintain economy, induces an increase in the number of infected people, then the connections between cities must be reduced (the more red the edges are, the less people are allowed to move).

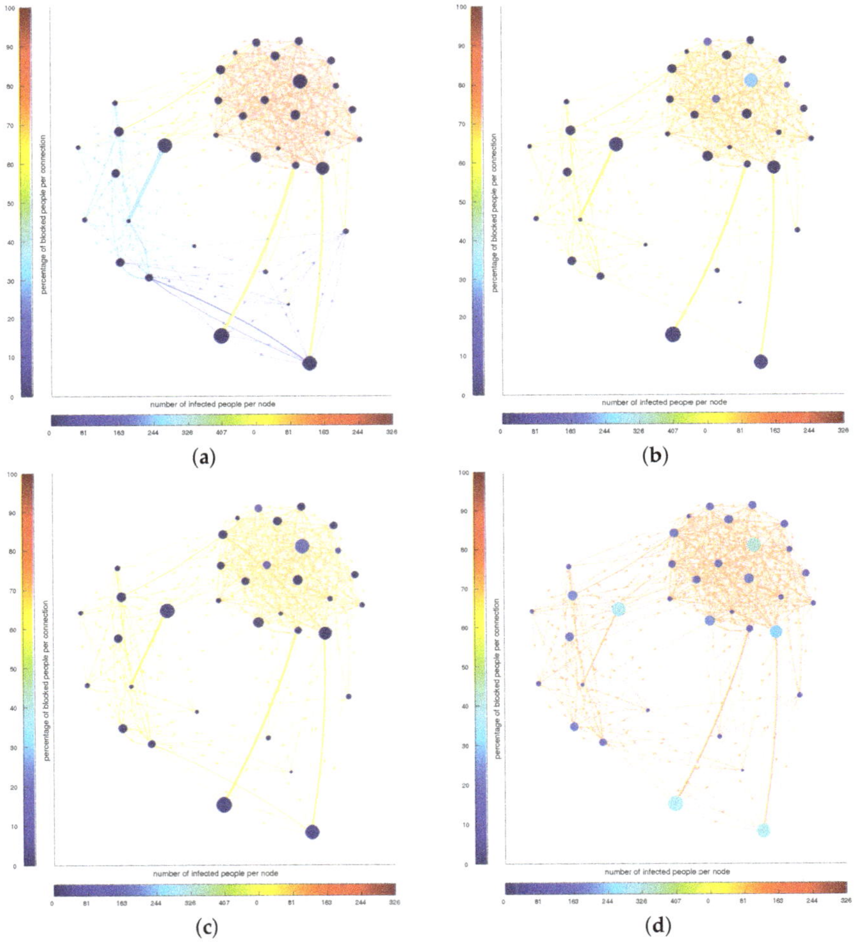

Figure 2. (**a**–**d**): Evolution of the strategy and infected people in Scenario B for timestamps $t = 0, 20, 40, 179$, respectively. The color of the nodes represents the number of infected people in that node, while the color of the edges represents the percentage of people blocked.

4. Conclusions

The network perspective allows to relax the assumption of uniform random mixing and then we are able to model the population interaction patterns during epidemics. Moreover, a network-based model provides useful and important insights about the spread of a disease; such insights cannot be inferred using the classical model. We have focused on a SEIR meta-population model on a network in order to characterize the epidemic dynamics and to predict possible contagion scenarios.

A network model can encompass differences in the number of interactions between individuals in a population. For example, there exist some realities where people may live in small environments and they have relatively few contacts (work and/or social life), while others may live in dense and more populated centers, where the usage of public and maybe crowded public transportation is very common, they work in high-contact environments and they have a large number of interactions with many others outside of work. The classical SEIR model does not allow to include such heterogeneity, while a network model can easily encompass it. Furthermore, it is possible to adapt our model and, instead of a geographical distinction of the subgroups of individuals, different stratifications of the population can be considered.

We justified the model by introducing and analyzing some of its properties, in particular, we proved a threshold theorem involving both biological parameters and the topology of the network. In a future paper, we will consider both time-dependent parameters and a detailed analysis of the asymptotic behavior of the solution of the proposed model. We point out that our analysis can be applied to recent models, e.g., [26], where numerical and statistical but not analytical results are provided.

Only recent works, e.g., [28–30], started investigating the trade-off between epidemic and economic costs with some analysis. In order to take a new step in this direction, we have also identified an optimal control problem that considers the advantages and benefits that arise from the application of optimal targeted policies, which lock down the various groups in an inhomogeneous way. The focus is on the balance between economic loss and loss of life. The main economic damages consist of lost productivity due to illness and also in the forgone productivity contributions of the blocked subpopulations. The lives lost are estimated via the number of the infected individuals, assuming that these losses represent a constant percentage of the latter.

Some preliminary numerical tests are provided; more effective results could be obtained by considering a suitable fitting of the parameters and based on some particular topology of the network. These issues will be analyzed in a future paper involving a different source of data [32] and recent optimization tools [33,34].

Author Contributions: Conceptualization, G.A. and G.N.; methodology, G.A., A.B. and G.N.; software, G.A.; validation, G.A. and A.B.; formal analysis, G.A., A.B. and G.N.; investigation, G.A. and G.N.; resources, G.A., A.B. and G.N.; data curation, G.A. and A.B.; writing—original draft preparation, G.A., A.B. and G.N.; writing—review and editing, A.B.; visualization, A.B.; supervision, G.A., A.B. and G.N.; project administration, G.N.; funding acquisition, G.N. All authors have read and agreed to the published version of the manuscript.

Funding: This research received funding from PRECISION project.

Institutional Review Board Statement: Not applicable.

Informed Consent Statement: Not applicable.

Data Availability Statement: Data sharing not applicable.

Acknowledgments: We acknowledge people from ADAMSS center for the insights on the developing of the model.

Conflicts of Interest: The authors declare no conflict of interest.

References

1. Kermack, W.O.; McKendrick, A.G.; Walker, G.T. A contribution to the mathematical theory of epidemics. *Proc. R. Soc. London Ser. Contain. Pap. Math. Phys. Character* **1927**, *115*, 700–721. [CrossRef]
2. Anderson, R.; May, R. *Infectious Diseases of Humans: Dynamics and Control*; Oxford University Press: Oxford, UK, 1991.
3. Hethcote, H. Mathematics of infectious diseases. *SIAM Rev.* **2000**, *42*, 599–653. [CrossRef]
4. Capasso, V. *Mathematical Structures of Epidemic Systems*; Springer: Berlin/Heidelberg, Germany, 1993.
5. Foppa, I. *A Historical Introduction to Mathematical Modeling of Infectious Diseases: Seminal Papers in Epidemiology*; Academic Press: Cambridge, MA, USA, 2016; pp. 1–197.

6. Chowell, G.; Sattenspiel, L.; Bansal, S.; Viboud, C. Mathematical models to characterize early epidemic growth: A review. *Phys. Life Rev.* **2016**, *18*, 66–97. [CrossRef] [PubMed]
7. Metcalf, C.; Lessler, J. Opportunities and challenges in modeling emerging infectious diseases. *Science* **2017**, *357*, 149–152. [CrossRef]
8. Keeling, M.; Rohani, P. *Modeling Infectious Diseases in Humans and Animals*; Princeton University Press: Princeton, NJ, USA, 2011; pp. 1–368.
9. Garrett, K.; Mundt, C. Epidemiology in mixed host populations. *Phytopathology* **1999**, *89*, 984–990. [CrossRef]
10. Rimbaud, L.; Fabre, F.; Papaïx, J.; Moury, B.; Lannou, C.; Barrett, L.; Thrall, P. Models of Plant Resistance Deployment. *Annu. Rev. Phytopathol.* **2021**, *59*, 125–152. [CrossRef]
11. Jeger, M.; Pautasso, M.; Holdenrieder, O.; Shaw, M. Modelling disease spread and control in networks: Implications for plant sciences. *New Phytol.* **2007**, *174*, 279–297. [CrossRef] [PubMed]
12. Kenah, E.; Chao, D.; Matrajt, L.; Halloran, M.; Longini, I., Jr. The global transmission and control of influenza. *PLoS ONE* **2011**, *6*, e19515. [CrossRef]
13. Heesterbeek, H.; Anderson, R.; Andreasen, V.; Bansal, S.; DeAngelis, D.; Dye, C.; Eames, K.; Edmunds, W.; Frost, S.; Funk, S.; et al. Modeling infectious disease dynamics in the complex landscape of global health. *Science* **2015**, *347*, aaa4339. [CrossRef] [PubMed]
14. Nowzari, C.; Preciado, V.; Pappas, G. Analysis and Control of Epidemics: A Survey of Spreading Processes on Complex Networks. *IEEE Control Syst.* **2016**, *36*, 26–46. [CrossRef]
15. Arino, J.; Bauch, C.; Brauer, F.; Driedger, S.; Greer, A.; Moghadas, S.; Pizzi, N.; Sander, B.; Tuite, A.; Van Den Driessche, P.; et al. Pandemic influenza: Modelling and public health perspectives. *Math. Biosci. Eng.* **2011**, *8*, 1–20. [CrossRef] [PubMed]
16. Lessler, J.; Cummings, D. Mechanistic models of infectious disease and their impact on public health. *Am. J. Epidemiol.* **2016**, *183*, 415–422. [CrossRef] [PubMed]
17. Masuda, N.; Miller, J.; Holme, P. Concurrency measures in the era of temporal network epidemiology: A review. *J. R. Soc. Interface* **2021**, *18*, 20210019. [CrossRef]
18. Layan, M.; Dellicour, S.; Baele, G.; Cauchemez, S.; Bourhy, H. Mathematical modelling and phylodynamics for the study of dog rabies dynamics and control: A scoping review. *PLoS Neglected Trop. Dis.* **2021**, *15*, e0009449. [CrossRef] [PubMed]
19. Earn, D.; Dushoff, J.; Levin, S. Ecology and evolution of the flu. *Trends Ecol. Evol.* **2002**, *17*, 334–340. [CrossRef]
20. Riley, S.; Eames, K.; Isham, V.; Mollison, D.; Trapman, P. Five challenges for spatial epidemic models. *Epidemics* **2015**, *10*, 68–71. [CrossRef] [PubMed]
21. Pellis, L.; Ball, F.; Bansal, S.; Eames, K.; House, T.; Isham, V.; Trapman, P. Eight challenges for network epidemic models. *Epidemics* **2015**, *10*, 58–62. [CrossRef]
22. Miller, J.; Kiss, I. Epidemic spread in networks: Existing methods and current challenges. *Math. Model. Nat. Phenom.* **2014**, *9*, 4–42. [CrossRef]
23. Hanski, I.; Gilpin, M. Metapopulation dynamics: Brief history and conceptual domain. *Biol. J. Linn. Soc.* **1991**, *42*, 3–16. [CrossRef]
24. Hanski, I.; Gaggiotti, O. *Ecology, Genetics and Evolution of Metapopulations*; Academic Press: Cambridge, MA, USA, 2004; pp. 1–696. [CrossRef]
25. Boccaletti, S.; Latora, V.; Moreno, Y.; Chavez, M.; Hwang, D.U. Complex networks: Structure and dynamics. *Phys. Rep.* **2006**, *424*, 175–308. [CrossRef]
26. Gatto, M.; Bertuzzo, E.; Mari, L.; Miccoli, S.; Carraro, L.; Casagrandi, R.; Rinaldo, A. Spread and dynamics of the COVID-19 epidemic in Italy: Effects of emergency containment measures. *Proc. Natl. Acad. Sci. USA* **2020**, *117*, 10484–10491. [CrossRef] [PubMed]
27. Chung, F.R.K. *Spectral Graph Theory; Volume 92, CBMS Regional Conference Series in Mathematics*; Published for the Conference Board of the Mathematical Sciences, Washington, DC, USA; American Mathematical Society: Providence, RI, USA, 1997; p. xii+207.
28. Alvarez, F.E.; Argente, D.; Lippi, F. *A Simple Planning Problem for COVID-19 Lockdown*; Working Paper 26981: National Bureau of Economic Research: Cambridge, MA, USA, 2020. [CrossRef]
29. Miclo, L.; Spiro, D.; Weibull, J. *Optimal Epidemic Suppression under an ICU Constraint*; Working Paper 1111; Toulouse School of Economics: Toulouse, France, 2020.
30. Birge, J.R.; Candogan, Z.; Feng, Y. *Controlling Epidemic Spread: Reducing Economic Losses with Targeted Closures*; Working Paper 2020-57; Becker Frieman Institute Chicago: Chicago, IL, USA, 2020.
31. Godio, A.; Pace, F.; Vergnano, A. Seir modeling of the italian epidemic of SARS-CoV-2 using computational swarm intelligence. *Int. J. Environ. Res. Public Health* **2020**, *17*, 3535. [CrossRef] [PubMed]
32. Rivieccio, B.; Micheletti, A.; Maffeo, M.; Zignani, M.; Comunian, A.; Nicolussi, F.; Salini, S.; Manzi, G.; Auxilia, F.; Giudici, M.; et al. CoViD-19, learning from the past: A wavelet and cross-correlation analysis of the epidemic dynamics looking to emergency calls and Twitter trends in Italian Lombardy region. *PLoS ONE* **2021**, *16*, e0247854. [CrossRef] [PubMed]
33. Zanni, L.; Benfenati, A.; Bertero, M.; Ruggiero, V. Numerical Methods for Parameter Estimation in Poisson Data Inversion. *J. Math. Imaging Vis.* **2015**, *52*, 397–413. [CrossRef]
34. Benfenati, A.; Ruggiero, V. Inexact Bregman iteration with an application to Poisson data reconstruction. *Inverse Probl.* **2013**, *29*, 065016. [CrossRef]

mathematics

Article

A Comparative Numerical Study and Stability Analysis for a Fractional-Order SIR Model of Childhood Diseases

Mohamed M. Mousa [1,2] **and Fahad Alsharari** [1,3,*]

[1] Department of Mathematics, College of Sciences and Human Studies at Hotat Sudair, Majmaah University, Al-Majmaah 11952, Saudi Arabia; mm.mousa@mu.edu.sa or mohamed.youssef@bhit.bu.edu.eg
[2] Department of Basic Engineering Sciences, Faculty of Engineering at Benha, Benha University, Benha 13512, Egypt
[3] Department of Mathematics, College of Science and Arts, Jouf University, Gurayat 77455, Saudi Arabia
* Correspondence: f.alsharari@mu.edu.sa

Abstract: The objective of this work is to examine the dynamics of a fractional-order susceptible-infectious-recovered (SIR) model that simulate epidemiological diseases such as childhood diseases. An effective numerical scheme based on Grünwald–Letnikov fractional derivative is suggested to solve the considered model. A stability analysis is performed to qualitatively examine the dynamics of the SIR model. The reliability and robustness of the proposed scheme is demonstrated by comparing obtained results with results obtained from a fourth order Runge–Kutta built-in Maple syntax when considering derivatives of integer order. Graphical illustrations of the numerical results are given. The inaccuracy of some results presented in two studies exist in the literature have been clearly explained. Generalizing of the cases examined in another study, by considering a model with fraction-order derivatives, is another objective of this work as well.

Keywords: SIR model; fractional derivatives; Grünwald–Letnikov method; stability analysis

Citation: Mousa, M.M.; Alsharari, F. A Comparative Numerical Study and Stability Analysis for a Fractional-Order SIR Model of Childhood Diseases. *Mathematics* **2021**, *9*, 2847. https://doi.org/10.3390/math9222847

Academic Editors: Mihaela Neamţu, Eva Kaslik and Anca Rădulescu

Received: 30 September 2021
Accepted: 2 November 2021
Published: 10 November 2021

Publisher's Note: MDPI stays neutral with regard to jurisdictional claims in published maps and institutional affiliations.

Copyright: © 2021 by the authors. Licensee MDPI, Basel, Switzerland. This article is an open access article distributed under the terms and conditions of the Creative Commons Attribution (CC BY) license (https://creativecommons.org/licenses/by/4.0/).

1. Introduction

The investigation of the prevalence of the contagious diseases with the aid of mathematical modeling has turn into an essential means to realize epidemiological patterns diseases. A lot of procedures can be studied in the formulation of the mathematical model to reflect the characteristics in spread types of a disease. Generally, there are two main types:

i. Direct contact type such as childhood diseases [1–3], Ebola virus disease [4,5], HIV infection [6–9], and COVID-19 [10–14].
ii. Indirect contact type, that is because of the existence of disease carriers in the natural, such as, over mosquitoes and their aquatic stage [15–18].

Childhood infections are the highly popular form of infections contagious. These infections diseases such as chicken pox, mumps, measles, etc., to which infants start their life susceptible, and mostly develop within next 5 years. Because children are very close together with their peers, such diseases can prevalence in a fast way. Because the vaccination is believed to be the best efficient approach counter to such diseases, the progress of a framework that would develop the optimum vaccine treatment stage needed to avoid the prevalence of childhood diseases is necessary. The (susceptible–infected–recovered) SIR models are traditional models that have been used to illustrate many epidemiological diseases [1–3,10–14].

The considered version of SIR model is the model given in [1–3]. This model that applied and used to achieve a well understanding of exactly how the childhood disease spreads within populations of various individuals with time, as well as the risk of surges in the susceptible individuals. The SIR is a dynamical system that is consists of three coupled ordinary differential equations (ODEs) that refer to the time evolution of the next three individuals:

i. Susceptible group (S): this group is not infected but perhaps becomes infected due to the spread of the virus or stay susceptible.
ii. Infected group (I): This group has already been infected by the virus and may spread it to the susceptible group. An infected people may stay infected or may be removed from this group because of recovering or death.
iii. Removed group (R): This group has been vaccinated against the virus as well as recovered individuals with permanent immunity. The SIR model considers that the efficiency of the vaccination is one hundred percentage. The normalized dynamical system that describes the SIR model is given below [3]:

The normalized dynamical system that describes the SIR model which considered in [3] is given below:

$$\begin{aligned} D_t S(t) &= (1-p)\omega - \omega\, S(t) - \beta\, S(t)\, I(t), \\ D_t I(t) &= -(\sigma+\omega)\, I(t) + \beta\, S(t)\, I(t), \\ D_t R(t) &= p\,\omega - \omega\, R(t) + \sigma\, I(t), \end{aligned} \tag{1}$$

where $D = d/dt$, β and σ are the natural birth rate and the natural death rate, respectively; however, the childhood disease mortality rate is considered extremely low. The factor p is a fraction of population that have been vaccinated at the birth every year ($0 \leq p < 1$) with taking into consideration that the remainder of the citizens is susceptible. The factor ω describes the rate of infection of the susceptible citizens due to a contact with infected peoples. Here, the total population is normalized. Recently, a big attention has been drawn to examine mathematical models that defined by fractional differential systems specially in the field of epidemiological diseases [19–21]. The greatest important property of such models is their memory impact [22], which is not occurring in the differential systems of integer derivatives. This memory impact is quite described by the mobility of the differentiation order for the fractional derivatives and realized as an inherited property on viruses' genomes and strains (as mentioned in [21]), that is helpful for modeling epidemiological diseases. It is known that the SIR model is highly impacted by the initial data and the integer order SIR model may not perfectly explain the spread of the diseases because of the local nature of the integer order derivatives. The fractional-order derivative is non-local in its nature and depends on the initial data. Thus, for well understanding of the epidemiological disease's models, it is helpful to replace the integer order SIR model by a fractional-order one. In the present study, we will replace time derivatives in the system (1) with fractional time derivatives of order γ. In the last few years, many researchers simulated and analyzed various epidemiological diseases using fractional-order dynamical models [23–27]. In this work we generally analyze and simulate the dynamics of the fractional-order SIR model that has been studied in [1,2],

$$\begin{aligned} {}^{C}_{t_0}D^{\gamma}_{t} S(t) &= (1-p)\omega - \omega\, S(t) - \beta\, S(t)\, I(t), \\ {}^{C}_{t_0}D^{\gamma}_{t} I(t) &= -(\sigma+\omega)\, I(t) + \beta\, S(t)\, I(t), \\ {}^{C}_{t_0}D^{\gamma}_{t} R(t) &= p\,\omega - \omega\, R(t) + \sigma\, I(t), \end{aligned} \tag{2}$$

where ${}^{C}_{t_0}D^{\gamma}_{t}$ is the Caputo fractional-order derivative operator of order $\gamma \in (0, 1]$ which can be defined as,

$${}^{C}_{t_0}D^{\gamma}_{t} F(t) = (1/\Gamma(n-\gamma)) \left(\int_{t_0}^{t} (t-\tau)^{n-\gamma-1} \frac{d^n F(\tau)}{d\tau^n}\, d\tau \right), \tag{3}$$

when it effects on a continuous function F on an interval $[0, T]$, where t_0 is the start time, $n = \lfloor \gamma \rfloor + 1$ and $\lfloor \gamma \rfloor$ is the integer part of the fraction γ. The Caputo fractional-order derivative has advantages for solving initial value problems only when it applied with analytical or semi-analytical approaches. The Caputo derivative can be approximated by the Grünwald–Letnikov (GL) method using finite differences of the fractional order, similar to the Euler method, to handle numerical solutions of initial value problems. The GL method is proceeding iteratively but the sum in the scheme becomes longer and longer, which reflects the memory effect.

This paper is structured as follows. An explanation of the Grünwald–Letnikov fractional derivative representation is given in Section 2. In Section 3, the GL discretized scheme for the considered fractional-order SIR model is developed. In Section 4, the qualitative stability analysis of the fractional-order SIR model is investigated, and the free equilibrium fixed point of the disease is examined. Section 5 contains the numerical simulation of the present results including comparisons with results displayed in [1–3] and with results of the Maple built-in scheme of the fourth order Runge–Kutta (RK4) method along with graphical illustrations. A sensitivity analysis, which demonstrates how the model dynamics differently perform as the values of the model parameters changed, is including in Section 5 as well. Finally, in Section 6, conclusions and observations are provided.

2. Explanation of the Grünwald–Letnikov Fractional Derivatives

Here, some fundamental concepts of fractional Grünwald–Letnikov derivatives are presented. The form of GL fractional derivative with order $\gamma \in (0, 1]$ is expressed as:

$$ {}_{t_0}D_t^{\gamma} F(t) = \lim_{h \to 0} (1/h^{\gamma}) \sum_{j=0}^{(t-t_0)/h} (-1)^j \binom{\gamma}{j} F(t - jh), \tag{4} $$

where h is the calculations time step and $\binom{\gamma}{j} = \frac{\Gamma(\gamma+1)}{\Gamma(j+1)\Gamma(\gamma-j+1)}$ is binomial coefficients.

The explicit numerical approximation of the fractional derivative of order γ-th at the points kh, ($k = 1, 2, \ldots$) is defined by the next formula [28–30]:

$$ (k - (Z/h))_{t_0} D_{t_k}^{\gamma} F(t) \approx \frac{1}{h^{\gamma}} \sum_{j=0}^{k} c_j^{(\gamma)} F\left(t_{k-j}\right), \tag{5} $$

where Z is noted as memory length, $t_k = kh$ and $c_j^{(\gamma)}$, $(j = 0, 1, \ldots)$ are the binomial coefficients that easily computed from the following equations [28]:

$$ c_0^{(\gamma)} = 1, \quad c_j^{(\gamma)} = \left(1 - \frac{1+\gamma}{j}\right) c_{j-1}^{(\gamma)}, \quad (j = 1,\, 2,\, \ldots). \tag{6} $$

Therefore, a general numerical approximation of the fractional-order differential equation:

$$ {}_{t_0}D_t^{\gamma} F(t) = G(F(t), t), \tag{7} $$

where G is a nonlinear function, that can be written as,

$$ F(t_k) = G(F(t_{k-1}), t_{k-1}) h^{\gamma} - \sum_{j=v}^{k} c_j^{(\gamma)} F\left(t_{k-j}\right), \tag{8} $$

With the aid of the principle of the short memory, the summation index v in Equation (8) is considered as $v = 1$ for all $k < Z/h$ and $v = k - (Z/h)$ for $k > Z/h$. However, we set $v = 1$ for all k when the principle of the short memory is not used. Clearly, the consequence for this approximation makes us pay a penalty in the shape of a few inaccuracies. If $F(t) \leq M$, we may easily estimate the memory length Z, providing the needed accuracy ξ as [31]:

$$ Z \geq [M/(\xi \cdot |\Gamma(1-\gamma)|)]^{1/\gamma} \tag{9} $$

An investigation of the short memory impact and the differences between long and short memory were examined and illustrated in [30]. Generally, we can easily apply the GL fractional derivative approximation to systems of fractional-order in the following form,

$$ {}_{t_0}D_t^{\gamma} F_i(t) = G_i(F_1(t), F_2(t), \dots, F_N(t), t),\ i = 1, 2, \dots, N. \tag{10}$$

3. Grünwald–Letnikov Discretization of Fractional-Order SIR Model

For numerical simulation of the fractional-order SIR model described in the system (2), we will implement the introduced GL discretization technique to solve it. This approach is established on the fact that for many classes of continuous functions, the fractional-order derivatives of Grünwald–Letnikov and Caputo are equivalent [31]. Using the fractional GL numerical approximation in the Equation (8), the numerical solution of the system (2), when ${}_{t_0}^{C}D_t^{\gamma} \equiv {}_{t_0}D_t^{\gamma}$, can be obtained using the following iterative formulas,

$$\begin{aligned} S(t_k) &= [(1-p)\,\omega - \omega\, S(t_{k-1}) - \beta\, S(t_{k-1})\, I(t_{k-1})]h^{\gamma} - \sum_{j=v}^{k} c_j^{(\gamma)} S\left(t_{k-j}\right), \\ I(t_k) &= [-(\sigma+\omega)\, I(t_{k-1}) + \beta\, S(t_k)\, I(t_{k-1})]h^{\gamma} - \sum_{j=v}^{k} c_j^{(\gamma)} I\left(t_{k-j}\right), \\ R(t_k) &= [p\omega - \omega\, R(t_{k-1}) + \sigma\, I(t_k)]h^{\gamma} - \sum_{j=v}^{k} c_j^{(\gamma)} R\left(t_{k-j}\right), \end{aligned} \tag{11}$$

where, $k = 1, 2, 3, \dots, N$, for $N = T/h$ and T is the simulation time. The binomial coefficients $c_j^{(\gamma)}$ are calculated using the formula in Equation (6). Along this paper we set the start data (initial conditions) as follows: $S(t_0) = k_1$, $I(t_0) = k_2$, $R(t_0) = k_3$ and the start time $t_0 = 0$.

4. Stability Analysis of the Fractional-Order SIR Model

In this section, we examine the stability of the considered SIR model. We proceed to find the fixed points (steady solutions) that occur when ${}_{t_0}D_t^{\gamma}S(t) = {}_{t_0}D_t^{\gamma}I(t) = {}_{t_0}D_t^{\gamma}R(t) = 0$, and evaluate their stability. The fixed point that occurs at a long-term behavior of the fractional system (2) are $P_f = \left(S_f^* = 1-p,\ I_f^* = 0,\ R_f^* = p\right)$. This fixed point is corresponding to the disease-free equilibrium which is described by the nonexistence of the infected nodes. The Jacobian matrix for the SIR system (2) is defined as:

$$ J = \begin{bmatrix} -\beta I - \omega & -\beta S & 0 \\ \beta I & \beta S - \omega - \sigma & 0 \\ 0 & \sigma & -\omega \end{bmatrix}, \tag{12}$$

At disease free equilibrium fixed point, the eigenvalues relating to the matrix J are,

$$ \lambda_1 = \beta(1-p) - \omega - \sigma,\ \lambda_2 = -\omega,\ \lambda_3 = -\omega. \tag{13}$$

This disease-free equilibrium fixed point is asymptotically stable if all the eigenvalues in Equation (13) satisfy the following condition [32]:

$$ |\arg(\lambda_{1,2,3})| > \frac{\pi}{2}\gamma. \tag{14}$$

From Equations (13) and (14), the condition that guarantees the stability of SIR dynamical system at the disease-free equilibrium point is,

$$ \beta(1-p) < \omega + \sigma,\ \forall\, \gamma \in (0, 1]. \tag{15}$$

Therefore, the vaccination reproduction number V_r can be defined as,

$$ V_r = \frac{\beta(1-p)}{\omega + \sigma}. \tag{16}$$

If V_r is higher than a limit value, an infectious disease can propagate in a susceptible group. Moreover, the condition in Equation (15) shows that a critical vaccination fraction p_c can defined as,

$$p_c = 1 - \frac{\omega + \sigma}{\beta}. \tag{17}$$

It is worth to note that above this critical fraction, the disease-free equilibrium is stable i.e., $p > p_c$. Therefore, the vaccination fraction must be large enough in order to effectively avoid surge period of the disease.

The other fixed point of the dynamical system (2) is the point corresponding to the endemic equilibrium point, which is described by the existence of the infected nodes. This point is obtained as $P_e = \left(S_e^* = \frac{1-p}{V_r},\ I_e^* = \frac{\omega}{\beta}(V_r - 1),\ R_e^* = \frac{1}{\beta S_e^*}\left[\omega p + \sigma - \frac{\sigma}{\beta}(\omega + \sigma)\right]\right)$. It is clear that the endemic equilibrium will only occur when $V_r > 1$. At this point, the eigenvalues relating to the matrix J are obtained as,

$$\lambda_1 = -\omega,\ \lambda_{2,3} = -\frac{\omega V_r}{2}\left(1 \pm \sqrt{1 - 4\frac{(\omega + \sigma)}{\omega V_r}}\right). \tag{18}$$

From Equations (14) and (18), the endemic equilibrium point P_ϵ is asymptotically stable if the following condition is fulfilled,

$$1 < V_r \leq \frac{4}{\omega}(\omega + \sigma),\ \forall\, \gamma \in (0, 1]. \tag{19}$$

5. Numerical Results and Discussion

In the current section, we present numerical results with many visualizations to simulate the dynamics of the present fractional-order SIR model subjected to several fractional order γ, initial conditions and parameters. Along this section, we set the time step $h = 0.01$ for the fractional GL iterative scheme and for the Maple RK4 built-in scheme. Firstly, we consider the numerical values of the case considered in [1,2] to show the inaccuracy of the results displayed in these papers. As mentioned in [1,2], we assume:

$$k_1 = 1,\ k_2 = 0.5,\ k_3 = 0,\ p = 0.9,\ \beta = 0.8,\ \omega = 0.4,\ \sigma = 0.03. \tag{20}$$

According to the parameters in Equation (20), the critical vaccination fraction $p_c = 0.4625$. Therefore, this case falls under the disease-free equilibrium case at which the steady state solution asymptotically approaches to the following fixed point,

$$P_f = \left(S_f^* = 0.1,\ I_f^* = 0,\ R_f^* = 0.9\right) \tag{21}$$

Looking at the results that were presented in [1,2], we found that the graphical representations of the solutions are inaccurate and do not agree with the corresponding steady state solution that shown in Equation (21). In this regard, we compared our numerical results with the results presented in [1,2] for the susceptible group $S(t)$ only at several values of the fractional order γ. This comparison is displayed in Figure 1. For the validation purpose of the present numerical results, we plot the solution of Maple RK4 built-in scheme along with our solution using the GL scheme at $\gamma = 1$.

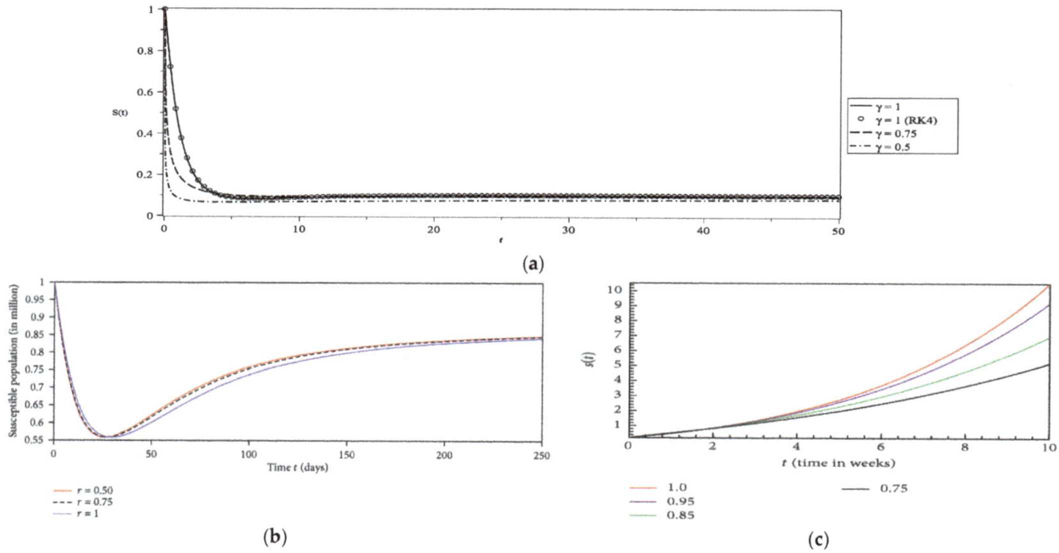

Figure 1. Comparison of $S(t)$ between (**a**) present results, (**b**) results of [1] and (**c**) results of [2].

From Figure 1, it is shown that the graphical representation of $S(t)$ in [1,2] is incorrect even in the transient region of the solution. From Figure 1c that was presented in [2], it can be shown that $S(0) = 0$ although the authors considered $S(0) = 1$ in their study. Moreover, the graphical representations of $I(t)$ and $R(t)$ that displayed in [1,2] are inaccurate as well because they do not agree with the corresponding fixed point in Equation (21). For $\gamma = 1$, the excellent agreement between the approximate solution using RK4 scheme and the GL scheme validates the approximate solution of GL scheme. Figure 1a illustrates that our approximate solution agrees with the steady state fixed point shown in Equation (21). This note is considered as another validation of the present results.

The rest of this section is related to the generalization of the cases presented in [3] to examine the analyze of those cases subject to various fractional orders, mainly $\gamma = 1$, $\gamma = 0.75$ and $\gamma = 0.5$ for the same initial conditions and parameters. Because of the considered SIR model encounters long interepidemic periods, we set the simulation time $T = 250$. For the purpose of the comparison with the results of [3], two versions of the solutions graphs were displayed. One at a small-time behavior $T = 10$, to show how the solutions at the transient region behave; and the other at a long-time behavior $T = 250$, to show how the solutions behave at long interepidemic period (steady state solution). Here, we set $\beta = 0.8$, $\omega = 0.4$ and $\sigma = 0.03$ for all the considered cases. According to these parameters, the critical value of the vaccination portion is obtained as $p_c = 0.4625$. The results of the RK4 method at $\gamma = 1$ are plotted by circle points ($\circ$).

- **Case 1**

We consider the initial conditions and case parameters as $k_1 = 1$, $k_2 = 0$, $k_3 = 0$ and $p = 0.9$. Here, we have a stable disease-free equilibrium case with a vaccination reproduction number $V_r = 0.186047$.

This case demonstrates the effect of high-level vaccination exposure on the initial individuals with no infective individuals. Figure 2 displays a comparison between present numerical results for S, I and R versus time for various fractional orders, $\gamma = 1$, $\gamma = 0.75$ and $\gamma = 0.5$, against the results obtained in [3] when $\gamma = 1$. It can be shown that there is an excellent agreement between the results of GL scheme, RK4 scheme and the obtained in at $\gamma = 1$. Furthermore, the graphical representation of the numerical results demonstrates that the solutions are monotone dependent on the order of the fractional derivative γ. The susceptible, infective, and removed individuals decrease as the fractional order γ decreases in both transient and steady state stage. Generally, as γ diminishes, the fixed equilibrium solution reduces as well. For this case, the disease has been successfully eradicated because of the effect of high-level vaccination fraction p. It is worth to notice that all individuals normally remain disease-free for all times.

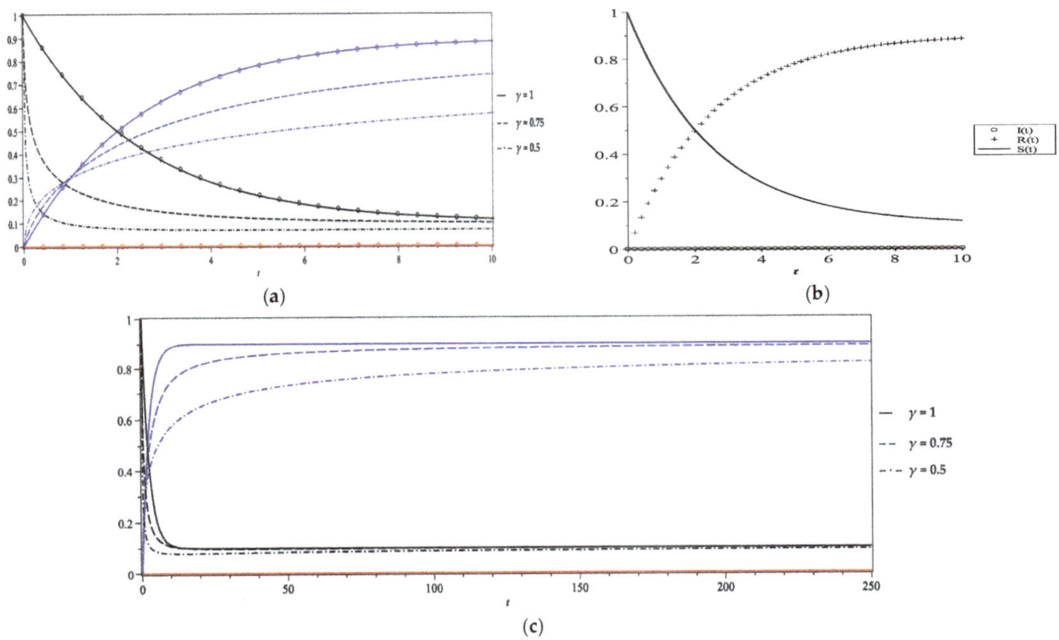

Figure 2. Behavior of various individual fractions (S in black, I in red and R in blue) versus time for Case 1 at different fractional order values (**a**) till $T = 10$, (**b**) results of [3] and (**c**) till $T = 250$.

- **Case 2**

In this case we make a slight change in the initial condition to study the influence of high-level vaccination exposure on the initial low-level infective individuals. Here we consider $k_1 = 0.8$, $k_2 = 0.2$, $k_3 = 0$ and $p = 0.9$. This case is also a stable disease-free equilibrium one, with a vaccination reproduction number $V_r = 0.186047$, in which the disease will be eradicated.

Figure 3 displays a comparison between present numerical results for susceptible, infective and removed individuals versus time for various fractional order γ, along with the results obtained in [3] when $\gamma = 1$. When considering an integer order, an excellent agreement between the results of GL scheme, RK4 scheme and the results obtained in [3] is realized. As shown from Figure 3, the susceptible and infective individuals reduce with the passage of time whereas the removed individual raises due to the appearance of recovered and vaccinated groups with long-lasting resistance against disease and hence the disease will be eradicated. Here, the disease-free equilibrium is reachable as soon as the vaccination amount level is more than the critical vaccination value p_c (i.e., $p > p_c$) as occurred in Case 1. The impact of the fractional order γ on the dynamic of the solutions is the same as its effect in Case 1.

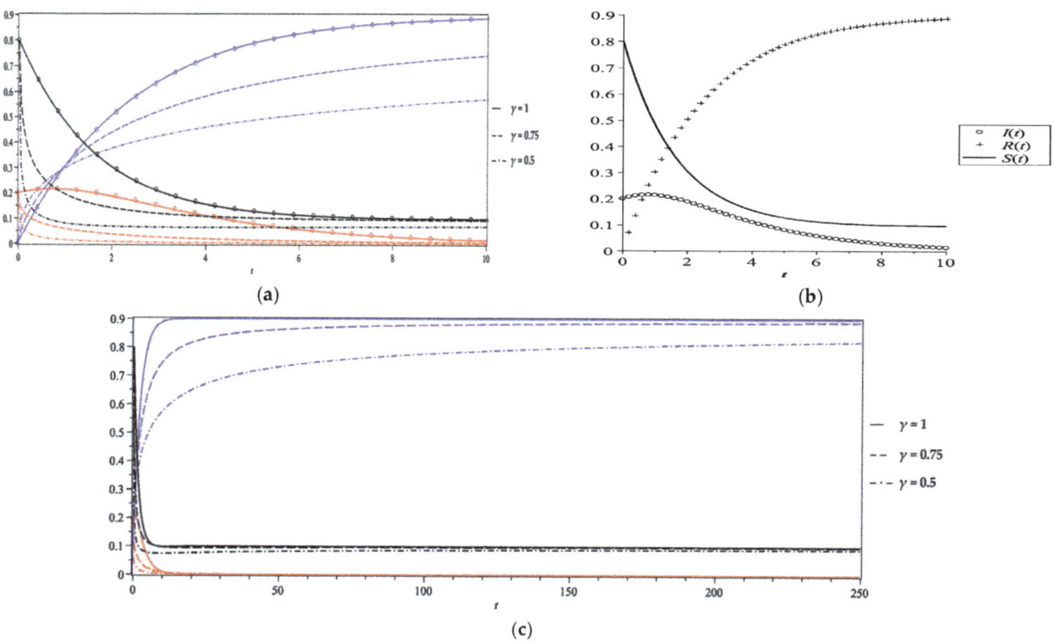

Figure 3. Behavior of various individual fractions (S in black, I in red and R in blue) versus time for Case 2 at different fractional order values (**a**) till $T = 10$, (**b**) results of [3] and (**c**) till $T = 250$.

- **Case 3**

Here we consider an endemic-equilibrium case by setting $k_1 = 0.8$, $k_2 = 0.2$, $k_3 = 0$ and $p = 0.3$. This case explains the impact of low-level vaccination exposure on the initial individuals with low-level infective population. Figure 4 shows a graphical representation of numerical results for the susceptible, infective, and removed population versus time for various fractional order γ, alongside the results obtained in [3]. As expected, when $\gamma = 1$, an excellent agreement between the obtained results and the results displayed in [3] is achieved.

From Figure 4, it can be notice that the number of susceptible individuals reduces, while the removed individuals increase with a small amount as time expands. Nevertheless, it is notable that the infective individuals will never vanish as time expands and hence the endemic case continues with no disease eradication. This proves that a disease-free equilibrium could not be accomplished when the vaccination amount is less than the critical vaccination value (i.e., $p < p_c$). In this case, the endemic equilibrium stays stable because a vaccination reproduction number $V_r = 1.302326$ satisfies the stability condition in Equation (19). In such an endemic-equilibrium case, the fractional order γ plays an influential role. The fractional order affects the susceptible, infective, and removed individuals by a different manner from disease-free equilibrium cases. When $\gamma < 1$, we found that as the fractional order γ decreases, the susceptible and infective groups rapidly decrease in the transient stage and then begin to increase once again in the steady state stage. In addition, for $\gamma < 1$, the susceptible individuals will be more than the amount for $\gamma = 1$ in the transient stage. While the infective individuals remain less than the amount for $\gamma = 1$ in both transient and steady state stages.

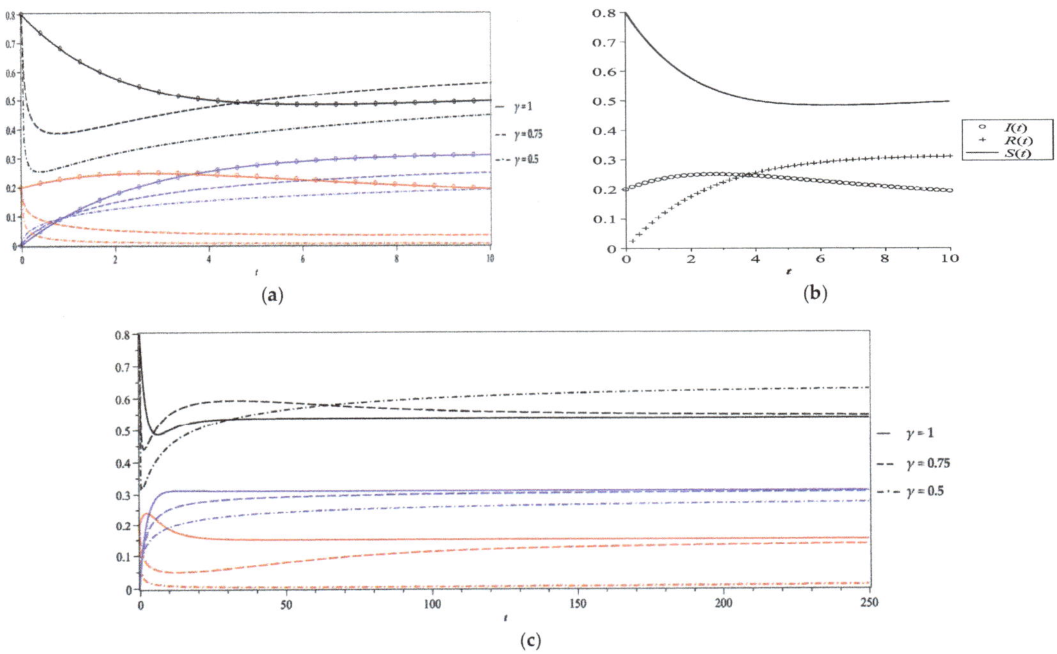

Figure 4. Behavior of various individual fractions (S in black, I in red and R in blue) versus time for Case 3 at different fractional order values (**a**) till $T = 10$, (**b**) results of [3] and (**c**) till $T = 250$.

- **Case 4**

Here we consider another endemic-equilibrium case by setting $k_1 = 0.8$, $k_2 = 0.2$, $k_3 = 0$ and $p = 0$. This case describes the influence of low-level vaccination exposure on the initial individuals with no vaccinated people. Figure 5 shows the results visualization for the susceptible, infective, and removed population with time for various values of γ, beside the results obtained in [3] for $\gamma = 1$.

As expected, an excellent agreement between our results and the results obtained in [3] is realized. Similar to Case 3, we have a stable endemic-equilibrium situation with a vaccination reproduction number $V_r = 1.860465$. Here, the number of susceptible individuals decreases while the infective individuals increase by the time for the steady state stage. The only role of the removed individuals is the very little fraction of the recovered population with long-lasting immunity. Here, the disease is quickly transported to the bulk of people. The impact of the fractional order γ on the dynamic of the solutions is similar in its impact in Case 3.

A sensitivity analysis of the considered cases is displayed in Table 1 which illustrates how the model dynamics differently behave as per parameters values changed. From the data presented in the table, it is clear that the numerical results are in excellent agreement with the results of the system fixed points in both disease-free equilibrium and endemic equilibrium case. For the disease-free equilibrium cases, the fraction of the infected population is vanishing for all values of the fractional order γ by the time and hence a disease has been eradicated. While for the endemic equilibrium cases, the fraction of the infected population decreases with the decreasing of the fractional order γ by the time, but the disease is not eradicated.

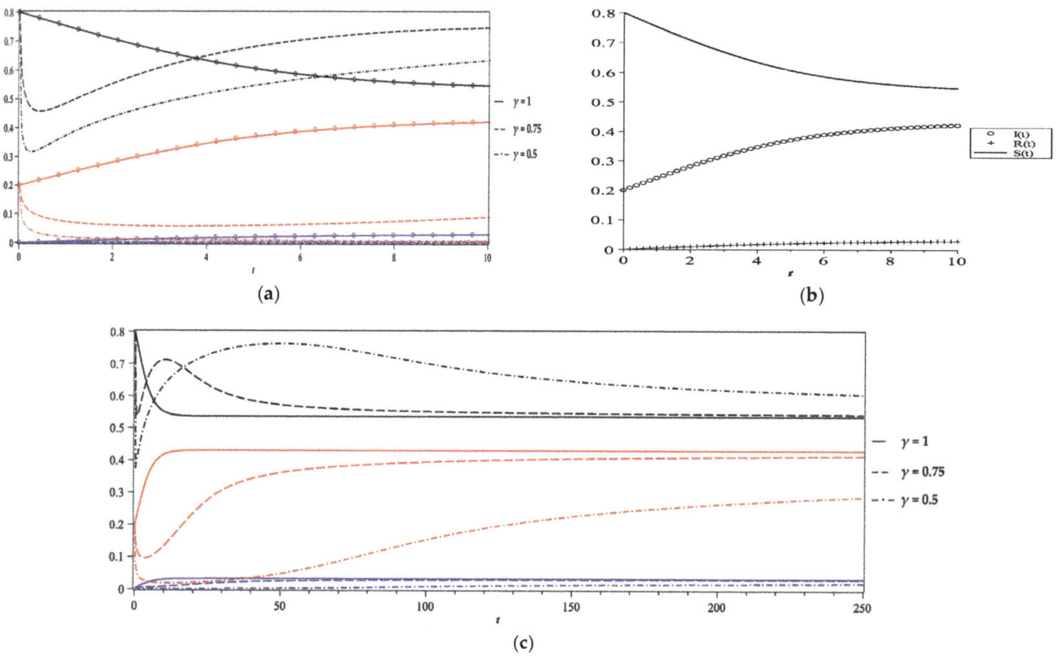

Figure 5. Behavior of various individual fractions (S in black, I in red and R in blue) versus time for Case 4 at different fractional order values (**a**) till $T = 10$, (**b**) results of [3] and (**c**) till $T = 250$.

Table 1. Impact of various parameter values on the population fractions.

Case	V_r	γ	**Fixed Point of the Stability Analysis**			**Steady State Equilibrium of the Numerical Solution at $T = 250$**			**Comment**
			S	I	R	S	I	R	
1	0.186	1 0.75 0.5	0.1	0.0	0.9	0.1000 0.0989 0.0914	0.0 0.0 0.0	0.8999 0.8897 0.8207	Removal of the disease
2	0.186	1 0.75 0.5	0.1	0.0	0.9	0.0999 0.0989 0.0913	0.0 0.0 0.0	0.9000 0.8897 0.8208	
3	1.302	1 0.75 0.5	0.5375	0.1512	0.3113	0.5375 0.5457 0.6277	0.1512 0.1363 0.0102	0.3113 0.3066 0.2742	No removal of the disease
4	1.860	1 0.75 0.5	0.5375	0.4302	0.0323	0.5375 0.5440 0.6063	0.4302 0.4140 0.2870	0.0323 0.0306 0.0188	

6. Conclusions

This paper deals with a new application of a Grünwald–Letnikov method for the achievement of a numerical simulation of a fractional-order SIR epidemiological diseases model. The significant advantage of the considered technique is that it can be used without system linearization or any other restrictions for any fractional-order dynamical system. A stability analysis is performed to qualitatively investigate the dynamics of the model in both disease-free and endemic equilibrium case. It is explained that the graphical representation presented in [1,2] are inaccurate. One of the issues that validates the present results is the excellent agreement between them with the results presented in [3] and RK4 results for integer order $\gamma = 1$. The excellent agreement between the asymptotic behavior of the solutions with the numerical results is another evident on the validation and the reliability of the considered numerical technique in studying such dynamical models in epidemiology. The proposed technique is reliable and accurate in simulating the behavior of the solutions for a long-time interval. The disease-free equilibrium is stable if the vaccination amount above a critical value p_c. The dynamics of the model is strongly dependent on the value of the fractional order γ. For the endemic equilibrium cases, the fraction of the infected population decreases by the reduction of the fractional order γ as time passes.

Author Contributions: Both authors contributed equally to this work. All authors have read and agreed to the published version of the manuscript.

Funding: This research was funded by Majmaah University, grant number R-2021-260.

Institutional Review Board Statement: Not applicable.

Informed Consent Statement: Not applicable.

Data Availability Statement: Not applicable.

Acknowledgments: The authors would like to thank the Deanship of Scientific Research at Majmaah University for supporting this work under Project Number No.: R-2021-260.

Conflicts of Interest: The authors declare that there are no conflicts of interest.

References

1. Ullah, A.; Abdeljawad, T.; Ahmad, S.; Shah, K.; Karapinar, E. Study of a fractional-order epidemic model of childhood diseases. *J. Funct. Spaces* **2020**, *2020*, 5895310. [CrossRef] [PubMed]
2. Haq, F.; Shahzad, M.; Muhammad, S.; Wahab, H.A.; Rahman, G.U. Numerical analysis of fractional order epidemic model of childhood diseases. *Discret. Dyn. Nat. Soc.* **2017**, *2017*, 4057089. [CrossRef]
3. Makinde, O.D. Adomian decomposition approach to a SIR epidemic model with constant vaccination strategy. *Appl. Math. Comput.* **2007**, *184*, 842–848. [CrossRef]
4. Rachah, A.; Torres, D.F.M. Mathematical modelling, simulation, and optimal control of the 2014 ebola outbreak in West Africa. Discret. *Dyn. Nat. Soc.* **2015**, *2015*, 842792.
5. GA, N.; MI, T.-E. A Mathematical model with quarantine states for the dynamics of Ebola virus disease in human populations. *Comput. Math. Methods Med.* **2016**, *2016*, 9352725.
6. Ogunlaran, O.M.; Noutchie, S.C. Mathematical model for an effective management of HIV infection. *Biomed. Res. Int.* **2016**, *2016*, 4217548. [CrossRef]
7. Djordjevic, J.; Silva, C.J.; Torres, D.F.M. A stochastic SICA epidemic model for HIV transmission. *Appl. Math. Lett.* **2018**, *84*, 168–175. [CrossRef]
8. Arenas, A.J.; González-Parra, G.; Naranjo, J.J.; Cogollo, M.; La Espriella, N.D. Mathematical analysis and numerical solution of a model of HIV with a discrete time delay. *Mathematics* **2021**, *9*, 257. [CrossRef]
9. Zafar, Z.U.A.; Rehan, K.; Mushtaq, M. HIV/AIDS epidemic fractional-order model. *J. Differ. Equ. Appl.* **2017**, *23*, 1298–1315. [CrossRef]
10. Omame, A.; Abbas, M.; Onyenegecha, C.P. A fractional-order model for COVID-19 and tuberculosis co-infection using Atangana–Baleanu derivative. *Chaos Solitons Fractals* **2021**, *153*, 111486. [CrossRef] [PubMed]
11. Okuonghae, D.; Omame, A. Analysis of a mathematical model for COVID-19 population dynamics in Lagos, Nigeria. *Chaos Solitons Fractals* **2020**, *139*, 110032. [CrossRef]
12. Cooper, I.; Mondal, A.; Antonopoulos, C.G. A SIR model assumption for the spread of COVID-19 in different communities. *Chaos Solitons Fractals* **2020**, *139*, 110057. [CrossRef] [PubMed]
13. Singh, H.; Srivastava, H.M.; Hammouch, Z.; Nisar, K.S. Numerical simulation and stability analysis for the fractional-order dynamics of COVID-19. *Results Phys.* **2021**, *20*, 103722. [CrossRef] [PubMed]
14. Yadav, S.; Kumar, D.; Singh, J.; Baleanu, D. Analysis and dynamics of fractional order Covid-19 model with memory effect. *Results Phys.* **2021**, *24*, 104017. [CrossRef]
15. Moreno, V.M.; Espinoza, B.; Bichara, D.; Holechek, S.A.; Castillo-Chavez, C. Role of short-term dispersal on the dynamics of Zika virus in an extreme idealized environment. *Infect. Dis. Model.* **2017**, *2*, 21–34. [CrossRef]
16. Ndaïrou, F.; Area, I.; Nieto, J.J.; Silva, C.J.; Torres, D.F.M. Mathematical modeling of Zika disease in pregnant women and newborns with microcephaly in Brazil. *Math. Methods Appl. Sci.* **2018**, *41*, 8929–8941. [CrossRef]
17. Ndaïrou, F.; Area, I.; Torres, D.F.M. Mathematical modeling of Japanese encephalitis under aquatic environmental effects. *Mathematics* **2020**, *8*, 1880. [CrossRef]

18. Angstmann, C.N.; Henry, B.I.; McGann, A.V. A fractional order recovery SIR model from a stochastic process. *Bull. Math. Biol.* **2016**, *78*, 468–499. [CrossRef]
19. Al-Sulami, H.; El-Shahed, M.; Nieto, J.J.; Shammakh, W. On fractional order dengue epidemic model. *Math. Probl. Eng.* **2014**, *2014*, 456537. [CrossRef]
20. Sweilam, N.H.; AL-Mekhlafi, S.M.; Baleanu, D. Optimal control for a fractional tuberculosis infection model including the impact of diabetes and resistant strains. *J. Adv. Res.* **2019**, *17*, 125–137. [CrossRef]
21. Ndaïrou, F.; Area, I.; Nieto, J.J.; Silva, C.J.; Torres, D.F.M. Fractional model of COVID-19 applied to Galicia, Spain and Portugal. *Chaos Solitons Fractals* **2021**, *144*, 110652. [CrossRef]
22. Du, M.; Wang, Z.; Hu, H. Measuring memory with the order of fractional derivative. *Sci. Rep.* **2013**, *3*, 3431. [CrossRef] [PubMed]
23. Lichae, B.H.; Biazar, J.; Ayati, Z. The fractional differential model of HIV-1 infection of CD4+ T-Cells with description of the effect of antiviral drug treatment, Comput. *Math. Methods Med.* **2019**, *2019*, 4059549. [CrossRef]
24. Danane, J.; Allali, K.; Hammouch, Z. Mathematical analysis of a fractional differential model of HBV infection with antibody immune response. *Chaos Solitons Fractals* **2020**, *136*, 109787. [CrossRef]
25. Li, Y.; Haq, F.; Shah, K.; Shahzad, M.; Rahman, G. Numerical analysis of fractional order Pine wilt disease model with bilinear incident rate. *J. Math. Comput. Sci.* **2017**, *17*, 420–428. [CrossRef]
26. Owolabi, K.M.; Atangana, A. Mathematical modelling and analysis of fractional epidemic models using derivative with exponential kernel. In *Fractional Calculus in Medical and Health Science*; CRC Press: Boca Raton, FL, USA, 2020; pp. 109–128.
27. El-Dessoky, M.M.; Khan, M.A. Modeling and analysis of an epidemic model with fractal-fractional Atangana-Baleanu derivative. *Alex. Eng. J.* **2022**, *61*, 729–746. [CrossRef]
28. Dorcak, L. Numerical models for the simulation of the fractional-order control systems. *arXiv* **2002**, arXiv:math/0204108v1.
29. Vinagre, B.M.; Chen, Y.Q.; Petráš, I. Two direct Tustin discretization methods for fractional-order differentiator/integrator. *J. Frankl. Inst.* **2003**, *340*, 349–362. [CrossRef]
30. Podlubny, I. *Fractional Differential Equations: An Introduction to Fractional Derivatives, Fractional Differential Equations, to Methods of Their Solution and Some of Their Applications*; Academic Press: New York, NY, USA, 1998.
31. Petráš, I. *Fractional-Order Nonlinear Systems: Modeling, Analysis and Simulation*; Springer: Berlin, Germany, 2011.
32. Tavazoei, M.S.; Haeri, M. A necessary condition for double scroll attractor existence in fractional-order systems. *Phys. Lett. A* **2007**, *367*, 102–113. [CrossRef]

mathematics

Article

Computational Study on the Dynamics of a Consumer-Resource Model: The Influence of the Growth Law in the Resource

Luis M. Abia [1], Óscar Angulo [2,*], Juan Carlos López-Marcos [1] and Miguel Ángel López-Marcos [1]

1 Departamento de Matemática Aplicada, Facultad de Ciencias, Universidad de Valladolid e IMUVa, 47011 Valladolid, Spain; lmabia@uva.es (L.M.A.); lopezmar@mac.uva.es (J.C.L.-M.); malm@mac.uva.es (M.Á.L.-M.)
2 Departamento de Matemática Aplicada, ETS de Ingenieros de Telecomunicación, Universidad de Valladolid e IMUVa, 47011 Valladolid, Spain
* Correspondence: oscar.angulo@uva.es; Tel.: +34-983-423000 (ext. 5835); Fax: +34-983-423-661

Abstract: The dynamics of a specific consumer-resource model for *Daphnia magna* is studied from a numerical point of view. In this study, Malthusian, chemostatic, and Gompertz growth laws for the evolution of the resource population are considered, and the resulting global dynamics of the model are compared as different parameters involved in the model change. In the case of Gompertz growth law, a new complex dynamic is found as the carrying capacity for the resource population increases. The numerical study is carried out with a second-order scheme that approximates the size-dependent density function for individuals in the consumer population. The numerical method is well adapted to the situation in which the growth rate for the consumer individuals is allowed to change the sign and, therefore, individuals in the consumer population can shrink in size as time evolves. The numerical simulations confirm that the shortage of the resource has, as a biological consequence, the effective shrink in size of individuals of the consumer population. Moreover, the choice of the growth law for the resource population can be selected by how the dynamics of the populations match with the qualitative behaviour of the data.

Keywords: size-structured population; numerical methods; characteristics method; growth laws; asymptotic behaviour; *Daphnia magna*

MSC: 92D25; 92D40; 65M25; 65M12; 35B40

Citation: Abia, L.M.; Angulo, Ó.; López-Marcos, J.C.; López-Marcos, M.Á. Computational Study on the Dynamics of a Consumer-Resource Model: The Influence of the Growth Law in the Resource. *Mathematics* **2021**, *9*, 2746. https://doi.org/10.3390/math9212746

Academic Editors: Mihaela Neamțu, Eva Kaslik and Anca Rădulescu

Received: 1 October 2021
Accepted: 26 October 2021
Published: 29 October 2021

Publisher's Note: MDPI stays neutral with regard to jurisdictional claims in published maps and institutional affiliations.

Copyright: © 2021 by the authors. Licensee MDPI, Basel, Switzerland. This article is an open access article distributed under the terms and conditions of the Creative Commons Attribution (CC BY) license (https://creativecommons.org/licenses/by/4.0/).

1. Introduction

Population dynamics is a very active field with different disciplines (epidemiology, ecology, etc.) in which a variety of theoretical studies and different numerical approaches arise. The purpose of this work is to analyse, from a numerical point of view, some aspects of the dynamics of a problem that describes the evolution of a *Daphnia magna* population, as an example of a consumer-resource model. As, in general, it is not feasible to obtain its dynamics from a theoretical point of view, the behaviour of the solutions to the problem is analysed with numerical techniques. The model is solved by means of an efficient numerical method adapted to the problem, which was proposed and analysed in [1]. It is a second-order method that has been used successfully to obtain numerical approximations to solutions of a wide range of physiologically structured population problems [1–3]. The numerical method provides the size dependence of the density of the consumer population and can afford numerical approximations even in the situation of unbounded density functions. With this help, the evolution of the consumer-resource model of the *Daphnia magna* is studied to evaluate the effect of different biological growth laws for the resource on the dynamics of the model. The method allows achieving certainty regarding different conclusions on the dynamics of the model.

Many population dynamics models can be formulated and analysed as a specific physiologically structured consumer-resource model. In de Roos and Persson [4], a trophic chain population model with a resource, a consumer, and a top predator is considered, and its stability analysis was performed in Sanchez and Getto [5]. Cuesta et al. [6] studied the stability of a problem for phytoplankton cell dynamics and allometry in the growth rate and a top predator. Pang et al. [7], also studied the stability of models structured by the age of infection for HIV transmission. The group of Lafferty approach to these models is through the formulation of a general complete macrostructured population model [8]. Aylaj and Noussair [9] analysed the convergence of a numerical method for the approximation to the solution of a multistage kinetic model of a physiologically structured population of insects with four stages in their life-cycle. The numerical integration for an erythropoiesis model with two dynamical growth factors (erythropoietin and glucocorticoids) was proposed in [10].

The pioneering work of Kooijman and Metz [11] was the beginning of the theoretical study of consumer-resource models. These authors presented a mathematical model for the development of an ectothermic population (*Daphnia magna*, water flea) in which the amount of food they are supplied with represented a regulatory mechanism for the population density. The model was length-structured and also included a stage of maturity: the difference between juveniles and adults in the population. A few years later, this model was studied by Thieme [12]. He described the population as a distributional solution of a classical partial differential equation formulation and explained heuristically why the well-posedness was a problem and to what extent it could be solved. He also included a dynamical environment in which individuals fed and that evolved with a general growth law in absence of a consumer. In [13], Diekmann et al. proved that, under appropriate assumptions, the local stability of a steady-state was determined by the spectral properties of the linearised semigroup. They applied this theory to the consumer-resource model that was transformed into a delay ordinary differential equation coupled with a birth law (Volterra functional formulation). They employed the age of individuals as the structuring variable with a finite life span and a finite delay in the description of the birth law. These results were extended to the infinite delay case in [14], which allowed them to study problems with infinite life span. The properties of the stability of the former model proposed by Kooijman and Metz, including the maturation stage, were studied in [15]. This work led to the theoretical support of [16], where de Roos et al. performed a numerical study of such equilibria. This analysis assumed positive continuously differentiable vital rates and only continuous at the value of maturity.

On the other hand, the numerical study of the dynamics of the model is also important because it gives an answer to problems that cannot be addressed theoretically. With respect to the research on the dynamics of the model, recent papers [17,18], tried to compute the roots of the characteristic equation numerically by means of pseudoespectral discretisation and then determine the stability properties. This last procedure was also applied to a more complex model that introduced to the system a superconsumer (predator) [5]. Moreover, suitable numerical methods were proposed to approximate the solution of the problem and were used to study its dynamics by means of a long-time integration. These studies concerned different cases that also included an unbounded consumer population [1–3,19].

Consumer-resource models with undefined sign growth rates for the consumer population still have to be considered theoretically concerning the existence, uniqueness, and regularity of solutions. Numerically, only [1–3] dealt with this problem but by employing a logistic growth law for the resource. The numerical method used in the simulations is well adapted to this situation in which the growth rate is allowed to change the sign, and therefore, individuals in the consumer population can shrink in size as time evolves.

The remainder of the paper is organised as follows. Section 2 describes a general consumer-resource model of a physiologically structured population and the growth laws for the resource considered in the numerical experiments. In this model, the maximum size of individuals in the population is not fixed and they are allowed to shrink. The main contribution is included in Section 3, where an extensive numerical experimentation, for the case of the *Daphnia magna*, is reported using vital rates and initial conditions selected from the literature. The simulations will show the complexity of the dynamics of both populations (consumer—*Daphnia magna*—and resource—algae) under different biological growth laws and parameter values in the model. The numerical experimentation is developed with an efficient numerical scheme, which is completely described in Appendix A to make the paper self-contained.

2. The *Daphnia magna* Model and the Growth of the Resource

The problem studied is in the form introduced by de Roos [19]: he did not consider a maturation stage to distinguish between juveniles and adults (therefore individuals started reproducing immediately after they were born), and for the first time, he described the growth rate of the consumer population with a function that had an undefined sign. This difficult issue, from a mathematical point of view, provided an answer to the modelisation of rare species that could diminish their size.

A size-structured consumer-resource model that represents the evolution of a single consumer population preying on a single resource (for instance, a *Daphnia magna* feeding on an algal population) is employed. In general, this model is composed of two nonlinear coupled problems. On the one hand, the evolution of the consumer population, which is structured by the length of the individuals. On the other hand, the evolution of a resource, which is seen as an unstructured population. Henceforth, the model is composed of a classical size-structured problem driven by a nonlinear hyperbolic partial differential equation (conservation law), a nonlocal boundary condition (the birth law), and an initial condition:

$$\begin{cases} u_t + (g(x,t,s(t))\,u)_x = -\mu(x,t,s(t))\,u, & x_0 < x < x_M(t), \quad t > 0, \\ g(x_0,t,s(t))\,u(x_0,t) = \displaystyle\int_{x_0}^{x_M(t)} \alpha(x,t,s(t))\,u(x,t)\,dx, & t > 0, \\ u(x,0) = u^0(x), \quad x_0 \le x \le x_M^0, \end{cases} \tag{1}$$

and by the following nonlinear initial value problem

$$\begin{cases} s'(t) = f(t,s(t),I(t)), & t > 0, \\ s(0) = s^0. \end{cases} \tag{2}$$

Here t and x represent, respectively, the time and the length of the consumer individuals, and x_0 and x_M^0 denote the minimum and maximum consumer individual length at the initial distribution, respectively. Functions $s(t)$ and $u(x,t)$ mean the available resource and the density of individuals with length x at time t, respectively. The solution of this evolution problem depends on the initial conditions $u^0(x)$, $x_0 \le x \le x_M^0$, and s^0, the vital rates that define the evolution of the consumer (growth, birth, and death), and the biological growth law of the resource. The fertility and mortality rates, α and μ, respectively, are non-negative functions. The growth rate, g, has the following special properties: it has an undetermined sign but $g(x_0,\cdot,\cdot) > 0$ at any time $t > 0$. These conditions on the growth rate mean that, on the one hand, individuals grow from their birth and, on the other hand, individuals in the population can shrink under food scarcity conditions. These vital rates are influenced by the time, the length of an individual and the available resource. Next, the biological growth law of the resource is represented as the sum of two terms $f(t,s,I) = f_S(t,s) + f_u(t,I)$ to fulfil the requirements of the further numerical experimentation. The first one, f_S, introduces the evolution of the resource without the consumer influence; the second one, f_u, takes into account the intake of resources by the consumer-population individuals through a quantity, $I(t)$, that represents its consumption along time. It is described with the following nonlocal term,

$$I(t) = \int_{x_0}^{x_M(t)} \gamma(x,t,s(t))\,u(x,t)\,dx, \quad t \ge 0, \tag{3}$$

where γ represents the intake of individuals of size x in the consumer population at time t and also depends on the available resource. Thus, the evolution of the individuals in the consumer population is affected by the competition for food through the vital rates, which depends on the available resource itself. The evolution of the feeding resource is modified by the individuals through (2).

Finally, the maximum individual length, $x_M(t)$, is not fixed and can evolve in a non-monotonical way with time due to the lack of restriction on the growth rate sign. Its evolution follows the corresponding characteristic curve. It is the solution of the next initial value problem,

$$\begin{cases} x'_M(t) = g(x_M(t), t, s(t)), \quad t > 0, \\ x_M(0) = x_M^0. \end{cases} \tag{4}$$

To close this section, different alternatives for the growth of the resource in the absence of a consumer, which is given by f_S, will be discussed. The analysis is focused on the growth phenomena with the introduction of three basic distributions which provide the main mathematical frameworks.

1. The exponential growth: $f_S(t,s) = r\,s, r \in \mathbb{R}$.
   This is the most basic and simplest growth function and, in essence, it is the growth relationship considered by Malthus (see [20], for instance). The parameter r is the specific growth rate. When $r \neq 0$, there is only an equilibrium for the differential equation associated: the trivial solution. If $r < 0$, the trivial equilibrium is an attractor. If $r > 0$, this solution is unbounded.
2. The chemostat growth function: $f_S(t,s) = r\,(K - s), r > 0, K > 0$.
   This is a modification of the exponential growth function with a limited increment that depends on the carrying capacity K. There is a stable nontrivial equilibrium that corresponds with the carrying capacity. When the $s^0 < K$ population increases, and the growth is slower because the available resources decrease. In the case $s^0 > K$, population decreases to the carrying capacity. It also can represent a population, given by s, that is growing up in an environment in which a constant quantity of feeding is provided (the carrying capacity) and r represents the dilution rate [20].
3. The Gompertz growth function: $f_S(t,s) = r\,s\,\log\left(\frac{K}{s}\right), r > 0, K > 0$.
   The Gompertz equation was formulated originally as a law of decreasing survivorship, but it has also been employed to model the growth of plants, tumours, and fisheries [20]. There is a limited carrying capacity. The population has a similar behaviour as when the biological growth is described with the well-known logistic growth law because there are two equilibria: the unstable trivial equilibrium and a stable nontrivial one.

Although there are other possibilities (logistic, power laws, Ricker's, Weibull or Gaussian distributions, etc., or diverse combinations among them) the results will show enough variability to consider that the discussion is enough with these choices. The logistic growth, which was studied in [1–3,19], deserves a special mention. Numerical simulations for this growth law will not be reported because the dynamics obtained with the numerical method are basically not different from the numerical results discussed for the Gompertz growth law.

3. Numerical Experimentation

As it was previously stated, the main interest is to discover the dynamics of the particular case introduced in [19] to describe the evolution of the *Daphnia magna* population, by using different growth laws for the resource. Therefore, most of the parameter data and functions that were employed in that work are considered in this study. The growth rate is given as $g(x,t,z) = g_0\left(\frac{z}{1+z} - x\right)$, the mortality rate function is chosen constant, $\mu(x,t,z) = \mu$, and the fertility rate is defined $\alpha(x,t,z) = \alpha\,\frac{z}{1+z}\,x^2$. The function that controls the way in which consumer individuals use the resource, in the definition of the nonlocal term (3), is considered $\gamma(x,t,z) = \frac{z}{1+z}\, x^2$. The functional that depends on the consumer is given by $f_u(t,I) = -I$. It has to be noted that the growth and fertility rates, and the nonlocal term weight function γ depend on the resource through the Micaelis–Menten law. Lastly, the parameters employed are $g_0 = 1$, $\mu = 0.1$, and $\alpha = 0.75$, as in [19].

The numerical method used in the simulations was analysed in [1] and, for the sake of completeness, is fully described in Appendix A. The second order of convergence was proven under enough regularity hypotheses on the function data. It should be pointed out that the following numerical results are performed beyond the limits of the numerical convergence theorem because the integration could be made on unbounded problems. On the one hand, the search of the asymptotic behaviour of the problem compels to employ an undetermined large time interval. On the other hand, some combination of the parameter values (as the one employed in [19]) makes the population density unbounded close to the maximum size $x_M(t)$, $t > 0$. This last case was explored in [3], and it arises from the relationship $g_0 > \mu = 0.1$, which makes the equilibrium solution unbounded when the size of individuals tends to the maximum size. It should be argued that this situation is fully admitted because most of the theoretical analysis made for the model asks for solutions on $\mathcal{L}^1([0,x_M(t)])$, $t > 0$ (see [13], for instance). Although the study performed in the following avoids this kind of unboundness, the numerical approximation given in [3] was based on modifying the removing procedure (see Equation (A3), Appendix A) to adapt it to the behaviour of the solution. Therefore, the value $g_0 = 0.075$ is employed, that ensures both the boundedness of the solution without affecting the behaviour of the model and the non-negativity of the resource at equilibrium.

In addition to the fixed parameters previously introduced in the vital functions, different experiments with diverse values of the parameters $r \in \mathbb{R}$, the specific growth rate of the resource, and $K \in \mathbb{R}^+$, the carrying capacity, are performed. In each trial, the following initial conditions are employed: $s^0 = \frac{K}{2}$ ($s^0 = 7$ in the case of Malthusian growth, where there is no declared carrying capacity), and

$$u^0(x) = \begin{cases} 100\,(\widetilde{x}_M - x)^{\beta}, & \text{if } 0 \le x \le \widetilde{x}_M, \\ 0, & \text{if } \widetilde{x}_M < x \le x_M^0, \end{cases} \tag{5}$$

where $\widetilde{x}_M = 0.875$ and $\beta = 0.5153$ are chosen to fulfil the compatibility between initial and boundary conditions.

If the growth rate of the resource is autonomous, the computation of the theoretical equilibrium of (1)–(4) is possible, due to the choice of the function data in this test problem (note that all the growth functions described in Section 2 are independent of time). Thus, when $f_S^*(s)$ is defined as the autonomous representation of $f_S(t,s)$, whenever a nontrivial steady state of (2), S^*, exists, the nontrivial equilibrium state for the coupled problem (the maximum size, consumer, and resource) is given by the following formulae [1]:

$$\begin{cases} x_M^* = \sqrt[3]{\dfrac{\mu\,(\mu+g_0)\,(\mu+2\,g_0)}{2\,\alpha\,g_0^2}}, \\ S^* = \dfrac{x_M^*}{1-x_M^*}, \\ u(x) = \begin{cases} \dfrac{\alpha}{g_0}\,\dfrac{1+S^*}{S^*}\, f_S^*(S^*)\left(1-\dfrac{x}{x_M^*}\right)^{\left(\frac{\mu}{g_0}-1\right)}, & \text{if } 0 \le x \le x_M^*, \\ 0, & \text{if } x_M^* < x \le x_M. \end{cases} \end{cases} \tag{6}$$

This nontrivial equilibrium state could be stable, unstable or even not exist. The initial values considered in (5) are different enough from this state to observe the convergence of the solution towards such a nontrivial equilibrium state when it happens.

The numerical integration is carried out in an undetermined finite time interval that varies in different circumstances (for instance, in the case of the Malthusian growth, the integration time is left free to arrive to extinction or an unbounded solution). In previous works [1], the convergence of the numerical approximation to the theoretical nontrivial equilibrium (when it exists) was stated. The good behaviour of the numerical scheme allows reaching the possible behaviours: unbounded solutions, trivial equilibrium (extinction of the consumer population), nontrivial equilibrium (coexistence of both populations), and periodical solutions (stable limit cycles). It is observed that it is enough to consider the discretisation parameters values as $h = k = 1.5625 \cdot 10^{-2}$ to describe the full asymptotic dynamics of the problem.

All in all, different scenarios of the growth of the resource in the absence of the population will be considered, and the influence of the consumer population on the dynamics will be discussed. Therefore, although the original work of de Roos [19] introduced a feeding resource that evolved with a logistic function, its behaviour is challenged with the three other new possibilities presented in the previous section.

3.1. Malthusian Growth

The first choice consists of considering an unlimited growth. The evolution of the resource is given by the Malthusian growth, with a constant intrinsic rate of growth, $f_S(t,s) = r\,s, r \in \mathbb{R}$. Therefore, the resource grows exponentially when $r > 0$ and declines to extinction when r is negative. It would only remain constant when births balance deaths ($r = 0$).

This kind of growth is introduced into the model. The main consequence expected would be that the consumer expedites the decline of the resources, due to the consumption effect produced by $f_u(t, I) = -I$. Therefore, it could be thought that $r = 0$ will remain the threshold value of the intrinsic rate of growth that separates the exponential growth from the extinction of the resource.

In order to validate this assumption, t_{up} and t_{down} are computed. t_{up} is the integration time at which the approximated value of the resource is higher than 10^{20}, a sufficiently large value to represent an exponential growth, i.e., an unbounded solution. t_{down} is the integration time at which the numerical approximation is lower than 10^{-20}, a small enough value to represent the extinction of the population. It is obvious that both values are mutually exclusive. In addition, the theoretical time values that correspond to these integration times but when there is no consumption of the resource are computed. The following formula is employed $t^*_{up} = \frac{1}{r} \log\left(\frac{10^{20}}{s^0}\right)$ or $t^*_{down} = \frac{1}{r} \log\left(\frac{10^{-20}}{s^0}\right)$. In Figure 1, the results obtained with the numerical experimentation are plotted. The intrinsic rate of growth r is in the set $[-1, 1]$ and r is drawn versus t_{up} and t^*_{up} or t_{down} and t^*_{down}, in a semilogarithmic scale. It is observed that the consumer influence is interpreted as a modification of the response speed. It decreases the time the resource takes to arrive to extinction and increases the time that the numerical approximation needs to reach the fixed upper bound. Furthermore, the value of the intrinsic growth at which the population remains constant disappears, and when $r = 0$, the resource tends to extinction slowly. This behaviour shows the small influence of the consumer population on the resource evolution with this kind of growth function. Finally, the evolution of the consumer population is closely related to the behaviour of the resource and it grows exponentially in the case of unlimited resources and decays to extinction with scarce resources.

This is a signal of the limitations of exponential growth that could describe accurately the growth for small values of time. After a certain period, other factors become involved and the growth is retarded and limited.

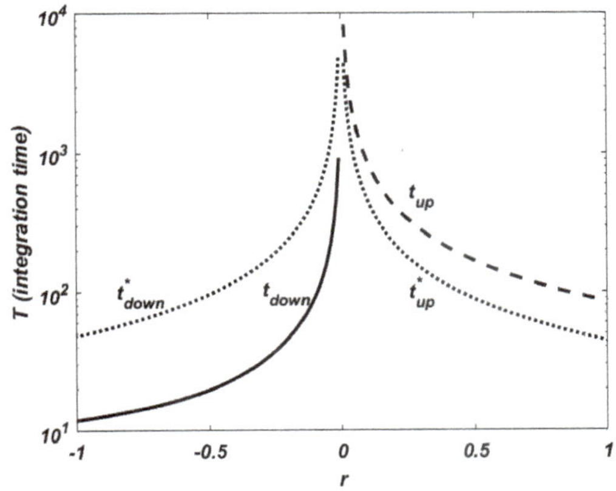

Figure 1. Malthusian growth. Behaviour of the numerical solution depending on r. t_{down} (solid line): time at which the approximated resource shows an exponential decay, its value is lower than 10^{-20} (extinction). t_{up} (dashed line): time at which it shows an exponential growth, its value is higher than 10^{20} (unbounded solution). Theoretical values in the absence of the consumer (dotted line): $t^*_{up} = \frac{1}{r}\log\left(\frac{10^{20}}{s^0}\right)$, $r > 0$, and $t^*_{down} = \frac{1}{r}\log\left(\frac{10^{-20}}{s^0}\right)$, $r < 0$.

3.2. Chemostat Growth

A growth law that circumvents some of the disadvantages of the Mathusian growth is introduced. It is clear that the effect of the limitation of the growth of the resource imposed by the carrying capacity of the environment affects its evolution. The chemostat growth law is a simple choice to modelise this situation. It is also called exponential confined growth law because the curve defined by the differential equation maintains the sign of the curvature along time as it approaches the carrying capacity; therefore, it is always confined to a region bounded by K. In terms of growth, it means that the resource rate of the growth is proportional to the available space. Thus, the dynamics of the resource in the absence of a consumer population are reduced to a stable nontrivial equilibrium that matches the carrying capacity.

An extensive numerical experimentation with different values of both parameters $r, K \in \mathbb{R}^+$ has been performed. It is observed that the size-structured consumer population defined in (1) modifies the nontrivial equilibrium to which the resource population tends to, but it does not modify its behaviour. Regardless of the value of parameters K and r employed, the resource grows very fast to a value close and below the carrying capacity of the environment (note that s^0 is lower than K, otherwise it would decrease) followed by a smooth transition to the stable nontrivial equilibrium.

Figures 2–5 confirm that statement. Three kinds of plots are shown. First, a 3D plot in which the density function $u(x,t)$ is drawn versus the size of individuals and time. Second, the evolution of the maximum size, the resource, and the total population of consumers, $x_M(t)$, $s(t)$, and $P(t) = \int_{x_0}^{x_M(t)} u(x,t)\,dx$, respectively, are represented versus time. This last quantity is computed numerically with the quadrature rule employed in the numerical scheme. In the last plot, the transition that these three quantities follow to the equilibrium is shown. It is a 3D plot in which the total population versus the resource and the maximum size, for every value of the time integration interval is drawn. Furthermore, the equilibrium value is shown.

This growth function confers stability to the equilibrium. However, the equilibrium reached by the resource population is modified by the action of the consumer population. It is observed that there is a threshold value of the carrying capacity K^* (whose value is $K^* = S^*$ in (6)), below which the consumer population tends to extinction, probably due to the lack of enough resources to survive (Allee effect type), and the resource tends to the carrying capacity. Therefore, the existence of a trivial equilibrium of the system corresponding to the extinction of the consumer population and a value equal to the carrying capacity for the resource population, can be assessed. This behaviour is shown in Figure 2, with $r = 4.5$ and $K = 3.4$.

However, if $K \geq K^*$, both populations tend to nontrivial steady states, which seem to be stable, and are given by the expressions in (6). It should be assumed that the trivial equilibrium for the consumer population becomes unstable and a new stable nontrivial equilibrium appears; then we can affirm the existence of a transcritical bifurcation point for the consumer population at $K = K^*$. The emergence of this nontrivial consumer equilibria modifies the behaviour of the resource (and the maximum size) because, in these conditions, it never reaches the carrying capacity and tends to the nontrivial equilibrium S^* (6). In this case, the nontrivial equilibrium reached by the system is modified and changes with the value of K. This coexistence pattern between both populations is shown in Figure 3 (experiment made with $r = 4.5$ and $K = 10$). Figure 4 shows that the same dynamics occur when the initial resource is larger than the carrying capacity, with the only novelty of a continuous decline to the asymptotic state. In this experiment the following values of parameters have been employed $r = 4.5$, $K = 10$, and $s^0 = 20$.

Finally, no other change is noted when the value of the carrying capacity K is augmented. It is only noticeable that the consumer population increases its size without any other related changes in the other variables $x_M(t)$ and $s(t)$. This is shown in a final experiment with $r = 4.5$ and $K = 20.2$, as shown in Figure 5.

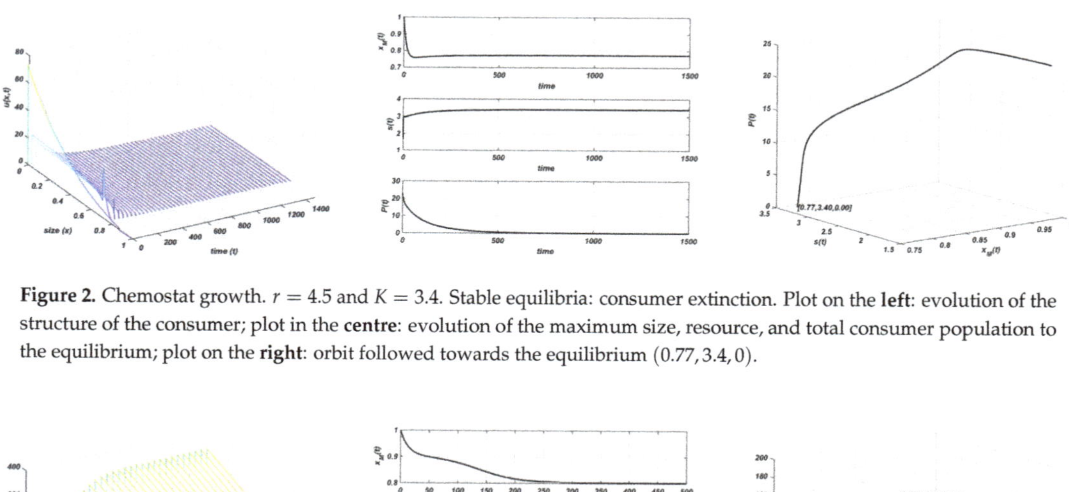

Figure 2. Chemostat growth. $r = 4.5$ and $K = 3.4$. Stable equilibria: consumer extinction. Plot on the **left**: evolution of the structure of the consumer; plot in the **centre**: evolution of the maximum size, resource, and total consumer population to the equilibrium; plot on the **right**: orbit followed towards the equilibrium $(0.77, 3.4, 0)$.

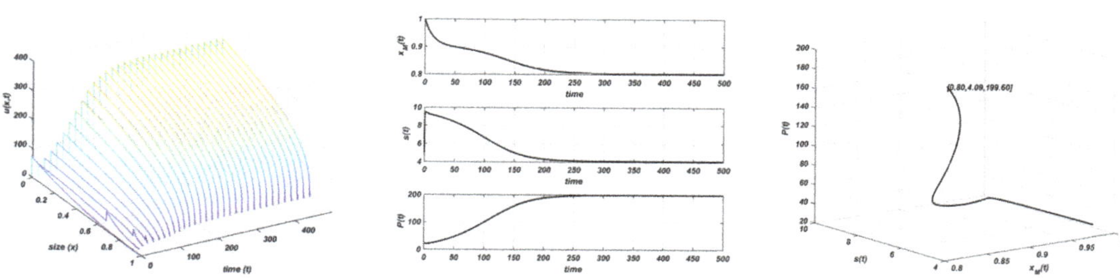

Figure 3. Chemostat growth. $r = 4.5$ and $K = 10$. Stable equilibria. Plot on the **left**: evolution of the structure of the consumer; plot in the **centre**: evolution of the maximum size, resource, and total consumer population to the equilibrium; plot on the **right**: orbit followed towards the equilibrium $(0.80, 4.09, 199.60)$.

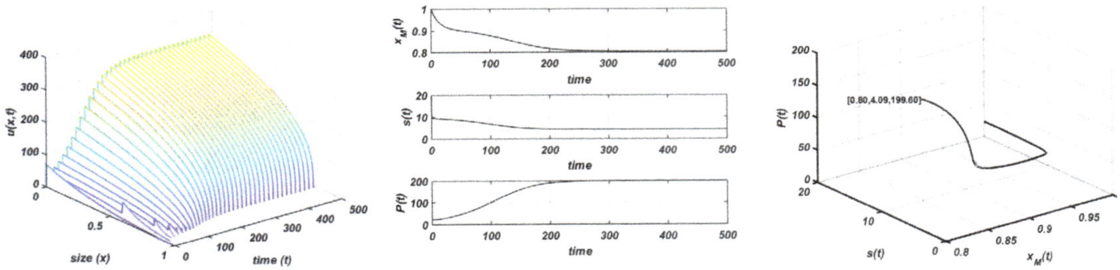

Figure 4. Chemostat growth. $r = 4.5$, $K = 10$, and $s^0 = 20$. Stable equilibria. Plot on the **left**: evolution of the structure of the consumer; plot in the **centre**: evolution of the maximum size, resource, and total consumer population to the equilibrium; plot on the **right**: orbit followed towards the equilibrium $(0.80, 4.09, 199.60)$.

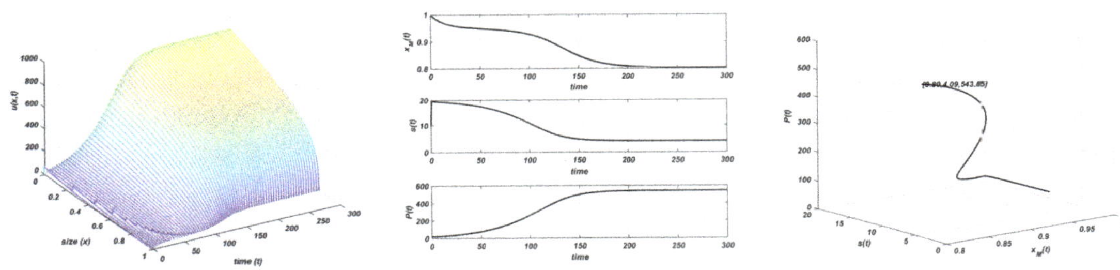

Figure 5. Chemostat growth. $r = 4.5$ and $K = 20.2$. Stable equilibria. Plot on the **left**: evolution of the structure of the consumer; plot in the **centre**: evolution of the maximum size, resource, and total consumer population to the equilibrium; plot on the **right**: orbit followed towards the equilibrium $(0.80, 4.09, 543.85)$.

3.3. Gompertz Growth

The last growth law in this study also presents an effect of limitation on the growth of the resource. Unlike the chemostat growth law, there are two possible theoretical equilibria in the system: an unstable trivial equilibrium state and a stable nontrivial equilibrium state that corresponds to the carrying capacity of the environment. Therefore, in this case, the evolution of the resource in the absence of the consumer population is towards the carrying capacity.

An exhaustive numerical experimentation with different values of both parameters $r, K \in \mathbb{R}^+$ has also been performed and the results are represented with similar figures as in the chemostat growth section. First, the evolution of the density function $u(x,t)$ is plotted versus the size of individuals and the time. The three plots in the middle of the figure represent the evolution of the maximum size, the resource, and the total population of consumers versus time, respectively. The right-hand size 3D plot in the figure shows the orbit that these three previously introduced quantities follow for every value in the time integration interval. When the system arrives to a stable equilibrium, its value is also shown.

Initially, a situation in which the consumer population does not affect the unstructured resource is observed. This happens when the carrying capacity is lower than a threshold value K^* (whose value is again $K^* = S^*$ in (6)). This case is similar to the corresponding one in the chemostat growth law. The resource approaches the carrying capacity while the consumer population tends to extinction, probably the consumer population does not find enough resources to survive. In Figure 6, the use of $r = 3$ and $K = 3.4$ allows showing how the whole system tends to a so-called stable trivial equilibrium state, $s = K$ and $u(x) = 0$, $x \in [0, x_M^*]$. Note that this situation happens regardless of the value of r.

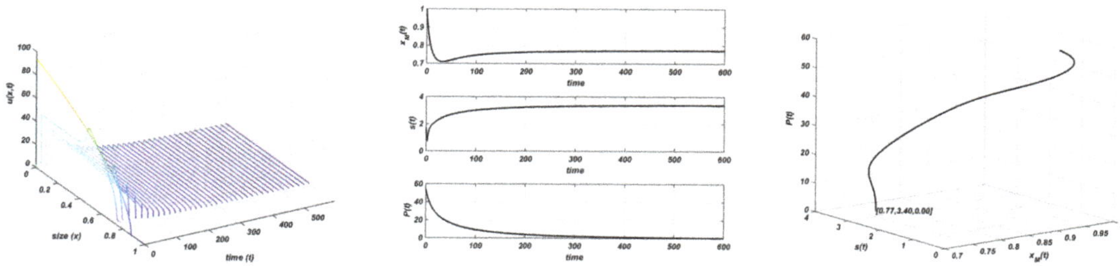

Figure 6. Gompertz growth. $r = 3$ and $K = 3.4$. Stable equilibria. Plot on the **left**: evolution of the structure of the consumer; plot in the **centre**: evolution of the maximum size, resource, and total consumer population to the equilibrium; plot on the **right**: orbit followed towards the equilibrium $(0.77, 3.40, 0)$.

When the carrying capacity increases, the resource does as well. As a consequence, the consumer population is able to survive, and when the consumer population appears in the system, the dynamics of the resource population change. Thus, when $K > K^*$, the resource population converges to the nonlinear steady-state given by S^* in (6), and the carrying capacity will never be a steady-state again. With respect to the consumer population, it tends to the nontrivial equilibrium $u^*(x)$, $x \in [0, x_M^*]$ given by (6). Therefore, the existence of a transcritical bifurcation at K^* can be stated for the consumer population: the trivial equilibrium becomes unstable and a nontrivial stable equilibrium emerges. In Figure 7, the parameter values employed are $r = 3$ and $K = 10$, and it is shown how both populations and the maximum size approach the stable steady-state given by (6). Thus, the consumer affects the resource, but the stability of the asymptotic state is maintained and, as a consequence, both populations coexist.

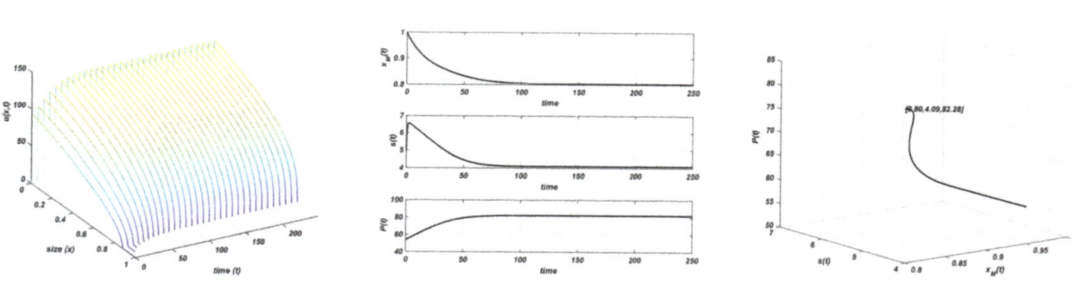

Figure 7. Gompertz growth. $r = 3$ and $K = 10$. Stable equilibria. Plot on the **left**: evolution of the structure of the consumer; plot in the **centre**: evolution of the maximum size, resource, and total consumer population to the equilibrium; plot on the **right**: orbit followed towards the equilibrium $(0.80, 4.09, 82.28)$.

Thus far, the behaviour is quite similar to what we have described before with the chemostat growth law. However, the stable nontrivial equilibrium lingers until the carrying capacity reaches a second threshold $K^{**} > K^*$, and then, it becomes unstable and a stable limit cycle emerges. The value of the threshold depends on r and decreases as r increases. Its existence was previously reported when the logistic growth law was employed [1]. Thus, the consumer population affects the behaviour of the resource and modifies the stability of its nontrivial equilibrium. We could describe K^{**} as a Hopf bifurcation.

The following figures are slight modifications of the others shown above: now the evolution of the involved quantities ($u(x,t)$, $x_M(t)$, $s(t)$, and $P(t)$) with respect to the full time-integration interval $[0, T]$ is not presented. Instead, the period, T_K, and the amplitude of the resource oscillations, A_K, of the corresponding limit cycle are computed numerically for each experiment. The time interval, in which the involved quantities are displayed, corresponds to $[0, 5\,T_K]$, i.e., five periods once the limit cycle reaches its stability. The right-hand picture is completely different. It represents a 3D plot of the stable limit cycle at which the evolutions of $x_M(t)$, $s(t)$, and $P(t)$ converge.

The following experiments are carried out with $r = 3$ and different values of the carrying capacity, $K > K^{**}$. In Figure 8, the announced behaviour for $K = 14.73$ is observed: the nontrivial equilibrium is unstable and a stable limit cycle emerges with $T_{14.73} = 15.26$ and $A_{14.73} = 1.14$. The nontrivial steady-state loses its stability, and both populations and the maximum size oscillate around the nontrivial equilibrium. It should be pointed out the free behaviour of the maximum size as was done in [1] with the logistic growth law.

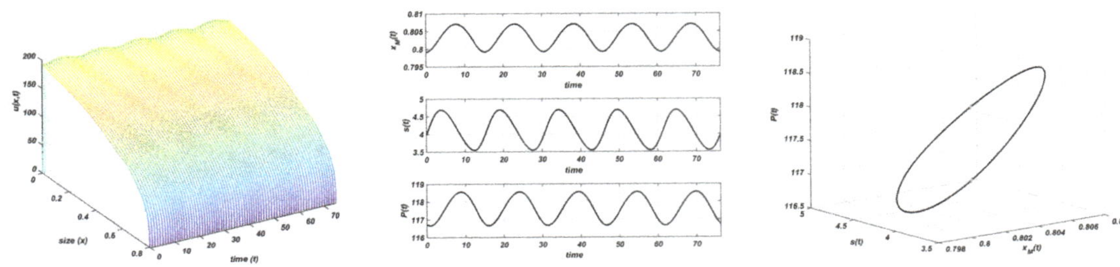

Figure 8. Gompertz growth. $r = 3$ and $K = 14.73$. Stable limit cycle. Plot on the **left**: evolution of the structure of the consumer; plot in the **centre**: evolution of the maximum size, resource, and total consumer population to the equilibrium; plot on the **right**: stable limit cycle.

Beyond this new threshold K^{**}, a complex dynamic, which has never been described in this kind of model so far, arises. It is shown in Figure 9, where simulations are made with $K = 15.21$, $K = 15.33$, $K = 15.45$, and $K = 15.93$. The instability of the nontrivial equilibrium and the convergence to a stable limit cycle is observed. However, the period of the oscillations increases from $T_{14.73} = 15.26$ to $T_{15.33} = 82.44$ and diminishes again to $T_{15.93} = 32.06$ (see Figure 9). This means that the period is doubled twice and, after that, it is also split twice. This is a common doubling period cascade. The amplitude of the oscillations (in the resource population) increases from $A_{14.73} = 1.14$ to $A_{15.93} = 10.04$.

Finally, there are no different situations beyond this value. In Figure 10, we show the numerical results we obtain with $K = 20.2$. The only statement is that the amplitude increases with K, $A_{20.2} = 14.45$. A final influence of parameter r reveals the increments of the consumer population (figure not shown).

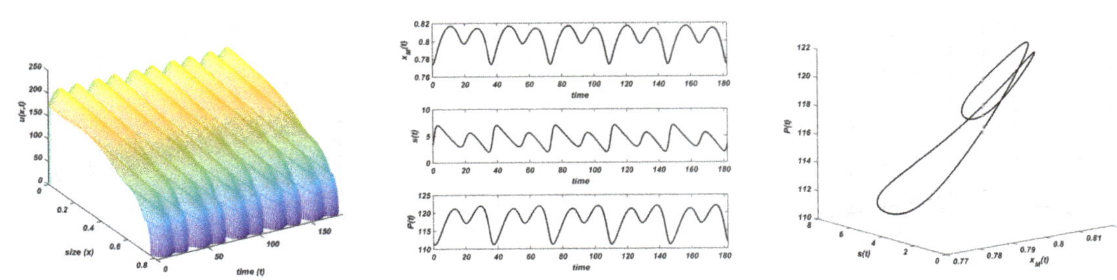

Figure 9. *Cont.*

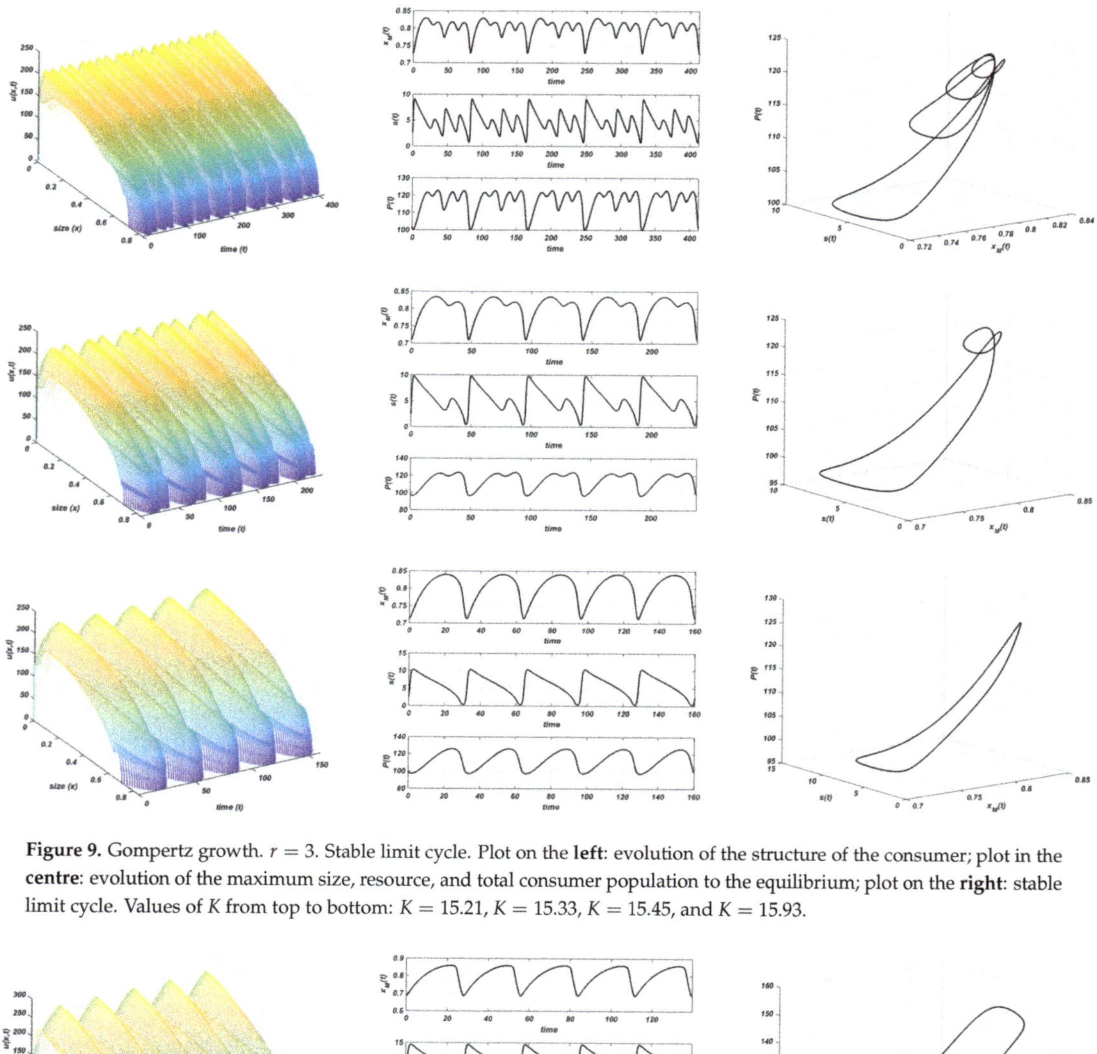

Figure 9. Gompertz growth. $r = 3$. Stable limit cycle. Plot on the **left**: evolution of the structure of the consumer; plot in the **centre**: evolution of the maximum size, resource, and total consumer population to the equilibrium; plot on the **right**: stable limit cycle. Values of K from top to bottom: $K = 15.21$, $K = 15.33$, $K = 15.45$, and $K = 15.93$.

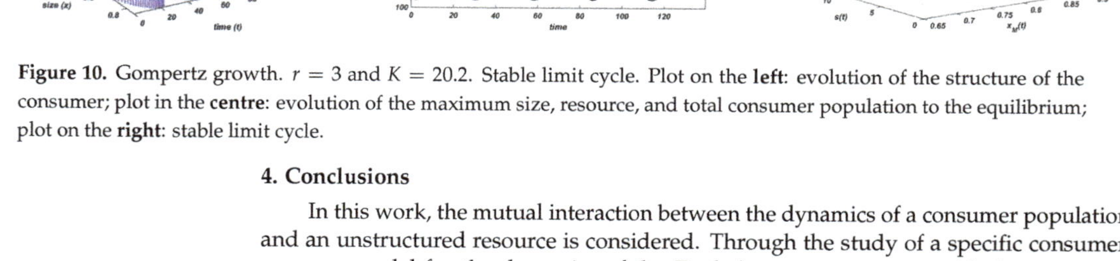

Figure 10. Gompertz growth. $r = 3$ and $K = 20.2$. Stable limit cycle. Plot on the **left**: evolution of the structure of the consumer; plot in the **centre**: evolution of the maximum size, resource, and total consumer population to the equilibrium; plot on the **right**: stable limit cycle.

4. Conclusions

In this work, the mutual interaction between the dynamics of a consumer population and an unstructured resource is considered. Through the study of a specific consumer-resource model for the dynamics of the *Daphnia* magna, we cope with the numerical difficulties due to an undefined sign growth law for the individual consumers. The numerical method employed is well adapted to this kind of size-structured population model: it efficiently provides the density function for the individuals in the consumer population without the numerical dissipation of first-order finite difference methods (the upwind, for instance) and the spurious oscillations from second-order finite difference methods.

Although existence, uniqueness, and regularity results about this model with a non-negative growth rate [12–15] have not been extended to the case of an undefined sign growth law yet, these properties are assumed under usual hypotheses that include enough smoothness on the function data. We focus on the numerical study of the long-time behaviour of both populations. In this case, a numerical method, that deals efficiently with the complex dynamics that arise, has been employed. With meaningful biological vital rates for the function data for the *Daphnia magna* model [19], the complex dynamics, which the model presents, are shown for each of three different growth laws.

The model behaves as a typical Malthusian population when this kind of growth law is chosen. The only influence of the consumer population is to slow down or to increase the speed at which the resource is unbounded or tends to extinction.

For the chemostat growth law, the consumer population does not influence the stable behaviour of the resource. On the one hand, the whole system stabilises the resource population towards the carrying capacity, and the consumer population tends to extinction, and on the other hand, the system stabilises towards the nontrivial steady-state given by (6). It should be pointed out that the consumer population increases with the increment of both parameters r and K.

With respect to the Gompertz growth law, the model exhibits many of the same properties as in the analysis of models that include the logistic growth law. In particular, several features have been presented, such as: stable and unstable equilibria, and stable periodical behaviours for different values of the parameters r and K of the resource growth law. However, in the extensive numerical experimentation with different values of r and K, a period cascade it is also obtained, short and long oscillations and short and long periods that have not been presented in this model before. These results could indicate the use of the Gompertz growth when other distributions do not match the real behaviour.

As the numerical simulations for each of the growth laws considered show, we can argue that the shortage of the resource has, as a biological consequence, not only the reduction in the total number of individuals of the stationary equilibria but also the effective shrink in size of the individuals of the population. This second effect has to be exclusively attributed to the undefined sign of the growth law of the consumer population. A future work in this direction is to confirm this effect as a general pattern for other consumer-resource population models in which an undefined sign growth rate is presented. Furthermore, we should expect some confirmation with real data from field experiments for these kinds of dynamics.

Author Contributions: All authors contributed equally to this article. All authors have read and agreed to the published version of the manuscript.

Funding: This research was funded in part by project MTM2017-85476-C2-1-P of the Spanish Ministerio de Economía y Competitividad, and European FEDER Funds, research grant RED2018-102650-T of the Spanish Ministerio de Economía y Competitividad, and research grant VA193P20 of the Junta de Castilla y León, Spain, and European FEDER funds (EU).

Institutional Review Board Statement: Not applicable.

Informed Consent Statement: Not applicable.

Data Availability Statement: Not applicable.

Acknowledgments: Authors are very grateful to the referees for their careful reading of the original manuscript and their aid to improve it.

Conflicts of Interest: The authors declare no conflict of interest. The funders had no role in the design of the study; in the collection, analyses, or interpretation of data; in the writing of the manuscript, or in the decision to publish the results.

Appendix A. Numerical Method

The description of the numerical scheme used in the simulations is introduced. This was built "ad hoc" for the problem in the current conditions in [1]. It also could be used in a long time integration searching for equlibria or periodicity.

The problem (1)–(4), ready to perform the numerical integration, is composed of four fully coupled problems:

$$\begin{cases} x'(t;t^*,x^*) = g(x(t;t^*,x^*),t,s(t)), & t \geq t^*, \\ x(t^*;t^*,x^*) = x^*, \end{cases}$$

$$\begin{cases} s'(t) = f(t,s(t),I(t)), & t > t^*, \\ s(t^*) = s^*, \end{cases}$$

$$u(x(t;t^*,x^*),t) = u(x^*,t^*)\exp\Big\{-\int_{t^*}^{t}\mu^*(x(\tau;t^*,x^*),\tau,s(\tau))d\tau\Big\}, \; t \geq t^*,$$

where $\mu^*(x,t,s(t)) = \mu(x,t,s(t)) + g_x(x,t,s(t))$,

$$g(x_0,t,s(t))\,u(x_0,t) = \int_{x_0}^{x_M(t)} \alpha(x,t,s(t))\,u(x,t)\,dx.$$

Their discretisation will provide numerical approximations for the physiological states, the evolution of the resource, and the solution at the grid points at each time level.

In the following, the equations of the numerical scheme are introduced. It is a second order method, based on the trapezoidal quadrature rule (both integral and differential versions). This is an implicit rule but an explicit predictor–corrector transformation is used to compute a first aproximation and, in the second stage, to obtain the right value (with higher order of convergence). The integration is made on a finite time interval $[0,T]$. Therefore, given $J \in \mathbb{N}$ and $\lambda \in \mathbb{R}^+$, the discretisation parameters are defined as $h = x_M^0/J$, $k = \lambda\, h$. Thus, the number of time levels are given by $N = [T/k]$, and the time discretisation levels are $t^n = n\,k$, $0 \leq n \leq N$.

Now, to perform the numerical integration, initial values are provided: the initial space discretisation, $\mathbf{X}^0$, $X_j^0 = j\,h$, $0 \leq j \leq J$, the initialisation of the resource s^0 and the initial condition on the initial grid, $\mathbf{U}^0$, $U_j^0 = u^0(X_j^0)$, $0 \leq j \leq J$. Then, the computation of the general step $\{\mathbf{X}^{n+1}, S^{n+1}, \mathbf{U}^{n+1}\}$, is built after the corresponding values at the previous time step $\{\mathbf{X}^n, S^n, \mathbf{U}^n\}$, $0 \leq n \leq N-1$, are known, by means of the following procedure. The equations of the first stage are given by,

$$\begin{cases} X_0^{n+1,*} = x_0, \quad X_{j+1}^{n+1,*} = X_j^n + k\,g(X_j^n,t^n,S^n), & 0 \leq j \leq J+n, \\ S^{n+1,*} = S^n + k\,f(t^n,S^n,\mathcal{Q}(\mathbf{X}^n,\gamma^n\cdot\mathbf{U}^n)), \\ U_{j+1}^{n+1,*} = U_j^n\,\exp\big\{-k\,\mu^*(X_j^n,t^n,S^n)\big\}, & 0 \leq j \leq J+n, \\ g(X_0^{n+1,*},t^{n+1},S^{n+1,*})\,U_0^{n+1,*} = \mathcal{Q}(\mathbf{X}^{n+1,*},\boldsymbol{\alpha}^{n+1,*}\cdot\mathbf{U}^{n+1,*}). \end{cases} \tag{A1}$$

The equations in the second stage are

$$\begin{cases} X_0^{n+1} = x_0, \quad X_{j+1}^{n+1} = X_j^n + \frac{k}{2}\left(g(X_j^n,t^n,S^n) + g(X_{j+1}^{n+1,*},t^{n+1},S^{n+1,*})\right), & 0 \leq j \leq J+n, \\ S^{n+1} = S^n + \frac{k}{2}\Big(f(t^n,S^n,\mathcal{Q}(\mathbf{X}^n,\gamma^n\cdot\mathbf{U}^n)) + f(t^{n+1},S^{n+1,*},\mathcal{Q}(\mathbf{X}^{n+1,*},\gamma^{n+1,*}.\mathbf{U}^{n+1,*}))\Big), \\ U_{j+1}^{n+1} = U_j^n\,\exp\Big\{-\frac{k}{2}\left(\mu^*(X_j^n,t^n,S^n) + \mu^*(X_{j+1}^{n+1,*},t^{n+1},S^{n+1,*})\right)\Big\}, & 0 \leq j \leq J+n, \\ g(X_0^{n+1},t^{n+1},S^{n+1})\,U_0^{n+1} = \mathcal{Q}(\mathbf{X}^{n+1},\boldsymbol{\alpha}^{n+1}\cdot\mathbf{U}^{n+1}). \end{cases} \tag{A2}$$

The discretisation of the nonlocal terms is made by means of a composite quadrature rule (in this case based on the trapezoidal one), $\mathcal{Q}(\mathbf{y}^n, \mathbf{V}^n) = \sum_{l=1}^{J+n} \frac{y_l^n - y_{l-1}^n}{2}(V_{l-1}^n + V_l^n)$.

The grid of the scheme is built following the characteristic curves and, therefore, a new node appears after each time level. For efficiency reasons, the number of grid nodes is kept constant by removing one grid node after every time step as in the following procedure: first, X_l^{n+1} such that

$$|X_{l+1}^{n+1} - X_{l-1}^{n+1}| = \min_{1 \le j \le J} |X_{j+1}^{n+1} - X_{j-1}^{n+1}|, \tag{A3}$$

is selected. Next, the selected node X_l^{n+1} is removed and the nodes are renamed. Thus, the quadrature is performed on this subgrid with a fixed number of nodes. Other versions of this procedure could be possible if the integration interval increases or decreases due to the variable maximum size. Finally, $\gamma^{n,*}$, $\boldsymbol{\alpha}^{n,*}$, γ^n, and $\boldsymbol{\alpha}^n$ are the vectors whose components are $\boldsymbol{\alpha}_j^{n,*} = \alpha(X_j^{n,*}, t^{n+1}, S^{n,*})$, $\gamma_j^{n,*} = \gamma(X_j^{n,*}, t^{n+1}, S^{n,*})$, $\boldsymbol{\alpha}_j^n = \alpha(X_j^n, t^n, S^n)$, and $\gamma_j^n = \gamma(X_j^n, t^n, S^n)$, respectively, $0 \le j \le J$; and the vector products in (A1) and (A2) are considered component-wise.

The analysis of this numerical method was performed in [1], where the corresponding smoothness properties demanded by this second-order numerical scheme were specified.

References

1. Angulo, O.; López-Marcos, J.C.; López-Marcos, M.A. Analysis of an efficient integrator for a size-structured population model with a dynamical resource. *Comput. Math. Appl.* **2014**, *68*, 941–961. [CrossRef]
2. Abia, L.M.; Angulo, O.; López-Marcos, J.C.; López-Marcos, M.A. Long-Time Simulation of a Size-Structured Population Model with a Dynamical Resource. *Math. Model Nat. Phenom.* **2010**, *5*, 1–21. [CrossRef]
3. Angulo, O.; López-Marcos, J.C.; López-Marcos, M.A. Numerical approximation of singular asymptotic states for a size-structured population model with a dynamical resource. *Math. Comput. Model.* **2011**, *54*, 1693–1698. [CrossRef]
4. de Roos, A.M.; Persson, L. Size-dependent life-history traits promote catastrophic collapses of top predators. *Proc. Nat. Acad. Sci. USA* **2002**, *99*, 12907–12912. [CrossRef] [PubMed]
5. Sánchez-Sanz, J.; Getto, P. Numerical bifurcation analysis of physiologically structured populations: Consumer–resource, cannibalistic and trophic models. *Bull Math. Biol.* **2016**, *78*, 1546–1584. [CrossRef] [PubMed]
6. Cuesta, J.A.; Delius, G.W.; Law, R. Sheldon spectrum and the plankton paradox: Two sides of the same coin—A trait-based plankton size-spectrum model. *J. Math. Biol.* **2017**, *76*, 67–96. [CrossRef] [PubMed]
7. Pang, J.; Chen, J.; Liu, Z.; Bi, P.; Ruan, S. Local and global stabilities of a viral dynamics model with infection-age and immune response. *J. Dyn. Diff. Equat.* **2019**, *31*, 793–813. [CrossRef]
8. Lafferty, K.; De Leo, G.; Briggs, C.; Dobson, A.; Gross, T.; Kuris, A. A general consumer-resource population model. *Science* **2015**, *349*, 854–857. [CrossRef] [PubMed]
9. Aylaj, B.; Noussair, A. Global weak solution for a multistage physiologically structured population model with resource interaction. *Nonlinear Anal. Real. World Appl.* **2010**, *11*, 1670–1684. [CrossRef]
10. Angulo, O.; Crauste, F.; López-Marcos, J.C. Numerical integration of an erythropoiesis model with explicit growth factor dynamics. *J. Comput. Appl. Math.* **2018**, *330*, 770–782. [CrossRef]
11. Kooijman, S.A.L.M.; Metz, J.A.J. On the dynamics of chemically stressed populations: The deduction of population consequences from effects on individuals. *Ecotoxicol. Environ. Saf.* **1984**, *8*, 254–274. [CrossRef]
12. Thieme, H.R. Well-posedness of physiologically structured population models for daphnia magna. *J. Math. Biol.* **1988**, *26*, 299–317. [CrossRef]
13. Diekmann, O.; Getto, P.; Gyllenberg, M. Stability and bifurcation analysis of volterra functional equations in the light of suns and stars. *SIAM J. Math. Anal.* **2007**, *39*, 1023–1069. [CrossRef]
14. Diekmann, O.; Gyllenberg, M. Equations with infinite delay: Blending the abstract and the concrete. *J. Differential. Equat.* **2012**, *252*, 819–851. [CrossRef]
15. Diekmann, O.; Gyllenberg, M.; Metz, J.A.J.; Nakaoka, S.; de Roos, A.M. Daphnia revisited: Local stability and bifurcation theory for physiologically structured population models explained by way of an example. *J. Math. Biol.* **2010**, *61*, 277–318. [CrossRef] [PubMed]
16. de Roos, A.M.; Metz, J.A.J.; Evers, E.; Leipoldt, A. A size dependent predator-prey interaction: Who pursues whom? *J. Math. Biol.* **1990**, *28*, 609–643. [CrossRef]
17. Breda, D.; Diekmann, O.; Gyllenberg, M.; Scarabel, F.; Vermiglio, R. Pseudospectral discretization of nonlinear delay equations: New prospects for numerical bifurcation analysis. *SIAM J. Appl. Dyn. Syst.* **2016**, *15*, 1–23. [CrossRef]
18. Breda, D.; Getto, P.; Sánchez-Sanz, J.; Vermiglio, R. Computing the eigenvalues of realistic Daphnia models by pseudospectral methods. *SIAM J. Sci. Comput.* **2015**, *37*, A2607–A2629. [CrossRef]

19. de Roos, A.M. Numerical methods for structured population models: The escalator boxcar train. *Numer. Methods Partial Differ. Equat.* **1988**, *4*, 173–195. [CrossRef]
20. Banks, R.B. *Growth and Diffusion Phenomena;* Springer: Berlin/Heidelberg, Germany, 1994.

mathematics MDPI

Article

Integrable Deformations and Dynamical Properties of Systems with Constant Population

Cristian Lăzureanu

Department of Mathematics, Politehnica University of Timişoara, Piața Victoriei 2, 300006 Timișoara, Romania; cristian.lazureanu@upt.ro

Abstract: In this paper we consider systems of three autonomous first-order differential equations $\dot{\mathbf{x}} = \mathbf{f}(\mathbf{x})$, $\mathbf{x} = (x, y, z)$, $\mathbf{f} = (f_1, f_2, f_3)$ such that $x(t) + y(t) + z(t)$ is constant for all t. We present some Hamilton–Poisson formulations and integrable deformations. We also analyze the case of Kolmogorov systems. We study from some standard and nonstandard Poisson geometry points of view the three-dimensional Lotka–Volterra system with constant population.

Keywords: Hamilton–Poisson systems; integrable deformations; Lotka–Volterra systems; Kolmogorov systems; stability; periodic orbits; heteroclinic orbits

MSC: 70K20; 70K42; 70K44

Citation: Lăzureanu, C. Integrable Deformations and Dynamical Properties of Systems with Constant Population. *Mathematics* **2021**, *9*, 1378. https://doi.org/10.3390/math9121378

Academic Editors: Mihaela Neamțu, Eva Kaslik and Anca Rădulescu

Received: 24 May 2021
Accepted: 11 June 2021
Published: 14 June 2021

Publisher's Note: MDPI stays neutral with regard to jurisdictional claims in published maps and institutional affiliations.

Copyright: © 2021 by the author. Licensee MDPI, Basel, Switzerland. This article is an open access article distributed under the terms and conditions of the Creative Commons Attribution (CC BY) license (https://creativecommons.org/licenses/by/4.0/).

1. Introduction

The dynamics of the interactions between the individuals of different categories of a population may be modeled by autonomous first-order differential equations. For a period of time, there are situations when the total size of the population can be assumed constant, that is, the model is a dynamical system with constant population. For instance, we recall here the SIR model [1], a model of the opiate-using career [2], and the NERA model of drug consumption in a given population [3]. Additionally, there are cases of high-dimensional Lotka–Volterra systems with constant population [4,5].

A dynamical system with constant population always has a constant of motion. In the three-dimensional case, if the system has a second constant of motion, then it admits a Hamilton–Poisson formulation. In addition, controlled versions of such a system that are also Hamilton–Poisson may be obtained by integrable deformations [6–8].

The paper is organized as follows. In Section 2, we particularize some known results about Hamilton–Poisson formulations and integrable deformations in the case of a three-dimensional dynamical system with constant population. In Section 3, we consider three-dimensional Kolmogorov systems. We deduce the general form of a Lotka–Volterra system with constant population and we present a Hamilton–Poisson formulation of it. We give integrable deformations that are also with a constant population. Then, we obtain the general form of a polynomial Kolmogorov system of degree 3 with constant population and a particular version of it, which is Hamilton–Poisson. In Section 4, we study the stability of the equilibrium states of the Lotka–Volterra system with a constant population restricted to $[0,\infty)^3$, the existence of periodic orbits, and we point out the heteroclinic orbits. We also study the properties of the energy-Casimir mapping associated with this system and their connections with the dynamics of the system.

2. Hamilton–Poisson Formulations and Integrable Deformations

In this section, we consider a system of three autonomous first-order differential equations:

$$\dot{x} = f_1(x,y,z)\ ,\ \dot{y} = f_2(x,y,z)\ ,\ \dot{z} = f_3(x,y,z), \tag{1}$$

where $\dot{u} = \frac{du}{dt}$, such that the function $H(x,y,z) = x + y + z$ is a constant of motion. In this case, we say that (1) is a three-dimensional dynamical system with constant population. System (1) has only one constant of motion, namely H, or there may exist another functionally independent constant of motion, denoted by C.

In the following, we particularize some known results about Hamilton–Poisson formulations and integrable deformations in the case of system (1). For the sake of completeness, we present some proofs. For details about Poisson geometry see, for example, [9–11].

Proposition 1. *Let $C \in C^1(P)$, where $P \subseteq \mathbb{R}^3$ is a C^∞ manifold, such that H and C are functionally independent on P, where $H(x,y,z) = x + y + z$. Then:*

(i) *System*

$$\dot{x} = C_z - C_y \,,\ \dot{y} = C_x - C_z \,,\ \dot{z} = C_y - C_x, \tag{2}$$

where $f_u := \frac{\partial f}{\partial u}$, is a dynamical system with constant population.

(ii) *C is also a constant of motion of system (2).*

(iii) *(P, Π_C, H) is a Hamilton–Poisson formulation of system (2), where the Poisson structure Π_C is given by*

$$\Pi_C = \begin{bmatrix} 0 & C_z & -C_y \\ -C_z & 0 & C_x \\ C_y & -C_x & 0 \end{bmatrix}, \tag{3}$$

and H is the Hamiltonian. Furthermore, C is a Casimir of the Poisson structure Π_C.

Following [12], we have:

Proposition 2. *Let $C, \nu \in C^1(P)$, where $P \subseteq \mathbb{R}^3$ is a C^∞ manifold. If H and C are functionally independent on P, where $H(x,y,z) = x + y + z$, then system*

$$\dot{x} = \nu(C_z - C_y) \,,\ \dot{y} = \nu(C_x - C_z) \,,\ \dot{z} = \nu(C_y - C_x), \tag{4}$$

is a dynamical system with constant population. Moreover, C is also a constant of motion and $(P, \nu\Pi_C, H)$ is a Hamilton–Poisson formulation of system (4), where Π_C is given by (3) and H is the Hamiltonian. Furthermore, C is a Casimir of the Poisson structure $\Pi_C^\nu := \nu\Pi_C$.

The next result follows by [13,14].

Proposition 3. *Let $C, \beta \in C^1(P)$, where $P \subseteq \mathbb{R}^3$ is a C^∞ manifold, $g \in \mathbb{R}$ and $H(x,y,z) = x + y + z$.*

(i) *If H and C are constants of motion of system (1) and they are functionally independent on P, then there is a differentiable function ν on an open and dense subset of P such that system (1) takes the form (4).*

(ii) *If H and $C + g\beta$ are functionally independent on P, then a family of integrable deformations of system (1) is given by:*

$$\dot{x} = f_1 + g\nu(\beta_z - \beta_y) \,,\ \dot{y} = f_2 + g\nu(\beta_x - \beta_z) \,,\ \dot{z} = f_3 + g\nu(\beta_y - \beta_x), \tag{5}$$

where g is a deformation parameter. Moreover, H and $C + g\beta$ are constants of motion of system (5).

(iii) *If H is the only constant of motion of system (1), then a family of deformations of system (1) for which H is also a constant of motion is given by (5).*

Proof. (i) By hypothesis, H and C are functionally independent on P, and thus at least one of the following inequalities

$$C_x \neq C_y \,,\; C_x \neq C_z \,,\; C_y \neq C_z$$

holds.

Let us consider $C_x \neq C_y$. Since H and C are constants of motion of system (1), it follows that

$$f_1 + f_2 + f_3 = 0 \,,\; C_x f_1 + C_y f_2 + C_z f_3 = 0.$$

Therefore $f_1 = \dfrac{C_z - C_y}{C_y - C_x} f_3$, $f_2 = \dfrac{C_x - C_z}{C_y - C_x} f_3$. Considering $\nu = \dfrac{f_3}{C_y - C_x}$, the conclusion follows.

(ii) From (i) we have $\nu \nabla H \times \nabla C = (f_1, f_2, f_3) = (\dot{x}, \dot{y}, \dot{z})$. Then $\nu \nabla H \times \nabla(C + g\beta) = \nu \nabla H \times \nabla C + g\nu \nabla H \times \nabla \beta = (f_1, f_2, f_3) + g\nu(\beta_z - \beta_y, \beta_x - \beta_z, \beta_y - \beta_x)$. Therefore system (5) is obtained. It is easy to see that H and $\tilde{C} = C + g\beta$ are constants of motion of system (5). Then, by Proposition 2, where C is replaced by $\tilde{C}$, (5) is a Hamilton–Poisson system. Moreover, if the deformation parameter g vanishes then system (5) becomes (1), as required.

(iii) Using (5), it immediately follows that $\dot{H} = 0$. □

We end this section with some considerations about the dynamics of a three-dimensional dynamical system with constant population.

Let (x_0, y_0, z_0) be an initial condition of the system with constant population (1) and $h = x_0 + y_0 + z_0$. Since H is a constant of motion, that is $H(x(t), y(t), z(t)) = H(x_0, y_0, z_0)$ for all t, it follows that $x(t) + y(t) + z(t) = h$ for all t. Therefore, it is clear that the dynamics of such a system takes place in the level surface $S = \{(x, y, z) \in \mathbb{R}^3 : x + y + z = h\}$. Moreover, if C is the second constant of motion, then the trajectories belong to the intersection of the level surfaces S and $\{(x, y, z) \in \mathbb{R}^3 : C(x, y, z) = C(x_0, y_0, z_0)\}$.

As A. J. Lotka himself highlighted, "Periodic phenomena play an important role in nature, both organic and inorganic" [15]. In our case, if H and C are functionally independent constants of motion and the intersection of the corresponding level surfaces is a closed curve, then the trajectory may be a periodic orbit. On the other hand, if H is the only one constant of motion of system (1) the Bendixon condition $\operatorname{div} \mathrm{f} \neq 0$ in a simply connected region of S precludes periodic solutions in that region (see, e.g., [16] and references therein).

3. A Particular Case: Polynomial Kolmogorov Systems

Let P_1, P_2, P_3 be three continuously differentiable functions on $\mathbb{R}^3$. Consider the differential equations

$$\dot{x} = xP_1(x, y, z) \,,\; \dot{y} = yP_2(x, y, z) \,,\; \dot{z} = zP_3(x, y, z), \tag{6}$$

called the three-dimensional ecological equations or Kolmogorov equations [17,18]. If P_1, P_2, P_3 are polynomial such that the maximum of their degrees is $k - 1$, then (6) is a three-dimensional polynomial Kolmogorov system of degree k. Particularly, if $k = 2$, then (6) is a three-dimensional Lotka–Volterra system. In this section, we deal with Kolmogorov systems of degree $k \in \{2, 3\}$ with constant population. More precisely, we deduce the general forms of such systems that admit Hamilton–Poisson formulations. We recall here that the Liouville, Darboux integrability, respectively, of Kolmogorov and Lotka–Volterra systems and their dynamical behavior have been widely investigated (see, e.g., [19–25] and references therein).

Proposition 4. *If $H(x, y, z) = x + y + z$ is a constant of motion of the three-dimensional Lotka–Volterra Equation (6), then*

$$P_1(x, y, z) = cy - bz \,,\; P_2(x, y, z) = -cx + az \,,\; P_3(x, y, z) = bx - ay, \tag{7}$$

where $a, b, c \in \mathbb{R}$.

Proof. Let $P_k(x,y,z) = a_k x + b_k y + c_k z + d_k, a_k, b_k, c_k, d_k \in \mathbb{R}, k \in \{1,2,3\}$. The condition $\dot{H} = 0$ implies $xP_1(x,y,z) + yP_2(x,y,z) + zP_3(x,y,z) = 0, \forall x,y,z \in \mathbb{R}$, or equivalently,

$$a_1x^2 + b_2y^2 + c_3z^2 + (b_1 + a_2)xy + (c_1 + a_3)xz + (c_2 + b_3)yz + d_1x + d_2y + d_4z = 0, \forall x,y,z \in \mathbb{R}.$$

Then $a_1 = b_2 = c_3 = d_1 = d_2 = d_3 = 0$, and denoting $b_1 = c, a_3 = b, c_2 = a$, the conclusion follows. □

Remark 1. *We obtained that a three-dimensional Lotka–Volterra system with constant population has the form*

$$\dot{x} = x(cy - bz)\ ,\ \dot{y} = y(-cx + az)\ ,\ \dot{z} = z(bx - ay),\ a,b,c \in \mathbb{R}. \tag{8}$$

Moreover, the function $C(x,y,z) = x^a y^b z^c$ *is a constant of motion of this system (see also [26]). In Section 4, we will study this system from some standard and nonstandard Poisson geometry points of view (for details about such approach see, e.g., [27]).*

The next result is a consequence of Propositions 2 and 3.

Proposition 5. *Let* $a, b, c > 0$.

(i) *The three-dimensional Lotka–Volterra system with constant population (8) has the Hamilton–Poisson formulation* (P, Π, H), *where* $P = (0,\infty)^3$, $H(x,y,z) = x + y + z$, *and the Poisson structure*

$$\Pi = \begin{bmatrix} 0 & cxy & -bxz \\ -cxy & 0 & ayz \\ bxz & -ayz & 0 \end{bmatrix}. \tag{9}$$

In addition, system (8) reads $(\dot{x}, \dot{y}, \dot{z}) = \nu \nabla H \times \nabla C$, *where* $C(x,y,z) = x^a y^b z^c$ *and the rescaling function* ν *is given by* $\nu(x,y,z) = x^{1-a}y^{1-b}z^{1-c}$.

(ii) *If* $\beta \in C^1(P)$ *such that* H *and* $C + g\beta$ *are functionally independent on* P, *then a family of integrable deformations of Lotka–Volterra system (8) is given by*

$$\begin{cases} \dot{x} = x(cy - bz) + gx^{1-a}y^{1-b}z^{1-c}(\beta_z - \beta_y) \\ \dot{y} = y(-cx + az) + gx^{1-a}y^{1-b}z^{1-c}(\beta_x - \beta_z) \\ \dot{z} = z(bx - ay) + gx^{1-a}y^{1-b}z^{1-c}(\beta_y - \beta_x) \end{cases}, \tag{10}$$

where $g \in \mathbb{R}$ *is a deformation parameter. Moreover, (10) is also a system with constant population and* $C + g\beta$ *is a constant of motion.*

Taking into account the form of system (10), it is natural to ask whether there are functions β such that (10) is a polynomial Kolmogorov system. The answer is affirmative. Indeed, let us consider $\beta(x,y,z) = x^a y^b z^c Q(x,y,z)$, where Q is an arbitrary polynomial of degree n. Then we obtain that system (10) becomes

$$\begin{cases} \dot{x} = x(cy - bz + g(cy - bz)Q + gyz(Q_z - Q_y)) \\ \dot{y} = y(-cx + az + g(-cx + az)Q + gxz(Q_x - Q_z)) \\ \dot{z} = z(bx - ay + g(bx - ay)Q + gxy(Q_y - Q_x)) \end{cases}, \tag{11}$$

which is a polynomial Kolmogorov system of degree $n + 2$. Moreover, by Proposition 5, it is a system with constant population for which $\tilde{C}(x,y,z) = x^a y^b z^c(1 + gQ(x,y,z))$ is a constant of motion.

In the sequel, we deduce the general form of a three-dimensional polynomial Kolmogorov system of degree 3 with a constant population. In addition, we present a family of such systems that are Hamilton–Poisson.

Proceeding as in the proof of Proposition 4, we obtain the following.

Proposition 6. *The family of three-dimensional polynomial Kolmogorov systems of degree 3 with constant population is given by*

$$\begin{cases} \dot{x} = x(cy - bz + a_1y^2 + a_2z^2 + a_3xy + a_4xz + a_5yz) \\ \dot{y} = y(-cx + az - a_3x^2 + a_6z^2 - a_1xy + a_7xz + a_8yz) \\ \dot{z} = z(bx - ay - a_4x^2 - a_8y^2 - (a_5 + a_7)xy - a_2xz - a_6yz) \end{cases} , a, b, c, a_k \in \mathbb{R}. \quad (12)$$

We note that system (12) is a deformation of the Lotka–Volterra system (8). Therefore, if this deformation is an integrable deformation, then system (12) has a second constant of motion and, consequently, it is a Hamilton–Poisson system.

Proposition 7. *A family of three-dimensional Hamilton–Poisson polynomial Kolmogorov systems of degree 3 with constant population is given by*

$$\begin{cases} \dot{x} = x(cy - bz + cqy^2 - brz^2 + cpxy - bpxz + ((c+1)r - (b+1)q)yz) \\ \dot{y} = y(-cx + az - cpx^2 + arz^2 - cqxy + ((a+1)p - (c+1)r)xz + aqyz) \\ \dot{z} = z(bx - ay + bpx^2 - aqy^2 + ((b+1)q - (a+1)p)xy + brxz - aryz) \end{cases} , \quad (13)$$

where , $a, b, c, p, q, r \in \mathbb{R}$. *Moreover,* $\tilde{C}(x, y, z) = x^a y^b z^c (1 + px + qy + rz)$ *is also a constant of motion of the above system.*

Proof. Using Proposition 6, it is obvious that (13) is a three-dimensional Kolmogorov system with constant population. In order to prove that (13) is a Hamilton–Poisson system, we show that it is a particular case of system (10). Let β be as in Proposition 5 such that system (10) becomes (12). We obtain

$$\begin{aligned} \beta_z - \beta_y &= x^a y^{b-1} z^{c-1}(a_1y^2 + a_2z^2 + a_3xy + a_4xz + a_5yz), \\ \beta_x - \beta_z &= x^{a-1} y^b z^{c-1}(-a_3x^2 + a_6z^2 - a_1xy + a_7xz + a_8yz), \\ \beta_y - \beta_x &= x^{a-1} y^{b-1} z^c(-a_4x^2 - a_8y^2 - (a_5 + a_7)xy - a_2xz - a_6yz), \end{aligned}$$

and

$$\begin{aligned} \beta_{xz} - \beta_{xy} &= x^{a-1} y^{b-1} z^{c-1}(aa_1y^2 + aa_2z^2 + aa_3xy + aa_4xz + aa_5yz + a_3xy + a_4xz), \\ \beta_{xy} - \beta_{yz} &= x^{a-1} y^{b-1} z^{c-1}(-ba_3x^2 + ba_6z^2 - ba_1xy + ba_7xz + ba_8yz - a_1xy + a_8yz), \\ \beta_{yz} - \beta_{xz} &= x^{a-1} y^{b-1} z^{c-1}(-ca_4x^2 - ca_8y^2 - c(a_5 + a_7)xy - ca_2xz - ca_6yz - a_2xz - a_6yz). \end{aligned}$$

Adding the above relations and equating the coefficients of the obtained polynomial to zero, we obtain

$$ca_4 = -ba_3, aa_5 = (c+1)a_6 - (b+1)a_8, ba_6 = -aa_2, ba_7 = (c+1)a_2 - (a+1)a_4, ca_8 = aa_1. \quad (14)$$

We denote $a_1 = cq, a_2 = -br, a_3 = cp$, where $p, q, r \in \mathbb{R}$. Using (14), we obtain that system (12) becomes (13). We also consider the function β given by $\beta(x, y, z) = x^a y^b z^c Q(x, y, z)$, where $Q(x, y, z) = p_1x + q_1y + r_1z$ such that $p = gp_1, q = gq_1, r = gr_1$. Now, on one the hand, function β fulfills the above conditions regarding partial derivatives. On the other hand, we have already seen that for such a function β, system (10) becomes (11), which also takes the particular form (13).

By Proposition 5 it follows that $\tilde{C}(x, y, z) = x^a y^b z^c (1 + px + qy + rz)$ is a constant of motion and, consequently, (13) is a Hamilton–Poisson system, as required. □

4. Dynamical Properties of the Three-Dimensional Lotka–Volterra System with Constant Population

In this section we study the dynamical properties of the three-dimensional Lotka–Volterra system with constant population (8) and their connections with the corresponding energy-Casimir mapping. Lotka–Volterra systems [28] are widely investigated from dynamical point of view (see, e.g., [29–31] and references therein).

As we have seen in Section 3, the three-dimensional Lotka–Volterra system with constant population is given by

$$\begin{cases} \dot{x} = x(cy - bz) \\ \dot{y} = y(-cx + az) \\ \dot{z} = z(bx - ay) \end{cases} . \tag{15}$$

In addition, the constants of motion of this system is

$$H(x,y,z) = x + y + z \, , \; C(x,y,z) = x^a y^b z^c. \tag{16}$$

In the following we consider $a, b, c > 0$ and we restrict our study on the set $[0, \infty)^3$, which is the region of ecological interest.

System (15) has the Hamilton–Poisson formulation given by Proposition 5. Therefore, the corresponding energy-Casimir mapping is given by

$$\mathcal{EC} : [0,\infty)^3 \to \mathbb{R}^2, \; \mathcal{EC}(x,y,z) = (x + y + z, x^a y^b z^c). \tag{17}$$

There are some connections between the dynamics of a Hamilton–Poisson system and the properties of the corresponding energy-Casimir mapping (see, e.g., [27,32–38]). Most of the connections are provided by the partition of the image of the energy-Casimir mapping given by the equilibrium states. In the abovementioned papers, the energy-Casimir mapping is defined on $\mathbb{R}^3$; thus, we are interested to see which of the properties remain true in our case.

Lemma 1. *Let $a, b, c > 0$. Then*

$$x^a y^b z^c \leq \frac{a^a b^b c^c}{(a+b+c)^{a+b+c}} (x + y + z)^{a+b+c} \, , \; \forall x, y, z \geq 0.$$

Moreover, the equality holds if and only if $\frac{x}{a} = \frac{y}{b} = \frac{z}{c}$.

Proof. If $xyz = 0$, then the inequality holds. Let $x, y, z > 0$. We apply Jensen's inequality for a real concave function $\varphi : A \subset \mathbb{R} \to \mathbb{R}$, namely

$$\varphi\left(\sum_{i=1}^{n} p_i x_i\right) \geq \sum_{i=1}^{n} p_i \varphi(x_i), \text{ for all } x_i \in A, p_i > 0, i \in \{1, 2, \ldots, n\}, \text{ where } \sum_{i=1}^{n} p_i = 1.$$

The equality holds if and only if $x_1 = x_2 = \ldots = x_n$ or φ is linear.

Let $\varphi : (0, \infty) \to \mathbb{R}, \varphi(t) = \ln t$ and $x_1 = \frac{x}{a}, x_2 = \frac{y}{b}, x_3 = \frac{z}{c}, p_1 = \frac{a}{a+b+c}, p_2 = \frac{b}{a+b+c}, p_3 = \frac{c}{a+b+c}$. Then the conclusion follows. □

Remark 2. *For $H(x,y,z) = x + y + z$ and $C(x,y,z) = x^a y^b z^c$, the inequality from Lemma 1 becomes*

$$C(x,y,z) \leq \frac{a^a b^b c^c}{(a+b+c)^{a+b+c}} H^{a+b+c}(x,y,z) \, , \; \forall x, y, z \geq 0, \tag{18}$$

and the equality holds if and only if $(x,y,z) = (aM, bM, cM), M \geq 0$.

Now, we can present the image of the energy-Casimir mapping (17).

Proposition 8. *The image of the energy-Casimir mapping* (17) *is the set*

$$Im(\mathcal{EC}) = \{(h_1,h_2) \in \mathbb{R}^2 : \; h_1 \geq 0, \; 0 \leq h_2 \leq \frac{a^a b^b c^c}{(a+b+c)^{a+b+c}} h_1^{a+b+c} \}. \tag{19}$$

Proof. By definition,

$$\begin{aligned} \mathrm{Im}(\mathcal{EC}) &= \{(h_1,h_2) \in \mathbb{R}^2 : \exists (x,y,z) \in [0,\infty)^3 \text{ such that } \mathcal{EC}(x,y,z) = (h_1,h_2)\} \\ &= \{(h_1,h_2) \in \mathbb{R}^2 : \exists (x,y,z) \in [0,\infty)^3 \text{ such that } x+y+z = h_1, x^a y^b z^c = h_2\}. \end{aligned}$$

Using Lemma 1, we obtain the conclusion. □

Remark 3. *The image of the considered energy-Casimir mapping is drawn in Figure 1. In the abovementioned papers, if* $Im(\mathcal{EC})$ *is a proper subset of* $\mathbb{R}^2$*, then its boundary is given by the images of the stable equilibrium states through* $\mathcal{EC}$*. Moreover, a partition of* $Im(\mathcal{EC})$ *is given by the images of all equilibrium states through* $\mathcal{EC}$*. If the image of an unstable family of equilibrium states through* $\mathcal{EC}$ *is a half-line, then there are heteroclinic orbits that connect pairs of such states. Furthermore, if* Σ^p *denotes an open subset of the partition of* $Im(\mathcal{EC})$ *that has dimension 2, then each point of* Σ^p *is the image of at least one periodic orbit through* $\mathcal{EC}$*. In our case, because* $\mathcal{EC}$ *is restricted to* $[0,\infty)^3$*, the image of the energy-Casimir mapping is a proper subset of* $\mathbb{R}^2$*, but without this restriction, it is* $\mathbb{R}^2$*. If the image of the energy-Casimir mapping is* $\mathbb{R}^2$*, the dynamics is more complicated, but we can expect to find periodic orbits and heteroclinic orbits.*

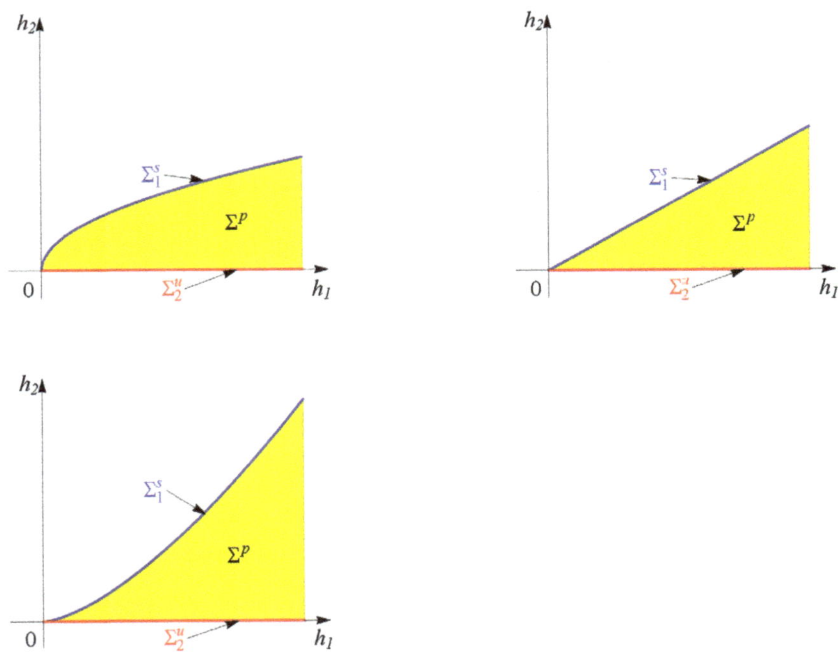

Figure 1. Image of the energy-Casimir mapping ($a+b+c<1, a+b+c=1$, and $a+b+c>1$ respectively).

In the following we study the abovementioned connections.

The equilibrium states of system (15) are given by the families

$$\mathcal{E}_1 = \{(aM, bM, CM) \mid M \in [0,\infty)\}\,,\ \mathcal{E}_2 = \{(M,0,0) \mid M \in [0,\infty)\}\,,$$
$$\mathcal{E}_3 = \{(0,M,0) \mid M \in [0,\infty)\}\,,\ \mathcal{E}_4 = \{(0,0,M) \mid M \in [0,\infty)\}. \tag{20}$$

Proposition 9. *If $M \neq 0$, the equilibrium states of system (15) that belong to the family $\mathcal{E}_1$ (20) are Lyapunov stable.*

Proof. Let $M > 0$. We use the Arnold method [39]. Consider the function

$$F(x,y,z) = C(x,y,z) - \lambda H(x,y,z) = x^a y^b z^c - \lambda(x+y+z)$$

and let $W = \ker dH(aM, bM, cM) = \mathrm{Span}_{\mathbb{R}}\{(1,0,0),(0,1,0)\}$. For $\lambda = a^a b^b c^c M^{a+b+c-1}$, we obtain:

(i) $dF(aM, bM, cM) = 0$;
(ii) $d^2F(aM, bM, cM)|_{W\times W} = -a^{a-1}b^{b-1}c^c M^{a+b+c-2}(b\,dx^2 + a\,dy^2)$, which is negative definite for all $M > 0$.

Therefore the equilibrium state (aM, bM, cM), $M > 0$ is nonlinearly stable. □

Proposition 10. *For every $M \in (0,\infty)$, the equilibrium states of system (15) that belong to the families $\mathcal{E}_i, i \in \{2,3,4\}$ (20) are unstable.*

Proof. Let J be the Jacobian matrix of system (15). Then

$$J(x,y,z) = \begin{bmatrix} cy - bz & cx & -bx \\ -cy & -cx + az & ay \\ bz & -az & bx - ay \end{bmatrix}. \tag{21}$$

The eigenvalues of $J(M,0,0)$ are given by $\lambda_1 = 0, \lambda_2 = -cM, \lambda_3 = bM$. Since $\lambda_3 > 0$, the equilibrium state $(M,0,0)$ is unstable for all $M > 0$. Analogously, we study the stability of the other equilibrium states. □

It remains to study the stability of $O(0,0,0)$, the common point of all families of equilibrium states.

Consider that system (15) is without the restriction $x, y, z \geq 0$. We choose the initial conditions $x(0) = 0, y(0) = -z_0, z(0) = z_0$, where $z_0 > 0$ is near to zero. Then we obtain the particular solution of system (15) given by $x(t) = 0, y(t) = \dfrac{-z_0}{1 - az_0 t}, z(t) = \dfrac{z_0}{1 - az_0 t}$, $t \in \left[0, \dfrac{1}{az_0}\right)$. Therefore, the equilibrium state $O(0,0,0)$ is unstable.

On the other hand, our system is restricted to $[0,\infty)^3$. The coordinate axes and the subspaces spanned by them, and also the first octant, are all invariant sets for a Lotka–Volterra system with constant population (15). Hence, populations that start non-negative remain non-negative. If V is a neighborhood of $O(0,0,0)$, then we say that $V \cap (0,\infty)^3$ is a positive neighborhood of $O(0,0,0)$.

The next result shows the behavior of a trajectory that starts in a positive neighborhood of $O(0,0,0)$.

Proposition 11. *For every positive neighborhood U of $O(0,0,0)$ there is a positive neighborhood V of $O(0,0,0)$ such that the trajectories $(x(t), y(t), z(t))$ of system (15) initially in V never leave U.*

Proof. Consider the initial conditions of system (15) given by $(x_0, y_0, z_0) \in (0,\infty)^3$. Thus $(x(t), y(t), z(t)) \in (0,\infty)^3$ for $t > 0$. We have to prove that for all $\varepsilon > 0$ there is $\delta > 0$ such that if $\sqrt{x_0^2 + y_0^2 + z_0^2} < \delta$, then $\sqrt{x(t)^2 + y(t)^2 + z(t)^2} < \varepsilon$ for $t > 0$.

For every $\varepsilon > 0$ let $\delta > 0$ such that $\delta < \frac{\varepsilon\sqrt{3}}{3}$. Then the following relations hold:

$$\sqrt{x(t)^2 + y(t)^2 + z(t)^2} \leq x(t) + y(t) + z(t) = x_0 + y_0 + z_0 \leq 3\sqrt{\frac{x_0^2 + y_0^2 + z_0^2}{3}} \leq \sqrt{3}\delta < \varepsilon,$$

for every $t > 0$, as required. □

Remark 4. *In the case when system* (15) *is restricted to* $[0, \infty)^3$, *Proposition* 11 *tells us that* $O(0, 0, 0)$ *behaves as a Lyapunov stable equilibrium state.*

Remark 5. *For every* $M \in [0, \infty)$, *the sets of images of the equilibrium states through* $\mathcal{EC}$ *are given by*

$$\begin{aligned}
\Sigma_1^s &= \{(aM + bM + cM, a^a b^b c^c M^{a+b+c}) : M > 0\} \\
&= \{(h_1, h_2) \in \mathbb{R}^2 : \ h_1 > 0, \ h_2 = \frac{a^a b^b c^c}{(a+b+c)^{a+b+c}} h_1^{a+b+c} \}, \\
\mathcal{EC}(0,0,0) &= (0,0), \\
\Sigma_2^u &= \{(M, 0) : M > 0\} = \{(h_1, h_2) \in \mathbb{R}^2 : \ h_1 > 0, \ h_2 = 0 \},
\end{aligned}$$

where "s" and "u" stand for stable and unstable, respectively. We note that in our case the boundary of $Im(\mathcal{EC})$ *is given by* $\Sigma_1^s \cup \Sigma_2^u \cup \{(0,0)\}$. *We also denote by* Σ^p *the interior of* $Im(\mathcal{EC})$ *(see Figure* 1*). Therefore,*

$$Im(\mathcal{EC}) = \Sigma_1^s \cup \{(0,0)\} \cup \Sigma_2^u \cup \Sigma^p. \tag{22}$$

In the next result we prove that there are periodic orbits of the considered system around the stable equilibrium states. In [40], the existence of periodic orbits of a similar system has been proven by another approach.

Proposition 12. *Let* $E_M^1 = (aM, bM, cM) \in \mathcal{E}_1$ *be such that* $M \in (0, \infty)$. *Then, for each sufficiently small* $\varepsilon \in \mathbb{R}_+^*$, *any integral surface*

$$\Sigma_\varepsilon^{E_M^1} : \ a^a b^b c^c M^{a+b+c-1}(x + y + z) - x^a y^b z^c - a^a b^b c^c M^{a+b+c}(a + b + c - 1) = \varepsilon^2$$

contains at least one periodic orbit $\gamma_\varepsilon^{E_M^1}$ *of system* (15) *whose period is close to* $\frac{2\pi}{M\sqrt{abc(a+b+c)}}$.

Proof. We apply a version of Moser's theorem in the case of a zero eigenvalue [41]. By (21), $J(aM, bM, cM)$ has the eigenvalues $\lambda_1 = 0$, $\lambda_{2,3} = \pm iM\sqrt{abc(a+b+c)}$ and the eigenspace corresponding to the eigenvalue zero has dimension 1. Consider the constants of motion $C_1(x, y, z) = x + y + z$ and $I(x, y, z) = a^a b^b c^c M^{a+b+c-1}(x + y + z) - x^a y^b z^c$. We obtain $W = \ker dC_1(aM, bM, cM) = \mathrm{Span}_{\mathbb{R}}\{(1, 0, 0), (0, 1, 0)\}$, $dI(aM, bM, cM) = 0$, and $d^2 I(aM, bM, cM)|_{W \times W} = a^{a-1} b^{b-1} c^c M^{a+b+c-2}(b\,dx^2 + a\,dy^2)$, which is positive definite for all $M > 0$.

By Theorem 2.1 from [41], for each sufficiently small $\varepsilon \in \mathbb{R}_+^*$, any integral surface $I(x, y, z) - I(aM, bM, cM) = \varepsilon^2$ contains at least one periodic solution of system (15) whose period is close to the period of the corresponding linear system around E_M^1, which finishes the proof. □

Another special orbit of a Hamilton–Poisson system is the heteroclinic orbit. In our case we have the following.

Proposition 13. *For each $M \in (0,\infty)$ there is a cycle of heteroclinic orbits $\mathcal{HE}_1 \cup \mathcal{HE}_2 \cup \mathcal{HE}_3$ that connect the unstable equilibrium states $(M,0,0), (0,0,M), (0,M,0)$ given by*

$$\mathcal{HE}_1(t) = \left(\frac{Mx_0}{(M-x_0)e^{bMt}+x_0}, 0, \frac{M(M-x_0)e^{bMt}}{(M-x_0)e^{bMt}+x_0}\right),$$
$$\mathcal{HE}_2(t) = \left(0, \frac{M(M-z_0)e^{zMt}}{(M-z_0)e^{aMt}+z_0}, \frac{Mz_0}{(M-z_0)e^{aMt}+z_0}\right),$$
$$\mathcal{HE}_3(t) = \left(\frac{M(M-y_0)e^{cMt}}{(M-y_0)e^{cMt}+y_0}, \frac{My_0}{(M-y_0)e^{cMt}+y_0}, 0\right),$$

where $t \in \mathbb{R}$ and $x_0, y_0, z_0 \in (0,M)$.

Proof. To obtain a parametric form of the heteroclinic orbit $\mathcal{HE}_1$, we first reduce our system from three degrees of freedom to one degree of freedom by using the level sets $H(x,y,z) = H(M,0,0)$ and $C(x,y,z) = C(M,0,0)$ and then integrate the resulting reduced differential equation. We have

$$x+y+z = M \text{ and } x^a y^b z^c = 0, a,b,c > 0.$$

Consider $y = 0$. Thus, $z = M - x$ and system (15) reduces to $\dot{x} = bx(x-M)$. Let $x_0 \in (0,M)$ be the initial condition of the above equation. We obtain the solution $x(t) = \dfrac{Mx_0}{(M-x_0)e^{bMt}+x_0}$ and then $\mathcal{HE}_1(t)$. We similarly obtain the other parametric forms of the heteroclinic orbits. □

Remark 6. *By geometrical point of view, the heteroclinic orbits given in Proposition 13 are open line segments with endpoints given by the unstable equilibrium states. Another approach in the study of the existence of heteroclinic orbits of a similar system has been used in [40].*

The dynamics of system (15) takes place at the intersection of the level sets $H(x,y,z) = h_1, C(x,y,z) = h_2$, where $(h_1,h_2) \in \mathrm{Im}(\mathcal{EC})$. The fiber of the energy-Casimir mapping $\mathcal{EC}$ corresponding to (h_1,h_2) is the set

$$\mathcal{F}_{(h_1,h_2)} = \left\{(x,y,z) \in [0,\infty)^3 \mid \mathcal{EC}(x,y,z) = (h_1,h_2)\right\}. \tag{23}$$

Then, in other words, the dynamics of system (15) takes place on the fibers of the corresponding energy-Casimir mapping. In the following we give a topological classification of these fibers and, as a consequence, we point out the connections between the partition of $\mathrm{Im}(\mathcal{EC})$ and the dynamics of the considered system. This classification is provided by the partition of the image of the energy-Casimir mapping (22).

Proposition 14. *Let $(h_1,h_2) \in Im(\mathcal{EC})$ (19) and $\mathcal{F}_{(h_1,h_2)}$ (23) be the corresponding fiber of the energy-Casimir mapping.*

(i) *If $(h_1,h_2) \in \Sigma_1^s$, then $\mathcal{F}_{(h_1,h_2)} = \{(aM,bM,cM)\}$, where $M = \dfrac{h_1}{a+b+c}$, that is, a stable equilibrium state. In addition, $\mathcal{F}_{(0,0)} = \{(0,0,0)\}$.*

(ii) *If $(h_1,h_2) \in \Sigma_2^u$, then $\mathcal{F}_{(h_1,h_2)}$ is the triangle with vertices at $E_M^2(M,0,0), E_M^3(0,M,0), E_M^4(0,0,M)$, where $M = h_1$, that is, three unstable equilibrium states and the cycle of heteroclinic orbits that connect them (see Proposition 13).*

(iii) *If $(h_1,h_2) \in \Sigma^p$, then $\mathcal{F}_{(h_1,h_2)} = \left\{(x,y,z) \in [0,\infty)^3 \mid x+y+z = h_1, x^a y^b z^c = h_2\right\}$, that is, a periodic orbit.*

Proof. (*i*) Let $(x, y, z) \in \mathcal{F}_{(h_1,h_2)}$ (23), $(h_1, h_2) \in \Sigma_1^s$, where Σ_1^s is given in Remark 5. Using Lemma 1, we obtain $\frac{x}{a} = \frac{y}{b} = \frac{z}{c}$. Since $x + y + z = h_1$, we obatin the conclusion.

(*ii*) We have $x^a y^b z^c = 0$ and $x + y + z = M$. Using Proposition 13 and Remark 6, the conclusion follows.

(*iii*) The dynamics of system (15) takes place at the intersection of the level sets $H(x, y, z) = h_1$ and $C(x, y, z) = h_2$, which is the closed curve drawn in Figure 2. On the other hand, by Proposition 12 we know that there are periodic orbits around the stable equilibrium (aM, bM, cM), for every $M > 0$. Hence, in this case, we conclude that the fiber is a periodic orbit. □

We notice that the level set $S_h = \{(x, y, z) \in \mathbb{R}^3 : x + y + z = h, x, y, z \geq 0\}, h > 0$ is locally foliated by the periodic orbits that expand from the stable equilibrium $\left(\frac{ah}{a+b+c}, \frac{bh}{a+b+c}, \frac{ch}{a+b+c}\right)$. Moreover, these orbits tend to the heteroclinic cycle given by the boundary of S_h (see Figure 2, right).

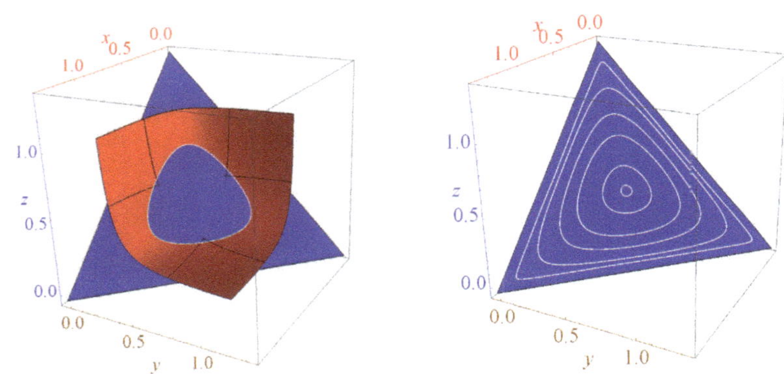

Figure 2. Left: The intersection of level sets: a periodic orbit. **Right**: A local foliation of the level set $S_h = \{(x, y, z) \in \mathbb{R}^3 : x + y + z = h, x, y, z \geq 0\}$ by periodic orbits around the stable equilibrium $\left(\frac{ah}{a+b+c}, \frac{bh}{a+b+c}, \frac{ch}{a+b+c}\right)$.

Remark 7. *Consider that system (15) describes the dynamics of the interactions between three categories of a constant population of total size h. After an interaction between individuals from distinct categories, they can change their category. At a moment t, we denote by $x(t), y(t), z(t)$ the number of individuals of these categories. At $t = 0$, the initial conditions are $x(0) = x_0, y(0) = y_0, z(0) = z_0$ such that $x_0, y_0, z_0 \in [0, h)$ and $x_0 + y_0 + z_0 = h$. The above analysis shows the following types of behaviors. First, if $x_0, y_0, z_0 \in (0, h)$, then we obtain a periodic behavior around the stable equilibrium $\left(\frac{ah}{a+b+c}, \frac{bh}{a+b+c}, \frac{ch}{a+b+c}\right)$. Second, if $z_0 = 0$, then $z(t) = 0$ for all $t > 0$ since the plane of equation $z = 0$ is an invariant set of our system. In this case, system (15) reduces to $\dot{x} = cxy, \dot{y} = -cxy$, and hence x increases in time and y decreases, that is $(x(t), y(t), 0)$ approaches $(h, 0, 0)$ in infinite time (the heteroclinic orbit $\mathcal{HE}_3$). Analogously, if $y_0 = 0$, then $(x(t), 0, z(t))$ approaches $(0, 0, h)$, and if $x_0 = 0$, then $(0, y(t), z(t))$ approaches $(0, h, 0)$. Finally, if (x_0, y_0, z_0) is an equilibrium state, then the number of individuals from each category stays unchanged.*

5. Conclusions

In this paper, Hamilton–Poisson formulations and integrable deformations of a three-dimensional dynamical system with constant population were presented. Particularly, general forms of some Kolmogorov systems with constant population were deduced. Using an integrable deformation of a Lotka–Volterra system with a constant population, a Hamilton–Poisson version of a Kolmogorov system of degree threes with a constant population were constructed.

The dynamical properties of the three-dimensional Lotka–Volterra system with a constant population were analyzed in connection with the associated energy-Casimir mapping. Such connections were reported in the case of other Hamilton–Poisson systems and, even if there is no a general result, the expected properties hold.

Funding: This research received no external funding.

Institutional Review Board Statement: Not applicable.

Informed Consent Statement: Not applicable.

Data Availability Statement: Not applicable.

Acknowledgments: We would like to thank the referees very much for their valuable comments and suggestions.

Conflicts of Interest: The author declares no conflict of interest.

References

1. Kermack, W.O.; McKendrick, A.G. A Contribution to the Mathematical Theory of Epidemics. *Proc. R. Soc. A Math.* **1927**, *115*, 700–721.
2. White, E.; Comiskey, C. Heroin epidemics, treatment and ODE modelling. *Math. Biosci.* **2007**, *208*, 312–324. [CrossRef] [PubMed]
3. Dauhoo, M.Z.; Korimboccus, B.S.N.; Issack, S.B. On the dynamics of illicit drug consumption in a given population. *IMA J. Appl. Math.* **2013**, *78*, 432–448. [CrossRef]
4. Bogoyavlenskij, O.I. Integrable Lotka—Volterra Systems. *Regul. Chaotic Dyn.* **2008**, *13*, 543–556. [CrossRef]
5. Kamp, P.H.V.; Kouloukas, T.E.; Quispel, G.R.W.; Tran, D.T.; Vanhaecke, P. Integrable and superintegrable systems associated with multi-sums of products. *Proc. R. Soc. A* **2014**, *470*, 20140481. [CrossRef] [PubMed]
6. Galajinsky, A. Remark on integrable deformations of the Euler top. *J. Math. Anal. Appl.* **2014**, *416*, 995–997. [CrossRef]
7. Lăzureanu, C. On the Hamilton–Poisson realizations of the integrable deformations of the Maxwell-Bloch equations. *C. R. Math.* **2017**, *355*, 596–600. [CrossRef]
8. Lăzureanu, C. Hamilton–Poisson Realizations of the Integrable Deformations of the Rikitake System. *Adv. Math. Phys.* **2017**, *2017*, 4596951. [CrossRef]
9. Libermann, P.; Marle, C.-M. *Symplectic Geometry and Analytical Mechanics*; D. Reidel: Dordrecht, The Netherlands, 1987.
10. Marsden, J.E.; Raţiu, T.S. *Introduction to Mechanics and Symmetry*, 2nd ed.; Texts in Applied Mathematics 17; Springer: New York, NY, USA, 1999.
11. Puta, M. *Hamiltonian Mechanical System and Geometric Quantization*; Kluwer Academic Publishers: Dordrecht, The Netherlands, 1993.
12. Tudoran, R.M. A normal form of completely integrable system. *J. Geom. Phys.* **2012**, *62*, 1167–1174. [CrossRef]
13. Gürses, M.; Guseinov, G.S.; Zheltukhin, K. Dynamical systems and Poisson structures. *J. Math. Phys.* **2009**, *50*, 112703. [CrossRef]
14. Lăzureanu, C. Integrable Deformations of Three-Dimensional Chaotic Systems. *Int. J. Bifurc. Chaos* **2018**, *28*, 1850066. [CrossRef]
15. Lotka, A.J. Analytical note on certain rhythmic relations in organic systems. *Proc. Natl. Acad. Sci. USA* **1920**, *6*, 410–415. [CrossRef] [PubMed]
16. Fečkan, M. A Generalization of Bendixson's Criterion. *Proc. Am. Math. Soc.* **2001**, *129*, 3395–3399. [CrossRef]
17. Kolmogorov, A. Sulla teoria di Volterra della lotta per l'esistenza. *G. Dell' Ist. Ital. Degli Attuari* **1936**, *7*, 74–80.
18. Sigmund, K. Kolmogorov and population dynamics. In *Kolmogorov's Heritage in Mathematics*; Charpentier, É., Lesne, A., Nikolski, N.K., Eds.; Springer: Berlin/Heidelberg, Germany, 2007.
19. Bianca, C.; Pappalardo, F.; Pennisi, M.; Ragusa M.A. Persistence Analysis in a Kolmogorov-type Model for Cancer-Immune System Competition. In Proceedings of the 11st International Conference of Numerical Analysis and Applied Mathematics (ICNAAM 2013), Rhodes, Greece, 21–27 September 2013; Volume 1558, pp. 1797–1800.
20. Diz-Pita, E.; Llibre, J.; Otero-Espinar, M.V.; Valls, C. The zero-Hopf bifurcations in the Kolmogorov systems of degree 3 in $\mathbb{R}^3$. *Commun. Nonlinear Sci. Numer. Simul.* **2021**, *95*, 105621. [CrossRef]
21. Grammaticos, B.; Moulin-Ollagnier, J.; Ramanic, A.; Strelcyn, J.-M.; Wojciechowski, S. Integrals of quadratic ordinary differential equations in $\mathbb{R}^3$: The Lotka–Volterra system. *Phys. A Stat. Mech. Its Appl.* **1990**, *163*, 683–722. [CrossRef]
22. Labrunie, S. On the polynomial first integrals of the (a, b, c) Lotka–Volterra system. *J. Math. Phys.* **1996**, *37*, 5539–5550. [CrossRef]
23. Liang, X.; Jiang, J.F. The dynamical behaviour of type-K competitive Kolmogorov systems and its application to three-dimensional type-K competitive Lotka–Volterra systems. *Nonlinearity* **2003**, *16*, 785–801. [CrossRef]
24. Llibre, J.; Ramirez, R.; Ramirez, V. Integrability of a class of N-dimensional Lotka–Volterra and Kolmogorov systems. *J. Differ. Equ.* **2020**, *269*, 2503–2531. [CrossRef]
25. Llibre, J.; Valls, C. Polynomial, rational and analytic first integrals for a family of 3-dimensional Lotka–Volterra systems. *Z. FÜr Angew. Math. Und Phys.* **2011**, *62*, 761–777. [CrossRef]

26. Cairó, L.; Feix, M.R. Families of invariants of the motion for the Lotka–Volterra equations: The linear polynomials family. *J. Math. Phys.* **1992**, *33*, 2440. [CrossRef]
27. Tudoran, R.M.; Aron, A.; Nicoară, Ş. On a Hamiltonian Version of the Rikitake System. *SIAM J. Appl. Dyn. Syst.* **2009**, *8*, 454–479. [CrossRef]
28. Lotka, A.J. *Analytical Theory of Biological Populations*; The Plenum Series on Demographic Methods and Population Analysis; Plenum Press: New York, NY, USA, 1998.
29. Diz-Pita, E.; Llibre, J.; Otero-Espinar, M.V. Phase portraits of a family of Kolmogorov systems depending on six parameters. *Electron. J. Differ. Equ.* **2021**, *2021*, 1–38.
30. Han, M.; Llibre, J.; Tian, Y. On the Zero-Hopf Bifurcation of the Lotka–Volterra Systems in $\mathbb{R}^3$. *Mathematics* **2020**, *8*, 1137. [CrossRef]
31. Llibre, J.; Martínez, Y.P. Dynamics of a family of Lotka–Volterra systems in $\mathbb{R}^3$. *Nonlinear Anal.* **2020**, *199*, 111915. [CrossRef]
32. Bînzar, T.; Lăzureanu, C. On some dynamical and geometrical properties of the Maxwell-Bloch equations with a quadratic control. *J. Geom. Phys.* **2013**, *70*, 1–8. [CrossRef]
33. Bînzar, T.; Lăzureanu, C. A Rikitake type system with one control. *Discret. Contin. Dyn. Syst. B* **2013**, *18*, 1755–1776. [CrossRef]
34. Lăzureanu, C. The Real-Valued Maxwell–Bloch Equations with Controls: From a Hamilton–Poisson System to a Chaotic One. *Int. J. Bifurc. Chaos* **2017**, *27*, 1750143. [CrossRef]
35. Lăzureanu, C. On a Hamilton–Poisson Approach of the Maxwell-Bloch Equations with a Control, Mathematical Physics. *Anal. Geom.* **2017**, *20*, 20.
36. Lăzureanu, C.; Bînzar, T. A Rikitake type system with quadratic control. *Int. J. Bifurc. Chaos* **2012**, *22*, 1250274. [CrossRef]
37. Lăzureanu, C.; Bînzar, T. Some geometrical properties of the Maxwell-Bloch equations with a linear control. In Proceedings of the XIII-th International Conference on Mathematics and its Applications, Timişoara, Romania, 1–3 November 2012; pp. 151–158.
38. Lăzureanu, C.; Bînzar, T. On a Hamiltonian version of controls dynamic for a drift-free left invariant control system on G_4. *Int. J. Geom. Methods Mod. Phys.* **2012**, *9*, 1250065. [CrossRef]
39. Arnold, V. Conditions for nonlinear stability of stationary plane curvilinear flows on an ideal fluid. *Dokl. Akad. Nauk.* **1965**, *162*, 773–777.
40. Llibre, J.; Zhang, X. Dynamics of Some Three-Dimensional Lotka—Volterra Systems. *Mediterr. J. Math.* **2017**, *14*, 126. [CrossRef]
41. Birtea, P.; Puta, M.; Tudoran, R.M. Periodic orbits in the case of a zero eigenvalue. *Comptes Rendus Math.* **2007**, *344*, 779–784. [CrossRef]

MDPI
St. Alban-Anlage 66
4052 Basel
Switzerland
Tel. +41 61 683 77 34
Fax +41 61 302 89 18
www.mdpi.com

Mathematics Editorial Office
E-mail: mathematics@mdpi.com
www.mdpi.com/journal/mathematics

www.ingramcontent.com/pod-product-compliance
Lightning Source LLC
Chambersburg PA
CBHW061810210926
51850CB00006B/659